Molecular Biology of Microbial Differentiation

Molecular Biology of Microbial Differentiation

Proceedings of the Ninth International Spore Conference,
Asilomar, California, 3-6 September 1984

Editors

James A. Hoch
Research Institute of Scripps Clinic
Scripps Clinic and Research Foundation
La Jolla, California 92037

Peter Setlow
Department of Biochemistry
The University of Connecticut Health Center
Farmington, Connecticut 06032

American Society for Microbiology
Washington, D.C. 1985

Library of Congress Cataloging in Publication Data

International Spores Conference (9th: 1984: Asilomar, Calif.)
  Molecular biology of microbial differentiation.

  Includes indexes.
  1. Microbial differentiation—Congresses. 2. Bacillus subtilis—Congresses. 3. Bacteria,
Sporeforming—Congresses. 4. Molecular biology—Congresses. 5. Bacteria—Physiology—
Congresses. I. Hoch, James A. II. Setlow, Peter. III. Title.
QR73.5.I58   1984   576'.13   85-1413

ISBN 0-914826-75-1   *55,862*

Molecular Biology of Microbial Differentiation

Contents

Preface

This book contains contributions from the participants at the Ninth International Spores Conference, held at the Asilomar Conference Center in Pacific Grove, Calif., 3–6 September 1984. The proceedings were organized by a program committee consisting of Roy Doi and Richard Losick, Co-Chairmen, with James Hoch, Peter Setlow, Issar Smith, and Linc Sonenshein. This meeting differed from several of those in the past in that the lectures were mainly concentrated in the area of the molecular biology of the sporulation and germination processes. The decision to focus the contents of the lectures was based upon the very rapid progress being made in this area of study of bacterial development. This book reflects the excellence of the presentations in this interesting and exciting area of microbial differentiation.

The meeting was made possible through the generous support of the National Institutes of Health, The National Science Foundation, and the U.S. Army Office of Research. Generous support from the following corporations was important to the success of this meeting: Abbott Laboratories of North Chicago, Bayer AG/Miles, Campbell's Soup, Cetus Corp., Dow Chemical Co., E. I. Du Pont de Nemours & Co., Ethicon, Gist-Brocades nv, Hoffmann-La Roche Inc., Merck Sharp & Dohme Research Laboratories, Monsanto, R. J. Reynolds, and Syntro Corp. We are especially grateful for this industrial support as it permitted many scientists with little federal support to participate and also made possible publication of this book.

Since the last meeting of two of our colleagues have passed away. Elizabeth B. Freese was a dedicated worker in the sporulation field and published on many aspects of the biochemistry of sporulation. Hans J. Rhaese was a provocative investigator in the area of nucleotide control of sporulation. We mourn the passing of these two friends.

The editors of this volume would like to express their gratitude to the authors submitting manuscripts to this book. Not only were the manuscripts of excellent quality and submitted (mostly) on time, but also they required very little editing on our part. This is the easiest editorial job either of us has ever had.

We also thank, on behalf of the meeting participants and spectators, the unselfish efforts of Roy H. Doi in the preparation for and execution of the meeting. His hard work and organizational talents were reflected in the excellence of the meeting.

JAMES A. HOCH
Division of Cellular Biology
Research Institute of Scripps Clinic
La Jolla, California 92037

PETER SETLOW
Department of Biochemistry
University of Connecticut Health Center
Farmington, Connecticut 06032

Molecular Cloning and Genetics of Sporulation and Germination Genes

Genes Controlling Development in *Bacillus subtilis*†

PATRICK J. PIGGOT[1] AND JAMES A. HOCH[2]

Microbiology Division, National Institute for Medical Research, Mill Hill, London, England,[1] and Division of Cellular Biology, Department of Basic and Clinical Research, Research Institute of Scripps Clinic, La Jolla, California 92037[2]

It is clear that one of the driving forces behind the study of the genetics of *Bacillus subtilis* is the quest to understand the simple differentiation cycle of sporulation and germination that this organism undergoes. The study of this process has resulted in the accumulation of vast amounts of knowledge of the physiology and genetics of postexponential-phase bacteria. It is this knowledge that will certainly be applicable to an understanding of microbial differentiation in various forms of procaryotes and eucaryotes and, it is hoped, will lend insights into the unique genetic mechanisms occurring during this phase of the life cycle.

Although it has only been a few years since the last compilation of a genetic map of this sort (25, 26), there have been substantial additions of new markers to the genetic map. This has resulted not only from a continuation of classical genetic studies using DNA-mediated transformation and transduction methods, but also from a minor explosion of molecular cloning of genes of interest in this organism. Several libraries of *B. subtilis* DNA have been constructed in bacteriophage λ (14), cosmid (4), and other vectors (32, 62), and these libraries have been a rich source of fragments of the chromosome carrying genes of interest. One of the more useful genetic tools used to study the genetic location of cloned genes is the so-called integrative vector (15). These vectors lack an origin of replication for *B. subtilis* and therefore cannot replicate in this organism but can integrate into the chromosome of the organism if a suitable region of homology, e.g., a cloned region of the genome, is cloned within the vector. In most cases such vectors include a chloramphenicol resistance gene which can be expressed in *B. subtilis* when integrated into the chromosome. In addition, all constructions in the vector are done in a permissive host such as *Escherichia coli*. Mapping of the location of the integrated plasmid allows one to determine the genetic location of the cloned gene of interest. Such methods have been particularly useful for genes that would have no obvious phenotype if mutated and for those genes whose genetics is not readily determined. In any case, the existence of integrative vectors provides a link between cloning and genetics of this organism.

Another powerful tool which will result in a rapid expansion of knowledge about the genetics of this organism is the transposon Tn917. The methodology for the use of this transposon to insert into the chromosome of *B. subtilis* has been painstakingly worked out by P. Youngman and associates. This system is described in detail in another paper (Youngman et al., this volume) and will not be described here. It is clear that transposon-mediated mutagenesis and transcriptional coupling will play a powerful role in our understanding of the mechanisms of gene expression in *B. subtilis*. Thus, it seems likely that within several years the density of genetic markers on the map as we presently know it should at least double or perhaps triple. Much of this will be the result of the discovery, in cloned DNA, of open reading frames of interest that can now be identified easily and manipulated. One may also speculate that within this time frame a significant portion of the genome will have been identified in cloned form and we should know substantially more about the molecular genetics of this organism.

In this article we have attempted to compile the known sporulation mutations, their genetic map position, information as to whether they have been cloned, and the number of complementation groups or open reading frames within each cloned fragment. A genetic map of *B. subtilis* showing both auxotrophic and developmental markers is presented in Fig. 1. A number of the markers on the map have not been compiled previously, and only those related to development are documented here. All of the developmental loci are inside the circle; the auxotrophic and other loci are outside the circle. The location of the origin is most likely to be very close to the *rrn0* operon, although its exact

† Publication no. 3784-BCR from the Research Institute of Scripps Clinic.

1

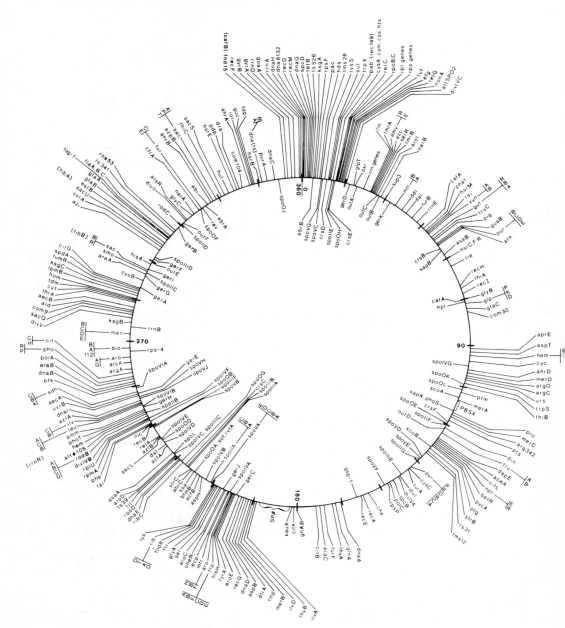

FIG. 1. Genetic map of the *B. subtilis* chromosome.

location has not been unequivocally determined (H. Yoshikawa, personal communication). The terminus of the map is located very close to the *gltA* and *gltB* loci (according to the data of Weiss and Wake [82] and Monteiro et al. [M. J. Monteiro, M. G. Sargent, and P. J. Piggott, J. Gen. Microbiol., in press]). In general, the order of markers is also shown; however, this order should not be taken as definitive. The markers were placed according to our best estimate from the published data. In some rare cases, three-factor crosses have been done for all of the markers within a region, but in the majority of the cases this is not true.

Table 1 lists the loci presently known to be associated with development in *B. subtilis*. Several of these loci have been cloned, and this information is also included. Of those cloned loci examined in detail by nucleic acid sequencing, some have been shown to be polycistronic.

TABLE 1. Sporulation and germination markers

Locus	Map position (°)	Comments
abrA	325	Antibiotic resistant. Mutation suppresses some of the pleiotropic phenotypes, but not Spo⁻, of spo0A mutations (76). May be the same as rev-4.
abrB	3	Antibiotic resistant. Major locus for mutations suppressing pleiotropic phenotypes of spo0 mutations without locus or allele specificity. Ribosome alterations have been observed in these mutants (77, 78). Same locus as cpsX (22, 23) and probably as absA, absB (36), and tolA (35).
abrC		Weak intragenic suppressors of spo0A (76).
catA	89	Overproduces extracellular protease; can sporulate in presence of glucose (37). Probably the same as scoC (47) and possibly the same as hpr (27).
cpsX		See abrB.
crsA	222	Carbon source-resistant sporulation; resistant to novobiocin and acridine orange during sporulation (74). Mutation in rpoD (R. H. Doi, personal communication); has similar phenotype to rvt mutations (69).
crsB	55	Requires high glucose for sporulation (74).
crsC	221	Carbon source-resistant sporulation (74).
crsD	8	As crsC.
crsE	11	Carbon source-resistant sporulation (74); maps in rpoBC operon; rfm-11 suppresses crsE; stv std mutants show partial resistance to catabolites in sporulation (75).
crsF	118	As crsC (74).
gdh		Structural gene for glucose dehydrogenase. Cloned on a phage vector (80).
gerA	290	Defective in germination response to alanine and related amino acids (50, 64, 79). Isolated from λ clone banks (49); contains several genes (A. Moir, personal communication).
gerB	314	Defective in germination response to the combinations of glucose, fructose, asparagine, and KCl (50).
gerC	201	Temperature-sensitive germination in alanine (50, 79). Has not been separated from linked mutations in the original isolate.
gerD	16	Defective germination in a range of germinants (50, 58).
gerE	253	Defective germination in a range of germinants. May be a spore coat defect. Cloned in SPβ (48, 50).
gerF	301	Defective germination in a range of germinants (50, 58).
gerG	294	Mutant lacks phosphoglycerate kinase activity (17). Germinates poorly in alanine; sporulates poorly (60).
gerH	247	Defective germination in a range of germinants (58).
gerI	297	Defective germination in a range of germinants (58).
gerJ	207	Defective germination in a range of germinants; map order gerJ-aroC-mtr-spoVIA-aroB2-trpC2. Allele gerJ51 (also called tzm) is present in many laboratory strains (81).
gerK	32	Defective germination response to glucose (34).
outA	21	Blocked in outgrowth after RNA, protein, and DNA synthesis has started; previous designation gspIV (2, 20, 58).
outB	28	Blocked in outgrowth before most macromolecular synthesis has started; previous designation gsp-81 (1, 18, 58).
outC	27	Blocked in outgrowth after RNA and protein synthesis has started, but before synthesis of DNA; previous designation gsp-25 (2, 58).
outD	122	Blocked in outgrowth; protein and DNA synthesis reduced; previous designation gsp-1 (19, 58).
outE	300	Blocked in outgrowth; RNA synthesis normal, protein synthesis reduced, and DNA synthesis prevented; previous designation gsp-42 (2, 58).
outF	316	Blocked in outgrowth; RNA and protein synthesis reduced and DNA synthesis prevented; previous designation gsp-4 (2, 58).
rev-4	324	Suppressor of some of the pleiotropic effects (but not asporogeny) of spo0 mutations; suppresses effect on sporulation of various drug resistance mutations (67, 68). May be the same as abrA.
rvtA	218	Suppressor of sporulation defect in spo0B, spo0E, and spo0F mutants (69). May be the same as sof-1.
rvt		Mutations causing the same phenotype as rvtA mutations but not mapping in the rvtA region (69).
sapA	114	Mutations overcome sporulation phosphatase-negative phenotype of early blocked spo mutants. Not definitely separate from phoS locus defined by mutations that cause constitutive phosphatase production (59). Deletions of region have PhoS phenotype (54).

Continued

TABLE 1—*Continued*

Locus	Map position (°)	Comments
sapB	56	Mutations overcome sporulation phosphatase-negative phenotype of early blocked *spo* mutations (59).
sas	211	Weak intragenic suppressor mutations of *spoIIA* (85).
scoA	109	Protease and phosphatase overproduction; delayed spore formation (47).
scoB	129	Protease and phosphatase overproduction; delayed spore formation (13, 38).
sof-1	218	Suppressor of sporulation defects in *spo0B*, *spo0E*, and *spo0F* mutants (30a, 42). The *sof-1* mutation is an alteration in codon 12 of the *spo0A* gene. Probably the same as *rvtA*.
spo0A	218	Mutants are blocked at stage 0 (30, 46). Locus codes for a protein of 29,700 daltons as determined from nucleic acid sequence (F. A. Ferrari et al., this volume). Mutants exhibit a wide variety of pleiotropic phenotypes, possibly as a result of transcription defects from promoters under control of minor forms of RNA polymerase (21, 86). Transcribed during vegetative growth.
spo0B	241	Mutants are blocked at stage 0 (30). Locus codes for a protein of 22,500 daltons as well as another protein of unknown function on a polycistronic transcript (15a; J. Bouvier, P. Stragier, C. Bonamy, and J. Szulmajster, Proc. Natl. Acad. Sci. U.S.A., in press). Mutants have most of the phenotypes of mutants bearing *spo0A* mutations. Transcribed during vegetative growth.
spo0C	218	Mutations with less pleiotropic phenotypes now known to be missense alterations in the *spo0A* gene product (F. A. Ferrari et al., this volume).
spo0D	234	Single allele resulting in a stage 0 block of sporulation. Mapped but not further characterized (33).
spo0E	120	Oligosporogenous mutations giving a stage 0 block. Possibly more than one gene. Phenotypes are less pleiotropic than *spo0A*, *spo0B*, or *spo0F* mutations (11, 30).
spo0F	323	Cloned on phage and plasmid vectors (43). DNA sequence contains a single open reading frame for a protein of 19,055 daltons (70). Inhibits sporulation when present in multiple copies (43): five copies inhibit; two do not (P. J. Piggot et al., this volume).
spo0G	223	Single allele resulting in stage 0 block of sporulation. Maps in the region of *spo0A* locus but genetically distinct from *spo0A* (33).
spo0H	11	Stage 0 locus that codes for a protein of 22,000 daltons as determined from the sequence of the *B. licheniformis* cloned gene. This gene complements *spo0H* mutations of *B. subtilis*. Transcript contains long 5' and 3' untranslated regions (61). Least pleiotropic of *spo0* mutations (29, 53).
spo0J	352	A stage 0 locus consisting of two alleles that yield a phenotype similar to *spo0H* mutations (31). Map close to the *spo-CM1* mutation which may be complemented by certain phage infections (10, 77). Closely linked by transformation to the gyrase gene.
spo0K	101	Stage 0 mutation mapping very close to the tryptophanyl tRNA synthetase gene (11; J. A. Hoch, unpublished data).
spo0L	106	Uncharacterized allele giving a *spo0* phenotype and mapping near to *spo0K* but genetically distinct from it (Hoch, unpublished data; listed in references 25 and 26).
spoIIA	211	Mutant blocked at stage II of sporulation. Region cloned on plasmid (45, 57) and phage (65) vectors. Transcribed as a polycistronic unit (57). DNA sequence has three adjacent open reading frames coding for proteins of 13, 16, and 22 kilodaltons (16). Located very close to, but in a separate transcriptional unit from, *spoVA* (55). Transformation starts 1 to 1.5 h after the start of sporulation (Piggot et al., this volume).
spoIIB	244	(11)
spoIIC	296	Cloned on phage and plasmid vectors (3).
spoIID	316	(53)
spoIIE	10	Cloned on plasmid vector (M. Young, personal communication).
spoIIF	120	(31)
spoIIG	135	Analysis of temperature-sensitive mutant indicates that expression must be tightly controlled (84). Cloned in plasmid and phage vector (5, 9). DNA sequence has homology with *rpoD* gene (73).
spoIIIA	220	Blocked at stage III of sporulation (53).
spoIIIB	221	(53). Cloned on phage vector (41).
spoIIIC	227	Possibly the same as *spoIVC* (33, 56).
spoIIID	302	(33)

Continued

TABLE 1—*Continued*

Locus	Map position (°)	Comments
spoIIIE	142	(33)
spoIIIF	239	Map order *hemA-spoIIB-attφ105-rodB-divIVB-spo0B-pheA-nic-recB-spoIIIF-spoVB* (44).
spoIVA	204	Blocked at stage IV of sporulation. Linked to *trpC* by transformation (45, 54).
spoIVB	213	May be an allele of *spo0A* (11; Hoch, unpublished data).
spoIVC	227	Contains at least two cistrons (12). Linked to *aroD* by transformation (58).
spoIVD	233	(31)
spoIVE	234	(33)
spoIVF	242	Linked to *spo0B* by transformation (11, 44).
spoIVG	97	(53)
spoVA	211	Blocked at stage V of sporulation. Map order *spoIIA-spoVA-lys* (55). Cloned on a plasmid and on a phage (57, 65). Transcribed as a polycistronic unit with open reading frames for proteins of 23, 15, 16, 36 and 34 kilodaltons (P. Fort and J. Errington, J. Gen. Microbiol., in press).
spoVB	239	(31, 43)
spoVC	7	Cloned on plasmid vectors (51).
spoVD	133	Linked to *spoVE* by transformation (11, 31).
spoVE	133	(11, 31). Isolated in λ clone bank and subcloned on a plasmid (P. J. Piggot and K.-F. Chak, unpublished data).
spoVF	148	Probably the same as *dpa* (6, 58). Mutants form octanol- and chloroform-resistant, heat-sensitive spores (6, 58). Form heat-resistant spores in the presence of dipicolinic acid.
spoVG	6	Previously called 0.4-kb gene (24, 63). Transcription turned on within 30 min of the start of sporulation (66).
spoVH	251	(28)
spoVJ	250	(28)
spoVIA	255	Blocked at stage VI of sporulation. Map order *argA-spoVIA-gerE-leuA* (39).
spoVIB	247	Map order *citF-gerE-spoVH-spoVJ-ilvB-leuA-spoVIB-pheA* (40).
spoL	227	"Decadent" sporulation (7).
sprE	91	Structural gene for subtilisin E. Map order (*hpr, glyB*)-*sprE-metD* (71, 83).
ssa	218	Alcohol-resistant sporulation. Maps very close to *spo0A* (8). *rvt* mutations have same phenotype (69). Probably *rvtA = sof-1 = ssa = spo0A*.
"0.3-kb gene"		Turned on at a late stage (T_3 to T_4) of sporulation. Requires correct functioning of at least *spo0B*, *spoIIA*, *spoIIE*, and *spoIIIE*. Codes for a protein of 6,750 daltons (52, 72).

This is the case for *spo0B*, *spoIIA*, and *spoVA* and probably will hold for other loci as well. Thus, one should not equate loci with single genes. One of the surprising findings in the sequencing of sporulation genes is the relatively low molecular weight of their putative gene products. None of these gene products exceed 30,000 daltons for the *spo0A*, *spo0B*, *spo0F*, *spo0H*, and *spoIIA* loci. This is considerably less than the size of the majority of vegetative proteins or proteins made in sporulating cells. The significance of this observation is unclear.

The promise that cloning of sporulation genes will increase our knowledge of the regulation of such genes and give insights into the functions of our gene products is now being fulfilled. It is hoped that such studies will allow us to paint a comprehensive picture of the mechanisms controlling development at the cellular level.

This research was supported, in part, by Public Health Service grant GM19416 from the National Institute of General Medical Sciences.

LITERATURE CITED

1. Albertini, A. M., M. L. Baldi, E. Ferrari, E. Isneghi, M. T. Zambelli, and A. Galizzi. 1979. Mutants of *Bacillus subtilis* affected in spore outgrowth. J. Gen. Microbiol. 110:351–363.
2. Albertini, A. M., and A. Galizzi. 1975. Mutant of *Bacillus subtilis* with a temperature-sensitive lesion in ribonucleic acid synthesis during germination. J. Bacteriol. 124:14–25.
3. Anaguchi, H., S. Fukui, H. Shimotsu, F. Kawamura, H. Saito, and Y. Kobayashi. 1984. Cloning of sporulation gene *spoIIC* in *Bacillus subtilis*. J. Gen. Microbiol. 130:757–760.
4. Aubert, E., F. F. Fargette, A. Fouet, A. Klier, and G. Rapoport. 1982. Use of a bifunctional cosmid for cloning large DNA fragments of *B. subtilis*, p. 11–24. *In* A. T. Ganesan, S. Chang, and J. A. Hoch (ed.), Molecular cloning and gene regulation in bacilli. Academic Press, Inc., New York.
5. Ayaki, H., and Y. Kobayashi. 1984. Cloning of sporulation-related gene *spoIIG* in *Bacillus subtilis*. J. Bacteriol. 158:507–512.
6. Balassa, G., P. Milhaud, E. Raulet, M. T. Silva, and J. C. F. Sousa. 1979. A *Bacillus subtilis* mutant requiring dipicolinic acid for the development of heat resistant spores. J. Gen. Microbiol. 110:365–379.
7. Balassa, G., P. Milhaud, J. C. F. Sousa, and M. T. Silva. 1979. Decadent sporulation mutants of *Bacillus subtilis*. J. Gen. Microbiol. 110:381–392.
8. Bohin, J.-P., and B. Lubochinsky. 1982. Alcohol-resistant

sporulation mutants of *Bacillus subtilis*. J. Bacteriol. 150:944–955.

9. **Bonamy, C., and J. Szulmajster.** 1982. Cloning and expression of *Bacillus subtilis* spore genes. Mol. Gen. Genet. 188:202–210.

10. **Bramucci, M. G., K. M. Keggins, and P. S. Lovett.** 1977. Bacteriophage PMB12 conversion of the sporulation defect in RNA polymerase mutants of *Bacillus subtilis*. J. Virol. 24:194–200.

11. **Coote, J. G.** 1972. Sporulation in *Bacillus subtilis*. Characterization of oligosporogenous mutants and comparison of their phenotypes with those of asporogenous mutants. J. Gen. Microbiol. 71:1–15.

12. **Dancer, B. N., and J. Mandelstam.** 1981. Complementation of sporulation mutations in fused protoplasts of *Bacillus subtilis*. J. Gen. Microbiol. 123:17–26.

13. **Dod, B., G. Balassa, E. Raulet, and V. Jeannoda.** 1978. Spore control (*Sco*) mutations in *Bacillus subtilis*. II. Sporulation and the production of extracellular proteases and α-amylase by *Sco* mutants. Mol. Gen. Genet. 163:45–56.

14. **Ferrari, E., D. J. Henner, and J. A. Hoch.** 1981. Isolation of *Bacillus subtilis* genes from a Charon 4A library. J. Bacteriol. 146:430–432.

15. **Ferrari, F. A., A. Nguyen, D. Lang, and J. A. Hoch.** 1983. Construction and properties of an integrable plasmid for *Bacillus subtilis*. J. Bacteriol. 154:1513–1515.

15a.**Ferrari, F. A., K. Trach, and J. A. Hoch.** 1985. Sequence analysis of the *spo0B* locus reveals a polycistronic transcription unit. J. Bacteriol. 161:556–562.

16. **Fort, P., and P. J. Piggot.** 1984. Nucleotide sequence of the *Bacillus subtilis* 168 *spoIIA* locus. J. Gen. Microbiol. 130:2147–2153.

17. **Freese, E., Y. K. Oh, E. B. Freese, M. D. Diesterhaft, and C. Prasad.** 1972. Suppression of sporulation of *Bacillus subtilis*, p. 212–221. *In* H. O. Halvorson, R. Hanson, and L. L. Campbell (ed.), Spores V. American Society for Microbiology, Washington, D.C.

18. **Galizzi, A., A. M. Albertini, P. Plevani, and G. Cassani.** 1976. Synthesis of RNA and protein in a mutant of *Bacillus subtilis* temperature sensitive during spore germination. Mol. Gen. Genet. 148:159–164.

19. **Galizzi, A., F. Gorrini, A. Rollier, and M. Polsinelli.** 1973. Mutants of *Bacillus subtilis* temperature sensitive in the outgrowth phase of spore germination. J. Bacteriol. 113:1482–1490.

20. **Galizzi, A., A. G. Siccardi, A. M. Albertini, A. R. Amileni, G. Meneguzzi, and M. Polsinelli.** 1975. Properties of *Bacillus subtilis* mutants temperature sensitive in germination. J. Bacteriol. 121:450–454.

21. **Gilman, M. Z., and M. J. Chamberlin.** 1983. Developmental and genetic regulation of *Bacillus subtilis* genes transcribed by σ²⁸-RNA polymerase. Cell 35:285–293.

22. **Guespin-Michel, J. F.** 1971. Phenotypic reversion in some early blocked sporulation mutants of *Bacillus subtilis*: isolation and phenotype identification of partial revertants. J. Bacteriol. 108:241–247.

23. **Guespin-Michel, J. F.** 1971. Phenotypic reversion in some early blocked sporulation mutants of *Bacillus subtilis*. Genetic studies of polymyxin resistant partial revertants. Mol. Gen. Genet. 112:243–254.

24. **Haldenwang, W. B., C. D. B. Banner, J. F. Ollington, R. Losick, J. A. Hoch, M. B. O'Connor, and A. L. Sonenshein.** 1980. Mapping of a cloned gene under sporulation control by insertion of a drug resistance marker into the *Bacillus subtilis* chromosome. J. Bacteriol. 142:90–98.

25. **Henner, D. J., and J. A. Hoch.** 1980. The *Bacillus subtilis* chromosome. Microbiol. Rev. 44:57–82.

26. **Henner, D. J., and J. A. Hoch.** 1982. The genetic map of *Bacillus subtilis*, p. 1–33. *In* D. Dubnau (ed.), The molecular biology of the bacilli, vol. 1. Academic Press, Inc., New York.

27. **Higerd, T. B., J. A. Hoch, and J. Spizizen.** 1972. Hyperprotease-producing mutants of *Bacillus subtilis*. J. Bacteriol.

112:1026–1028.

28. **Hill, S. H. A.** 1983. *spoVH* and *spoVJ*—new sporulation loci in *Bacillus subtilis* 168. J. Gen. Microbiol. 129:293–302.

29. **Hoch, J. A., and J. L. Mathews.** 1973. Chromosomal location of pleiotropic negative sporulation mutations in *Bacillus subtilis*. Genetics 73:215–228.

30. **Hoch, J. A., and J. Spizizen.** 1969. Genetic control of some early events in sporulation of *Bacillus subtilis* 168, p. 112–120. *In* L. L. Campbell (ed.), Spores IV. American Society for Microbiology, Bethesda, Md.

30a.**Hoch, J. A., K. Trach, F. Kawamura, and H. Saito.** 1985. Identification of the transcriptional suppressor *sof-1* as an alteration in the *spo0A* protein. J. Bacteriol. 161:552–555.

31. **Hranueli, D., P. J. Piggot, and J. Mandelstam.** 1974. Statistical estimate of the total number of operons specific for *Bacillus subtilis* sporulation. J. Bacteriol. 119:684–690.

32. **Hutchison, K. W., and H. O. Halvorson.** 1980. Cloning of randomly sheared fragments from a φ105 lysogen of *Bacillus subtilis*. Gene 8:267–268.

33. **Ionesco, H., J. Michel, B. Cami, and P. Schaeffer.** 1970. Genetics of sporulation in *Bacillus subtilis* Marburg. J. Appl. Bacteriol. 33:13–24.

34. **Irie, R., T. Okamoto, and J. Fujita.** 1982. A germination mutant of *Bacillus subtilis* deficient in response to glucose. J. Gen. Appl. Microbiol. 28:345–356.

35. **Ito, J.** 1973. Pleiotropic nature of bacteriophage tolerant mutants obtained in early-blocked asporogeneous mutants of *Bacillus subtilis* 168. Mol. Gen. Genet. 124:97–106.

36. **Ito, J., G. Mildner, and J. Spizizen.** 1971. Early blocked asporogeneous mutants of *Bacillus subtilis* 168. I. Isolation and characterization of mutants resistant to antibiotic(s) produced by sporulating *Bacillus subtilis* 168. Mol. Gen. Genet. 112:104–109.

37. **Ito, J., and J. Spizizen.** 1973. Genetic studies of catabolite repression insensitive sporulation mutants of *Bacillus subtilis*. Colloq. Int. C.N.R.S. 227:81–82.

38. **Jeannoda, V., and G. Balassa.** 1978. Spore control (*Sco*) mutations in *Bacillus subtilis*. IV. Synthesis of alkaline phosphatase during sporulation of *Sco* mutants. Mol. Gen. Genet. 163:65–73.

39. **Jenkinson, H. F.** 1981. Germination and resistance defects in spores of a *Bacillus subtilis* mutant lacking a coat polypeptide. J. Gen. Microbiol. 127:81–91.

40. **Jenkinson, H. F.** 1983. Altered arrangement of proteins in the spore coat of a germination mutant of *Bacillus subtilis*. J. Gen. Microbiol. 129:1945–1958.

41. **Jenkinson, H. F., and J. Mandelstam.** 1983. Cloning of the *Bacillus subtilis lys* and *spoIIIB* genes in phage φ105. J. Gen. Microbiol. 129:2229–2249.

42. **Kawamura, F., and H. Saito.** 1983. Isolation and mapping of a new suppressor mutation of an early sporulation gene *spo0F* mutation in *Bacillus subtilis*. Mol. Gen. Genet. 192:330–334.

43. **Kawamura, F., H. Shimotsu, H. Saito, H. Hirochika, and Y. Kobayashi.** 1981. Cloning of *spo0* genes with bacteriophage and plasmid vectors in *Bacillus subtilis*, p. 109–113. *In* H. S. Levinson, A. L. Sonenshein, and D. J. Tipper (ed.), Sporulation and germination. American Society for Microbiology, Washington, D.C.

44. **Lamont, I. L., and J. Mandelstam.** 1984. Identification of a new sporulation locus, *spoIIIF*, in *Bacillus subtilis*. J. Gen. Microbiol. 130:1253–1261.

45. **Liu, H.-M., K.-F. Chak, and P. J. Piggot.** 1982. Isolation and characterization of a recombinant plasmid carrying a functional part of the *Bacillus subtilis spoIIA* locus. J. Gen. Microbiol. 128:2805–2812.

46. **Michel, J. F., and B. Cami.** 1969. Selection de mutants de *Bacillus subtilis* bloqués au debut de la sporulation—nature des mutations selectionnées. Ann. Inst. Pasteur (Paris) 116:3–18.

47. **Milhaud, P., G. Balassa, and J. Zucca.** 1978. Spore control (*Sco*) mutations in *Bacillus subtilis*. I. Selection and genetic mapping of *Sco* mutants. Mol. Gen. Genet.

163:35–44.

48. **Moir, A.** 1981. Germination properties of a spore coat-defective mutant of *Bacillus subtilis*. J. Bacteriol. **146**:1106–1116.

49. **Moir, A.** 1983. The isolation of λ transducing phages carrying the *citG* and *gerA* genes of *Bacillus subtilis*. J. Gen. Microbiol. **129**:303–310.

50. **Moir, A., E. Lafferty, and D. A. Smith.** 1979. Genetic analysis of spore germination mutants of *Bacillus subtilis* 168: the correlation of phenotype with map location. J. Gen. Microbiol. **165**:165–180.

51. **Moran, C. P., R. Losick, and A. L. Sonenshein.** 1980. Identification of a sporulation locus in cloned *Bacillus subtilis* deoxyribonucleic acid. J. Bacteriol. **142**:331–334.

52. **Ollington, J. F., and R. Losick.** 1981. A cloned gene that is turned on at an intermediate stage of spore formation in *Bacillus subtilis*. J. Bacteriol. **147**:443–451.

53. **Piggot, P. J.** 1973. Mapping of asporogenous mutations of *Bacillus subtilis*: a minimum estimate of the number of sporulation operons. J. Bacteriol. **114**:1241–1253.

54. **Piggot, P. J., and R. S. Buxton.** 1982. Bacteriophage PBSX-induced deletion mutants of *Bacillus subtilis* 168 constitutive for alkaline phosphatase. J. Gen. Microbiol. **128**:663–669.

55. **Piggot, P. J., K.-F. Chak, H.-M. Liu, and H. Lencastre.** 1983. *Bacillus subtilis*: genetics and spore formation, p. 163–166. *In* D. Schlessinger (ed.), Microbiology—1983. American Society for Microbiology, Washington, D.C.

56. **Piggot, P. J., and J. G. Coote.** 1976. Genetic aspects of bacterial endospore formation. Bacteriol. Rev. **40**:908–962.

57. **Piggot, P. J., C. A. M. Curtis, and H. Lencastre.** 1984. Use of integrational vectors to identify a polycistronic transcriptional unit required for differentiation of *Bacillus subtilis*. J. Gen. Microbiol. **130**:2123–2136.

58. **Piggot, P. J., A. Moir, and D. A. Smith.** 1981. Advances in the genetics of *Bacillus subtilis* differentiation, p. 29–39. *In* H. S. Levinson, A. L. Sonenshein, and D. J. Tipper (ed.), Sporulation and germination. American Society for Microbiology, Washington, D.C.

59. **Piggot, P. J., and S. Y. Taylor.** 1977. New types of mutations affecting formation of alkaline phosphatase by *Bacillus subtilis* in sporulation conditions. J. Gen. Microbiol. **102**:69–80.

60. **Prasad, C., M. Diesterhaft, and E. Freese.** 1972. Initiation of spore germination in glycolytic mutants of *Bacillus subtilis*. J. Bacteriol. **110**:321–328.

61. **Ramakrishna, N., E. Dubnau, and I. Smith.** 1984. The complete DNA sequence and regulatory regions of the *Bacillus licheniformis* spo0H gene. Nucleic Acids Res. **12**:1779–1790.

62. **Rapport, G., A. Klier, A. Billault, F. Fargette, and R. Dedonder.** 1979. Construction of a colony bank of *E. coli* containing hybrid plasmids representative of the *Bacillus subtilis* 168 genome. Expression of functions harbored by the recombinant plasmids in *B. subtilis*. Mol. Gen. Genet. **176**:239–245.

63. **Rosenbluh, A., C. D. B. Banner, R. Losick, and P. C. Fitz-James.** 1981. Identification of a new developmental locus in *Bacillus subtilis* by construction of a deletion mutation in a cloned gene under sporulation control. J. Bacteriol. **148**:341–351.

64. **Sammons, R. L., A. Moir, and D. A. Smith.** 1981. Isolation and properties of spore germination mutants of *Bacillus subtilis* 168 deficient in the initiation of germination. J. Gen. Microbiol. **124**:229–241.

65. **Savva, D., and J. Mandelstam.** 1984. Cloning of the *Bacillus subtilis* spoIIA and spoVA loci in phage φ105DI:1t. J. Gen. Microbiol. **130**:2137–2145.

66. **Segall, J., and R. Losick.** 1977. Cloned *Bacillus subtilis* DNA containing a gene that is activated early during sporulation. Cell **11**:751–761.

67. **Sharrock, R. A., and T. Leighton.** 1981. Intergenic suppressors of temperature-sensitive sporulation in *Bacillus subtilis* are allele non-specific. Mol. Gen. Genet. **183**:532–537.

68. **Sharrock, R. A., and T. Leighton.** 1982. Suppression of defective-sporulation phenotypes by the *Bacillus subtilis* mutation *rev4*. Mol. Gen. Genet. **186**:432–438.

69. **Sharrock, R. A., S. Rubinstein, M. Chan, and T. Leighton.** 1984. Intergenic suppression of spo0 phenotypes by the *Bacillus subtilis* mutation *rvtA*. Mol. Gen. Genet. **194**:260–264.

70. **Shimotsu, H., F. Kawamura, Y. Kobayashi, and H. Saito.** 1983. Early sporulation gene spo0F: nucleotide sequence and analysis of gene product. Proc. Natl. Acad. Sci. U.S.A. **80**:658–662.

71. **Stahl, M. L., and E. Ferrari.** 1984. Replacement of the *Bacillus subtilis* subtilisin structural gene with an in vitro derived deletion mutation. J. Bacteriol. **158**:411–418.

72. **Stephens, M. A., N. Lang, K. Sandman, and R. Losick.** 1984. A promoter whose utilization is temporally regulated during sporulation in *Bacillus subtilis*. J. Mol. Biol. **176**:333–348.

73. **Stragier, P., J. Bouvier, C. Bonamy, and J. Szulmajster.** 1984. A developmental gene product of *Bacillus subtilis* homologous to the sigma factor of *Escherichia coli*. Nature (London) **312**:376–378.

74. **Sun, D., and I. Takahashi.** 1982. Genetic mapping of catabolite-resistant mutants of *Bacillus subtilis*. Can. J. Microbiol. **28**:1242–1251.

75. **Sun, D., and I. Takahashi.** 1984. A catabolite-resistance mutation is localized in the *rpo* operon of *Bacillus subtilis*. Can. J. Microbiol. **30**:423–429.

76. **Trowsdale, J., S. M. H. Chen, and J. A. Hoch.** 1978. Genetic analysis of phenotypic revertants of spo0A mutants in *Bacillus subtilis*: a new cluster of ribosomal genes, p. 131–135. *In* G. Chambliss and J. C. Vary (ed.), Spores VII. American Society for Microbiology, Washington, D.C.

77. **Trowsdale, J., S. M. H. Chen, and J. A. Hoch.** 1979. Genetic analysis of a class of polymyxin resistant partial revertants of stage 0 sporulation mutants of *Bacillus subtilis*: a map of the chromosomal region near the origin of replication. Mol. Gen. Genet. **173**:61–70.

78. **Trowsdale, J., M. Shiflett, and J. A. Hoch.** 1978. New cluster of ribosomal genes in *Bacillus subtilis* with regulatory role in sporulation. Nature (London) **272**:179–181.

79. **Trowsdale, J., and D. A. Smith.** 1975. Isolation, characterization, and mapping of *Bacillus subtilis* 168 germination mutants. J. Bacteriol. **123**:83–95.

80. **Vasantha, N., B. Uratani, R. F. Ramaley, and E. Freese.** 1983. Isolation of a developmental gene of *Bacillus subtilis* and its expression in *Escherichia coli*. Proc. Natl. Acad. Sci. U.S.A. **80**:785–789.

81. **Warburg, R. J., and A. Moir.** 1981. Properties of a mutant of *Bacillus subtilis* 168 in which spore germination is blocked at a late stage. J. Gen. Microbiol. **124**:243–253.

82. **Weiss, A. S., and R. G. Wake.** 1983. Restriction map of DNA spanning the replication terminus of the *Bacillus subtilis* chromosome. J. Mol. Biol. **171**:119–137.

83. **Wong, S.-L., C. W. Price, D. S. Goldfarb, and R. H. Doi.** 1984. The subtilisin E gene of *Bacillus subtilis* is transcribed from a σ^{37} promoter in vivo. Proc. Natl. Acad. Sci. U.S.A. **81**:1184–1188.

84. **Young, M.** 1976. The use of temperature-sensitive mutants to study gene expression during sporulation in *Bacillus subtilis*. J. Bacteriol. **126**:928–936.

85. **Yudkin, M. D., and L. Turley.** 1980. Suppression of asporogeny in *Bacillus subtilis*. Allele-specific suppression of a mutation in the spoIIA locus. J. Gen. Microbiol. **121**:69–78.

86. **Zuber, P., and R. Losick.** 1983. Use of a *lacZ* fusion to study the role of the spo0 genes of *Bacillus subtilis* in developmental regulation. Cell **35**:275–283.

Molecular Cloning and Nucleotide Sequence of the *spo0A* Locus and Its Mutations

FRANCO A. FERRARI,† KATHLEEN TRACH, DOMINIQUE LE COQ, AND JAMES A. HOCH

Division of Cellular Biology, Department of Basic and Clinical Research, Research Institute of Scripps Clinic, La Jolla, California 92037

The initiation of sporulation appears to be controlled by at least nine genes called *spo0* genes. The majority of the alleles of these *spo0* genes map in five loci that are unlinked on the *Bacillus subtilis* chromosome; the other four loci are defined by one or two alleles at each locus (6, 12). If we assume that sporulation is triggered by some internal metabolic signal that is transmitted to the transcription machinery to initiate the transcription of sporulation-specific genes, then *spo0* mutants may be envisioned as being blocked in any one of these processes, including the formation of the metabolic signal or the mechanism by which this signal is transmitted to the transcription machinery. Mutations in the transcription machinery itself also might be specific for sporulation. In addition to being defective in sporulation, stage 0 mutants have large pleiotropic effects on the synthesis of a wide variety of gene products that appear concomitantly with the sporulation process (2). The most pleiotropic phenotypes are exhibited by the mutations in the *spo0A* locus. In this report we examine the sequence of the *spo0A* locus and of some of its mutations that have profound effects on these phenotypes.

RESULTS

The *spo0A* locus was identified and isolated from a library of *B. subtilis* DNA cloned in the λ vector Charon 4A (3). The insert within one of these clones was subjected to extensive restriction enzyme analysis, and subfragments of the inserts were cloned within the integrative vector pJH101 (4). Using these subcloned fragments, we were able to identify those regions of the original cloned insert that contained transforming activity for *spo0A* mutations. Figure 1 shows the location of alleles in the *spo0A* locus, as defined by the subfragments cloned in the plasmids indicated at the bottom of the figure. All of the alleles except two were contained within the *Hpa*I to *Eco*RI fragment of 840 base pairs (bp). None of the mutations was found to the left of the *Hpa*I site. Two mutations were found in the *Eco*RI to *Hinc*II fragment of 660 bp to the right

of the majority of the alleles. These alleles to the right include the *spo0A9V* mutation, which is the classically less pleiotropic mutation in the *spo0A* locus (2). It should be noted that integration of the plasmids pJF1377 and pJF2042 by Campbell-type recombination in Spo⁺ strains results in a Cmʳ Spo⁻ phenotype. This result suggests that the fragments contained within these plasmids are included within the *spo0A* transcription unit. None of the other plasmids gives this phenotype when integrated as described.

To verify the location of the deletion 204, which covers many of the *spo0A* alleles, chromosomal DNAs from the wild-type strains and a strain bearing this deletion were digested with restriction endonucleases *Eco*RI and *Bgl*II, electrophoresed on agarose gels, transferred to nitrocellulose, and probed with a radioactive fragment DNA extending from the *Eco*RI site leftward to a *Bal*I site just left of the indicated *Cla*I site in Fig. 1. The results of this analysis, shown in Fig. 2, indicate that the *Eco*RI to *Bgl*II fragment is 0.53 kilobase in the Spo⁺ strain and 0.35 kilobase in the deletion strain. Thus, the deletion is approximately 180 bp long and is verified to be between the *Eco*RI and *Bgl*II sites.

Sequence of the *spo0A* locus. The entire *spo0A* locus and its flanking regions were sequenced by use of the Maxam and Gilbert technique (9). The potential open reading frames derived from this sequence are shown in Fig. 3. A large open reading frame extending from the *Hpa*I site through the *Eco*RI site, from left to right as shown in Fig. 3, was present. Since the majority of the *spo0A* mutations were located between the *Hpa*I and *Eco*RI restriction sites, we concluded that this major open reading frame must include the *spo0A* locus. In the region between the *Eco*RI site and the *Hinc*II site, no open reading frame of any substantial size was evident. This region contains both the *spo0A9V* and *spo0A153* alleles. One sizable open reading frame with the potential to code for a protein of about 17,000 molecular weight was uncovered in the region immediately after the *Bal*I restriction site and it stopped after the *Cla*I site.

Transcription of the *spo0A* locus. High-resolution S1 mapping (1) was used to determine the sites of initiation and termination of transcription for the *spo0A* locus in this region. mRNA

---

† Present address: Syntro Corp., San Diego, CA 92121.

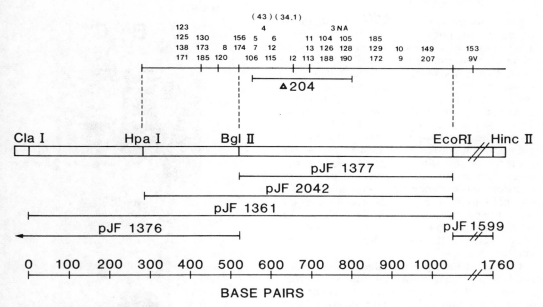

FIG. 1. Genetic and restriction maps of the *spo0A* locus. The top line is the genetic map of the region showing the restriction fragments in which the alleles fall. All the alleles above the deletion 204 bar are contained within the deletion. The plasmids have the indicated amount of DNA cloned in pJH101. Plasmid pJF1376 extends to the next *Eco*RI site, which is approximately 5.3 kilobases to the left of the *Bgl*II site.

fractions from logarithmically growing cells were hybridized to restriction fragments end labeled at the *Eco*RI and *Bgl*II sites. The site of transcription initiation was found to be approximately 270 bp upstream to the left of the *Bgl*II site. Termination of transcription was approximately 130 bp downstream to the right of the *Eco*RI site. This termination site (Fig. 4) has a typical terminatorlike structure preceding it that is capable of forming a hairpin loop with a calculated ΔG (25°C) = -19.2 kcal (ca. -80.4 kJ)/mol (15).

A putative promoter sequence ahead of the transcription start site has been found that resembles closely those observed for sigma-37-dependent promoters of *spoVC* (11), *spoVG* (10), and *sprE* (16) (Fig. 5). The -10 region of *spo0A* is identical to the -10 region of *spoVC* in seven of nine positions. The -35 region is identical to the -35 region of *spoVC* in four of six positions. In addition, an 11-bp sequence overlapping the putative -35 region of *spo0A* was found in the promoter region upstream from the *spo0F* coding sequence (14). If one aligns this sequence with known sigma-37 promoters, a good homology is found, suggesting that this may indicate a sigma-37-dependent promoter for the *spo0F* locus. We believe that the translation start for the *spo0A* gene is the GTG codon at position +51 (Fig. 4). This codon is commonly used for translation starts in *B. subtilis*, and it is preceded by a strong ribosome binding site,

GGAGG, located an optimal distance upstream. Earlier, we thought that the ATG codon downstream from this in the same frame at position +135 was the start of translation (4a). However, the location of the *sof-1* allele described below at position +86 suggests that the GTG codon is the actual start of translation. The protein specified by this sequence is 29,691 daltons.

Identification of the *spo0C* alleles. It has been a question for some time whether the *spo0A9V* allele was a less pleiotropic mutation of the *spo0A* locus or whether it actually was a mutation in a second locus linked to *spo0A* that has a different phenotype from *spo0A* mutations. To resolve this question, we cloned both the *spo0A9V* and *spo0A153* mutations and determined their sequence by the Maxam and Gilbert technique (9). The results of these analyses showed that both alleles were in the same base of the 11th codon preceding the translation stop codon of the *spo0A* gene (Fig. 6). Although these mutations reside in the same base of the same codon, they result in different substitutions of amino acids in the protein. Thus, both mutations give rise to a missense protein that is partially active, resulting in some of the phenotypes of the complete *spo0A* alleles.

Genetic location of the *sof-1* mutation. Two groups of investigators have recently isolated suppressors of *spo0F* mutations that allow sporulation to occur in the presence of a defective *spo0F* gene product (8, 13). In addition, such suppressors are active on strains carrying *spo0B* or *spo0E* mutations. Both of these mutations, *sof-1* and *rvt11*, were found to map in the region of the chromosome between *lys-1* and *aroD* and close to the *spo0A* locus. To map the *sof-1* mutation with respect to the *spo0A* locus, we used the plasmids described in Fig. 1 to determine the location of the *sof-1* allele. A strain, UOT0550 (*spo0FΔS sof-1*), was transformed by each of these integrative vectors, and among the chloramphenicol-resistant transformants the presence of the wild-type allele of *sof-1* was assayed by determining whether any of the Cm^r transformants were now Spo$^-$. Plasmids pJF1376 and pJF1361 were capable of transforming the strain to Spo$^-$. This locates the *sof-1* allele in the overlapping region in these plasmids, that is, the region between the *Cla*I and *Bgl*II restriction sites.

Cloning of the *sof-1* mutation. To clone the *sof-1* mutation, a transformant from a cross with plasmid pJF1361 was extracted and its chromosomal DNA was subjected to digestion with restriction endonuclease *Eco*RI. This *Eco*RI digest was then self-ligated with T4 ligase, and the ligation mixture was used to transform *Escherichia coli*. Among the chloramphenicol-resistant *E. coli* strains that arose from this transformation, one plasmid was found that was identical to plasmid pJF1361 in restriction map and was capable of transforming the *sof* phenotype to strains bearing mutations in the *spo0F* gene. Thus, this plasmid contains the *sof-1* allele rather than the wild-type allele in the DNA between the *Cla*I and *Bgl*II sites. Sequence analysis of the insert in this plasmid, pJH2074, revealed only a single base change (Fig. 7), and this change results in a transversion of a T to a G in the sequence ATAATC, resulting in the ATAAGC. This sequence codes for part of the amino-terminal end of the *spo0A* gene (Fig. 8), and the mutation results in a substitution of lysine for asparagine in the 12th position of the *spo0A* gene product. No other differences from the wild type were observed in this sequence.

DISCUSSION

The *spo0A* locus is believed to code for a protein of 29,691 daltons, as determined from the nucleic acid sequence of the locus. Because we have been unable to obtain an intact *spo0A* gene in either *E. coli* or *B. subtilis* on a plasmid, we have been unable to confirm this by minicell-type experiments. By use of gene fusion experiments we have been able to show that the copy number of the *spo0A* gene product is approximately 10 to 100 copies per cell, and this gene product is maximally expressed during the vegetative phase of growth, declining as the cells enter sporulation (unpublished data). This is

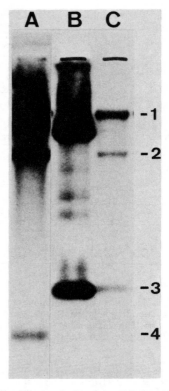

FIG. 2. Southern analysis of the deletion 204-carrying strain. Chromosomal DNAs from the wild-type strain (lane C) and a strain carrying the Δ204 mutation (lane A) or the plasmid pJF1377 (lane B) were digested with *Bgl*II and *Eco*RI restriction nucleases. The digests were electrophoresed in agarose, transferred to nitrocellulose membranes, and hybridized to nick-translated plasmid pJF1361. Band 1 is a 5.3-kilobase *Eco*RI-*Bgl*II fragment starting at the *Bgl*II site in the *spo0A* gene and ending at an *Eco*RI site to the left (see Fig. 1) of the gene. Band 2 is the *Eco*RI-*Eco*RI fragment containing the 9V allele to the right of the *spo0A* gene. This fragment appears because the probe is contaminated with labeled DNA homologous to this region. Bands 3 and 4 are the *Bgl*II-*Eco*RI fragments from the *spo0A* gene.

consistent with our previous observations that the *spo0A* gene products are vegetative products and lends support to our working hypothesis of many years that these gene products are dispensable vegetative functions that sense the metabolic state of the cell and regulate the decision as to whether to divide or to begin sporulation.

Perhaps one of the most interesting results of this study is the finding that the *sof-1* suppressor is a missense mutation in the amino-terminal portion of the *spo0A* protein. Since the *sof-1* suppressor is capable of suppressing mutations in the *spo0B*, *spo0E*, and *spo0F* genes, it seems clear that the *spo0B*, *spo0E*, and *spo0F* gene

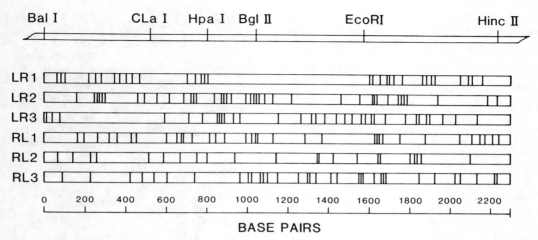

FIG. 3. Open reading frames in the sequence of the *spo0A* region. Each line represents a stop codon in the left to right (LR) or right to left (RL) direction.

```
          -67                           -37                              -7    +1
AGAATAACAAAGATATGCCACTAATATTGGTGATTATGATTTTTTTAGAGGGTATATAGCGGTTTTGTCGAATGTAAA

           12                            42                              72
CATGTAGCAAGGGTGAATCCTGTTAACTACATTTGGGGAGGAAGAAACGTGGAGAAAATTAAAGTTTGTGTTGCTGAT
                                        fMetGluLysIleLysValCysValAlaAsp

           90                           120                             150
GATAATCGAGAGCTGGTAAGCCTGTTAAGTGAATATATAGAAGGACAGGAAGACATGGAAGTGATCGGCGTTGCTTAT
AspAsnArgGluLeuValSerLeuLeuSerGluTyrIleGluGlyGlnGluAspMetGluValIleGlyValAraTyr

          168                           198                             228
AACGGACAGGAATGCCTGTCGCTGTTTAAAGAAAAAGATCCCGATGTGCTCGTATTAGATATTATTATGCCGCATCTA
AsnGlyGlnGluCysLeuSerLeuPheLysGluLysAspProAspValLeuValLeuAspIleIleMetProHisLeu

          246                           BglII                          306
GACGGGACTTGCGGTTTTAGAGAGGCTGAGGGAATCAGATCTGAAAAAACAGCCGAATGTCATTATGCTGACAGCCTTT
AspGlyLeuAlaValLeuGluArgLeuArgGluSerAspLeuLysLysGlnProAsnValIleMetLeuThrAlaPhe

          324                           354                             384
GGGCAGGAAGATGTCACGAAAAAGGCCGTCGATTTAGGCGCGTCCTACTTTATTCTCAAACCGTTTGATATGGAAAAC
GlyGlnGluAspValThrLysLysAlaValAspLeuGlyAlaSerTyrPheIleLeuLysProPheAspMetGluAsn

          402                           432                             462
CTTGTCGGCCATATCCGCCAGGTCAGCGGAAATGCCAGCAGTGTGACGCATCGTGCGCCATCATCGCAAAGCAGTATT
LeuValGlyHisIleArgGlnValSerGlyAsnAlaSerSerValThrHisArgAlaProSerSerGlnSerSerIle

          480                           510                             540
ATACGCAGCAGCCAGCCTGAACCAAAGAAGAAAAATCTCGACGCGAGCATCACAAGCATTATCCATGAAATCGGCGTC
IleArgSerSerGlnProGluProLysLysLysAsnLeuAspAlaSerIleThrSerIleIleHisGluIleGlyVal

          558                           588                             618
CCAGCCCATATTAAAGGCTATCTCTATCTGCGCGAAGCAATCTCAATGGTATACAATGACATCGAATTGCTCGGCAGC
ProAlaHisIleLysGlyTyrLeuTyrLeuArgGluAlaIleSerMetValTyrAsnAspIleGluLeuLeuGlySer

          636                           666                             696
ATTACAAAAGTCCTCTATCCGGACATCGCCAAAAAATTTAACACAACCGCAAGCCGTGTAGAAAGAGCGATCCGCCAT
IleThrLysValLeuTyrProAspIleAlaLysLysPheAsnThrThrAlaSerArgValGluArgAlaIleArgHis

          714                           744                             774
GCAATTGAAGTGGCATGGAGCAGAGGAAACATTGATTCCATTTCCTCGTTGTTTGGTTATACTGTCAGCATGACAAAA
AlaIleGluValAlaTrpSerArgGlyAsnIleAspSerIleSerSerLeuPheGlyTyrThrValSerMetThrLys

          792                           EcoRI   822                      852
GCTAAACCTACCAACAGTGAATTCATTGCAATGGTTGCGGATAAGCTGAGGTTAGAGCATAAGGCTTCTTAAACATGA
AlaLysProThrAsnSerGluPheIleAlaMetValAlaAspLysLeuArgLeuGluHisLysAlaSer OC

          870                           900                             930  T
GCTTATTAAGTGGTCATTAAATCAAACGTCTTTTATTTATTAGTTTGCGCTGATAAATAGGAGGCGTTTTGTTTTGGG
          >>>>>>>>>>>>>>>>>>>>>      <<<<<<<<<<<<<<<<<<<<<<
```

FIG. 4. Nucleotide sequence of the *spo0A* locus. Transcription starts at nucleotide +1 and stops at nucleotide 933.

```
         -35                         -10
spo0A  ATTA TGATTT TTTTAGAGGGTATATAGC GGTTTTGTC GAATGTAA

spoVC  TTCG AGGTTT AAATCCTTATCGTTATG  GGTATTGTT TGTAATA

spoVG  GAGC AGGATT TCAGAAAAAATCGT      GGAATTGAT ACACTAA

sprE   GAAT AGTCTT TTAAGTAAGTCTACT     CGAATTTTT TTA

spo0F  ATTA TGATTT TCGTCAAAAGTAAGC     AGTATTGTA TG
```

FIG. 5. Sequence of putative spo0A promoter compared with known sigma-37-dependent promoters. The underlined nucleotides are common to spo0A and spo0F genes.

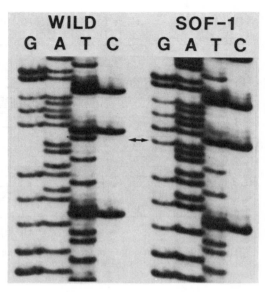

FIG. 7. Autoradiograph of a portion of the sequencing gel for the ClaI-EcoRI fragment isolated from wild-type and sof-1 strains.

products are unnecessary for sporulation if the spo0A product has been altered by missense mutation. This suggests that the role of the spo0B, spo0E, and spo0F gene products is to modulate the activity of the spo0A gene product, perhaps by activating it from an inactive form in response to environmental signals. These gene products could act either in a sequential or a concerted fashion to effect this activity. Thus, the sof-1 mutation is envisioned to give rise to an altered spo0A gene product that no longer requires the interaction of these spo0 genes in order to be active.

This model impinges on recent results of Gilman and Chamberlin (5) on transcription from promoters controlled by the sigma-28 form of RNA polymerase. These investigators found that sigma-28-dependent promoters required the spo0A, spo0B, spo0E, and spo0F gene products in order to be transcribed. This transcription could be restored in spo0B, spo0E, and spo0F strains by the introduction of the rvtA1 allele, which we now are virtually certain is the equivalent of the sof-1 mutation of spo0A. This suggests that the important gene product for transcription from sigma-28-controlled promoters is the spo0A gene product. One might envision the spo0A gene product as a cotranscriptional factor that interacts with RNA polymerase molecules containing minor forms of sigma factors. Alternatively, it is still possible that the spo0A gene product and the 28,000-dalton sigma factor are identical.

These studies have shown also that the spo0A9V mutation and a similar mutation,

spo0A153, are missense alterations in the carboxyl-terminal region of the protein. The spo0A9V allele is a classic allele that is less pleiotropic than most spo0A mutations and for some time was thought to be in the locus spo0C rather than spo0A. The finding that the spo0A9V allele is actually part of the spo0A gene simply indicates that the phenotypes attributed to spo0A9V are those of a partially active spo0A protein that loses the highly pleiotropic phenotypes of mutations in this locus, but nevertheless is unable to sporulate. Several years ago, it was shown that the introduction of rifampin, rif, mutations in spo0A9V strains resulted in the conversion of the less pleiotropic phenotype to a phenotype more characteristic of the other alleles at this locus (7). One interpretation of this result would be that the spo0A gene product interacts directly with RNA polymerase and that mutation in a polymerase subunit simply compensates for mutation in the spo0A gene. A

```
      ECORI

168  GAA TTC ATT GCA ATG GTT GCG DAT AAG CTG AGG TTA GAG CAT AAG GCT TCT TAA
     GLU PHE ILE ALA MET VAL ALA ASP LYS LEU ARG LEU GLU HIS LYS ALA SER  OC

153  GAA TTC ATT GCA ATG GTT GAG GAT AAG CTG AGG TTA GAG CAT AAG GCT TCT TAA
     GLU PHE ILE ALA MET VAL GLU ASP LYS LEU ARG LEU GLU HIS LYS ALA SER  OC

9V   GAA TTC ATT GCA ATG GTT GTG GAT AAG CTG AGG TTA GAG CAT AAG GCT TCT TAA
     GLU PHE ILE ALA MET VAL VAL ASP LYS LEU ARG LEU GLU HIS LYS ALA SER  OC
```

FIG. 6. Sequence of mutations at the carboxyl terminus of the spo0A gene. The box indicates the amino acid changes in each mutant.

```
+1
AACATGTAGCAAGGGTGAATCCTGTTAACTACATTTGGG

GAGGAAGAAAC GTG GAG AAA AAT AAA GTT TGT
            fMet Glu Lys Ile Lys Val Cys
```

```
wild     GTT GCT GAT GAT AAT CGA GAG CTG GTA AGC
         Val Ala Asp Asp Asn Arg Glu Leu Val Ser

sof-1    GTT GCT GAT GAT AAG CGA GAG CTG GTA AGC
         Val Ala Asp Asp Lys Arg Glu Leu Val Ser
```

FIG. 8. Amino-terminal portion of the *spo0A* gene.

model of this kind would be consistent with the hypothesis advanced above.

Obtaining the sequence of the *spo0A* gene product allowed us to compare it with two amino acid sequences of known proteins and thus to determine whether the *spo0A* gene product was related in any way to any protein of known function. Upon comparison of this protein with other proteins in the National Institutes of Health Sequence Data Bank, a relationship was found between this protein and the *ompR* gene product of *E. coli* (17). Comparison of the two sequences reveals a conserved region encompassing 97 amino acids, beginning about 44 amino acids from the amino terminus of both proteins (Fig. 9). In this region, 36 of the residues are identical and 26 are conservative or favored changes. In the portion of both proteins amino terminal to this region, 17 of 43 residues are identical and 4 of 43 are conserved (data not shown). In contrast, the carboxyl portion of each molecule shows no identity above random chance. The function of *ompR* gene product in *E. coli* is to regulate at the transcriptional level

the production of *ompF* and *ompC* matrix proteins of the outer membrane of *E. coli*. There is some evidence that *ompR* regulates several genes other than these two genes, and it seems possible that *ompR* may be more than just a repressor of the *ompF* and *ompC* genes (C. Higgins, personal communication). It will be of interest to see whether these proteins are related in function as well as sequence.

SUMMARY

The *spo0A* locus has been cloned and sequenced, and the product of this locus has been determined to be a protein of 29,691 daltons. Several mutations in this region have been cloned and sequenced. It has been found that the *spo0A9V* and the *spo0A153* mutations are mutations in the 10th codon preceding the carboxyl terminus of the *spo0A* protein and that both mutations result in different missense mutations from alterations in the same base of this codon. Thus, the *spo0A* locus appears to consist of a single gene product, not two genes, *spo0A* and *spo0C*, as previously thought. The *sof-1* mutation has been localized by genetic studies to the region of the *spo0A* locus. Cloning and sequencing of the *sof-1* mutation reveals that this mutation results in a change in the 12th codon of the *spo0A* protein, giving rise to a missense mutation at this position. This result indicates that the *spo0B*, *spo0E*, and *spo0F* gene products are unnecessary for efficient sporulation in the presence of an altered *spo0A* gene product. A hypothesis is advanced that the function of the *spo0B*, *spo0E*, and *spo0F* gene products is to

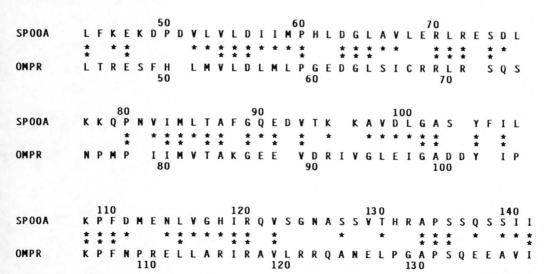

FIG. 9. Comparison of the amino acid sequence of the *spo0A* protein and the *ompR* protein. The numbers above and below the line are residue numbers for *spo0A* and *ompR*, respectively. Identical residues are denoted by two stars; conserved or favored residue changes, by one star.

modulate the activity of the *spo0A* gene product by a concerted or sequential mechanism. It is suggested that the *spo0A* gene product interacts directly with the transcription machinery. The primary amino acid structure of the *spo0A* protein resembles closely, in one region, that of the *ompR* gene product of *E. coli*.

LITERATURE CITED

1. **Berk, A. J., and P. A. Sharp.** 1977. Sizing and mapping of early adenovirus mRNAs by gel electrophoresis of S1 endonuclease digested hybrids. Cell **12**:721–732.
2. **Brehm, S. P., S. P. Staal, and J. A. Hoch.** 1973. Phenotypes of pleiotropic-negative sporulation mutants of *Bacillus subtilis*. J. Bacteriol. **115**:1063–1070.
3. **Ferrari, E., D. J. Henner, and J. A. Hoch.** 1981. Isolation of *Bacillus subtilis* genes from a Charon 4A library. J. Bacteriol. **146**:430–432.
4. **Ferrari, F. A., A. Nguyen, D. Lang, and J. A. Hoch.** 1983. Construction and properties of an integrable plasmid for *Bacillus subtilis*. J. Bacteriol. **154**:1513–1515.
4a. **Ferrari, F. A., K. Trach, J. Spence, E. Ferrari, and J. A. Hoch.** 1984. Cloning and sequence analysis of the *spo0A* locus of *Bacillus subtilis*, p. 323–331. *In* A. T. Ganesan and J. A. Hoch (ed.), Genetics and biotechnology of bacilli. Academic Press, Inc., New York.
5. **Gilman, M. Z., and M. J. Chamberlin.** 1983. Developmental and genetic regulation of *Bacillus subtilis* genes transcribed by sigma-28-RNA polymerase. Cell **35**:285–293.
6. **Hoch, J. A.** 1976. Genetics of bacterial sporulation. Adv. Genet. **18**:69–99.
7. **Ikeuchi, T., K. Babasaki, and K. Kurahashi.** 1979. Genetic evidence for possible interaction between ribonucleic acid polymerase subunit and the *spo0C* gene product of *Bacillus subtilis*. J. Bacteriol. **139**:327–332.
8. **Kawamura, F., and H. Saito.** 1983. Isolation and mapping of a new suppressor mutation of an early sporulation gene *spo0F* mutation in *Bacillus subtilis*. Mol. Gen. Genet. **192**:330–334.
9. **Maxam, A. M., and W. Gilbert.** 1980. Sequencing end-labeled DNA with base-specific chemical cleavages. Methods Enzymol. **65**:499–560.
10. **Moran, C. P., Jr., N. Lang, C. D. B. Banner, W. G. Haldenwang, and R. Losick.** 1981. Promoter for a developmentally regulated gene in *Bacillus subtilis*. Cell **25**:783–791.
11. **Moran, C. P., Jr., N. Lang, and R. Losick.** 1981. Nucleotide sequence of a *Bacillus subtilis* promoter recognized by *Bacillus subtilis* RNA polymerase containing σ^{37}. Nucleic Acids Res. **9**:5979–5990.
12. **Piggot, P. J., and J. G. Coote.** 1973. Genetic aspects of bacterial endospore formation. Bacteriol. Rev. **40**:908–962.
13. **Sharrock, R. A., S. Rubinstein, M. Chan, and T. Leighton.** 1984. Integenic suppression of the *spo0* phenotypes by the *Bacillus subtilis* mutation *rvtA*. Mol. Gen. Genet. **194**:260–264.
14. **Shimotsu, H., F. Kawamura, Y. Kobayashi, and H. Saito.** 1983. Early sporulation gene *spo0F*: nucleotide sequence and analysis of gene product. Proc. Natl. Acad. Sci. U.S.A. **80**:658–662.
15. **Tinoco, I., Jr., P. N. Borer, B. Dengler, M. D. Levine, O. C. Uhlenbeck, D. M. Crothers, and J. Gralla.** 1973. Improved estimation of secondary structure in ribonucleic acids. Nature (London) New Biol. **246**:40–41.
16. **Wong, S.-L., C. W. Price, D. S. Goldfarb, and R. W. Doi.** 1984. The subtilisin E gene of *Bacillus subtilis* is transcribed from a σ^{37} promoter *in vivo*. Proc. Natl. Acad. Sci. U.S.A. **81**:1184–1188.
17. **Wurtzel, E. T., M.-Y. Chou, and M. Inouye.** 1982. Osmoregulation of gene expression. I. DNA sequence of the *ompR* gene of the *ompB* operon of *Escherichia coli* and characterization of its gene product. J. Biol. Chem. **257**:13685–13691.

Analysis of the Control of *spo* Gene Expression in *Bacillus subtilis*

P. J. PIGGOT, J. W. CHAPMAN, AND C. A. M. CURTIS

Division of Microbiology, National Institute for Medical Research, Mill Hill, London NW7 1AA, England

About 50 genetic loci are known to be specifically involved in formation of spores by *Bacillus subtilis* (14, 16). There is evidence that some of these loci are expressed in a dependent sequence (1, 22), although this does not imply that all are expressed in a single sequence. How the whole network of locus expression is organized remains poorly understood. In attempting to unravel the control network, it is instructive to consider a gene, X, that is switched on during sporulation and has the function of switching on some other gene, Y. Such genes must surely exist to account for the sequences that are observed. However, it seems inherently unlikely that X is an isolated gene with the sole function of switching on one other gene—if so, why have X at all? Rather, it seems likely either that X would have more than the function of switching on one other gene or that it could be part of a polycistronic operon which has additional functions. We have been pursuing the latter possibility by analyzing cloned DNA from the *spoIIA–spoVA* region, where recombination frequencies and phenotypes had suggested two, or more, genes in close proximity (13). Initially, a plasmid, pHM2, carrying part of the *spoIIA* locus was isolated from the clone bank of Rapoport et al. (12, 19). The plasmid-borne portion of the *spoIIA* region complemented some, but not all, chromosomal *spoIIA* mutations, from which it was concluded that the locus contained more than one gene (12); it was inferred that pHM2 contained the *spoIIA* promoter. We have now cloned the rest of the locus and have established that it is transcribed as a polycistronic unit. We are analyzing how its transcription is controlled by using *lac* fusions.

The approach outlined above is now being supplemented by analysis of the effects of gene copy number. This has been applied to *spoIIA*. We are also interested in genes that are necessary for sporulation, but which inhibit it when present in high copy number; the first published example of this was *spo0F* (10), and it has also been reported for *spoVG* (2). Clearly, such "copy number" genes must be fairly precisely controlled to allow spore formation. They may indeed be pivotal to the control of the whole process. Even if this is not so, analysis of the effects of gene copy number is a powerful analytical tool.

EXPERIMENTAL PROCEDURES

Strains. The *B. subtilis* strains used were BR151 (*metB10 lys-3 trpC2*), SL566 (*spo0A34 phe-12 rif-2 tal-1*), SL989 (*recE4 metB10 lys-3*), SL964 (*spo0B136 metC3 tal-1*), SL1013 (*spoIIA69 lys-3 trpC2*), SL1060 (*spoIIA69 trpC2 recE4*), and JH649 (*spo0F221 phe-1 trpC2*). The *Escherichia coli* strains were DH1 (8) and SL2034, a Dam⁻ strain whose origin is obscure. Plasmids were maintained in DH1 or SL2034 (when they were to be restricted with *Bcl*I) unless otherwise stated.

lac **fusions.** Methods used were essentially those described by Donnelly and Sonenshein (5). Plasmid pDB1, kindly provided by A. L. Sonenshein, has the same promoter-cloning site as pCED6 (5), but is unable to replicate in *B. subtilis* and expresses chloramphenicol, rather than kanamycin, resistance in *B. subtilis*.

Copy number measurements. The copy number relative to the chromosome of autonomously replicating plasmids was estimated by quantitation of ethidium bromide fluorescence by use of gel photography (17, 18).

Measurement of copy number of pPP41 integrated into the chromosome utilized Southern (21) hybridization of DNA restricted with *Bgl*II and fractionated by electrophoresis in 0.3% agarose gels. The 12-kilobase (kb) plasmid pPP41 contained no *Bgl*II sites so that successive tandem repeats of pPP41 in the chromosome gave a progressively larger *Bgl*II fragment hybridizing to a probe of DNA from within the cloned region of pPP41: zero copies, 6.3 kilobase pairs (kbp); one copy, 18.3 kbp; two copies, 30.3 kbp, etc.

Sporulation in liquid culture. The modified Schaeffer system described by Leighton and Doi (11) was used for spore formation. Sporulation was monitored by phase-contrast microscopy.

All other experimental procedures have been described previously (15).

RESULTS

Cloning the rest of the *spoIIA* locus. By Southern hybridization, with pHM2 (12) used as a probe, it was deduced that a *Bgl*II fragment of

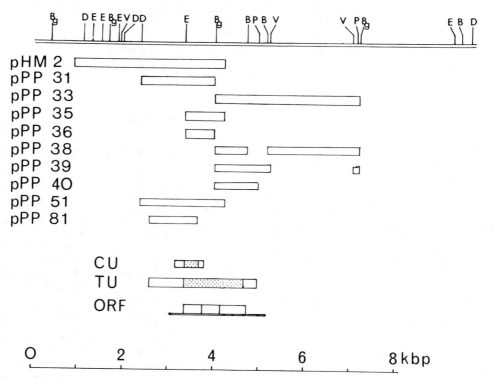

FIG. 1. Restriction map of the *spoIIA* region of the chromosome. Above the map the approximate locations of mutations *spo-69*, *spo-37*, and *spo-1* are indicated. Below the map, the regions cloned in different plasmids are shown. pPP31, pPP33, pPP38, pPP39, and pPP40 have pJAB1 as vector; pPP35, pPP36, and pPP51 have pJH101 as vector; pPP81 has pDB1 as vector. The minimum (stippled area) and maximum sizes of the *spoIIA69* complementing unit (CU; 3) and of the transcriptional unit (TU) are also indicated, as are the open reading frames (ORF) identified in the DNA sequence (7). The scale in kbp is shown at the bottom of the figure. Restriction sites are: B, *Bcl*1; Bg, *Bgl*II; D, *Hind*III; E, *Eco*RI; P, *Pst*I; V, *Eco*RV.

chromosomal DNA of about 3.3 kbp encompassed the portion of the *spoIIA* locus not contained in pHM2. Accordingly, 168 Trp⁺ DNA was cut with *Bgl*II, and fragments of about this size were fractionated by electrophoresis in 0.8% agarose and extracted by the method of Dretzen et al. (6). They were then cloned into the *Bcl*1 site of the vector pJAB1, and clones were selected as tetracycline-resistant transformants of the Dam⁻ *E. coli* strain SL2034. From the transformants one was isolated that harbored a plasmid, pPP33, containing the rest of the *spoIIA* locus (15) and a portion of the adjacent *spoVA* locus (J. Errington, personal communication). The restriction map of the *spoIIA* region, deduced from the maps of pHM2 and pPP33, is shown in Fig. 1. The locations of *spoIIA* mutations are also indicated; these were determined from the ability of various subclones to recombine with them so as to give *spo*⁺ recombinants. Analysis of the chromosome of BR151 by South-

ern hybridization confirmed that the DNA had not been rearranged on cloning.

Size of the *spoIIA* transcriptional unit. The approximate size of the *spoIIA* transcriptional unit was determined by using clones of portions of the *spoIIA* region in integrational plasmid vectors. The vectors cannot replicate in *B. subtilis* (they are maintained in *E. coli* where they can replicate). They cannot transform *B. subtilis* unless they have *B. subtilis* DNA cloned in them; when selection is made for a vector determinant, the whole plasmid integrates into the chromosome at the region of homology by a Campbell-like mechanism. When the homologous region is entirely within a chromosomal transcriptional unit, the transcriptional unit is inactivated (Fig. 2); if the homologous region contains even one end of a transcriptional unit, the unit is not inactivated. A series of integrational plasmids carrying different portions of *spoIIA* (Fig. 1) were used to transform the *spo*⁺

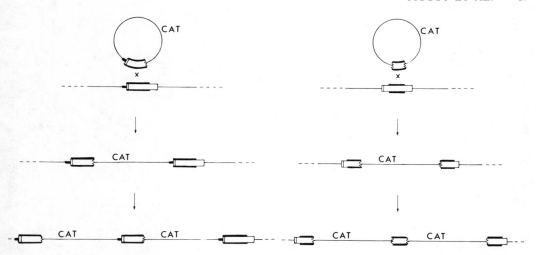

FIG. 2. Diagram illustrating the integration of plasmids into the *B. subtilis* chromosome by a Campbell-like mechanism. The region of homology between plasmid and chromosome is indicated by thick lines. A complete transcriptional unit (TU) is shown as a rectangle; incomplete TUs are indicated by serrated ends to the rectangle. The bottom part of the diagram shows the structure upon tandem duplication of the integrated plasmid. The location of the vector-determined chloramphenicol resistance is indicated by CAT. On the left side integration of a plasmid having one end of the TU leads to a complete TU in the chromosome. On the right side, integration of a plasmid whose region of homology is entirely within the TU results in loss of the complete TU.

strain BR151, selecting for vector-determined chloramphenicol resistance. Plasmids pPP31, pPP33, pPP39, pPP40, pPP51, and pPP81 did not disrupt sporulation, whereas plasmids pPP35 and pPP36 did (Table 1). Analysis by Southern hybridization confirmed that integration was by a Campbell-like mechanism (15). From these results it is deduced that there is a transcriptional unit with upper and lower size limits shown in Fig. 1. The lower limit is substantially larger than the cistron defined by complementation of *spoIIA* mutations (3) and covers all known *spoIIA* point mutations. This indicates that the transcriptional unit is polycistronic and is the only transcriptional unit in the *spoIIA* locus. As pPP33 and pPP39 do not disrupt sporulation on transformation and do not in-

TABLE 1. Transformations of BR151 to chloramphenicol resistance

Donor	Transformants[a]	
	Spo⁻	Spo⁺
pPP31	0	136
pPP33	0	712
pPP35	340	0
pPP36	325	0
pPP39	0	1,640
pPP40	0	108
pPP51	0	508
pPP81	0	206
None	0	0

[a] The numbers are colonies obtained from 0.2 ml of transformed culture.

clude all of the adjacent *spoVA* locus, it is concluded that the *spoVA* locus is not in the *spoIIA* transcriptional unit. A subclone of pPP33, pPP38, gives further information on this point. It differs from pPP33 only in the deletion of a 0.43-kbp *Bcl*I piece from the middle of the cloned region. Chloramphenicol-resistant transformants of BR151 with this plasmid have either a SpoII (30%) or a SpoV (70%) phenotype. This fits with the left-hand *Bgl*II–*Bcl*I portion of the insert (Fig. 1) being in the *spoIIA* transcriptional unit and indicates that the right-hand *Bcl*I–*Bgl*II portion (Fig. 1) is entirely within the *spoVA* transcriptional unit; thus the two transcriptional units are close together, but separate.

Analysis of the DNA sequence of the *spoIIA* locus confirms the in vivo analysis outlined above. The locus consists of three adjacent open reading frames on the same strand (7). Their position corresponds to the transcriptional unit (Fig. 1), with the position of the first open reading frame corresponding to the cistron in pHM2 that was delimited by Tn*1000* mutagenesis and subcloning (13). The direction of reading confirms that the cloned region on pHM2 contains the *spoIIA* promoter.

lac **fusion of** *spoIIA*. We have fused the *spoIIA* region to the *E. coli* β-galactosidase structural gene (*lacZ*), using a transcriptional fusion vector, pDB1, kindly provided by A. L. Sonenshein. pDB1 is an integrational vector for *B. subtilis* and is maintained in *E. coli*. It has a unique *Hin*dIII site for cloning promoters. The *Hin*dIII fragment containing the *spoIIA* pro-

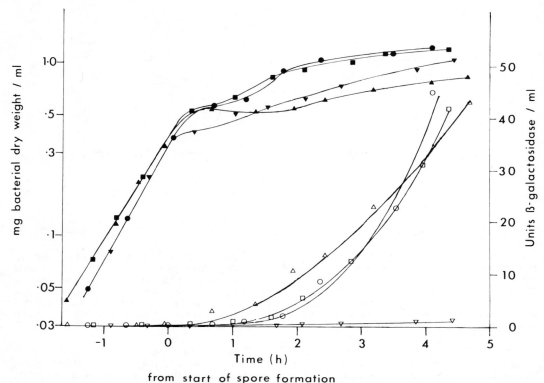

FIG. 3. Production of β-galactosidase in liquid sporulation cultures. Bacterial growth is shown with filled symbols, β-galactosidase activity with empty symbols. BR151 transformed with pPP81 and grown at 5 (●, ○) and (50 ■, □) μg of chloramphenicol per ml. SL566 transformed with pPP81 and grown at 5 (▼, ▽) and 50 (▲, △) μg of chloramphenicol per ml. β-Galactosidase units are the increase in E_{410nm} in 1 h with o-nitrophenyl galactoside as substrate. Enzyme activity is expressed as units per milliliter of culture except for BR151 (pPP81)-Cm50, where it is expressed as units per 0.2 ml.

moter was unsuitable for this purpose. We used instead an AhaIII fragment of 1.05 kbp. This was isolated from a plasmid subclone of pHM2. It includes part of open reading frame 1 and extends about 700 bp upstream from the translation start of open reading frame 1. HindIII linkers were ligated to the ends of the fragment, and it was then cloned into the HindIII site of the positive selection vector pJAB1; from there it was cloned into pDB1, creating pPP81. Analysis confirmed that pPP81 contained the required insert (Fig. 1).

Strain BR151 was transformed to chloramphenicol resistance with pPP81. The transformants, selected on nutrient agar containing 5 μg of chloramphenicol per ml, had β-galactosidase activity as indicated by hydrolysis of the chromogenic substrate 5-bromo-4-chloro-indolyl galactoside (X-gal) incorporated into the agar. (A plasmid having the same insert as pPP81, but in the opposite orientation, had no β-galactosidase activity.) Transformant clones were then assayed for enzyme activity during growth and sporulation in liquid culture (Fig. 3). Significant

synthesis of β-galactosidase did not occur until about 1.5 h after the start of sporulation.

pPP81 was transformed into a spo0A mutant, again selecting for resistance to 5 μg of chloramphenicol per ml, but using the protoplast transformation system of Chang and Cohen (4). A transformant clone was picked and monitored for β-galactosidase activity during growth and sporulation in liquid culture. No significant activity was detected (Fig. 3).

Effect of spoIIA copy number. It is known that plasmids integrated into the B. subtilis chromosome by a Campbell-like mechanism can form tandem repeats (24). Further, the number of repeats can be increased by raising the level of the drug for which the plasmid confers resistance (24). Accordingly, pPP81 transformant clones were grown on agar containing 30, or 50, μg of chloramphenicol per ml. Cultures were then assayed for enzyme activity during growth and sporulation. This procedure increased the amount of enzyme made by the spo+ strain. More strikingly, it overcame the effect of the spo0A mutation on β-galactosidase synthesis

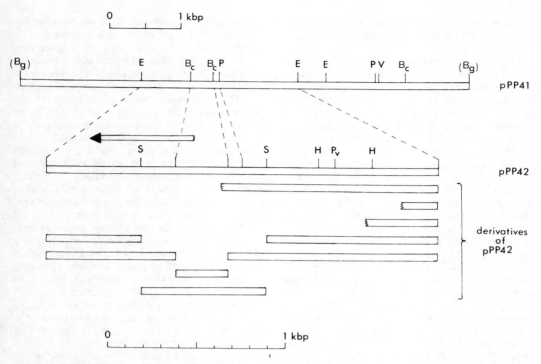

FIG. 4. Restriction map of the *spo0F* region cloned on pPP41. The lower part of the figure shows the maps of pPP42 and its derivatives drawn on an expanded scale (at the bottom) compared with that of pPP41 (at the top). Serrated ends indicate that these are deleted derivatives of pPP42 and the end has only been located approximately. The position of the *spo0F* open reading frame (20) is indicated above the map of pPP42, with the arrow showing the direction of reading. All the derivatives of pPP42 allowed sporulation. Restriction sites are: Bc, *Bcl*I; Bg, *Bgl*II; E, *Eco*RI; H, *Hin*dIII; P, *Pst*I; Pv, *Puv*II; S, *Sst*I; V, *Eco*RV. The *Bgl*II sites at the ends of pPP41 are in parentheses, as they are presumed to be present on the chromosome but are lost on cloning into pJAB1 to give pPP41.

(Fig. 3). In preliminary experiments qualitatively similar results have been obtained with *spo0B*, *spo0C*, and *spo0F* mutants. Increasing the copy number did not impair sporulation of *spo+* strains or restore the ability to sporulate to *spo0* strains. In no case did expression become constitutive; rather, the enzyme still began to be synthesized after the start of sporulation.

Analysis of *spo0F*. To clone *spo0F*, we first established that fragments of 6.0 to 6.6 kbp resulting from *Bgl*II digestion of chromosomal DNA had transforming activity for *spo0F*. Fragments of this size were then cloned into the *Bcl*I site of pJAB1, and a plasmid, pPP41, was isolated that had *spo0F* transforming activity (Fig. 4). Comparison of the restriction map of pPP41 with that of the *spo0F* chromosomal region (obtained by Southern hybridization) confirmed that no rearrangements had taken place during cloning. Moreover, the 2.2-kbp *Eco*RI fragment obtained from pPP41 was identical in size to the *spo0F* fragment cloned by Kawamura et al. (10); this fragment contains the entire *spo0F* gene (20). It was subcloned from pPP41 into the

*Eco*RI site of pHV33, and the resulting plasmid, pPP42, was stably maintained in *E. coli*. In *B. subtilis* it was also stably maintained as an autonomous plasmid, provided selection for chloramphenicol resistance was applied. It suppressed the ability of strain SL989 to sporulate, confirming the observations of Kawamura et al. (10). When selection for chloramphenicol resistance was removed, plasmid pPP42 was rapidly lost and, concomitantly, the ability to form spores was restored.

Multiple copies of the 0.7-kbp *Sst*I piece that includes part of the *spo0F* structural gene and extends more than 400 bp upstream from the *spo0F* translation start site (20) did not suppress sporulation. Thus, presence of the *spo0F* promoter region alone did not prevent sporulation. Various subclones and deletion derivatives of pPP42 were tested, and the ability to suppress sporulation correlated with the presence of multiple copies of the intact *spo0F* gene. It is tentatively concluded that the copy number effect is caused by the *spo0F* gene product. However, the results are also consistent with the effect being caused by a region extending for more than 400 bp upstream from the *spo0F* open reading frame.

The copy number of pPP42 in *B. subtilis* SL989 growing exponentially in media containing 5 μg of chloramphenicol per ml was about 14 per chromosome, assuming a chromosome size of 3×10^9 daltons (23). This was clearly an average value for a heterogeneous population; some bacteria in the population even contained no plasmid. It is not known how autonomous plasmid replication relates to chromosome replication. Moreover it is not known what happens to plasmids during sporulation either when chromosome replication stops or when the bacterium divides unequally. We therefore concentrated our effort on the integratable plasmid pPP41 to obtain more precise information on the effects of gene copy number.

B. subtilis BR151 was transformed with pPP41, and recombinants were selected on nutrient agar containing 5 μg of chloramphenicol per ml; these were Spo⁺. A single transformant clone was used to inoculate Penassay broth containing 5 or 30 μg of chloramphenicol per ml. It was grown overnight at 35°C and plated onto nutrient agar containing the same levels of antibiotic. All the colonies at 5 μg/ml were brown and opaque; all examined contained 80%, or more, spores. All those at 30 μg/ml were unpigmented and translucent; all examined contained <0.1% spores. Clones were then grown in Penassay broth containing the same levels of chloramphenicol, and samples were taken to measure copy number. Bacteria in the culture grown at 5 μg of chloramphenicol per ml contained a single copy of pPP41 (and so two copies of *spo0F*). Those grown at 30 μg of chloramphenicol per ml contained four copies of pPP41 (and so five copies of *spo0F*) integrated by a Campbell-like mechanism. In both cases less than 10% of the population had different copy numbers from those indicated. The inhibition of sporulation was not caused by the level of chloramphenicol, nor by the replication of the vector, as comparable experiments with *spoIIA* clones did not inhibit sporulation. Thus, it was caused by the cloned region of pPP41. This included additional DNA to the *spo0F* gene (notably *ctrA*). However, in view of our results with pPP42 and subclones from it, and also the results of Kawamura et al. (10), it seems likely that it is the five copies of *spo0F* that inhibit sporulation.

DISCUSSION

Analysis of the *spoIIA* locus with integrational plasmids has established that the locus is transcribed in vivo as a single polycistronic unit and is transcribed separately from the adjacent *spoVA* locus. The use of integrational plasmid

vectors in this way is novel. It depends on the demonstration that, when such vectors had *B. subtilis* DNA cloned in them, they integrated into the *B. subtilis* chromosome at the region of homology by a Campbell-like mechanism on selection for the vector-determined drug resistance (15). It provides the first demonstration of a polycistronic operon being involved in spore formation. However, there is no reason to think that *spoIIA* is exceptional in this regard.

We do not yet know how expression of *spoIIA* is regulated or how it regulates expression of other sporulation genes. To analyze the former matter, strains were constructed in which transcription of *spoIIA* led to β-galactosidase synthesis. Transcription of *spoIIA* started about 1.5 h after the start of sporulation and was prevented by *spo0* mutations. Increasing the copy number of *spoIIA-lac* fusions in a *spo⁺* strain increased the level of β-galactosidase synthesis severalfold. One way to interpret this is to assume that a positive activator controls *spoIIA* expression and is present in excess when a single *spoIIA* gene is present. Increasing the gene copy number causes more expression by utilizing more activator. Presumably, such an activator can be titrated out by sufficient copies, and careful quantitation of gene copy number against gene expression may indicate whether this is the case or whether the gene is being expressed at the maximal rate without the involvement of an activator.

Increasing the copy number of *spoIIA-lac* fusions in *spo0* strains overcame the block in β-galactosidase synthesis caused by *spo0* mutations. The easiest interpretation for such behavior is to assume that increasing copy numbers titrate out a repressor and allow gene expression. This could imply that *spo0* genes function to remove such a repressor. Again, careful quantitation may help to establish such an interpretation and to integrate it with the considerations in the previous paragraph. Strikingly, in all circumstances in which *spoIIA* was transcribed (as monitored by β-galactosidase synthesis), transcription started about 1.5 h after the start of sporulation. It would seem that the determinants of this temporal control are separate from the *spo0* loci. It is hoped that a search for mutants expressing *spoIIA* constitutively may identify the temporal control(s).

The *spo0F* locus exhibits a much more striking example of the effect of copy number (10). The effect appears to require the intact gene or a large region upstream from the gene. Our results also bring to light the extreme sensitivity of sporulation to *spo0F* in that five copies of the gene prevent sporulation although two do not. The mechanism for this effect is unknown. Kawamura and Saito (9) have recently described a second site mutation, *sof*, that completely suppresses all *spo0F* mutations, including deletions; *sof* is in *spo0A* (J. A. Hoch, personal communication). It will be interesting to see whether high copy numbers of the wild-type *spo0F* gene can prevent sporulation in the presence of *sof*. The proximity of *spo0F* to the origin of chromosome replication means that its gene dosage relative to a marker near the terminus (such as *spo0A*) would be two- to fourfold higher during exponential growth than during sporulation; this clearly must be taken into account in interpreting the effects of *spo0F* copy number. It is even possible that *spo0F* is a gene that couples the state of chromosome replication to the start of sporulation.

SUMMARY

Integratable plasmids carrying different portions of the *B. subtilis spoIIA* locus were used to establish that the locus is a polycistronic unit. A plasmid has been constructed in which expression of *E. coli* β-galactosidase comes under the control of the *spoIIA* promoter. It was shown that *spoIIA* transcription commences about 90 min after the start of sporulation and ordinarily requires the functioning of *spo0* loci. The presence of multiple copies of the *spoIIA-lacZ* fusion increases the level of β-galactosidase in a *spo*⁺ strain and allows enzyme expression in *spo0* strains; expression remains subject to sporulation control. This indicates complex regulation of *spoIIA* expression.

In an analysis of the effects of gene dosage it was shown that five copies of the *spo0F* region prevent spore formation. The relationship of such "copy number" genes and of polycistronic operons to the whole network of gene expression during sporulation was discussed.

LITERATURE CITED

1. **Balassa, G.** 1969. Biochemical genetics of bacterial sporulation. 1. Unidirectional pleiotropic interactions among genes controlling sporulation in *Bacillus subtilis*. Mol. Gen. Genet. **104**:73–103.
2. **Banner, C. D. B., C. P. Moran, and R. Losick.** 1983. Deletion analysis of a complex promoter for a developmentally regulated gene from *Bacillus subtilis*. J. Mol. Biol. **168**:351–365.
3. **Chak, K.-F., H. Lencastre, H.-M. Liu, and P. J. Piggot.** 1982. Facile *in vivo* transfer of mutations between the *Bacillus subtilis* chromosome and a plasmid harboring homologous DNA. J. Gen. Microbiol. **128**:2813–2816.
4. **Chang, S., and S. N. Cohen.** 1979. High frequency transformation of *Bacillus subtilis* protoplasts by plasmid DNA. Mol. Gen. Genet. **168**:111–115.
5. **Donnelly, C. E., and A. L. Sonenshein.** 1984. Promoter-probe plasmid for *Bacillus subtilis*. J. Bacteriol. **157**:965–967.
6. **Dretzen, G., M. Bellard, P. Sassone-Corsi, and P. Cham-**

bon. 1981. A reliable method for the recovery of DNA fragments from agarose and acrylamide gels. Anal. Biochem. **112**:295–298.
7. **Fort, P., and P. J. Piggot.** 1984. Nucleotide sequence of sporulation locus, *spoIIA*, in *Bacillus subtilis*. J. Gen. Microbiol. **130**:2147–2153.
8. **Hanahan, D.** 1983. Studies on transformation of *Escherichia coli* with plasmids. J. Mol. Biol. **166**:557–580.
9. **Kawamura, F., and H. Saito.** 1983. Isolation and mapping of a new suppressor mutation of an early sporulation gene *spo0F* mutation in *Bacillus subtilis*. Mol. Gen. Genet. **192**:330–334.
10. **Kawamura, F., H. Shimotsu, H. Saito, H. Hirochika, and Y. Kobayashi.** 1981. Cloning of *spo0* genes with bacteriophage and plasmid vectors in *Bacillus subtilis*, p. 109–113. In H. S. Levinson, A. L. Sonenshein, and D. J. Tipper (ed.), Sporulation and germination. American Society for Microbiology, Washington, D.C.
11. **Leighton, T. J., and R. H. Doi.** 1971. The stability of messenger ribonucleic acid during sporulation in *Bacillus subtilis*. J. Biol. Chem. **246**:3189–3195.
12. **Liu, H.-M., K.-F. Chak, and P. J. Piggot.** 1982. Isolation and characterization of a recombinant plasmid carrying a functional part of the *Bacillus subtilis spoIIA* locus. J. Gen. Microbiol. **128**:2805–2812.
13. **Piggot, P. J., K.-F. Chak, H.-M. Liu, and H. Lencastre.** 1983. *Bacillus subtilis*: genetics and spore formation, p. 163–166. In D. Schlessinger (ed.), Microbiology—1983. American Society for Microbiology, Washington, D.C.
14. **Piggot, P. J., and J. G. Coote.** 1976. Genetic aspects of bacterial endospore formation. Bacteriol. Rev. **40**:908–962.
15. **Piggot, P. J., C. A. M. Curtis, and H. Lencastre.** 1984. Use of integrational plasmid vectors to demonstrate the polycistronic nature of a transcriptional unit (*spoIIA*) required for sporulation of *Bacillus subtilis*. J. Gen. Microbiol. **130**:2123–2136.
16. **Piggot, P. J., A. Moir, and D. A. Smith.** 1981. Advances in the genetics of *Bacillus subtilis* differentiation, p. 29–39. In H. S. Levinson, A. L. Sonenshein, and D. J. Tipper (ed.), Sporulation and germination. American Society for Microbiology, Washington, D.C.
17. **Projan, S. J., S. Carleton, and R. P. Novick.** 1983. Determination of plasmid copy number by fluorescence densitometry. Plasmid **9**:182–190.
18. **Pulleyblank, D. E., M. Shure, and J. Vinograd.** 1977. The quantitation of fluorescence by photography. Nucleic Acid Res. **4**:1409–1418.
19. **Rapoport, G., A. Klier, A. Billault, F. Fargette, and R. Dedonder.** 1979. Construction of a colony bank of *E. coli* containing hybrid plasmids representative of the *Bacillus subtilis* 168 genome. Expression of functions harbored by the recombinant plasmids in *B. subtilis*. Mol. Gen. Genet. **176**:239–245.
20. **Shimotsu, H., F. Kawamura, Y. Kobayashi, and H. Saito.** 1983. Early sporulation gene *spo0F*: nucleotide sequence and analysis of gene product. Proc. Natl. Acad. Sci. U.S.A. **80**:658–662.
21. **Southern, E.** 1979. Gel electrophoresis of restriction fragments. Methods Enzymol. **68**:152–176.
22. **Waites, W. M., D. Kay, I. W. Dawes, D. A. Wood, S. C. Warren, and J. Mandelstam.** 1970. Sporulation in *Bacillus subtilis*. Correlation of biochemical events with morphological changes in asporogenous mutants. Biochem. J. **118**:667–676.
23. **Wake, R. G.** 1980. How many chromosomes in the *Bacillus subtilis* spore, and of what size? Spore Newsl. **7**:21–26.
24. **Young, M.** 1984. Gene amplification in *Bacillus subtilis*. J. Gen. Microbiol. **130**:1613–1621.

Bacillus subtilis Genes Involved in Spore Outgrowth

MAURIZIO GIANNI,[1] FRANCESCO SCOFFONE,[2] AND ALESSANDRO GALIZZI[1]

Dipartimento di Genetica e Microbiologia, "A. Buzzati-Traverso,"[1] and Istituto di Genetica Biochimica ed Evoluzionistica, Consiglio Nazionale delle Ricerche,[2] Pavia, Italy

The isolation of *Bacillus subtilis* mutants which are temperature sensitive only in spore outgrowth has proved the presence of functions specific to this stage of the bacterial life cycle (4). Thus, the process of spore outgrowth is not a mere resumption of vegetative life; rather, it appears to consist of a controlled sequence of steps, whose regulation is still incompletely known. Overall, in several attempts and using different enrichment procedures, we have isolated 12 mutants, belonging to five loci, which are temperature sensitive during outgrowth. Genetic analysis and physiological characterization of the mutant strains have failed, so far, to give an answer to the main questions: (i) are genes involved in spore outgrowth expressed only during outgrowth, and (ii) if there is a differential expression of genes, at what level is regulation taking place and through which mechanisms? With the aim of defining more precisely the role of different genes in spore outgrowth, we have tried a direct approach to the problem—cloning some of the genes involved in the process. This should enable us to analyze the genes at the level of the DNA sequence and to study gene expression and regulation at the molecular level.

Here we present data about the molecular cloning and preliminary characterization of genes involved in spore outgrowth. The four clones which will be described are all derived from the library of *B. subtilis* DNA in phage lambda Charon 4A constructed by Ferrari, Henner, and Hoch (2). To single out the desired clones, we followed two strategies. First, we analyzed the DNA from the library for the ability to transform outgrowth temperature-sensitive mutants to ts^+. With this procedure we have isolated one clone with the *outB* marker. The second method consisted of screening the library by means of hybridization to ^{32}P-labeled RNA prepared from outgrowing spores in the presence of a large excess of cold competitor RNA extracted from vegetative cells. Three clones were isolated by the latter procedure.

EXPERIMENTAL PROCEDURES

Construction of an ordered collection from the lambda library. To each well of a sterile microtiter tray, 100 µl of TYD medium (2) containing 1% agarose was added. After the agarose had solidified, 33 µl of top agarose (TYD medium containing 0.4% agarose and the host strain DP50) was added. Single plaques from the lambda library were picked into wells with a toothpick. Each set of plaques was replicated onto square petri dishes with a sterile plastic replicator. The transfer was done through an intermediate microtiter tray whose wells contained 30 µl of TM buffer (10 mM Tris-hydrochloride, pH 7.4; 10 mM MgCl₂). By these procedures large patches of lysis were obtained.

Extraction of RNA from outgrowing spores. Spores were inoculated into nutrient broth. After 12 or 24 min of incubation at 37°C, further growth was stopped by adding sodium azide and rapid chilling. The cells were collected by centrifugation, the pellet was resuspended in DNase buffer (10 mM sodium acetate, pH 5.3; 50 mM NaCl; 1 mM MgCl₂; 50 µg of DNase per ml) and was sonicated in the presence of glass beads. The RNA was extracted with water-saturated phenol.

In vitro labeling of RNA. One microgram of RNA was ethanol precipitated; the pellet was dried and then dissolved in 5 µl of 50 mM Tris (pH 9.5). The RNA was partially hydrolyzed in a sealed capillary tube, immersed in boiling water for 2 min, and then cooled in ice. The content of the capillary was transferred to a tube containing 5 µl of buffer (50 mM Tris, pH 9.5; 20 mM MgCl₂; 10 mM dithiothreitol), 20 µCi of [γ-^{32}P]ATP, and 1 U of polynucleotide kinase. After incubation at 37°C for 30 min, the reaction was stopped by the addition of 200 µl of Tris-EDTA buffer and heating at 70°C for 5 min. The unincorporated nucleotide was removed by gel filtration. The specific activity obtained was between 2×10^6 and 4×10^6 cpm/µg of RNA.

RESULTS AND DISCUSSION

Gene *outB*. The involvement of gene *outB*, mapping between *amyE* and *aroI906*, in spore outgrowth was demonstrated by the isolation of a temperature-sensitive mutant (*outB81*). Screening the Charon 4A library for clones capable of transforming a mutant strain from temperature sensitivity to ts^+ gave a positive result for five clones that, from later analysis, turned out to be identical. The positive clone λ C4BsG40 contains a *B. subtilis* DNA insert of 14

FIG. 1. Electron micrograph of a heteroduplex molecule between λ C4BsG40 and λ C4BsG40Δ20. The lower panel is a drawing of the heteroduplex. The bar represents 5 kbp.

kb (kilobase pairs), present on a single *Eco*RI fragment. Because of the size of the insert, we searched for the presence of additional markers linked to *outB* by transformation. The DNA from λ C4BsG40 was capable of transforming an *aroI906* mutant to prototrophy and *amyE* strains to ability to hydrolyze starch. Since *outB* maps between *amyE* and *aroI*, this result indicates that the entire *outB* gene is present in the λ clone. To compare and orient the genetic map

with a restriction map of the cloned fragment, we have isolated a deletion mutant of λ C4BsG40. The mutant was obtained by screening the survivors of a chelating agent treatment (8). The deletion involves about 4.0 kb. Restriction analysis and physical mapping by heteroduplex analysis in an electron microscope place the deletion at the junction between the *B. subtilis* DNA insert and the right arm of λ DNA (Fig. 1). The mutant phage DNA was still effective in trans-

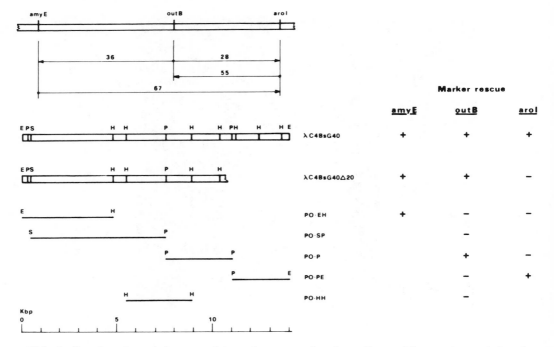

FIG. 2. Genetic and restriction map of the region surrounding the *outB* gene. The genetic map is based on two- and three-factor crosses in transduction, and the distances are expressed as 100 minus the percentage of cotransduction. The lower part of the figure reports the restriction map of λ C4BsG40, the extent of its deletion derivative, and the subfragments cloned in various plasmids. The results of the marker rescue experiments are summarized at the right of each clone. The restriction sites are summarized as follow: E, *Eco*RI; H, *Hind*III; P, *Pst*I; S, *Sal*I.

formation experiments involving the *amyE* and *outB* markers, whereas it was completely inactive when used to select for Aro⁺.

More data about the physical location of *outB* were obtained by subcloning several fragments of the insert into suitable plasmids (pBR322 and pBR325) and by testing for marker rescue by transformation (Fig. 2). The only plasmid that gave transformation to Aro⁺ was pO · PE, carrying the *Pst*I-*Eco*RI fragment of 2.8 kb to the right end of the insert, as represented in Fig. 2. Transformation to *ts*⁺ of strain *outB81* was

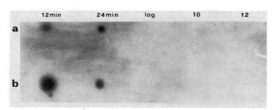

FIG. 3. Dot hybridization of nick-translated pO · P to RNA extracted at various stages of growth. RNA extracted from outgrowing spores (at 12 and 24 min), and from cells during logarithmic growth and at t_0 and t_2 was spotted on Gene Screen filters and probed with ^{32}P-labeled pO · P DNA. Samples of top row contained 6.8 μg of RNA; those of bottom row, 34 μg.

obtained with pO · P, whereas pO · HH, whose insert partially overlaps the one present in pO · P, was totally ineffective. Therefore, the *outB81* mutation must lie between the right *Hind*III site of pO · HH and the *Hind*III nearest to the right *Pst*I site of pO · P. The 3.5-kb *Pst*I fragment of pO · P was recloned in a *B. subtilis* replicative plasmid (pHV14) and assayed for complementation. Transformants whose spores were *ts*⁺ were obtained when the recipient was *rec* proficient, whereas no *ts*⁺ transformant was observed with competent cells of an *outB81 recE* strain. From these results we infer that the *Pst*I fragment of 3.5 kb does not contain the complete *outB* gene. Since the plasmid pO · P has at least a portion of the *outB* gene, it could be used as a probe to study the expression of the gene during the cell cycle. Total RNA was extracted from spores 12 and 24 min after initiation of germination and outgrowth and from cells during logarithmic growth, and at t_0, t_2, and t_4. The RNA was spotted on Gene Screen filters (New England Nuclear Corp.) and probed with nick-translated pO · P. The only samples that gave positive results were the ones obtained from outgrowing spores (Fig. 3). We conclude that the *outB* gene is transcribed only during outgrowth. This is in agreement with the physi-

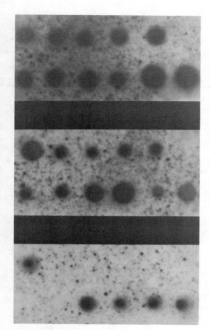

FIG. 4. Hybridization-competition experiment. DNA extracted from 10 lambda clones was spotted (1.7 μg) on nitrocellulose paper. After denaturation it was hybridized to RNA from outgrowing spores (12 min) and labeled in vitro with [γ-^{32}P]ATP and polynucleotide kinase. The amount of RNA added was 1 μg at a specific activity of 10^7 cpm/μg. Top panel: no competitor RNA added. Middle panel: 500 μg of unlabeled vegetative RNA was added. Bottom panel: 1,000 μg of unlabeled vegetative RNA was added. To the top right corner of each filter, DNA from the λ Charon 4A vector was spotted. The sample at the bottom right is DNA from λ C4BsG40. The four positive clones are (from left to right, top to bottom): λ C4BsUG2, IG9, UF7, and BA7.

ological data which confine the temperature-sensitive period to be beginning of outgrowth (1).

Screening the lambda library by hybridization-competition. We have screened the library of *B. subtilis* DNA in λ Charon 4A by hybridization to labeled RNA from outgrowing spores in the presence of a 500-fold excess of unlabeled vegetative RNA (7). By this procedure we hoped to isolate clones with sequences expressed during outgrowth. To simplify the screening from the original library, we prepared an ordered collection of clones in microtiter trays. Each set of clones were then replicated on square petri dishes to obtain patches of lysis, whose DNA was blotted on nitrocellulose filters. The RNA from outgrowing spores (12 min of incubation in nutrient broth) was labeled in vitro with [γ-^{32}P]ATP and polynucleotide kinase. From the original screening we chose 10 positive clones

which hybridized to a significant extent. Subsequent experiments with higher concentrations of vegetative competitor RNA showed that only four of the clones could stand a more stringent analysis. Figure 4 reports such an experiment with increasing concentrations of competitor RNA. The four positive clones were named BA7, IG9, UF7, and UG2. The sizes of their inserts were 16.5, 12.6, 11.4, and 11.1 kb, respectively. As expected, as a result of the average size of the insert in λ Charon 4A, all clones were transcribed during vegetative growth as well. A preliminary restriction map of the *B. subtilis* DNA inserts is reported in Fig. 5. From the restriction maps it appears that two clones, λ C4BsIG9 and λ C4BsBA7, have a common region. The overlap was verified by Southern blot and hybridization to the nick-translated *Eco*RI fragments of 1.8 and 7.8 kb (unpublished data).

Interestingly enough, λ C4BsG40, the clone identified by transformation in the same library, was not detected in the screening with the [^{32}P]RNA probes. A possible explanation is that the clone was absent from the ordered collection employed. Alternatively, the screening may not be sensitive enough for sequences expressed at a low level. Two clones, λ C4BsIG9 and λ C4BsUF7, were further analyzed. Fragments from the insert were subcloned in pJH101 (3), and the recombinant plasmids were used to transform *B. subtilis* competent cells, selecting for chloramphenicol-resistant transformants. Since pJH101 is unable to replicate in *B. subtilis*, transformants arise by integration of the plasmid into the chromosome at the site of homology (5, 6). This enabled us to genetically map the cloned sequences by means of the chloramphenicol resistance determinant (6). The insert in clone λ C4BsIG9 is linked in transduction to *pyrA*, *dnaP*, *dnaA*, and *glnA*. The order inferred from two-factor crosses and map distances was *pyrA-dnaP-cam-dnaA-glnA*. In the same region we have previously mapped gene *outD*, by means of a temperature-sensitive mutant. Neither clone IG9 or clone BA7 was capable of transforming strains with the *outD* mutation to *ts*$^+$. The insert derived from clone λ C4BsUF7 is linked in transduction to *pyrA* and in transformation to *pycA*.

Time of expression of the cloned sequences. The cloned sequences of λ C4BsIG9 and λ C4BsUF7, as well as their derivatives in pJH101, were nick translated and used to probe RNA extracted at various times of growth and spotted on Gene Screen filters. The aims of the experiment were to determine the time of expression of the cloned sequences and eventually to define the region transcribed during outgrowth. The DNA from clone IG9 hybridizes to RNA extracted from

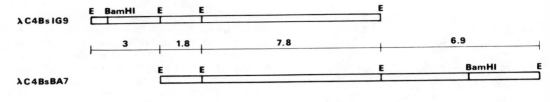

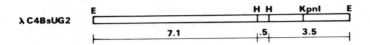

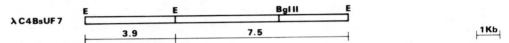

FIG. 5. Restriction map of the four clones isolated by hybridization-competition.

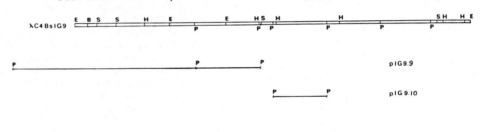

B = Bam HI
H = Hind III
P = Pst I
S = Sst I

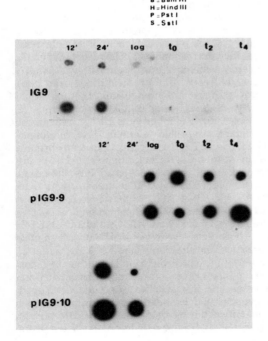

FIG. 6. Dot hybridization of nick-translated DNA from λ C4BsIG9 and plasmids derived from it to RNA extracted at various stages of growth. The top panel represents a detailed restriction map of λ C4BsIG9 and derived plasmids pIG9-9 and pIG9-10. The bottom panel is an autoradiograph of the dot hybridization experiment. Two quantities (6.8 μg, top row; 24 μg, bottom row) of RNA extracted from outgrowing spores (12 and 24 min) and cells in log phase and at t_0, t_2, and t_4 were spotted on Gene Screen filters and probed with ^{32}P-labeled DNA.

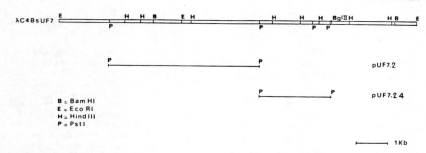

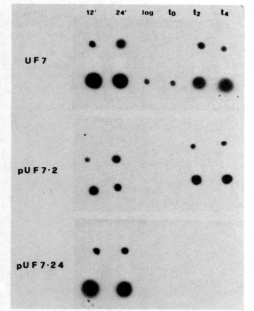

FIG. 7. Dot hybridization of nick-translated DNA from λ C4BsUF7 and plasmids derived from it to RNA extracted at various stages of growth. The top panel represents a detailed restriction map of λ C4BsUF7 and derived plasmids pUF7-2 and pUF7-24. The bottom panel is an autoradiograph of the dot hybridization experiment (for details see legend to Fig. 6).

outgrowing spores and, to a lesser extent, to RNA from vegetative cells and cells in the process of sporulation (Fig. 6). One of the plasmid derivatives (pIG9-9), with a large insert embracing the left half of the cloned DNA, hybridizes strongly to RNA from vegetative and sporulating cells and does not hybridize to RNA from outgrowing spores. Plasmid pIG9-10, with a 1.7-kb insert, gives a strong signal only with RNA extracted from outgrowing spores. Thus, the region transcribed during outgrowth is in the *Eco*RI fragment of 7.8 kb common to both λ C4BsIG9 and λ C4BsBA7. The results obtained with λ C4BsUF7 and its derivatives are shown in Fig. 7. The DNA from the original clone hybridizes to all RNA samples, even though to a lesser extent to RNA from vegetative cells. One of the plasmids (pUF7-2) hybridizes to RNA from outgrowing spores and to RNA derived from sporulating cells. Plasmid pUF7-24, whose *B. subtilis* DNA insert is adjacent to the one present in pUF7-2, only gives hybrids with outgrowth RNA. We conclude that the sequence of

interest should be in the central portion of the lambda clone.

Further experiments with different subfragments should enable us to define more precisely the extent of the region specifically transcribed during spore outgrowth. The same type of experiments, i.e., genetic mapping via plasmid integration and dot hybridization to unlabeled RNA, will be performed with the third clone isolated (λ C4BsUG2).

In conclusion, we have succeeded in cloning four DNA sequences of *B. subtilis* whose expression appears to be temporally programmed. The molecular basis for this regulation is under investigation.

SUMMARY

From the Charon 4A library of Ferrari et al. (2) we have isolated four genes involved in *B. subtilis* spore outgrowth. One of the genes (*outB*) was previously identified by a mutation whose phenotypic effect is temperature-sensitive spore outgrowth. Three genes have been obtained by screening the library by hybridization with labeled RNA from outgrowing spores in the presence of a large excess of unlabeled vegetative RNA. Fragments of the cloned DNA were subcloned in pJH101. By plasmid integration and PBS1 transduction, the chromosomal loci were mapped. Using DNA probes derived from the cloned sequences in dot hybridization to RNA extracted from cells at different developmental stages, we have shown that the corresponding transcripts are present only during the outgrowth phase.

We thank Lee Ferrari for reading the manuscript.

The work was partially supported by Consiglio Nazionale delle Richerche (Rome) grant 83.01990.04. M.G. was a recipient of a fellowship from the Ministero degli Esteri (Italy).

LITERATURE CITED

1. **Albertini, A. M., and A. Galizzi.** 1975. Mutant of *Bacillus subtilis* with a temperature-sensitive lesion in ribonucleic-cacid synthesis during germination. J. Bacteriol. **124**:14–25.

2. **Ferrari, E., D. J. Henner, and J. A. Hoch.** 1981. Isolation of *Bacillus subtilis* genes from a Charon 4A library. J. Bacteriol. **146**:430–432.

3. **Ferrari, F. A., A. Nguyen, D. Lang, and J. A. Hoch.** 1983. Construction and properties of an integrable plasmid for *Bacillus subtilis*. J. Bacteriol. **154**:1513–1515.

4. **Galizzi, A., A. M. Albertini, M. L. Baldi, E. Ferrari, E. Isnenghi, and M. T. Zambelli.** 1978. Genetic studies of spore germination and outgrowth in *Bacillus subtilis*, p. 150–157. *In* G. Chamblis and J. C. Vary (ed.), Spores VII. American Society for Microbiology, Washington, D.C.

5. **Galizzi, A., F. Scoffone, G. Milanesi, and A. M. Albertini.** 1981. Integration and excision of a plasmid in *Bacillus subtilis*. Mol. Gen. Genet. **182**:99–105.

6. **Haldenwang, W. G., C. D. B. Banner, J. F. Ollington, R. Losick, J. A. Hoch, M. B. O'Connor, and A. L. Sonenshein.** 1980. Mapping a cloned gene under sporulation control by insertion of a drug resistance marker into the *Bacillus subtilis* chromosome. J. Bacteriol. **142**:90–98.

7. **Mangiarotti, G., S. Chung, C. Zuker, and H. F. Lodish.** 1981. Selection and analysis of cloned developmentally-regulated *Dictyostelium discoideum* genes by hybridization-competition. Nucleic Acids Res. **9**:947–963.

8. **Parkinson, J. S., and R. J. Huskey.** 1971. Deletion mutants of bacteriophage lambda. Isolation and initial characterization. J. Mol. Biol. **56**:369–384.

Functions of *spo0F* Upstream Region for the Sporulation of *Bacillus subtilis*

HIROFUMI YOSHIKAWA, HIDETAKA SONE, TATSUYA SEKI, FUJIO KAWAMURA, AND HIUGA SAITO

Institute of Applied Microbiology, University of Tokyo, Bunkyo-Ku, Tokyo 113, Japan

We have cloned a 2.2-kilobase pair (kb) DNA fragment carrying a functional sporulation gene, *spo0F*, of *Bacillus subtilis* by means of "prophage transformation" (5, 6) and determined the sequence of 1,162 base pairs (bp), including an open reading frame which codes for a 19,000-dalton polypeptide (7). In the course of constructing plasmids we unexpectedly found that, when pUB110-derived high-copy plasmids carrying the 2.2-kb DNA fragment were introduced into *B. subtilis* cells, these plasmids markedly inhibited sporulation of the host cells (6). When the plasmid-harboring cells, which are otherwise sporulation proficient, were incubated on a sporulation medium, they segregated sporeforming cells at a frequency of around 10^{-5}. These sporeforming cells contained smaller plasmids than the parental ones, which carried deletions of various sizes extending into the 2.2-kb insert. In several instances the coding region, the ribosome binding site, and the putative promoters of the *spo0F* gene remained intact, but the upstream sequences were extensively deleted. This sporulation inhibition by the *spo0F* gene propagated in the high-copy-number plasmids appears similar to that reported for *spoVG* by Banner et al. (1). The latter workers carried out in vitro deletion analyses and found that the upstream A+T-rich box was essential not only for sporulation inhibition but also for development of *spoVG* function.

Here we describe the effects of deletion and addition mutations in the region upstream of the *spo0F* gene on sporulation inhibition and sporulation function of this gene.

EXPERIMENTAL PROCEDURES

Microorganisms and plasmids. *B. subtilis* strains used are listed in Table 1.

The parental plasmids, pUBSF4 and pUBSF13 (6.7 kb; Fig. 1), both of which consist of pUB110 and a 2.2-kb fragment carrying the functional *spo0F* gene, were constructed as described previously (6). Derived plasmids are shown in Fig. 3.

Measurement of sporulation frequency. Cells were inoculated into sporulation medium (nutrient broth modified by Schaeffer) and shaken at

37°C for 20 h. In the case of plasmid-carrying strains, 5 μg of kanamycin was added to 1 ml of the medium for selection of plasmid-carrying cells. After incubation, samples were spread over tryptose blood agar base (TBAB; Difco) plates either directly to count viable cells or after heating at 90°C for 10 min to count sporulating cells.

Construction of upstream deletion mutations. A number of plasmids carrying the *spo0F* upstream deletion mutations were constructed from pUBSF13 DNA digested with HindIII. After elimination of the short DNA fragment, the large DNA fragment was digested with nuclease Bal-31 to various extents, and synthetic BamHI linkers were ligated to the blunt ends left after Bal-31 digestion. Cleavage with BamHI generated "sticky" ends at the linker sites and also cleaved pUB110 DNA at a unique BamHI site (0.8 kb away from the inserted DNA fragment) (Fig. 1). Ligation at low DNA concentration joined the BamHI linkers at the digested ends of the *spo0F* fragment to the BamHI site in the pUB110 DNA. A collection of deletion-derivative plasmids was thus generated. *B. subtilis* strain UOT-0277 was transformed by these plasmids, and kanamycin-resistant colonies were picked. Plasmid DNA was prepared from these individual transformants, and nucleotide sequences at the junction sites were determined. Some of these plasmids were designated the pOFΔ5' series.

Unless otherwise stated, we will refer to the position of individual bases in the upstream region of *spo0F* gene by their distances from the initiation codon of the gene as shown in Fig. 2.

In vitro mutagenesis in the internal region: minor mutations. pOFΔ3'-1 DNA was digested with BclI, PstI, or SacI. Each digest was treated with the Klenow fragment of DNA polymerase I in the presence of four deoxynucleoside triphosphates to generate blunt ends, and the termini were joined by ligation at a low DNA concentration. By these treatments a four-nucleotide insertion occurred as a result of BclI digestion, and a four-nucleotide deletion occurred as a result of PstI or SacI treatment; all changes were in the upstream region of the *spo0F* gene. These plasmids were designated pOFΔ3'-B1, -P1, and -S1, respectively.

TABLE 1. *B. subtilis* strains used

Strain	Genetic marker	Reference for *spo* gene
UOT-0277	*hisA1 metB5 nonB1(hsrM1) recE4*	*spo⁺*
UOT-0532	*trpC2 metB5 leuA8 nonB1 spo0F77*	Y. Kobayashi (unpublished data)
UOT-0779	*trpC2 metB5 nonB1 spo0F221*	JH649; J. Hoch (3)
UOT-0789	*trpC2 metB5 leuA8 nonB1 spo0FΔS (φ105wt)*	Deletion
UOT-0820	*trpC2 leuA8 nonB1 spo0FH1*	Insertion

PstI insertion. A large foreign DNA fragment was inserted into the unique *Pst*I site in the *spo0F* upstream region. pUBSF13 DNA was digested with *Pst*I and ligated with ρ11 phage DNA which had been digested with the same restriction enzyme. One of the resulting plasmids, named p0FP2, had a 1,600-bp insert at the *Pst*I site.

HpaI insertion. A foreign DNA fragment was inserted into the unique *Hpa*I site at the 26th base downstream from the initiation codon of the *spo0F* gene. The procedure was the same as that described above, except that the restriction enzyme was *Hpa*I and the foreign DNA was φ29 bacteriophage DNA. The resulting plasmid, p0FH1, had a 600-bp insert at the *Hpa*I site.

Introduction of *spo0F* gene fragments into the genome of the temperate phage φ105. Phage

φCM is a derivative of φ105 which has the chloramphenicol acetyltransferase gene from plasmid pC194 and a unique *Bam*HI site for cloning foreign DNA fragments (details will be described elsewhere; H. Miyachi and T. Seki, unpublished data). The *spo0F* DNA fragments from a series of p0F deletion plasmids were isolated by double digestion with *Bam*HI and *Eco*RI. The *Eco*RI termini were converted to *Bam*HI "sticky" ends by using a synthetic adapter and subsequent digestion with *Bam*HI. These fragments were then ligated with *Bam*HI-digested φCM, and prophage transformations were carried out with UOT-0789 carrying the φ105 prophage as recipient. Chloramphenicol-resistant transformants were selected and analyzed further to determine whether the *spo0F* fragments had inserted into the prophage genome. Since the chromosomal *spo0F* allele in strain UOT-0789 (*spo0FΔS*) deleted the DNA portion homologous to the *spo0F77* mutation site, cellular DNA was isolated from strains with *spo0F* DNA fragments inserted in the prophage genome, and their ability to transform the *spo0F77* strain UOT-0532 (Y. Kobayashi, unpublished data) to Spo⁺ was examined.

RESULTS AND DISCUSSION

Effect of *spo0F* coding region on sporulation inhibition. We examined the possibility that sporulation inhibition by the *spo0F* gene propagated on a high-copy-number plasmid was due to the production of gene products in excess. Since the unique *Hpa*I cleavage site of the *spo0F*-carrying plasmids was in the coding region of *spo0F*, 26 bp downstream from the initiation codon, *Hpa*I was used to linearize pUBSF4 DNA. *Bam*HI linkers were attached to the resulting blunt ends

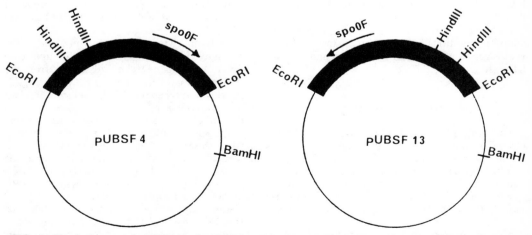

FIG. 1. Physical maps of pUBSF4 and pUBSF13. Thick lines represent 2.2-kb *spo0F* DNA fragment, and thin lines represent 4.5-kb vector plasmid pUB110.

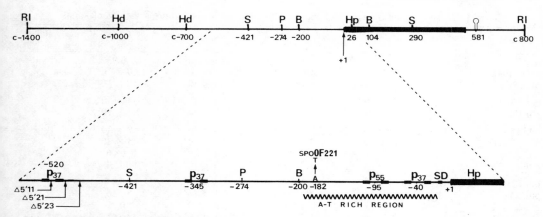

FIG. 2. *Eco*RI DNA fragment of 2.2 kb containing the *spo0F* gene. Negative numbers indicate distances from the first letter of the initiation codon of the *spo0F* coding frame. Thick line represents the coding frame. Semithick lines indicate ribosome binding site (S-D sequence) and −10 and −35 boxes of putative promoters for σ^{55} and σ^{37} RNA polymerases (4, 8). Arrows indicate the extent of upstream deletions created by Bal-31 treatments. A stem-and-loop signal corresponds to potential terminator. Abbreviations for restriction sites: RI, *Eco*RI; Hd, *Hin*dIII, S, *Sac*I; P, *Pst*I; B, *Bcl*I; Hp, *Hpa*I.

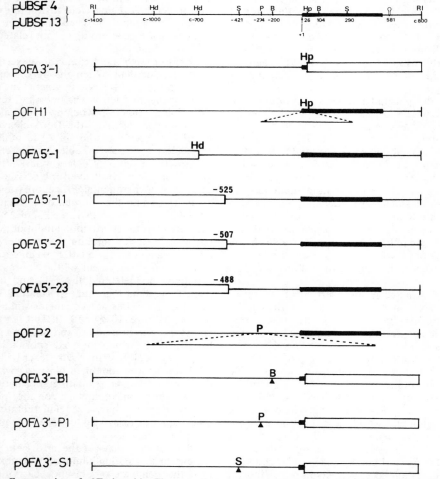

FIG. 3. Construction of p0F plasmids. The open lines represent deletions, and the additional lines indicated as triangles represent insertions. Only relevant restriction sites and base pair numbers, which are used to construct each plasmid, are shown. Abbreviations are the same as in Fig. 2. Only the relevant regions of the *spo0F* fragment are shown.

TABLE 2. Sporulation frequency of sporulation-proficient hosts harboring various plasmids

Plasmid	Mutation	Viable cells	Spores	S/V^a	Inhibition
pUB110		9.3×10^7	8.7×10^7	0.93	−
pUBSF4		4.4×10^7	4.5×10^2	1.0×10^{-5}	+
pUBSF13		1.9×10^8	8.8×10^3	4.6×10^{-5}	+
pOFΔ3′-1	Coding region	6.2×10^8	8.0×10^2	1.3×10^{-6}	+
pOFH1	Coding region	4.9×10^7	1.0×10^4	2.0×10^{-4}	+
pOFΔ5′-1	Upstream deletion	7.8×10^7	3.8×10^3	4.9×10^{-5}	+
pOFΔ5′-11	Upstream deletion	2.5×10^8	8.7×10^3	3.4×10^{-5}	+
pOFΔ5′-21	Upstream deletion	3.6×10^8	2.4×10^8	0.66	−
pOFΔ5′-23	Upstream deletion	1.7×10^8	1.6×10^8	0.94	−
pOFP2	Upstream insertion	3.2×10^8	1.4×10^8	0.44	−
pOFΔ3′-B1	Upstream addition	2.0×10^8	1.1×10^8	0.55	−
pOFΔ3′-P1	Upstream deletion	1.2×10^8	3.8×10^7	0.32	−
pOFΔ3′-S1	Upstream deletion	1.8×10^8	1.1×10^8	0.61	−
pOF221	*spo0F221*	9.1×10^7	2.3×10^7	0.25	−

^{a} Sporulation frequency is shown as the ratio of sporeforming cells to viable cells, S/V.

with T4 ligase, and then both the linker site and the unique *Bam*HI site derived from the parental plasmid were cleaved with *Bam*HI, giving two fragments. The large DNA fragment was purified and ligated end to end, giving rise to pOFΔ3′-1 (5.1 kb) which had lost a 1.6-kb fragment including most of the *spo0F* gene's coding region (Fig. 3).

When the plasmid pOFΔ3′-1 was introduced into otherwise sporulation-proficient host cells, sporulation was completely inhibited (Table 2). The plasmids carried by essentially all spores which did appear had some sort of deletion. The difference in the frequencies of sporulation between cells carrying pUBSF4 and pOFΔ3′-1 may reflect the difference in deletion frequencies of these two plasmids; i.e., plasmid pOFΔ3′-1 was more stable than pUBSF4, probably as a result of the smaller size of the former plasmid. Indeed, plasmid pOFΔ3′-1 is the most stable among the series of *spo0F*-carrying plasmids and was, therefore, used in many of the following experiments.

pOFH1, which had an insertion of 600 bp of foreign DNA at the *Hpa*I site in the *spo0F* coding region (Fig. 3), also exhibited sporulation inhibition and frequently gave rise to deleted plasmids (Table 2).

Although the complete gene product was not essential for sporulation inhibition, it is possible that the N-terminal peptide exerts the inhibition, as discussed by Banner et al. (1). Therefore, we constructed a series of plasmids in which the coding region was wholly lost. For this purpose pUBSF4 DNA cleaved with *Hpa*I was further digested with Bal-31. The digested DNA was attached to *Bam*HI linkers and circularized after *Bam*HI digestion, as described above. Although the results are preliminary, one of these plas-

mids, in which the upstream 50 bases together with the coding region were lost, still exhibited sporulation inhibition. Therefore, it is evident that the *spo0F* gene product is not related to this inhibitory function.

Extent of *spo0F* upstream region exerting sporulation inhibition. When the *spo0F* upstream region in the 2.2-kb DNA fragment was cleaved with various restriction enzymes and tested for its inhibitory function when propagated in high-copy-number plasmids, the DNA cleaved with *Hin*dIII still retained the ability to inhibit host sporulation (pOFΔ5′-1; Table 2). Therefore, we constructed a series of deletion-derivative plasmids from *Hin*dIII-cleaved pUBSF13 as described in Experimental Procedures and introduced them into the sporulation-proficient host UOT-0277. As shown in Table 2, pOFΔ5′-21 and -23 were inert for sporulation inhibition, whereas pOFΔ5′-11 exerted inhibition. However, the inhibitory function of the latter plasmid appeared to be "leaky" since all of the spores formed contained nondeleted pOFΔ5′-11 and upon repeated incubation in the sporulation medium they formed spores at frequencies similar to the parental culture (10^{-4} or 10^{-5}). This feature is in contrast to the situation of sporulation inhibition by pOFΔ5′-1, which allowed sporulation at a frequency of 4.9×10^{-5}. As described above, the latter spores contained plasmids which had deletions no longer exhibiting sporulation inhibition. Thus, one boundary for the inhibitory region is around base 525, and tentatively the long sequence from −50 to −515 is required for this phenomenon.

Effect of mutations in the upstream, internal regions on sporulation inhibition. We next examined the behavior of plasmids carrying mutations in the region between −50 and −525 upstream from the *spo0F* coding region. Four types of plasmids were constructed as described in Experimental Procedures. The plasmid p0FP2 having a large insertion of 1,600 bp of foreign DNA was found to be inert for sporulation inhibition (Table 2). Unexpectedly, more minor modifications, such as the four-base insertion or deletion introduced at the *Bcl*I, *Pst*I, and *Sac*I sites of p0FΔ3'-1, also abolished the sporulation inhibition (Table 2).

In addition to mutations constructed in vitro, we also examined a spontaneous mutant of *spo0F*, *spo0F221*. The mutation site of *spo0F221* was located on the *Bcl*I to *Bcl*I DNA fragment in the 2.2-kb DNA (Fig. 2) by transformation experiments (H. Shimotsu, unpublished data). Base sequence analysis of this DNA fragment from the mutant revealed a base change of A to T at −182 (Fig. 2). The mutant allele in UOT-0779 (*spo0F221*) was incorporated into a plasmid by the "plasmid-gene conversion" method devised by Canosi et al. (2). When the competent cells of strain UOT-0779 were transformed with pUBSF13 monomer DNA, kanamycin-resistant transformants on TBAB plates formed colonies of two types: rough (1.3×10^3/ml) and smooth (1.4×10^4/ml). The smooth colonies were characteristic of sporulation-inhibited cells harboring intact pUBSF13, and the rough colonies seemed to be of the *spo0* mutant type. By testing the transformation capacity of DNA from rough colonies with *spo0F221* and *spo0F77* mutants as recipients, the cells in the rough colonies were found to harbor *spo0F221*-carrying plasmids. This plasmid was termed p0F221. When this plasmid was introduced into the sporulation-proficient host, UOT-0277, the plasmid was found to be inert for sporulation inhibition (Table 2). This result was surprising in two respects: (i) only a single base change in the *spo0F* upstream region diminished sporulation inhibition capacity, and (ii) the upstream mutation itself caused the mutant *spo0F221* phenotype.

Behavior of upstream mutant alleles when present as a single copy. The combined effects of the *spo0F221* allele on sporulation inhibition and sporulation function seemed similar to the effect of promoter deletion mutants of *spoVG* (1). Therefore, we tested the sporulation function of a number of the upstream mutants described above. The upstream-deleted *spo0F* alleles (named *spo0FΔ5'-1, -11, -21*) in p0FΔ5'-1, -11, and -21 were inserted into φCM as described in Experimental Procedures, and then UOT-0789 (sporulation deficient by *spo0FΔS*: deletion of *Sac*I to *Sac*I DNA portion) was infected with these φCM derivatives to form chloramphenicol-resistant lysogens. Interestingly, all the lysogens carrying *spo0F* mutant alleles in their prophages were sporulation deficient, whereas the lysogen

TABLE 3. Complementation test of *spo0F* mutants[a]

Phage	Origin of *spo0F* allele	Sporulation
φ105 *spo0F*		+
φCM *spo0FΔ5'-1*	p0FΔ5'-1	−
φCM *spo0FΔ5'-11*	p0FΔ5'-11	−
φCM *spo0FΔ5'-21*	p0FΔ5'-21	−

[a] Host strain: UOT-0820.

from the same host infected with φ105 *spo0F*, which carried the intact 2.2-kb *spo0F* DNA fragment, was sporulation proficient. It thus seems likely that the upstream mutations diminish the functional expression of the *spo0F* gene.

Complementation test between upstream deletion and coding frame mutations. The function of the *spo0F* upstream region was assumed to be either regulation of *spo0F* expression or production of an essential sporulation factor(s) other than *spo0F* product. To answer this question, the following complementation test was carried out. The temperate φCM phage described above, carrying *spo0F* upstream mutant alleles, was used to infect *B. subtilis* UOT-0820, which had a mutant allele, *spo0FH1*, from p0FH1 on the bacterial chromosome. As shown in Table 3, the lysogenization did not restore the sporulation activity, suggesting the absence of any complementation effect.

It seems likely from these results that the upstream region of the *spo0F* gene plays an important role in regulating the expression of this gene. This regulatory function may be correlated with sporulation inhibition when the *spo0F* gene is propagated on a high-copy-number plasmid. As discussed by Banner et al. (1) for the *spoVG* gene, we also favor the postulate that amplification of *spo0F* titrates a regulatory factor(s) that controls the expression of other sporulation-controlling genes.

Banner et al. (1) described the importance of an upstream A+T-rich box for the function of the *spoVG* promoter. Although we cannot find the distinctive A+T-rich box in the *spo0F* upstream region, there are many regions which are A+T rich as compared with the average value of around 57% A+T for the whole genome: −1 to −100, 71%, −101 to −201, 74%. More upstream regions appear to be rather poor in A+T: −201 to −300, 55%; −301 to −400, 52%; −401 to −500, 49%. Although it is still premature to suggest any specific control mechanism, sporulation genes appear to be under controls which are different from those seen with other highly studied procaryote genes.

SUMMARY

When a 2.2-kb DNA fragment carrying a functional *spo0F* gene of *B. subtilis* was propagated in high-copy-number plasmids, sporulation of the host *B. subtilis* cells was completely inhibited. The inhibition-exerting portion of this DNA was studied by generating mutations in vitro by treatments such as Bal-31 nuclease digestion and in vitro deletion and insertion. The results of these studies indicated that an intact, approximately 500-bp region upstream from the initiation codon of the *spo0F* gene was essential for the inhibition, whereas the coding region was unnecessary. Furthermore, when those genes which had modifications only in the upstream noncoding region were introduced into *spo0F*-deficient hosts, they could not restore the ability of the host to sporulate, and complementation tests for sporulation between upstream mutations and a coding frame mutation were negative. From these data we assume that the 500-bp upstream region plays an important role(s) in the control of *spo0F* expression.

This study was supported, in part, by a Special Promotion Study Grant-in-Aid from the Ministry of Education, Science and Culture in Japan.

ADDENDUM IN PROOF

Recently we found a mistake in the previously published DNA sequence: one C should be added at the −271 position of the sequence in Fig. 1. By this correction, a leftward open reading frame consisting of 125 codons was found in the upstream region. We therefore cannot exclude the possibility that this new open frame is the structure gene of *spo0F*.

LITERATURE CITED

1. **Banner, D. B., C. P. Moran, Jr., and R. Losick.** 1983. Deletion analysis of a complex promoter for a developmentally regulated gene from *Bacillus subtilis*. J. Mol. Biol. **168:**351–365.
2. **Canosi, U., A. Iglesias, and T. A. Trautner.** 1981. Plasmid transformation in *Bacillus subtilis*: effects of insertion of *Bacillus subtilis* DNA into plasmid pC194. Mol. Gen. Genet. **181:**434–440.
3. **Hoch, J. A., and J. L. Mathews.** 1973. Chromosomal location of pleiotropic negative sporulation mutations in *Bacillus subtilis*. Genetics **73:**215–228.
4. **Johnson, W. C., C. P. Moran, Jr., and R. Losick.** 1983. Two RNA polymerase sigma factors from *Bacillus subtilis* discriminate between overlapping promoters for a developmentally regulated gene. Nature (London) **302:**800–804.
5. **Kawamura, F., H. Saito, H. Hirochika, and Y. Kobayashi.** 1980. Cloning of sporulation gene, *spo0F*, in *Bacillus subtilis* with ρ11 phage vector. J. Gen. Appl. Microbiol. **26:**363–373.
6. **Kawamura, F., H. Shimotsu, H. Saito, H. Hirochika, and Y. Kobayashi.** 1981. Cloning of *spo0* genes with bacteriophage and plasmid vectors in *Bacillus subtilis*, p. 109–113. *In* H. S. Levinson, A. L. Sonenshein, and D. J. Tipper (ed.), Spores VIII. American Society for Microbiology, Washington, D.C.
7. **Shimotsu, H., F. Kawamura, and Y. Kobayashi, and H. Saito.** 1983. Early sporulation gene *spo0F*: nucleotide sequence and analysis of gene product. Proc. Natl. Acad. Sci. U.S.A. **80:**658–662.
8. **Wang, P.-Z., and R. H. Doi.** 1984. Overlapping promoters transcribed by *Bacillus subtilis* σ^{55} and σ^{37} RNA polymerase holoenzymes during growth and stationary phases. J. Biol. Chem. **259:**8619–8625.

Progress in the Molecular Genetics of Spore Germination in *Bacillus subtilis* 168

ANNE MOIR,[1] IAN M. FEAVERS,[1] AAMIR R. ZUBERI,[1] RACHEL L. SAMMONS,[2] IAN S. ROBERTS,[2] JEFFREY R. YON,[2] EDITH A. WOLFF,[2†] AND DEREK A. SMITH[2]

Department of Microbiology, Sheffield University, Sheffield S10 2TN,[1] and Department of Genetics, Birmingham University, Birmingham B15 2TT,[2] United Kingdom

The overall aim of this work is to achieve an understanding of the bacterial spore germination process at the molecular level. Over the past 12 years more than 100 mutants of *Bacillus subtilis* whose spores germinate abnormally have been isolated, and progress in their study has been reviewed at intervals (25a, 29, 38). Spores of most mutants respond poorly or not at all to specific germinants (25, 42). Some are resistant to inhibitors of germination (26a). Spore germination phenotypes are not selectable, but genetic analysis was greatly aided by the facility to distinguish between Ger$^+$ and Ger$^-$ colonies with tetrazolium dye (25). Thus, by use of transduction and transformation procedures, spore germination mutations have been assigned to 13 different regions of the chromosome (*gerA* to *gerM* inclusive), the numbers varying from single representatives to in excess of 60 at different locations. In general, mutations at any one locus give rise to mutants with similar germination phenotypes. For example, some mutants are blocked in response to specific germinants (*gerA*, *gerB*, *gerC*, *gerK*; 13, 29), whereas others, including *gerD* (R. J. Warburg, A. Moir, and D. A. Smith, J. Gen. Microbiol., in press), *gerE*, *gerF*, and *gerJ* (46), germinate poorly in all germinants. Some mutants (*gerE* and *gerJ*) are blocked at a later stage of germination (23, 46). The *gerE* mutant produces spores with a defective spore coat (18, 23); two other mutant loci affecting spore coat and germination properties have been designated *spoVIA* and *spoVIB* (16, 17). The *gerJ* mutants are also abnormal in their sporulation (45). Finally, the *gerL* mutants (26a) may possess an altered cell membrane.

The many phenotypically and genetically distinguishable *ger* mutants identify genes whose products are necessary (directly or indirectly) for spore germination and facilitate the cloning of such genes. One approach has already been exploited; the Ger phenotype is not selectable in *B. subtilis* or, of course, detectable in *Escherichia coli*. This problem was circumvented by screening banks of partially *Eco*RI-digested *B. subtilis* DNA cloned in λ vectors for phages which would complement *E. coli* strains with the same enzyme deficiencies as those caused by mutations in genes closely linked to *ger* genes of *B. subtilis*. In this way the *citG* (fumarase) gene (24) and *hisA* and *ilv-leu* genes (44) were isolated, and the cloned fragments were screened for *ger* genes. The *gerA* region was present in representative λ *citG* phages, but the *gerF*, *gerH*, and *gerE* genes were absent from λ *hisA* and λ *ilv* or λ *leu* phages (although the phages could be useful as hybridization probes).

The purpose of this review is twofold. First, it is to describe the successful exploitation of the *citG* phages in the subcloning of fragments of the *gerA* region which has permitted complementation analysis and some DNA base sequencing of that region. Second, it is to report an alternative cloning strategy exploiting transposon Tn*917* (50, 51). With this strategy *ger* mutations have been induced, and the DNA close to their transposon insertion sites in three different regions of the chromosome has been isolated.

EXPERIMENTAL PROCEDURES

Bacterial strains. Derivatives of *B. subtilis* 168 and *E. coli* used in this work either have already been described (7, 24, 25, 28, 36, 47, 50) or are described in the text.

Media. Routine propagation of *E. coli* and bacteriophage λ was as previously described (24, 24a, 27). Media for growth and sporulation of *B. subtilis* strains were those used by Moir et al. (25). Liquid CCY medium (48) was used for spore formation where specified.

Preparation of *B. subtilis* spores. The method used has been described (25). Germination studies followed the method of Sammons et al. (36).

Scoring of the germination phenotype of *B. subtilis* colonies. The tetrazolium overlay technique (25) was used. Wild-type spores germinate and reduce 2,3,5-triphenyl tetrazolium chloride to give a red coloration; colonies of germination mutants reduce the dye more slowly or not at all.

Genetic techniques. Transduction of *B. subtilis* (15) and transformation of competent *B. subtilis* cells (1) were by standard techniques. Recombinants were selected on minimal medium (1),

---

† Present address: Genetic Systems Corporation, Seattle, WA.

with lactate as carbon sources for Cit$^+$ selection (34). Protoplasts were transformed and regenerated by the method of Chang and Cohen (6). Tn917 mutagenesis of *B. subtilis* was as described by Youngman et al. (50). Tn1000 mutagenesis of plasmids in *E. coli* was essentially as described by Guyer (11).

Plasmids and subcloning. Fragments of *B. subtilis* DNA were cloned into pHV33, a shuttle vector able to replicate in *B. subtilis* and *E. coli* (30). Constructions are described in more detail elsewhere (24a; A. R. Zuberi, I. M. Feavers, and A. Moir, manuscript in preparation). The cloning of a 4.3-kilobase (kb) fragment from the *red* region was by ligation of the DNA, purified from an agarose gel, into the EcoRI site of pUC9 (43). The generation in vivo of plasmids carrying DNA adjacent to sites of Tn917 insertion was as described by Youngman et al. (51).

Preparation of chromosomal, phage, and plasmid DNA. Standard procedures were used (24, 35, 41). Small-scale plasmid preparations (4, 14) were used for screening and for mapping of Tn1000 insertions.

DNA labeling and hybridization. Plasmid DNA was labeled with [α-^{32}P]dCTP by nick translation (33) and was used in Southern blots (39) or plaque hybridization assays (3).

DNA sequencing. DNA sequencing by the dideoxy chain termination method (37) was carried out as described elsewhere (22) with the use of M13 phage derivatives M13mp10 and M13mp11 and the synthetic 17-mer primer. The computer program FRAMESCAN (40), which also predicts regions of coding DNA by a statistical analysis of codon usage, was used to identify open reading frames. Analysis of hydropathy values of the translated amino acid sequence followed the program of Kyte and Doolittle (20).

RESULTS

Analysis of the *citG-gerA* Region

Complementation studies. The *citG* region of *B. subtilis* cloned in λgtWES was shown to carry at least part of the nearby *gerA* region, since the phage DNA could repair the *gerA1* mutation by recombination (24). Mutants in *gerA*, of which a large number have been isolated (25, 36, 42), were altered in their response to alanine and related amino acid germinants, but germinated normally in a glucose, fructose, asparagine, and KCl mixture. Some *gerA* mutants (*gerA38* and *gerA44*) showed an altered concentration dependence for alanine-induced germination (36). The locus may represent the region encoding a germination receptor common to L-alanine and its analogs. The *gerA1* mutation was chosen for the recombination experiment because it was the least closely linked to *citG* of a range of *gerA* mutations tested (36).

Restriction fragments were subcloned into pHV33 for complementation studies. Subclones were constructed in *E. coli* and introduced into *E. coli* or *B. subtilis* as required. A restriction map of the region and the extent of inserts in some of the constructed plasmids are shown in Fig. 1.

The position of *citG* in the cloned DNA was established by examining the properties of plasmids carrying specific EcoRI fragments. The 5.1-kb fragment could repair the *citG4* mutation to *cit*$^+$, but there was an additional requirement for the 1.55-kb fragment for complementation of both *E. coli* and *B. subtilis* fumarase defects. Tn1000 (γδ) mutagenesis further defined the position of the gene, and expression studies in *E. coli* confirmed that the gene encodes a protein of M_r ca. 49,000 and extends across the boundary of the 1.55- and 5.1-kb EcoRI fragments (24a).

To study complementation of *gerA* in *B. subtilis*, *recE4* derivatives of 10 *gerA* mutant strains were constructed by congression, and protoplasts of these were transformed with a range of plasmid subclones. The tetrazolium overlay test was used to score the germination phenotype of transformants, a red or pink coloration being positive (compared with the white color of all the *gerA* mutants) and indicating complementation. In general, the development of color in complementing clones was slower than that of the *ger*$^+$ control. Particularly in cases where larger plasmids had been introduced (e.g., pAAM6, 14.8 kb), the transformants were not stable, and complementing and noncomplementing sectors were seen in the same clone, even though selection for the plasmid-encoded drug resistance was maintained. In the case of the more stable plasmids, where segregation was not visible and homogeneous spore preparations could be prepared, the germination behavior of spore suspensions was intermediate between that of mutant and wild type.

Despite these reservations, the results obtained identified three *gerA* complementation groups (Fig. 1; Zuberi et al., in preparation). At this stage the region of DNA sufficient to complement a group of mutants is defined as a complementation unit; DNA sequencing is required to determine the number of genes present in each unit. The *citG*-proximal complementation unit, intact in pAAM2 and pAAM53 but not in pAAM14, complemented *gerA11*, *gerA2*, *gerA7*, *gerA12*, *gerA16*, and *gerA61* mutations (group I). Another group of mutations, *gerA44*, *gerA3*, and *gerA38*, were complemented by both pAAM53 and pAAM14 (group II). Complementation group III is identified by *gerA1*, which

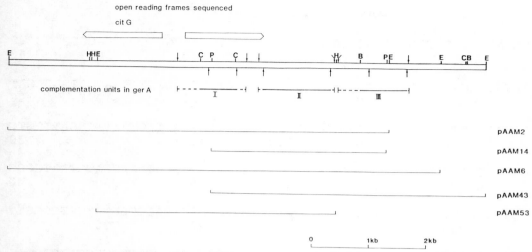

FIG. 1. Summary map of the *citG-gerA* region of the *B. subtilis* chromosome. The restriction map of an 8.3-kb segment, comprising four adjacent *Eco*RI fragments, is shown. Restriction sites are: B, *Bgl*II; C, *Cla*I; E, *Eco*RI; H, *Hin*dIII; P, *Pst*I. Below the map is shown the extent of the cloned DNA carried by five pHV33-derived plasmids. The estimates of minimum and maximum extents of complementation units I, II, and III in *gerA* are included immediately below the genetic map. The positions of Tn*1000* inserts which have contributed to these estimates are shown as arrows; those drawn below the map destroy the complementation activity of a particular unit, and those drawn above the map do not. The positions of *citG* and *gerA* open reading frames deduced from DNA sequence are indicated.

was not complemented by pAAM2, but was by pAAM6 and pAAM43.

Transposon mutagenesis of the *gerA* region. Insertion of Tn*1000* usually accompanies the mobilization of pBR322-derived plasmids by F (11). This approach was used (Zuberi et al., in preparation) to locate *gerA* complementation units on the basis of their inactivation by Tn*1000* insertion. The complementation of *gerA* mutants by plasmids carrying the entire *gerA* region, such as pAAM6, was unstable. Since Tn*1000* insertion increases plasmid size by 5.7 kb and further decreases stability, two smaller and more stably complementing plasmids were used for insertional mutagenesis.

Plasmid pAAM43 (11.2 kb) is a more stable derivative of pAAM10 recovered from *B. subtilis* and has the same restriction map as its parent (Fig. 1). It complements mutants in groups II and III, giving a stable pink tetrazolium color. Plasmid pAAM53 (11.3 kb) carries complementation units I and II and gives a high proportion of complementing, but segregating, transformant clones.

These plasmids were introduced into RB308 (8), an F⁺ *E. coli* strain, and transferred to an F⁻ strain, RB85, by conjugation. In these experiments only 40 to 60% of transferred plasmids carried Tn*1000*, similar to results obtained with a different pHV33 derivative (8). Tn*1000* inserts in pAAM43 and pAAM53 were mapped by restriction analysis, and selected plasmids were

introduced into *B. subtilis* strain AM168 (*gerA11 recE4 thrA5*), AM170 (*gerA44 recE4*), or AM093 (*gerA1 recE4 thrA5*), representatives of groups I, II, and III, respectively, as appropriate. Although the addition of Tn*1000* to either of the plasmids decreased the apparent stability in *B. subtilis*, complementation could still be scored. The map positions of inserts defining minimum and maximum extents of complementation units are indicated in Fig. 1. Inserts in either of the two complementation units in pAAM53, or in pAAM43, did not interfere with the function of the other complementation unit carried.

To define whether the left-hand end of complementation unit III extended through the *Bgl*II site (Fig. 1), a mutation was introduced by filling the recessed 3′ ends of *Bgl*II-linearized DNA of pAAM3 and then ligating the blunt ends to give a 4-base pair insertion. This mutation, *ger-570*, when transferred onto the *B. subtilis* chromosome, conferred a GerA phenotype. A *gerA570 recE4* strain was constructed, complementation by the plasmids described above was tested, and the mutation was assigned to complementation group III.

It has been shown that a plasmid carrying only DNA to the left of the *citG*-proximal *Cla*I site can repair *gerA11* by recombination (Zuberi et al., in preparation). Complementation unit I must therefore extend through this *Cla*I site.

The genetic evidence reveals three *gerA* complementation groups, implying that there are at

```
                                    ◄─────── citG
                            GTA              GGAGG
CCCATGGTGTCTCGTTCAATTCTGTATTCCATTTATGTATC CCTCC ATAACGGTTGCTTC
       10        20        30        40        50        60


AATAGATCATAATTCGCCAATGACACAAAAAATCCTTTTTATTTTTCGATCTTGGCAGAT
       70        80        90        100       110       120


TTTAATTTTAATTCAAGATATACGCGGGAATCCTTTGAATATTTGTATTTTGGATTTGTC
       130       140       150       160       170       180


CTGCCGTACTCCTTTGATGTATAAGAGACATCATAATGCTGTTCAAATGATCTAATGAGC
       190       200       210       220       230       240


TGTTCAATTTCTTCTTGTTTTCCAGCTAGCCTTACCTGTGCCACTTCGTATCCTCCGCTT
       250       260       270       280       290       300


TCATAAAAAAACTTACTCCATGAAATACTCTAACACAGTATATCATTTTTTTAACAGGAA
       310       320       330       340       350       360

                                    gerA ──────────►
                                        MetGluGlnThrGluPheLys
AAGATAACCTCTACTAAGGTTTTGGATAA GAGGTGA CCTCATTGGAACAAACAGAGTTTA
       370       380       390       400       410       420
```

FIG. 2. DNA sequence of the *citG-gerA* intergenic region. The noncoding strand of the first *gerA* gene is shown. The positions of likely ribosome binding sites and translation starts of *citG* and gene I of *gerA* are indicated.

least three adjacent *gerA* genes. If only three, they are of approximately average size, as the region extends over approximately 4 kb. The genetic data suggest that all three units can be expressed separately, at least at a sufficient level to give some complementation. The clustering of three alanine-specific germination genes is likely to have some functional significance. Perhaps the gene products act as a multiprotein complex; they might be expressed coordinately, but are unlikely to form a polycistronic operon. The analysis of all genes in the *gerA* locus, and of their expression in vivo and in vitro is necessary before firm conclusions can be drawn.

Sequence studies of the *citG-gerA* region. Expression of the DNA cloned in *E. coli* did not reveal any proteins which could correspond to the three *gerA* genes, although it permitted identification of the *citG* gene product (24a). DNA sequencing provides an opportunity to analyze possible genes and their predicted protein products and to assign the probable location of regulatory sequences, especially when results are integrated with the genetic evidence for the location of complementation units. So far, a 3.8-kb segment including *citG* and part of *gerA* has been sequenced on both strands (J. S. Miles and J. R. Guest, manuscript in preparation; I. M. Feavers, J. S. Miles, and A. Moir, manuscript in preparation). The open reading frame corresponding in position to *citG*, in the ex-

pected orientation and encoding a protein of the expected size, is shown in Fig. 1. It is preceded by a ribosome binding site and several sequences which show some homology with recognized *B. subtilis* promoter consensus sequences; experiments are currently in progress to localize the functional promoter (J. S. Miles, personal communication).

The sequence to the right of *citG* (Fig. 1 and 2) contains an open reading frame with a favored codon usage which commences 369 base pairs from the *citG* initiator codon, is transcribed in the opposite direction to *citG*, and could encode a protein of 466 amino acids (calculated M_r 51,943). This reading frame corresponds to complementation unit I of *gerA* (Fig. 1). The termination codon is, however, some 200 to 250 bp downstream of the estimated limit of this complementation unit. It is possible that Tn*1000* inserts toward the C terminus may not have destroyed the function of the protein; small inaccuracies in mapping inserts may also have contributed to the discrepancy. Part of the region sequenced is shown in Fig. 2. Upstream of the proposed TTG start (UUG is recognized as a functional initiator codon in gram-positive bacteria [21] and is found in several gram-positive genes sequenced [32]) lies a sequence complementary to seven bases near the 3' terminus of 16S rRNA. This would provide a strong ribosome binding site of $\Delta G = -16$ kcal (2, 21).

The intergenic region between *citG* and *gerA* is likely to contain the promoter for the first *gerA* gene, as well as the *citG* promoter. Possible *citG* promoters have been examined elsewhere (Miles and Guest, in preparation). The region has a very high adenine-plus-thymine content, like many promoter regions. Upstream of *gerA* (between residues 320 and 360), it is possible to identify partial homologies with several recognized consensus sequences of *B. subtilis* promoter regions. This could be coincidental, given the very high adenine-plus-thymine content of the sequences being compared. The transcriptional starts must be identified in vivo before promoters can be assigned in the region.

From the deduced amino acid sequence an overall amino acid composition has been obtained (Table 1). Analysis of relative hydropathy values along the sequence (Fig. 3) reveals that the protein contains an internal region, extending approximately from residues 233 to 430, which is of a largely hydrophobic character. It includes several segments of at least 20 residues that have average hydropathy values well above +1.6, characteristics common to membrane-spanning segments of known membrane-bound proteins (20). Strikingly, the hydrophobic region contains no lysine, whereas the rest of the protein contains 36, many of which occur in clusters. The presence of this distinct and com-

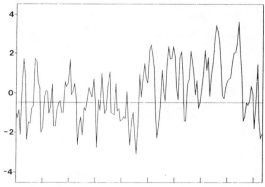

FIG. 3. Hydropathy profile of the polypeptide encoded by *gerA* gene I, computed as described by Kyte and Doolittle (20). The average hydropathy values of seven-residue segments, scanning from N to C terminus and moving three residues at time, are plotted at their midpoints. Hydrophobic regions have a high average hydropathy. The horizontal line represents the average hydropathy of a large number of sequenced proteins. Divisions along the *x* axis correspond to 50-amino acid residues.

paratively hydrophobic domain within the protein is intriguing. It will be interesting to determine whether this gene product might, for example, contain an anchoring membrane-bound domain or be found in a multimeric complex with other hydrophobic proteins.

The progress made in the molecular analysis of *gerA* has its basis in the extensive genetic studies previously undertaken. New genetic approaches designed to provide a strategy for cloning genes of interest are being exploited in an attempt to clone other *ger* loci.

Exploitation of Tn917

Isolation and general properties of new mutants. A Tn917-induced germination mutant (strain 5188), derived from strain PY79, was designated *gerM96* (E. A. Wolff, M.Sc. Thesis, University of Birmingham, Birmingham, England, 1983). The mutation was 54 and 82% cotransduced with *argA* and *ilvB*, respectively, with the use of phage PBS1 and selection for macrolide-lincosamide-streptogramin B resistance (MLS[r]). In further three-point crosses *gerM* was cotransformed with *citF83* (54%), *ilvB* (37%), and *leuC2* (14%), the results suggesting the order *citF-gerM-ilvB-leuC* (Fig. 4). The linkages of *gerM* to *citF* and *ilvB* were lower than might have been expected for a mutation between these two markers, but this is not surprising since the transposon insertion will increase the relative distance between flanking markers and may well inhibit crossing over in its vicinity. For these reasons it is difficult to assess the

TABLE 1. Amino acid composition of the *gerA* gene I product deduced from the nucleotide sequence

Amino acid	Amino acid composition of the entire protein		Amino acid composition between residues 233 and 430	
	No.	% by wt	No.	% by wt
Ala	30	4.10	17	5.65
Arg	18	5.41	8	5.85
Asn	21	4.61	5	2.67
Asp	20	4.43	4	1.62
Cys	1	0.21	0	0.00
Gln	14	3.45	5	3.00
Glu	30	7.45	7	4.23
Gly	23	2.53	13	3.47
His	4	1.06	2	1.28
Ile	42	9.15	20	10.59
Leu	59	12.85	31	16.41
Lys	36	8.88	0	0.00
Met	9	2.27	5	3.07
Phe	23	6.51	13	8.95
Pro	23	4.30	10	4.54
Ser	34	5.70	21	8.56
Thr	28	5.45	7	3.31
Trp	1	0.36	1	0.87
Tyr	14	4.40	6	4.58
Val	36	6.87	23	10.67
Total	466	100	198	100

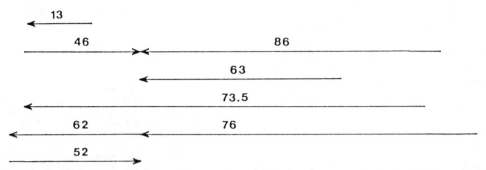

FIG. 4. Genetic map of the *gerM* region. Arrows point to selected markers; numbers indicate 100 − percent cotransformation.

location of *gerM* relative to *gerE*, which also maps in this region (23, 25).

The *gerM96* strain differed from *gerE36* mutants in forming lysozyme-resistant spores. Like *gerE* and *gerJ* mutants (23, 46), however, the germination of *gerM96* spores in L-alanine and in an L-asparagine, glucose, fructose, and KCl mixture appeared to be blocked at an intermediate stage. Binding of germinant (characterized by C_{50} values; 36) and loss of heat resistance were normal, but loss of absorbance of spore suspensions ceased at a level only 50% of that lost by wild-type spores. The partially germinated spores were phase gray and remained so for over 60 min. In Penassay broth they eventually became phase dark and outgrew normally.

A further two MLS^r mutations included in strain PY79 by the transposon were 11 and 3% cotransformed with *rpsE* and *rpoB*, respectively. *gerD* mutations are located in this region (25; R. Warburg et al., in press); the representative mutation *gerD48* is 31% cotransformed with *rpsE* and 6% with *rpoB*. On this basis the two mutations were designated *gerD97* (strain 5189) and *gerD98* (strain 5190). As a result of uncertainty about the effect of the transposon on linkages in transformation, it is not clear how close *gerD97* and *gerD98* are to *gerD48*. Germination of spores of *gerD97* and *gerD98* in L-alanine or an L-alanine, glucose, fructose, and KCl mixture was slower than that of wild type, the effect being more pronounced for *gerD97* than for *gerD98*. The phenotypes of these strains are similar to those of other *gerD* mutants.

Finally, an association between a Tn917-induced spore pigmentation abnormality and germination deficiency has been established. The

site of the insertion resulting in the formation of red colonies, designated the *red* locus (50), was 40% cotransformed with *hisA* and 19% cotransformed with *gerF*. Pigment appeared at approximately stage IV of sporulation, depending upon sporulation medium and temperature. Optimum production was at 30°C on potato glucose yeast extract agar or in L-broth; spores were not pigmented if prepared in CCY liquid medium at 30, 37, or 42°C. Fully pigmented *red*::Tn917 spores prepared on PGYEA at 30°C were defective at a late stage in germination. In response to both L-alanine and an L-alanine, glucose, fructose and KCl mixture, their loss of A_{580} was only 50% of that of wild-type spores and they became phase gray. The response of less pigmented *red*::Tn917 spores prepared on the same medium at 37 or 42°C increases to 80 to 90% of wild type, whereas those prepared in CCY germinate normally. Spores of *red*::Tn917 prepared by any of the above procedures were lysozyme resistant.

Molecular analysis of the new mutants. By use of the rescue vector method described by Youngman et al. (in press) chromosomal fragments on either side of the *gerM* Tn917 insertion have been cloned. This method exploits the plasmids pTV20 and pTV21Δ2 which were constructed by inserting a pBR322-derived replicon, pHW9, carrying the ampicillin (Amp) resistance gene of pBR322 and a chloramphenicol (Cm) resistance gene expressed in *B. subtilis* in opposite orientations near the middle of Tn917. The original Tn917 insertions in the *B. subtilis* chromosome can be replaced by plasmid DNA by homologous recombination between Tn917 sequences. The chromosomal DNA can then be digested

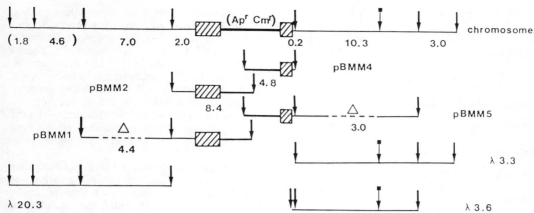

FIG. 5. Plasmids and λ Charon 4 phages carrying DNA from the *germ* region of the chromosome. The heavy line represents pHW9 DNA (carrying Cm^r and Ap^r markers) inserted in either orientation between the ends of Tn*917* (cross-hatched) as a result of recombination between the chromosomal copy of Tn*917* and pTV20 or pTV21Δ2. Plasmids derived from the insertion of pHW9 using pTV21Δ2 are shown on the right of the insertion and those using pTV20 are on the left. The relative order of the fragments of λ20.3 is uncertain. A possible order of fragments deduced from the clones is indicated on the chromosomal map. Fragment sizes are given in kb. Plain arrows indicate *Eco*RI sites, and the arrow with a square indicates a *Bam*HI site.

with restriction enzymes, for example, with *Eco*RI, which cuts once (between Cm^r and Amp^r) in the plasmid. Ligation of *Eco*RI fragments then generates new ampicillin-resistant plasmids which carry DNA from either side of the Tn*917* insertion depending on whether pTV20 or pTV21Δ2 was originally inserted.

Four such plasmids carrying DNA from either side of the *germ96* insertion were obtained by this procedure (Fig. 5). Plasmids pBMM1 and pBMM5 were used to probe a λ Charon 4A bank (9) for phage carrying homologous DNA sequences and hybridized with 3 and 33 plaques, respectively, of 800 tested. The three "pBMM1" phages (designated λ20.3) were identical and carried a *B. subtilis* insert consisting of three *Eco*RI fragments of approximately 1.8, 4.6, and 7.0 kb (Fig. 5). Restriction maps of 10 of the "pBMM5" phages showed that each carried a 10.3-kb *Eco*RI fragment; in addition, one (phage λ3.3) had a 3.0-kb and another (phage λ3.6) a 0.2-kb *Eco*RI fragment. The 10.3-kb fragment hybridized to the 3.0-kb fragment of pBMM5, which is probably a deletion product of this fragment since it hybridized with a 10.3-kb *Eco*RI fragment of the chromosome. pBMM4 hybridized with a 2.1-kb chromosomal *Eco*RI fragment, suggesting that this was the site of Tn*917* insertion. None of the phages examined so far carries such a fragment. The 7.0-kb fragment of λ20.3 hybridizes with the 4.4-kb fragment of pBMM1, which is therefore likely to contain a deletion. As predicted, none of the phages could repair the *germ* mutation by transformation. Nevertheless, pBMM4 and pBMM2 between

them should contain the entire fragment, and attempts are being made to construct an intact *germ* locus from these two plasmids.

It is important to establish that the additional *Eco*RI fragments of pBMM1 and pBMM5 do come from the *germ* region. This has been shown for the 3.0-kb fragment by the following experiments. pBMM5 and phage λ3.6 were digested with *Eco*RI and the fragments were subcloned into pACYC184 (5), forming plasmids pBMM7, pBMM8, and pBMM9. pBMM7 carried the 4.8-kb pBMM5 fragment, pBMM8 carried the 3.0-kb fragment of pBMM5, and pBMM9 carried both the 0.2- and 10.3-kb fragments of phage λ3.6. These plasmids were used to transform strain 4980 (*trpC2 leuD117*) to tetracycline resistance (Tc^r), and the Tc^r was then mapped by transformation in relation to *ilvB* and *citF*. Derivatives of strain 4980 harboring pBMM7, pBMM8, and pBMM9 yielded, respectively, 2, 16, and 3% cotransformation of Tc^r with *ilvB*. Only the derivative harboring pBMM9 was tested with *citF*; 19% cotransformation of Tc^r was obtained. In the *citF*$^+$ strain cotransformed to Tc^r in which pBMM9 had integrated, the cotransfer of *citF* and *leuD* was reduced from 26.5 to 4%, probably as a result of integration of the plasmid between these markers. Although both the plasmids and phage tested carry DNA from the *germ* region, none transformed *citF*, *gerE*, or *ilvB*. Recently, *citF* and *gerE* have been cloned on the *E. coli* plasmid pCT1047 (S. Hasnain and C. M. Thomas, personal communication). Experiments are in progress to determine whether the phages described above show

TABLE 2. Plasmids carrying DNA from the *gerD* region

Plasmid	Derivation	*Eco*RI fragment sizes (kb)[a]
pBMD1	pTV20 inserted into *gerD97*	9, 2.6
pBMD2	pTV20 inserted into *gerD97*	9, 4, 3.8, 2.9
pBMD6	pTV20 inserted into *gerD97*	5.6[b]
pBMD3	pTV21Δ2 inserted into *gerD97*	4.5, 11
pBMD4	pTV20 inserted into *gerD98*	9.5
pBMD5	pTV21Δ2 inserted into *gerD98*	5

[a] In each case the first fragment carries the ampicillin resistance gene, replication functions, and part of the transposon (6.5 kb for pTV20, 4.6 kb for pTV21Δ2).

[b] The 5.6-kb fragment is a partially deleted form of the 9-kb fragment of pBMD1 and pBMD2.

any homology with the *B. subtilis* DNA carried by this plasmid or whether it hybridizes with other Charon 4A phages.

Similar analyses of *gerD97* and *gerD98* have not progressed so far, but they are encouraging. Five plasmids (pBMD1 to pBMD5 inclusive), carrying chromosomal DNA from either side of both the *gerD97* and *gerD98* transposon insertions have been prepared (Table 2). None of the plasmids could repair *gerD48* in congression experiments. The plasmids pBMD4 and pBMD6 both hybridized to a 4.1-kb *Eco*RI fragment of the wild-type *B. subtilis* 168 chromosome. This suggests that the transposon may have inserted into the same 4.1-kb *Eco*RI fragment in *gerD97* and *gerD98*. In this case, many of the plasmids must have suffered deletions, since their restriction maps are very different. Labeled pBMD1 and pBMD3 DNA is being used to probe for homologous inserts in the λ Charon 4A library.

The presence in bacteria of plasmid pPL733, which contains a *B. subtilis* 2.7-kb *Eco*RI fragment, resulted in colonies with a brown pigmentation (26). The fragment exerts a promoter activity in postexponential growth, and its presence results in extracellular protease production. The fragment originates from a region 80 to 85% cotransduced with *hisA1* by PBS1 phage. Three λ Charon 4A phages with homology to the fragment have been isolated from a *B. subtilis* gene library (26). Since the map locations of *red*::Tn*917* and the 2.7-kb fragment were similar and red and brown colony pigmentations are alike, homology between the insertion and the fragment was tested. Transformation experiments were carried out with strain PY152 (*trp*

thr red::Tn*917*) as recipient, a low concentration of donor DNA from a *thr*[+] but otherwise isogenic strain, and congressing concentrations of pPL733 and the three λ Charon 4A phages (generously supplied by P. S. Lovett and S. Mongkolsuk). Thr[+] recombinants were screened for MLS sensitivity. All three λ phages congressed the strain to MLS[s], at a frequency of 5 to 10%, but the plasmid did not. Bacteria from the MLS[s] colonies did not produce red pigment sporulation, and their spores germinated normally.

Restriction analysis of the three λ Charon 4A phages showed that they carry overlapping *Eco*RI fragments spanning a region of 25 kb (Fig. 6). The common 4.3-kb fragment, adjacent to the previously characterized 2.7-kb fragment, has been subcloned into pUC9 (to give pKAS501) and a restriction map has been constructed (Fig. 6b). This fragment congresses MLS[s] and therefore corresponds to the site of the *red*::Tn*917* insertion. It should now be possible to determine precisely the location of this gene on the fragment, determine the nature of its control, and investigate any interaction with a gene(s) on the 2.7-kb fragment. The phages described here will extend the probes available for further analysis of the *his-gerF* region.

DISCUSSION

The molecular analysis of the first region to be cloned, *gerA*, has revealed an interesting degree of complexity, with three adjacent complementation units, each corresponding to at least one gene. The *gerA* region has been thought to encode an alanine receptor (36); the role of each gene must now be considered. Their products could function separately in the transmission of the alanine germination stimulus in the spore, or they may interact to form a multisubunit receptor complex. There are few physiological distinctions among the many *gerA* mutants studied. Mutants (*gerA38* and *gerA44*) which appear to lower the affinity of the spore for L-alanine map in group II. The polypeptide encoded by this gene could be the site of alanine binding. All the mutations in groups I and III inactivate alanine-stimulated germination at all concentrations of alanine, although several of the mutations in I are conditional on the temperature of spore germination. These groups could identify genes for other essential subunits in a receptor complex. DNA sequencing is in progress; the long open reading frame, which corresponds reasonably well with the expected limits of complementation unit I, would encode a polypeptide of several distinct domains, including one which is markedly hydrophobic in character. Inhibitor studies (49) and biophysical measurements of

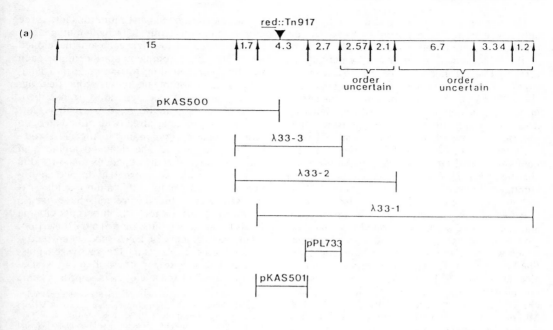

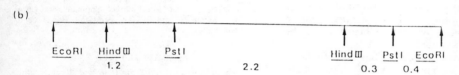

FIG. 6. (a) Physical map of the *red*::Tn917 region of the *B. subtilis* chromosome determined from *Eco*RI digests of plasmids and λ Charon 4 phages with overlapping fragments. pKAS500 was obtained from an insertion of pTV21Δ2 into *red*::Tn917. Arrows indicate *Eco*RI sites. (b) Restriction map of the 4.3-kb *Eco*RI fragment cloned in pKAS501. No *Bam*HI or *Xho*I sites were detected. All sizes are given in kb.

spore membrane properties (31) have implicated a hydrophobic environment, probably at the membrane, as the site involved in the triggering of germination. The predicted properties of this polypeptide would be consistent with this hypothesis; it could act, for example, as a membrane-traversing protein with at least one other domain, more hydrophilic in character, thus anchored to the membrane. As RNA or protein products of the *gerA* region have not yet been identified, conclusions must be drawn with caution. Expression of the *gerA* genes, perhaps fused to an *E. coli* promoter, will be attempted, to provide independent evidence of the gene products and their characteristics; biochemical and immunological studies may help to identify the location and function of the *gerA* germination proteins in the spore.

The germination apparatus is present in the mature spore, and the *gerA* genes are probably expressed during sporulation. It will now be possible to investigate the timing of transcription of individual *gerA* genes, to attempt to identify promoter regions, and to compare controls with those of other genes expressed during sporulation (10, 19). A similar study of other germination loci, for example, those whose function is required for the common response to a wide range of germinant and those blocked at later stages in the germination sequence, will inevitably contribute to the overall understanding of germination.

The transposon Tn917 has been used to generate *ger* mutations, some in previously defined loci and some in new locations. These Tn917 insertion mutants have provided the opportunity for recovery of DNA from these loci, on either side of the point of insertion. The plasmids recovered have frequently suffered deletion, however, and this has limited the usefulness of the DNA recovered for further molecular analysis. Moreover, each clone can contain only part of the gene into which the insertion has occurred. This latter limitation could be overcome by inserting the plasmid carrying cloned DNA into a wild-type chromosome and excising the plasmid complete with an intact cloned gene by use of an appropriate restriction enzyme.

The use of the plasmids containing DNA on either side of the transposon as hybridization probes of a gene library should permit recovery of the intact gene. In this work, λ derivatives carrying DNA from regions near Tn917 insertion sites have been recovered. In the case of gerM96, cloned DNA exactly corresponding to the site of insertion has not yet been identified, and the necessary fragment will be sought from other libraries or cloned from chromosomal DNA.

Extensive, though not all contiguous, segments of cloned DNA from the hisA-red region and the citF-gerE-gerM-ilv-leu region have been identified. In both regions a correlation of physical and genetic maps will be possible; a number of genes mapping in these areas could be present and are being tested. The cloned citF-gerE region is being used as a probe to seek phages which will overlap also with the fragments identified near gerM. Other cloning procedures may also be useful; a 1.5-kb BclI fragment which complements the gerE36 mutation has been cloned in a φ105 vector (W. James and J. Mandelstam, personal communication). The gerE gene may encode a regulatory gene required for expression of late sporulation genes (18); closer examination of the gene organization in this region will now be possible by using the various clones available.

A spore germination defect has been found in the red mutant of Youngman et al. (50). Inactivation of the red gene can lead to the accumulation late in sporulation of an uncharacterized red-brown pigment, and spores formed under these conditions are blocked at an intermediate stage of germination. It could be that the pigmented compound has some inhibitory effect on events occurring during sporulation or germination, or perhaps that the overproduction of pigment and germination deficiency are independent effects of a failure of a regulatory system controlled by the red gene. The relationship between red and the adjacent EcoRI fragment which, when cloned in a plasmid, results in pigment overproduction would be relevant to an understanding of the function of the red gene.

These clones should permit the analysis of germination genes which are very different in function from gerA, but are essential nonetheless for the completion of the alanine-triggered germination response within the spore.

SUMMARY

Genetic studies of spore germination mutants have identified at least 13 ger loci. The first ger region to be cloned, gerA, has been subcloned from λ citG phages into the shuttle vector pHV33. Complementation analysis using plasmids carrying different extents of cloned DNA has permit-

ted classification of 10 gerA mutations into three complementation groups. Tn1000 mutagenesis has been used to localize each on the cloned DNA. The three complementation units, each probably corresponding to one gene, are adjacent, but their complementation properties suggest that they may be separately transcribed. DNA sequencing of the region, which is currently in progress, has already identified an open reading frame corresponding to the gerA complementation unit I. The deduced amino acid sequence suggests that it encodes a 466-residue polypeptide with an internal hydrophobic domain of approximately 190 amino acids. Attempts to clone other ger regions have used a plasmid recovery system which permits cloning of DNA on either side of a Tn917 insertion. Extensive segments of DNA have been recovered, in some cases using the recovered plasmids as hybridization probes of libraries. Detailed analysis of these clones is in progress.

This work was supported by the Science and Engineering Research Council in the form of research grants to A.M. and D.A.S. and studentships to A.R.Z., I.S.R., and J.R.Y. We are most grateful to the donors of bacterial strains, plasmids, and phages, for the technical support of Debbie Sutton in Sheffield and Sue Hayley, Hazel Howell, Joyce Yeomans, and Carol Price in Birmingham, and for the secretarial assistance of Muriel Hinchliffe.

LITERATURE CITED

1. **Anagnostopoulos, C., and J. Spizizen.** 1961. Requirements for transformation in Bacillus subtilis. J. Bacteriol. **81**:741–746.
2. **Band, L., and D. J. Henner.** 1984. Bacillus subtilis requires a "stringent" Shine-Dalgarno region for gene expression. DNA **3**:17–21.
3. **Benton, W. D., and R. W. Davis.** 1977. Screening λ gt recombinant clones by hybridization to single plaques in situ. Science **196**:180–182.
4. **Birnboim, H. C., and J. Doly.** 1979. A rapid alkaline extraction procedure for screening recombinant plasmid DNA. Nucleic Acids Res. **7**:1513–1524.
5. **Chang, A. C. Y., and S. N. Cohen.** 1978. Construction and characterization of amplifiable multicopy DNA cloning vehicles derived from the PISA cryptic plasmid. J. Bacteriol. **134**:1141–1156.
6. **Chang, S., and S. N. Cohen.** 1979. High frequency transformation of Bacillus subtilis protoplasts by plasmid DNA. Mol. Gen. Genet. **168**:111–115.
7. **Dabbs, E. R.** 1983. Mapping of the genes for Bacillus subtilis ribosomal proteins S9, S11 and BL27 by means of antibiotic resistant mutants. Mol. Gen. Genet. **191**:295–300.
8. **De Lencastre, H., K. F. Chak, and P. J. Piggot.** 1983. Use of Escherichia coli transposon Tn1000 (γδ) to generate mutations in Bacillus subtilis DNA. J. Gen. Microbiol. **129**:3203–3210.
9. **Ferrari, E., D. J. Henner, and J. A. Hoch.** 1981. Isolation of Bacillus subtilis genes from a Charon 4A library. J. Bacteriol. **146**:430–432.
10. **Fliss, E. R., and P. Setlow.** 1984. Complete nucleotide sequence and start sites for transcription and translation of the Bacillus megaterium protein C gene. J. Bacteriol. **158**:809–813.
11. **Guyer, M. S.** 1978. The γδ sequence of F is an insertion sequence. J. Mol. Biol. **126**:347–365.
12. **Henner, D. J., and J. A. Hoch.** 1980. The Bacillus subtilis

chromosome. Microbiol. Rev. **44:**57–82.

13. **Irie, R., T. Okamoto, and Y. Fujita.** 1982. A germination mutant of *Bacillus subtilis* deficient in response to glucose. J. Gen. Appl. Microbiol. **28:**345–354.

14. **Ish-Horowicz, D., and J. F. Burke.** 1981. Rapid and efficient cosmid vector cloning. Nucleic Acids Res. **9:**2989–2998.

15. **Jamet, C., and C. Anagnostopoulos.** 1969. Etude d'une mutation tres faiblement transformable au locus de la threonine desaminase de *Bacillus subtilis*. Mol. Gen. Genet. **105:**225–242.

16. **Jenkinson, H. F.** 1981. Germination and resistance defects in spores of a *Bacillus subtilis* mutant lacking a coat polypeptide. J. Gen. Microbiol. **127:**81–91.

17. **Jenkinson, H. F.** 1983. Altered arrangement of proteins in the spore coat of a germination mutant of *Bacillus subtilis*. J. Gen. Microbiol. **129:**1945–1958.

18. **Jenkinson, H. F., and H. Lord.** 1983. Protease deficiency and its association with defects in spore coat structure, germination and resistance properties in a mutant of *Bacillus subtilis*. J. Gen. Microbiol. **129:**2727–2737.

19. **Johnson, W. C., C. P. Moran, and R. Losick.** 1983. Two RNA polymerase sigma factors from *Bacillus subtilis* discriminate between overlapping promoters for a developmentally regulated gene. Nature (London) **302:**800–804.

20. **Kyte, J., and R. F. Doolittle.** 1982. A simple method for displaying the hydropathic character of a protein. J. Mol. Biol. **157:**105–132.

21. **McLaughlin, J. R., C. L. Murray, and J. C. Rabinowitz.** 1981. Unique features in the ribosome binding site sequence of the Gram positive *Staphylococcus aureus* β-lactamase gene. J. Biol. Chem. **256:**11283–11291.

22. **Messing, J.** 1983. New M13 vectors for cloning. Methods Enzymol. **101:**20–77.

23. **Moir, A.** 1981. Germination properties of a spore-defective coat mutant of *Bacillus subtilis* 168. J. Bacteriol. **148:**1106–1116.

24. **Moir, A.** 1983. The isolation of λ transducing phages carrying the *citG* and *gerA* genes of *Bacillus subtilis*. J. Gen. Microbiol. **129:**303–310.

24a.**Moir, A., I. M. Feavers, and J. R. Guest.** 1984. Characterization of the fumarase gene of *Bacillus subtilis* 168 cloned and expressed in *Escherichia coli* K12. J. Gen. Microbiol. **130:**3009-3017.

25. **Moir, A., E. Lafferty, and D. A. Smith.** 1979. Genetic analysis of spore germination mutants of *Bacillus subtilis* 168: the correlation of phenotype with map location. J. Gen. Microbiol. **111:**165–180.

25a.**Moir, A., and D. Smith.** 1985. The genetics of spore germination in *Bacillus subtilis*, p. 89–100. *In* G. J. Dring, D. J. Ellar,' and G. W. Gould (ed.), Fundamental and applied aspects of bacterial spores. Academic Press, Inc., New York.

26. **Mongkolsuk, S., Y. Chang, R. B. Reynolds, and P. S. Lovett.** 1983. Restriction fragments that exert promoter activity during post-exponential growth of *Bacillus subtilis*. J. Bacteriol. **155:**1399–1406.

26a.**Morse, R. W., and D. A. Smith.** 1985. Mutations in *Bacillus subtilis* 168 affecting the inhibition of spore germination by a barbiturate, p. 101–115. *In* G. J. Dring, D. J. Ellar, and G. W. Gould (ed.), Fundamental and applied aspects of bacterial spores. Academic Press, Inc., New York.

27. **Murray, N. E., P. Manduca de Ritis, and L. A. Foster.** 1973. DNA targets for the *Escherichia coli* K restriction system analysed genetically in recombinants between phages phi80 and lambda. Mol. Gen. Genet. **120:**261–281.

28. **Ohne, M., B. Rutberg, and J. A. Hoch.** 1973. Genetic and biochemical characterization of mutants of *Bacillus subtilis* defective in succinate dehydrogenase. J. Bacteriol. **115:**738–745.

29. **Piggot, P. J., A. Moir, and D. A. Smith.** 1981. Advances in the genetics of *Bacillus subtilis* differentiation, p. 29–39. *In* H. S. Levinson, A. L. Sonenshein, and D. J. Tipper (ed.), Sporulation and germination. American Society for Microbiology, Washington, D.C.

30. **Primrose, S. B., and S. D. Ehrlich.** 1981. Isolation of plasmid deletion mutants and study of their instability. Plasmid **6:**193–201.

31. **Racine, F. M., J. F. Skomurski, and J. C. Vary.** 1981. Alterations in *Bacillus megaterium* QM B1551 spore membrane with acetic anhydride and L-proline, p. 224–227. *In* H. S. Levinson, A. L. Sonenshein, and D. J. Tipper (ed.), Sporulation and germination. American Society for Microbiology, Washington, D.C.

32. **Ramakrishna, N., E. Dubnau, and I. Smith.** 1984. The complete DNA sequence and regulatory regions of the *Bacillus licheniformis spo0H* gene. Nucleic Acids Res. **12:**1779–1790.

33. **Rigby, P. W. J., M. Dieckmann, C. Rhodes, and P. Berg.** 1977. Labelling deoxyribonucleic acid to high specific activity *in vitro* by nick translation with DNA polymerase I. J. Mol. Biol. **113:**237–251.

34. **Rutberg, B., and J. A. Hoch.** 1970. Citric acid cycle: gene-enzyme relationships in *Bacillus subtilis*. J. Bacteriol. **104:**826–833.

35. **Sammons, R. L., and C. Anagnostopoulos.** 1982. Identification of a cloned DNA segment at a junction of chromosome regions involved in rearrangements in the *trpE26* strains of *Bacillus subtilis*. FEMS Lett. **15:**265–268.

36. **Sammons, R. L., A. Moir, and D. A. Smith.** 1981. Isolation and properties of spore germination mutants of *Bacillus subtilis* 168 deficient in the initiation of germination. J. Gen. Microbiol. **124:**229–241.

37. **Sanger, F., S. Nicklen, and A. R. Coulson.** 1977. DNA sequencing with chain-terminating inhibitors. Proc. Natl. Acad. Sci. U.S.A. **74:**5463–5467.

38. **Smith, D. A., A. Moir, and R. L. Sammons.** 1978. Progress in genetics of spore germination in *Bacillus subtilis*, p. 158–163. *In* G. Chambliss and J. C. Vary (ed.), Spores VII. American Society for Microbiology, Washington, D.C.

39. **Southern, E. M.** 1975. Detection of specific sequences among DNA fragments separated by gel electrophoresis. J. Mol. Biol. **98:**503–517.

40. **Staden, R., and A. D. McLachlan.** 1982. Codon preference and its use in identifying protein coding regions in long DNA sequences. Nucleic Acids Res. **10:**141–156.

41. **Thomas, C. M.** 1981. Complementation analysis of replication and maintainance functions of broad host range plasmids RK2 and RP1. Plasmid **5:**277–291.

42. **Trowsdale, J., and D. A. Smith.** 1975. Isolation, characterization, and mapping of *Bacillus subtilis* 168 germination mutants. J. Bacteriol. **123:**83–95.

43. **Vieira, J., and J. Messing.** 1982. The pUC plasmids, an M13mp7-derived system for insertion mutagenesis and sequencing with synthetic universal primers. Gene **19:**259–268.

44. **Walton, D. A., A. Moir, R. Morse, I. Roberts, and D. A. Smith.** 1984. The isolation of λ phage carrying DNA from the histidine and isoleucine-valine regions of the *Bacillus subtilis* chromosome. J. Gen. Microbiol. **130:**1577–1586.

45. **Warburg, R. J.** 1981. Defective sporulation of a spore germination mutant of *Bacillus subtilis* 168, p. 98–100. *In* H. S. Levinson, A. L. Sonenshein, and D. J. Tipper (ed.), Sporulation and germination. American Society for Microbiology, Washington, D.C.

46. **Warburg, R. J., and A. Moir.** 1981. Properties of a mutant of *Bacillus subtilis* 168 in which spore germination is blocked at a late stage. J. Gen. Microbiol. **124:**243–253.

47. **Ward, J. B., Jr., and S. A. Zahler.** 1973. Genetic studies of leucine biosynthesis in *Bacillus subtilis*. J. Bacteriol. **116:**719–726.

48. **Wilkinson, B. J., D. J. Ellar, I. R. Scott, and M. A. Koncewicz.** 1977. Rapid, chloramphenicol-resistant activation of membrane electron transport on germination of *Bacillus* spores. Nature (London) **266:**174–176.

49. **Yasuda-Yasaki, Y., S. Namiki-Kanie, and Y. Hachisuka.**

1978. Inhibition of germination of *Bacillus subtilis* spores by alcohols, p. 113–116. *In* G. Chambliss and J. C. Vary (ed.), Spores VII. American Society for Microbiology, Washington, D.C.

50. **Youngman, P. J., J. B. Perkins, and R. Losick.** 1983. Genetic transposition and insertional mutagenesis in *Ba-* *cillus subtilis* with the *Streptococcus faecalis* transposon Tn917. Proc. Natl. Acad. Sci. U.S.A. **80:**2305–2309.

51. **Youngman, P. J., J. B. Perkins, and R. Losick.** 1984. A novel method for the rapid cloning in *Escherichia coli* of *Bacillus subtilis* chromosomal DNA adjacent to Tn*917* insertions. Mol. Gen. Genet. **195:**424–433.

Use of Tn917-Mediated Transcriptional Gene Fusions to *lacZ* and *cat-86* for the Identification and Study of *spo* Genes in *Bacillus subtilis*

PHILIP YOUNGMAN, JOHN B. PERKINS, AND KATHLEEN SANDMAN

Department of Cellular and Developmental Biology, The Biological Laboratories, Harvard University, Cambridge, Massachusetts 02138

Gene fusion-generating insertion elements, such as the Mu*d* (*lac* Ap^r) derivatives of Casadaban and his co-workers (4, 5) and conceptually similar derivatives of Tn5 (1; L. Kroos and D. Kaiser, Proc. Natl. Acad. Sci. U.S.A., in press), have proved to be extremely powerful and versatile tools for the identification and study of regulated genes in gram-negative bacteria. Such transposon derivatives carry a promoterless copy of the *Escherichia coli lac* operon (or some other gene or operon that encodes an easily scored or assayed gene product) very close to one of their terminal inverted repeats, oriented such that insertions into a chromosomal gene can place expression of the transposon-borne, promoterless genes under control of the promoter for the interrupted gene. This makes it possible to obtain gene fusions as an automatic consequence of generating an insertional mutation and therefore substantially reduces the work required to take advantage of the many ways in which gene fusions can facilitate studies of gene regulation.

In the present work, we describe several derivatives of the gram-positive transposon Tn917 (12) that generate insertion-mediated transcriptional gene fusions in a similar manner. Some of these derivatives carry promoterless copies of the *E. coli lacZ* gene, some carry promoterless copies of the *Bacillus pumilus cat-86* gene, and some carry both genes as a tandem pair (and thus can mediate simultaneous *lac* and *cat* fusions). Methods are presented for the use of these transposons to identify and analyze developmentally regulated genes of *B. subtilis* whose products are involved in the process of spore formation (*spo* genes). The methods are general in nature, however, and might readily be applied to the study of many kinds of regulated genes in any gram-positive species in which Tn917 can cause insertional mutations.

EXPERIMENTAL PROCEDURES

Bacterial strains, plasmids, and genetic methods. *B. subtilis* hosts for plasmid constructions and transposition experiments were the 168 derivatives BD170 (6) or PY79 (P. Youngman, J. B. Perkins, and R. Losick, Plasmid, in press). *E. coli* hosts for the cloning of DNA adjacent to Tn917 insertions were MM294 (10) or HB101 (3). Culture media, preparation of transformation-competent bacteria, and selections for MLS^r (resistance to macrolide, lincosamide, and streptogramin B antibiotics), Cm^r (chloramphenicol resistance), Tc^r (tetracycline resistance), and Ap^r (ampicillin resistance) were as described previously (13, 14; Youngman et al., in press). pTV1 (13; J. B. Perkins and P. Youngman, Plasmid, in press) and pTV8 (Youngman et al., in press) were described previously, and the construction of pTV32, pTV51, pTV52, pTV53, pTV54, and pTV55 will be described in detail elsewhere.

Maintenance of pE194-derived replicons and selections for transposition. To avoid enriching for chromosomal insertions of Tn917 derivatives in bacteria that harbor transposon-containing pE194-derived replicons, the bacteria were maintained and subcultured at 32°C in the presence of a drug against which the plasmid conferred resistance by virtue of a marker present outside transposon sequences. For example, pTV1-containing bacteria were maintained at 32°C in the presence of both chloramphenicol (5 μg/ml) and MLS drugs (erythromycin, 1 μg/ml; lincomycin, 25 μg/ml). To select for transpositions, pTV1-containing bacteria were streaked for single colonies, and a 32°C liquid culture in Luria-Bertani broth (2) was started from a fresh single colony. (The absence of a chromosomal Tn917 insert in bacteria from such a colony may be confirmed by passaging at 45°C in the absence of drugs and then verifying loss of both Cm^r and MLS^r.) The 32°C culture was allowed to reach a Klett value of 60 to 70 and then was diluted at least 1:100 into 47°C Luria-Bertani medium containing only MLS drugs (again, 1 μg of erythromycin and 25 μg of lincomycin per ml) and allowed to grow overnight (15 to 25 h). To confirm that the enrichment for transpositions in such an overnight outgrowth culture has been successful, the percentage of bacteria that contain chromosomal insertions should be determined by comparing MLS^r CFU per milliliter at both 32 and 47°C. Because we frequently find

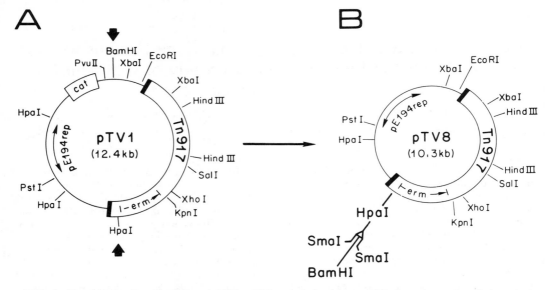

FIG. 1. Restriction maps of pTV1 and pTV8. pTV8 was derived from pTV1 through a series of alterations, including (i) the removal from pTV1 of its *cat*-containing *Hpa*I-*Pvu*II fragment, (ii) the removal from pTV1 of a *Bam*HI site (indicated by an arrow), and (iii) the replacement of one of the *Hpa*I sites in pTV1 (indicated by an arrow) with a small DNA "linker" that contains *Bam*HI and *Sma*I sites. These alterations are explained in detail elsewhere (Youngman et al., in press).

that insert-containing bacteria represent a substantial fraction of the outgrowth population (5 to 50%), but not the majority, we routinely dilute an outgrowth sample once again (1:50) into 47°C MLS drug-containing broth and allow growth to reach late log phase. Bacteria from the second outgrowth may be concentrated, quick frozen in small portions, and stored indefinitely at −70°C for future use.

Use of MUG to identify Tn917 lac insertions in genes active during sporulation. Insert-containing bacteria were spread onto Difco sporulation medium agar (13) at a dilution calculated to produce approximately 500 colonies per plate. After the colonies had developed to their full size (approximately 24 h at 37°C), plates were sprayed with a fine mist of 4-methylumbelliferyl-β-D-galactoside (MUG) solution (Sigma) by use of a chromatography atomizer. The MUG solution was prepared at an initial concentration of 10 mg/ml in dimethyl sulfoxide and then diluted 1:10 with distilled water to increase the volume. This produced a cloudy suspension that was transferred to the atomizer and immediately sprayed onto agar plates, covering their surfaces as evenly as possible. A total of approximately 0.3 to 0.5 ml of diluted MUG suspension was applied to each plate. After a period of 10 to 15 min, which allowed excess liquid to be absorbed into the agar, plates were illuminated with a hand-held long-wavelength UV light source (which produced a major emission band at 366 nm) to excite fluorescence.

RESULTS AND DISCUSSION

Tn917 derivatives that mediate transcriptional gene fusions. In principle, any transposon can be made to function like the widely used gene fusion-generating derivatives of the transposon phage Mu. The key requirement is to identify a site near one end of the transposon where foreign DNA (such as a promoterless *lacZ* or *cat-86* gene) can be inserted without interfering with transposition. Such a site was discovered in Tn917, a *Hpa*I site centered 240 base pairs (bp) from the terminal inverted repeat at the *erm*-proximal end of the transposon (Fig. 1A), and this site was "replaced" with a small DNA "linker" that contained *Bam*HI and *Sma*I restriction sites (Fig. 1B), to facilitate the insertion of foreign DNA (Youngman et al., in press). By using these *Bam*HI and *Sma*I sites, promoterless copies of the *E. coli lacZ* gene or the *B. pumilus cat-86* gene, or both, were inserted by ligation, in an orientation that makes their expression dependent upon transcription entering the transposon across its *erm*-proximal inverted repeat. This produced a family of transposon-bearing plasmids capable of generating various kinds of transposition-mediated gene fusions, including pTV32 (Fig. 2A) and pTV51 (Fig. 2B), which mediate fusions to *lacZ*, pTV52 (Fig. 2C), which mediates fusions to *cat-86*, and pTV53 (Fig. 2D), which simultaneously mediates fusions to both *lacZ* and *cat-86*.

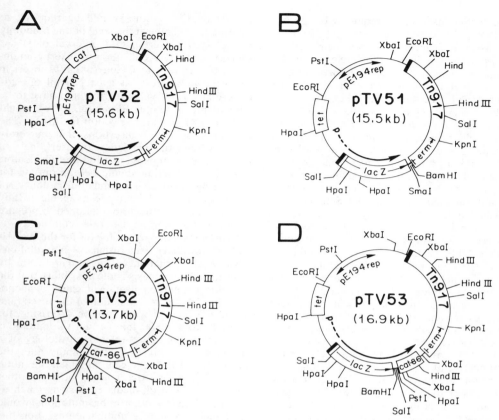

FIG. 2. Restriction maps of pTV32, pTV51, pTV52, and pTV53. These plasmids were derived from pTV8 (Fig. 1) through insertions at its *Bam*HI or *Sma*I sites of restriction fragments from pZΔ326 (P. Zuber, unpublished data) or pPL703 (11) that contained promoterless copies of the *E. coli lacZ* gene or the *B. pumilus cat-86* gene. A weak promoter in plasmid sequences outside the transposon (labeled "P" and marked with an arrow that indicates the direction and extent of the transcription it directs) makes bacteria that contain pTV32, pTV51, or pTV53 blue on X-gal plates and makes bacteria that contain pTV52 or pTV53 resistant to low levels of chloramphenicol.

The construction of all of the plasmids shown in Fig. 2 was significantly simplified by the chance existence of a promoter, in plasmid sequences just outside of the transposon, oriented to direct a low level of transcription across the *erm*-proximal inverted repeat of the transposon. Because of this promoter, bacteria that contain pTV32, pTV51, and pTV53 are moderately blue in the presence of X-gal, and bacteria that contain pTV52 and pTV53 are resistant to low levels of chloramphenicol (<5 μg/ml). Transpositions of the Tn*917* derivatives from these various plasmids into chromosomal sites separate the transposons from the plasmid-associated promoter, of course, and result in detectable expression of the transposon-borne *lacZ* or *cat-86* genes only when the Tn*917* derivatives become inserted in the appropriate orientation into a chromosomal transcription unit.

Insertional mutagenesis with Tn*917* derivatives: considerations of scale and randomness. pTV1, pTV8, pTV32, pTV51, pTV52, and pTV53 all depend for autonomous replication in *B. subtilis* upon plasmid *rep* functions derived ultimately from the *Staphylococcus aureus* replicon pE194 (9). All such plasmids are temperature sensitive for replication in *B. subtilis*, exhibiting progressively decreasing copy numbers with increasing temperatures above 32°C, and failing entirely to replicate at temperatures above 45°C (7). This makes it possible to recover transpositions of Tn*917* into chromosomal sites simply by subjecting bacteria that contain transposon-bearing pE194 replicons to a selection for MLS^r (the phenotype conferred by *erm*) at 47°C. For example, the direct plating of a 32°C culture of pTV1-containing bacteria at 47°C in the presence of erythromycin and lincomycin yields chromosomal inserts at a frequency of approximately 5×10^{-5} per viable count (at 32°C). An analysis of such inserts has revealed a significantly nonrandom distribution: we estimate that perhaps as many as 99% of all inserts are clustered in several "hot spot" regions of the chromosome (P. J. Youngman, J. B. Perkins, and K. Sandman *in* A. T. Ganesan and J. A. Hoch, ed., *Genetics and Biotechnology of Bacilli*, in press; S. A. Zahler, personal communication). The remaining 1% or so of insertions appear to be highly random, however, and permit Tn917 to be used effectively for insertional mutagenesis in all regions of the chromosome (Youngman, Perkins, and Sandman, in press; K. Sandman, unpublished data). Interestingly, the addition of even as much as 4.5 kb of extra DNA to Tn917 at its *Hpa*I site (as in the case of pTV53) causes no detectable reduction in transposition frequency (P. Youngman, unpublished data).

Selection in liquid culture has proved to be the most convenient method for recovering chromosomal insertions on a large scale, but such selections are only successful when appropriate allowances are made for the persistence of MLS^r in the "background" population. Even though pE194-derived replicons are completely replication defective at 47°C, approximately three cell generations are required to deplete the plasmid copy number to one per cell. Moreover, even after an *erm*-bearing plasmid is lost from the cell, a low level of drug resistance persists for a few generations as a result of the continued presence in plasmidless segregants of some *erm*-encoded ribosome methyltransferase and some previously methylated ribosomes. These factors allow substantial proliferation of a plasmid-containing cell shifted to 47°C even if that cell lacks a chromosomal Tn917 insert. Thus, a 32°C culture of plasmid-containing bacteria at a Klett value not exceeding 70 must be diluted at least 100-fold for the insert-containing population to emerge through the background growth after overnight incubation at 47°C in the presence of erythromycin and lincomycin.

In preparing a population of insert-containing bacteria with the objective of recovering an insertional mutation in a particular genetic locus, considerations of scale are critical. A 32°C culture of plasmid-containing bacteria at a Klett value less than 70 will have a titer not greater than 5×10^7 CFU/ml. If chromosomal inserts exist at a frequency of 5×10^{-5}/CFU, insert-containing bacteria will have a titer of 2.5×10^3/ml. However, if only 1% of these insertions are distributed with a high degree of randomness, the titer of bacteria having a randomly placed insertion will be on the order of only 2.5×10^1/ml. If the *B. subtilis* genome is taken to be approximately 3×10^3 kb in size, then at least 120 ml of 32°C culture would be required to saturate the genome at the level of one randomly placed Tn917 insert per kb interval of DNA. Thus, to prepare a collection of insert-containing bacteria likely to include a Tn917 insert in any gene of interest, at least 500 ml of 32°C culture at a Klett value of 70 would have to be diluted into a 50-liter fermentor for overnight growth at 47°C under selection for MLS^r. The collection of Tn917 insertions that currently serves as our principal source of insertional *spo* mutations was prepared in a 100-liter fermentor. This same collection should be appropriate for recovering many other kinds of mutants, of course, and it will be made available to any investigator in the field not equipped to prepare a similar collection on the scale required.

Construction of partial diploids for the analysis of Tn917-mediated gene fusions. A potential disadvantage of obtaining a *lac* or *cat* fusion by insertional mutagenesis is the possibility that the insertional mutation itself will cause a phenotype that affects the regulation of the fused (and disrupted) gene. It would not be surprising, for example, to discover that certain *spo* mutations distort the sporulation-related physiology of a cell in such a way as to profoundly alter the expression of the very gene in which the mutation occurred, and the analysis of many other kinds of transposon-mediated fusions (such as ones generated by auxotrophic mutations) would be subject to similar complications. For this reason, we have developed simple procedures that permit the study of Tn917-mediated fusions in partial diploids where the insertional mutation of interest is complemented by an intact copy of the interrupted gene.

These procedures take advantage of a method recently devised to facilitate the cloning in *E. coli* of *B. subtilis* DNA adjacent to Tn917 insertions (14). Derivatives of Tn917 were constructed, such as the one contained in pTV21Δ2 (Fig. 3), in which an intact copy of a pBR322-derived cloning vector, pHW9 (8), has been inserted by ligation into a site near the center of the transposon. pHW9 carries a *cat* gene of gram-positive origin which can serve as a selectable marker in *B. subtilis*. When linearized pTV21Δ2 DNA is used to transform to Cm^r *B. subtilis* bacteria that already contain a chromosomal Tn917 insert (a *spo*::Tn917 mutant, for example), the pBR322 sequences become integrated into the chromosomal copy of the transposon by homologous recombination.

Precisely the same strategy may be used to integrate pBR322 sequences (along with the *cat* gene) at the site of a Tn917 *lac*-generated insertional *spo* mutation, as diagrammed in Fig. 4A. Chromosomal DNA extending in either direction from the integrated pBR322 sequences may

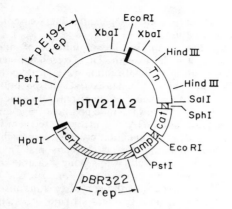

FIG. 3. Restriction map of pTV21Δ2 (see 14).

then be cloned into *E. coli*, with an Ap^r selection, simply by transforming with insert-containing DNA that was first digested with certain restriction enzymes and then ligated at a dilute concentration (to circularize the fragments). Several restriction enzymes, including *Hind*III, *Sph*I, and *Xba*I, have recognition sites in the *lac*-distal arm of the transposon, but not in the

lac-proximal arm. Clones generated with such enzymes (as illustrated in Fig. 4A with *Hind*III) thus extend into chromosomal DNA "upstream" from the transposon terminal inverted repeat spanned by the transcriptional fusion. If no recognition site for the enzyme used occurs between the promoter of the fused gene and the transposon, the cloned DNA will include the entire gene fusion. When such plasmid clones are transformed back into *B. subtilis* (with a selection for Cm^r), they integrate into the recipient genome by "Campbell recombination," regenerating an intact copy of the insertionally disrupted gene and returning the *lac* fusion to the chromosome in a single step (Fig. 4B). Even in the unlikely event that the interval between the transposon and the promoter for the *lac* fusion contains recognition sites for *Hind*III, *Sph*I, and *Xba*I, partial rather than complete digests with these enzymes should in most cases generate a plasmid clone with the structure appropriate for obtaining a partial diploid with restored gene function when the cloned DNA is transformed back into *B. subtilis*.

Use of pTV53 as a probe to identify and study regulated promoters. Transposon derivatives that generate simultaneous transcriptional fusions to

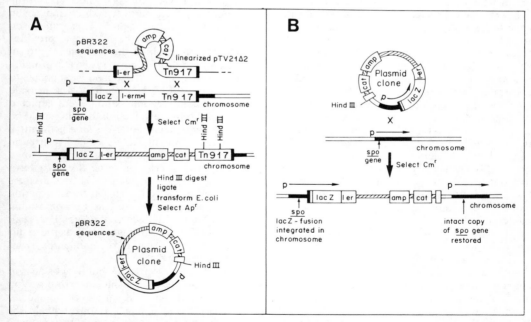

FIG. 4. Construction of partial diploids for the analysis of gene fusions generated by Tn917-mediated insertional mutagenesis. Bacteria already containing a Tn917-mediated *lacZ* fusion are first transformed to Cm^r with linearized pTV21Δ2 DNA (A, top). Transformants that result from the indicated marker-replacement recombination event are MLS^s and have pBR322 sequences integrated at the site of the gene fusion. DNA including the entire gene fusion can then be cloned into *E. coli* (A, bottom), as explained in the text. A *B. subtilis* partial diploid that contains the fusion of interest as well as an intact copy of the insertionally disrupted gene (B) may then be obtained by transforming "wild-type" *B. subtilis* bacteria to Cm^r with the *E. coli* plasmid clone.

lacZ and *cat-86*, such as the derivative contained in pTV53 (Fig. 2D), offer certain special advantages for identifying and manipulating regulated genes. The *lacZ* gene product (β-galactosidase) can be detected in colonies (by the development of blue color on X-gal plates, for example) even when it is made only by cells in stationary phase. The *cat-86* gene product (a chloramphenicol acetyltransferase) confers a selectable phenotype (Cm^r), but only when it is present in growing cells. A transposon that mediates simultaneous fusions to *lacZ* and *cat-86* thus provides a way of specifically identifying transcription units whose activity is restricted to stationary phase: only insertions into such transcription units should result in bacteria that are sensitive to chloramphenicol but that make blue colonies on X-gal plates.

Tandem fusions to *lacZ* and *cat-86* offer other kinds of advantages for studying regulated genes that are active during the vegetative phase. The production of β-galactosidase provides for a sensitive assessment of gene activity in colonies on agar plates (blueness on X-gal), and the *cat-86* gene makes it possible to select or enrich for certain kinds of mutations that affect gene fusions. In combination, *lacZ* and *cat-86* should permit a versatile genetic analysis of fusions involving a wide range of regulated genes.

Use of special β-galactosidase substrates for the identification of promoters that become active in sporulating bacteria. Another way of using *lacZ* fusion-generating transposons to identify transcription units that become active during sporulation is to take advantage of β-galactosidase substrates that can be applied to plates after colonies have developed and whose hydrolysis by β-galactosidase can be detected with sufficient sensitivity to provide a "real time" indication of gene activity. Probably the best substrate of this kind (D. Gelfand, personal communication) is MUG, which yields a highly fluorescent hydrolysis product. The amount of MUG hydrolyzed within 10 min of being sprayed onto a plate (even by a colony expressing a *lacZ* fusion at moderately low levels) is very easy to detect. Thus, when *B. subtilis* bacteria that contain Tn*917 lac* insertions at "random" chromosomal sites are plated for single colonies on sporulation-inducing media and treated with MUG after colonies have developed to their full size (when most of the bacteria are no longer growing and sporulation is in progress), brightly fluorescent colonies reflect insertions into transcription units that are induced to high levels of expression during some phase of sporulation. As with the *cat-lac* screen of pTV53-generated insertions discussed above, this approach has the important advantage of being able to identify stationary phase-specific transcription units regardless of whether they encode gene products essential for sporulation.

Interconversion by recombination of drug markers associated with Tn*917*-mediated gene fusions and insertional mutations: a strategy for systematic pairwise epistasis tests of *spo* mutations. The remarkably efficient natural capacity of *B. subtilis* bacteria to take up linear fragments of transforming DNA and incorporate them by homologous recombination creates the possibility of manipulating existing chromosomal Tn*917* inserts in many useful ways. One example has already been discussed: the integration of pBR322 sequences at the site of existing insertional mutations by transformation with pTV21Δ2 (Fig. 4A). Similar strategies actually make it possible to convert ordinary Tn*917*-generated insertional mutations into *lacZ* fusions, *cat-86* fusions, or tandem *lacZ*–*cat-86* fusions, by using plasmids pTV54 (Fig. 5A) and pTV55 (Fig. 5B), in combination with some of the plasmids shown in Fig. 2. Any Tn*917*-generated insertional mutation in which the *erm*-proximal end of Tn*917* is oriented toward the promoter of the interrupted gene should be appropriate for such manipulations.

In the various *lacZ* fusion-generating derivatives of Tn*917* described above, 278 base pairs of transposon DNA remains to the "left" of the inserted *lacZ* gene, and over 5 kb of transposon DNA remains to the "right." This transposon DNA on either side of the *lacZ* gene provides homology to recombine the *lacZ* gene into an ordinary chromosomal copy of Tn*917* in much the same way that transposon sequences to either side of pBR322 DNA in pTV21Δ2 provide homology for the recombination event depicted in Fig. 4A. So that a selection for Cm^r could be used to integrate a promoterless *lacZ* gene into a chromosomal copy of Tn*917*, a small restriction fragment carrying a *cat* gene (with its own promoter) was inserted into the *Bam*HI site of pTV51 (Fig. 2B) to produce pTV54 (Fig. 5A). As diagrammed in Fig. 6 (panel A), transformation of competent cells that contain a chromosomal Tn*917* insertion with linearized pTV54 DNA can yield stably Cm^r transformants only as the result of a "marker-replacement" recombination event that integrates *lacZ*.

pTV55 is similar to pTV54, but it lacks an *erm* gene. As a consequence, using linearized pTV55 DNA to recombine *lacZ* into a chromosomal copy of Tn*917* (Fig. 6, panel B, top) yields MLS^s transformants. Thus, after a *spo*::Tn*917* mutation has been converted into a *lacZ* fusion by transformation to Cm^r (and MLS^s) with pTV55 DNA, different *spo*::Tn*917* mutations can be transformed into the fusion-containing strain with a selection for MLS^r. This should make it possible in a very easy way to combine different

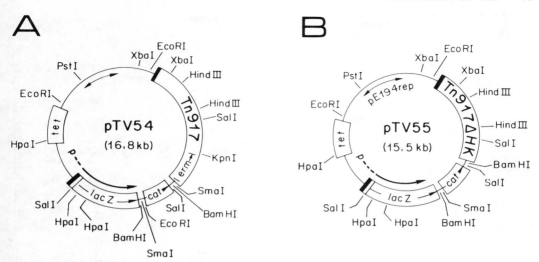

FIG. 5. Restriction maps of pTV54 and pTV55. pTV54 was obtained by inserting a BamHI-generated, cat-containing restriction fragment from pMI1101 (Youngman et al., in press) into the BamHI site of pTV51 (Fig. 2). pTV55 was obtained as a result of in vivo recombination between pTV54 and pTV13 (H.-P. Bieman and J. B. Perkins, unpublished data; a derivative of pTV5 [Youngman et al., in press] in which the erm gene has been deleted and a cat gene has been inserted).

spo mutations in the same cell and to examine the effect of one on the expression of the other by measuring the activity of a lacZ fusion. Such epistasis tests should make it possible to sort spo genes systematically into regulation classes on the basis of their expression response to mutations in various regulatory loci.

Finally, if a strain harboring a Cm^r MLS^s lacZ fusion (obtained by transformation with pTV55) were transformed to MLS^r with linearized pTV53 DNA (Fig. 6, panel B, bottom), the promoter-containing cat gene would be removed, and a cat-86 gene fusion would be created in the same recombination event. The capacity to replace one kind of fusion with another in vivo or to interconvert by recombination the drug markers associated with particular gene fusions permits a highly flexible analysis of Tn917-mediated gene fusions in B. subtilis and takes full advantage of the natural genetic tractability of this species.

SUMMARY

Derivatives of the gram-positive transposon Tn917 have been constructed that carry promoterless copies of the E. coli lacZ or the B. pumilus cat-86 gene, or both genes as a tandem pair. Insertions of these transposon derivatives into chromosomal genes can effect transcriptional gene fusions, placing expression of the promoterless, transposon-borne lacZ or cat-86 genes under control of the promoters (and other regulatory elements) of the interrupted chromo-

somal transcription units. Such gene fusions permit easy and quantitative assays for gene activity in liquid culture (both lacZ and cat-86), provide a sensitive semiquantitative assessment of gene activity in colonies of bacteria on plates (lacZ), and in certain cases can provide selections for some kinds of mutations that affect the regulation of the interrupted transcription units (cat-86).

Tn917-mediated fusions involving spo genes can be recovered in four ways. The first is simply to generate insertional spo mutations with Tn917 lac or Tn917 cat. In half of such insertional mutations, the transposon will be in an orientation that activates the transposon-borne lacZ or cat-86 genes for expression. The second is to generate a population of bacteria containing Tn917 cat-lac insertions at many different chromosomal sites and then to screen them on sporulation-inducing agar media for ones that produce blue colonies in the presence of X-gal, yet test as being sensitive to chloramphenicol. The third is to screen such an insert-containing population for bacteria that yield fluorescent colonies on sporulation-inducing media after an application of MUG. The fourth is to convert ordinary spo::Tn917 insertional mutations into lacZ fusions by recombination in vivo.

Tn917-mediated spo gene fusions obtained through these various approaches should represent an important tool in future work for defining and studying the many classes of regulated genes which together comprise the complex program of temporally regulated gene activity associated

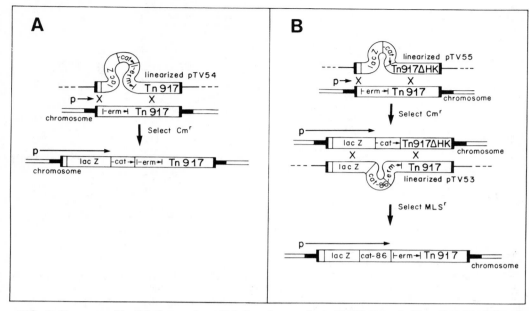

FIG. 6. Recovery of *lacZ* fusions and *cat-86* fusions form ordinary Tn*917*-generated insertional mutations by homologous recombination. When bacteria already containing a Tn*917*-generated insertional mutation are transformed to Cm^r with linearized pTV54 DNa (A), the indicated recombination event integrates a promoterless *lacZ* gene into the chromosomal copy of Tn*917*. If the original Tn*917* insertion is in the appropriate orientation, a *lacZ* fusion will result. A *lacZ* fusion can be produced in a similar manner by transforming with linearized pTV55 DNA (B, top), and the fusion-containing transformants become MLS^s in the process. A tandem *lacZ–cat-86* fusion can be obtained by transforming MLS^s *lacZ* fusion-containing bacteria to MLS^r with linearized pTV53 DNA (B, bottom).

with the differentiation of vegetative *B. subtilis* bacteria into dormant endospores.

We thank P. Zuber, M. Igo, P. Lovett, S. Zahler, and D. Dubnau for generous gifts of strains and plasmids.

This work was carried out in the laboratory of R. Losick at Harvard University and was supported by National Science Foundation grant PCM-8208349 to R.L. and P.Y., a postdoctoral fellowship to J.B.P. from the National Institutes of Health, and a predoctoral fellowship to K.S. from the National Science Foundation.

LITERATURE CITED

1. **Bellofatto, V., L. Shapiro, and D. A. Hodgson.** 1984. Generation of a Tn5 promoter probe and its use in the study of gene expression in *Caulobacter crescentus*. Proc. Natl. Acad. Sci. U.S.A. **81:**1035–1039.
2. **Bertani, G.** 1951. Studies on lysogenesis. I. The mode of phage liberation by lysogenic *Escherichia coli*. J. Bacteriol. **62:**293–300.
3. **Boyer, H. W., and R. Roulland-Dussoix.** 1968. A complementation analysis of the restriction and modification of DNA in *Escherichia coli*. J. Mol. Biol. **41:**459–472.
4. **Casadaban, M. J., and J. Chou.** 1984. *In vivo* formation of gene fusions encoding hybrid β-galactosidase proteins in one step with transposable Mu-*lac* transducing phage. Proc. Natl. Acad. Sci. U.S.A. **81:**535–539.
5. **Casadaban, M. J., and S. N. Cohen.** 1979. Lactose genes fused to exogenous promoters in one step using a Mu-*lac* bacteriophage: *in vivo* probe for transcriptional control sequences. Proc. Natl. Acad. Sci. U.S.A. **76:**4530–4533.
6. **Dubnau, D., R. Davidoff-Abelson, and I. Smith.** 1969. Transformation and transduction in *Bacillus subtilis*: evidence for separate modes of recombinant formation. J. Mol. Biol. **45:**155–179.
7. **Gryczan, T. J., J. Hahn, S. Contente, and D. Dubnau.** 1982. Replication and incompatibility properties of plasmid pE194 in *Bacillus subtilis*. J. Bacteriol. **152:**722–735.
8. **Horinouchi, S., and B. Weisblum.** 1982. Nucleotide sequence and functional map of pC194, a plasmid that specifies inducible chloramphenicol resistance. J. Bacteriol. **150:**815–825.
9. **Iordanescu, S.** 1976. Three distinct plasmids originating in the same *Staphylococcus aureus* strains. Arch. Roum. Pathol. Exp. Microbiol. **35:**111–118.
10. **Meselson, M., and R. Yuan.** 1968. DNA restriction enzyme from *Escherichia coli*. Nature (London) **217:** 1110–1114.
11. **Mongkolsuk, S., Y.-W. Chiang, R. B. Reynolds, and P. S. Lovett.** 1983. Restriction fragments that exert promoter activity during postexponential growth of *Bacillus subtilis*. J. Bacteriol. **155:**1399–1406.
12. **Tomich, P. K., F. Y. An, and D. B. Clewell.** 1980. Properties of erythromycin-inducible transposon Tn*917* in *Streptococcus faecalis*. J. Bacteriol. **141:**1366–1374.
13. **Youngman, P. J., J. B. Perkins, and R. Losick.** 1983. Genetic transposition and insertional mutagenesis in *Bacillus* with *Streptococcus faecalis* transposon Tn917. Proc. Natl. Acad. Sci. U.S.A. **80:**2305–2309.
14. **Youngman, P., J. B. Perkins, and R. Losick.** 1984. A novel method for the rapid cloning in *Escherichia coli* of *Bacillus subtilis* chromosomal DNA adjacent to Tn917 insertions. Mol. Gen. Genet. **195:**424–433.

Use of Cloned *spoIIA* and *spoVA* Probes to Study Synthesis of mRNA in Wild-Type and Asporogenous Mutants of *Bacillus subtilis*

DEMETRIS SAVVA† AND JOEL MANDELSTAM

Microbiology Unit, Department of Biochemistry, University of Oxford, Oxford OX1 3QU, United Kingdom

The formation of endospores in *Bacillus subtilis* involves about 50 known genetic loci (6), and the total number of loci may be considerably larger (7). The biochemical and biophysical properties of mutants blocked at various stages of sporulation led to the suggestion that the genetic loci, presumed to constitute operons, form a dependent sequence in which the expression of an operon depends on the expression of the previous operon in the sequence (11). The simplest model is a linear sequence in which one of the translation products of each operon activates or derepresses the next one (12); this could occur either by the protein product acting as such or, alternatively, by its having an enzymatic activity that generated a small molecule with a regulatory function. Other models have been considered, such as the parallel transcription of operons (3), the control of transcription through changes in sigma factors of RNA polymerase (see, e.g., 10), and the possibility that the sequence might be branched (14).

It was subsequently shown that the phenotypic characteristics of the later stages from about stage IV onwards did not, in fact, reflect an order of gene transcription but depended on the self-assembly of proteins that had been synthesized earlier (4, 8). From these last reports it appears that nearly all the operons involved in sporulation are transcribed during the first four stages, but it still remains to be determined whether the sequence during this phase of development is linear or branched, whether the control is negative or positive, and whether the regulators are large or small molecules.

The advent of cloning techniques and their successful application to a variety of sporulation genes means that a start can now be made toward answering the first of these questions. This can be done by using cloned genes, or portions of them, as probes to detect the appearance of the corresponding mRNA molecules.

The present paper describes experiments that made use of a recombinant phage derived from φ105, carrying both the *spoIIA* and the *spoVA*

loci (17). The DNA sequences of the two loci have been determined (5; P. Fort and J. Errington, J. Gen. Microbiol., in press), and a restriction map of the recombinant phage and the 7-kilobase-pair insert carrying the two operons is shown in Fig. 1. Restriction fragments lying totally within each locus were prepared from this phage and were used as probes to discover the times of transcription of *spoIIA* and *spoVA* in the wild type and also to determine whether transcription occurred in a variety of sporulation mutants.

EXPERIMENTAL PROCEDURES

Growth of bacteria. Cultures of *B. subtilis* 168 or of Spo⁻ mutants were grown and resuspended in sporulation medium by the method of Sterlini and Mandelstam (18); samples (100 ml) were removed at different times after resuspension for the isolation of RNA.

Isolation of RNA. Cultures (100 ml) were chilled by the addition of 20 ml of frozen resuspension medium and were centrifuged for 3 min on a bench centrifuge. The cell pellet was resuspended in 1 ml of cold lysozyme solution (10 mg/ml in 50 mM Tris hydrochloride, pH 8.0) and immediately centrifuged for 30 s in a microcentrifuge. The pellet was resuspended in 0.5 ml of cold lysozyme solution and frozen in liquid nitrogen. After thawing, the cells were added to 3 ml of guanidinium isothiocyanate solution (6 M guanidinium isothiocyanate, 5 mM sodium citrate [pH 7.0], 0.1 M β-mercaptoethanol, 0.5% Sarkosyl), and the RNA was isolated by the method of Maniatis et al. (13).

RNA pellets were dissolved in water and stored at −70°C; concentrations were estimated by measuring the absorbance of suitably diluted samples at 260 nm.

Gel electrophoresis. Samples of RNA (8 μg) were mixed with an equal volume of loading buffer (20% [wt/vol] glycerol, 50 mM boric acid, 5 mM sodium tetraborate, 10 mM sodium sulfate, 25 mM methyl mercuric hydroxide, 0.02% [wt/vol] bromophenol blue) and were electrophoresed on 1.2% agarose gels in the presence of 5 mM methyl mercuric hydroxide (1). Transfer of the RNA from the gel onto nitrocellulose filters

---

† Present address: Department of Physiology and Biochemistry, University of Reading, Whiteknights, Reading, Berks., United Kingdom.

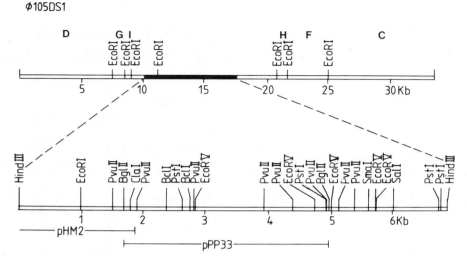

FIG. 1. Simplified restriction map of bacteriophage φ105DS1 (showing only the *Eco*RI sites) and a detailed restriction map of the cloned *B. subtilis* DNA (solid line) (17). The parts of the insert present in plasmids pHM2 (9) and pPP33 (15) are also shown. The *Eco*RI fragments of φ105DS1 which are identical to those of the vector (φ105DI:1t) are lettered as described by Bugaichuk et al. (2).

was performed as described previously (17), with the use of 20× SSC (3 M NaCl, 0.3 M sodium citrate).

Electrophoresis of DNA and isolation of DNA restriction fragments from agarose gels was performed as described previously (17).

DNA-RNA hybridizations. Probe DNA (200 ng; see Results) was labeled by nick translation (16) with the use of [α-^{32}P]dCTP (3,000 Ci/mmol; Amersham International).

Two types of hybridization were performed: (i) "Northern" hybridizations with nitrocellulose filters prepared as described above and (ii) dot hybridizations with nitrocellulose filters onto which the RNA samples (4 μg) were spotted.

Nitrocellulose filters were baked for 2 h at 80°C under vacuum, placed in the minimum volume of prehybridization solution (50% [wt/vol] formamide, 0.9 M NaCl, 5 mM EDTA, 50 mM sodium phosphate [pH 7.7], 0.1% [wt/vol] sodium dodecyl sulfate, 0.02% [wt/vol] bovine serum albumin, 0.02% [wt/vol] Ficoll, 0.02% [wt/vol] polyvinyl pyrrolidone, 0.01% [wt/vol] salmon sperm DNA), and incubated in a shaking water bath at 43°C for 3 h. The nick-translated probe DNA was denatured by boiling for 5 min and added to the prehybridization solution together with dextran sulfate (final concentration, 10%, wt/vol). Hybridization was allowed to proceed for 18 h in a shaking water bath at 43°C. Filters were washed in two changes (100 ml each for 15 min each) of 0.36 M NaCl–2 mM EDTA–20 mM sodium phosphate [pH 7.7]–0.1% [wt/vol] sodium dodecyl sulfate followed by two changes (100 ml each for 15 min) of 0.018 M NaCl–0.1

mM EDTA–1 mM sodium phosphate (pH 7.7)–0.1% (wt/vol) sodium dodecyl sulfate. After drying, the filters were autoradiographed at −70°C.

RESULTS

Choice of hybridization probes. The restriction map of a recombinant phage derived from φ105 and containing the *spoIIA* and *spoVA* loci (17) is shown in Fig. 1. On the basis of the DNA sequences of the *spoIIA* and *spoVA* loci (5; Fort and Errington, in press), a number of restriction fragments that lie totally within each locus were chosen for use as probes to detect their transcripts. In most cases these restriction fragments were isolated from low-melting-point agarose gels, but in a few cases they were subcloned in one of the pUC plasmids (19). The probes used for the detection of the *spoIIA* and *spoVA* mRNAs were the 0.67-kilobase-pair *Eco*RI-*Bgl*II fragment and the 1.89-kilobase-pair *Eco*RV-*Eco*RV fragment, respectively (these are shown in Fig. 2); identical results were obtained with a number of other probes in each locus.

Timing of transcription of *spoIIA* and *spoVA* operons. RNA isolated from *B. subtilis* 168 at hourly intervals after resuspension in sporulation medium (18) was electrophoresed on agarose gels containing methyl mercuric hydroxide, transferred to nitrocellulose filters, and hybridized to the different nick-translated probe DNAs; the specific activity of the probes was between 10^7 and 5×10^7 cpm/μg of DNA.

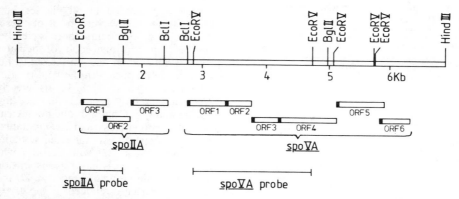

FIG. 2. Genetic map of the *B. subtilis* DNA in bacteriophage φ105DS1 (5; Fort and Errington, in press). Solid boxes indicate ribosome binding sites, and open boxes indicate the open reading frames (ORF). Also shown are some restriction enzyme sites and the fragments used as probes for the hybridization experiments.

The results of these hybridizations (Fig. 3) show that neither of the two loci studied appeared to be transcribed during vegetative growth, as evidenced by the absence of hybridization between any of the probes and RNA isolated immediately after the resuspension of the cells in sporulation medium (i.e., time zero [t_0]).

The *spoIIA* probe was found to hybridize weakly to RNA isolated at 1 h after resuspension (i.e., t_1) and strongly to RNA isolated at later times (Fig. 3a). Therefore, it appears that the mRNA for *spoIIA* starts being synthesized about 1 h after the initiation of sporulation and

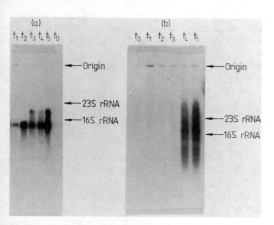

FIG. 3. Northern hybridizations indicating the timing of transcription for the *spoIIA* (a) and *spoVA* (b) loci. Each track contained 8 μg of RNA isolated from *B. subtilis* 168 at different times after resuspension; these are indicated as t_0, t_1, t_2, t_3, t_4, and t_5. The positions of the 23S and 16S rRNA from *B. subtilis* are shown by arrows; phage lambda DNA digested with *Hind*III was also used to provide molecular-weight markers (not shown).

that it is present for at least the next 4 h. The DNA probe hybridized to only one mRNA species, the size of which was about 1.4 kilobases; this is in agreement with earlier results suggesting that the *spoIIA* locus is a polycistronic operon (15).

With the *spoVA* probe, weak hybridization was obtained at t_3, but hybridization was strong with RNA samples isolated 4 or 5 h after resuspension (Fig. 3b); later experiments indicated that the *spoVA* transcript is present at $t_{3.5}$. However, it proved impossible to establish the exact size of the mRNA as a result of smearing of the samples. This was presumably due to the degradation of this messenger molecule.

Transcription of *spoIIA* and *spoVA* mRNA in Spo⁻ mutants. A number of Spo⁻ mutants were used to isolate RNA at 2.5 and 5 h after resuspension in sporulation medium. Table 1 shows the results of both the dot and Northern hybridizations between these RNA samples and the *spoIIA* and *spoVA* probes.

As a control, a mutant, *spoIIA4*, was used. This mutant has a large deletion covering the whole of the *spoIIA* locus and most of the *spoVA* locus, and as expected, it did not produce transcripts for either of the loci. The results also indicate that the *spoIIA* mRNA was present in all the stage 0 and stage II mutants studied; however, only trace amounts of *spoIIA* mRNA were present in the stage 0 mutants. All the mutants blocked at later stages of sporulation which were tested were found to produce normal amounts of the *spoIIA* mRNA. None of the mutants studied produced any *spoIIA* mRNA at t_0.

The *spoVA* mRNA was not detectable in any stage 0 mutants except *spo0A* and *spo0E*, which produced it in trace amounts. Similar small amounts were detected in all the mutants tested

TABLE 1. Synthesis of the mRNA molecules for *spoIIA* and *spoVA* in Spo⁻ mutants

Mutation	*spoIIA* mRNA[a,c]	*spoVA* mRNA[b,c]
spo0A43	t	t
spo0B136	t	−
spo0C9	t	−
spo0D8	t	−
spo0E11	t	t
spo0F221	t	−
spo0G14	t	−
spo0H17	t	−
spo0J93	t	−
spo0K141	t	−
spoIIA4	−	−
spoIIA69	+	t
spoIIB592	+	t
spoIIC298	+	t
spoIID68	+	t
spoIIE20	+	t
spoIIG55	+	t
spoIIIA65	+	t
spoIIIB2	+	t
spoIIIF590	+	t
spoIVA67	+	t
spoIVC133	+	t
spoVA89	+	+
spoVC285	+	+
spoVIC1149	+	+
gerE36	+	+

[a] The RNA was isolated at $t_{2.5}$.
[b] The RNA was isolated at t_5.
[c] The presence and absence of mRNA is indicated by + and −, respectively; t, trace amounts.

which were blocked before stage V, whereas those at stage V, or later, produced normal amounts.

DISCUSSION

The results of the Northern blot experiments (Fig. 3) show that *spoIIA* begins to be transcribed at about t_1, whereas transcription of *spoVA* is clearly seen only after t_3; both loci remain switched on until at least t_5. It is also apparent from the size of the *spoIIA* transcript that the mRNA molecule is polycistronic, as would be expected from the published nucleotide sequence (5). The mRNA for *spoVA* appears to be less stable and gives a continuous spread of material varying in size between 500 and about 4,000 bases. The upper value is close to that expected (4.03 kilobases) if the six genes in this locus (Fort and Errington, in press) also constitute an operon.

All the *spo0* mutants tested produced the *spoIIA* messenger molecule in greatly diminished amounts, though *spo0A*, *spo0E*, and *spo0J* generated significantly more than the other *spo0* mutants (results not shown). By contrast, the message is produced in all classes of *spoII* mutants tested. However, in all of these, and indeed in the *spo0* mutants, there is clearly some form of transcriptional control because the mRNA is not synthesized until about t_1, i.e., the same time as it appears in the wild type.

The transcription of *spoVA* is clearly different. The mRNA is produced in trace amounts in *spo0A* and *spo0E* mutants and not measurably in any of the other classes of *spo0* mutants. This should be compared with the results obtained by Zuber and Losick (22) with *spoVG*, which is transcribed very early after resuspension and whose expression is unaffected in *spo0J* mutants and is only partially reduced in *spo0C*, *spo0E*, and *spo0F* mutants. The mRNA corresponding to *spoVA* was found in only trace amounts in all the *spoII*, *spoIII*, and *spoIV* mutants tested. It is, however, formed in mutants blocked later in sporulation such as *spoVC* and *spoVIC*, and also in *gerE*.

The fact that the mRNA for *spoIIA* is present in all classes of *spoII* mutants could mean that the idea of a single linear sequence is an oversimplification. Alternatively, it could mean that *spoIIA* is expressed before all the other *spoII* loci. This would entail the supposition that it is misleading to base a "pleiotropic hierarchy" (14, 20) on the occurrence or nonoccurrence of such secondary biochemical events as the formation of alkaline phosphatase or DNase.

What is shown clearly by these experiments is (i) that the *spoIIA* and *spoVA* loci are subject to transcriptional control, (ii) that they generate polycistronic messages, and (iii) that the data, though far from complete, indicate that there is a degree of dependence in the expression of sporulation genes. Thus, the messenger molecules for *spoIIA* and for *spoVA* are expressed feebly or not at all in mutants blocked at earlier stages. Therefore, in a general way, these results support the model of a dependent sequence (12), which is also borne out by the patterns of protein synthesis during sporulation (21).

SUMMARY

The *spoIIA* and *spoVA* loci have been cloned, and DNA restriction fragments that lie totally within the two loci have been used as probes to study their transcription by means of DNA-RNA hybridizations. The results provide evidence for transcriptional control, the existence of operons, and some form of dependent sequence during sporulation.

This work was supported by the Science and Engineering Research Council.

LITERATURE CITED

1. **Bailey, J. M., and N. Davidson.** 1976. Methylmercury as a reversible denaturing agent for agarose gel electropho-

resis. Anal. Biochem. **70**:75–85.

2. **Bugaichuk, U. D., M. Deadman, J. Errington, and D. Savva.** 1984. Restriction enzyme analysis of *Bacillus subtilis* bacteriophage φ105 DNA. J. Gen. Microbiol. **130**:2165–2167.

3. **Coote, J. G., and J. Mandelstam.** 1973. Use of constructed double mutants for determining the temporal order of expression of sporulation genes in *Bacillus subtilis*. J. Bacteriol. **114**:1254–1263.

4. **Dion, P., and J. Mandelstam.** 1980. Germination properties as marker events characterizing later stages of *Bacillus subtilis* spore formation. J. Bacteriol. **141**:786–792.

5. **Fort, P., and P. J. Piggot.** 1984. Nucleotide sequence of sporulation locus *spoIIA* in *Bacillus subtilis*. J. Gen. Microbiol. **130**:2147–2153.

6. **Henner, D. J., and J. A. Hoch.** 1980. The *Bacillus subtilis* chromosome. Bacteriol. Rev. **44**:57–82.

7. **Hranueli, D., P. J. Piggot, and J. Mandelstam.** 1974. Statistical estimate of the total number of operons specific for *Bacillus subtilis* sporulation. J. Bacteriol. **119**:684–690.

8. **Jenkinson, H. F., W. D. Sawyer, and J. Mandelstam.** 1981. Synthesis and order of assembly of spore coat proteins in *Bacillus subtilis*. J. Gen. Microbiol. **123**:1–16.

9. **Liu, H.-M., K. F. Chak, and P. J. Piggot.** 1982. Isolation and characterization of a recombinant plasmid carrying a functional part of the *Bacillus subtilis spoIIA* locus. J. Gen. Microbiol. **128**:2805–2812.

10. **Losick, R., and J. Pero.** 1981. Cascades of sigma factors. Cell **25**:582–584.

11. **Mandelstam, J.** 1969. Regulation of bacterial spore formation. Symp. Soc. Gen. Microbiol. **19**:377–402.

12. **Mandelstam, J.** 1976. Bacterial sporulation: a problem in the biochemistry and genetics of a primitive developmental system. Proc. R. Soc. London Ser. B **193**:89–106.

13. **Maniatis, T., E. F. Fritsch, and J. Sambrook.** 1982. Molecular cloning. Cold Spring Harbor Laboratory, Cold Spring Harbor, N.Y.

14. **Piggot, P. J., and J. G. Coote.** 1976. Genetic aspects of bacterial endospore formation. Bacteriol. Rev. **40**: 908–962.

15. **Piggot, P. J., C. A. M. Curtis, and H. de Lencastre.** 1984. Use of integrational plasmid vectors to demonstrate the polycistronic nature of a transcriptional unit (*spoIIA*) required for sporulation of *Bacillus subtilis*. J. Gen. Microbiol. **130**:2123–2136.

16. **Rigby, P. W. J., M. Dieckmann, C. Rhodes, and P. Berg.** 1977. Labelling deoxyribonucleic acid to high specific activity *in vitro* by nick translation with DNA polymerase I. J. Mol. Biol. **113**:237–251.

17. **Savva, D., and J. Mandelstam.** 1984. Cloning of the *Bacillus subtilis spoIIA* and *spoVA* loci in phage φ105DI:1t. J. Gen. Microbiol. **130**:2137–2145.

18. **Sterlini, J. M., and J. Mandelstam.** 1969. Commitment to sporulation in *Bacillus subtilis* and its relationship to development of actinomycin resistance. Biochem. J. **113**:29–37.

19. **Vieira, J., and J. Messing.** 1982. The pUC plasmids, an M13mp7-derived system for insertion mutagenesis and sequencing with synthetic universal primers. Gene **19**:259–268.

20. **Young, M., and J. Mandelstam.** 1979. Early events during bacterial endospore formation. Adv. Microb. Physiol. **20**:103–162.

21. **Yudkin, M. D., H. Boschwitz, Y. Lorch, and A. Keynan.** 1982. Changes in the pattern of protein synthesis during the first three hours of sporulation in *Bacillus subtilis*. J. Gen. Microbiol. **128**:2165–2177.

22. **Zuber, P., and R. Losick.** 1983. Use of a *lacZ* fusion to study the role of the *spo0* genes of *Bacillus subtilis* in developmental regulation. Cell **35**:275–283.

Small, Acid-Soluble Spore Proteins of *Bacillus*: Products of a Sporulation-Specific, Multigene Family

EDWARD R. FLISS, MICHAEL J. CONNORS, CHARLES A. LOSHON, EVERARDO
CURIEL-QUESADA, BARBARA SETLOW, AND PETER SETLOW

Department of Biochemistry, University of Connecticut Health Center, Farmington, Connecticut 06032

From 10 to 20% of the protein in dormant spores of various *Bacillus* species is degraded to amino acids in the first minutes of spore germination (12). The amino acids produced by this degradation support much of the protein synthesis that occurs early in spore germination, as dormant spores lack many of the enzymes required for amino acid biosynthesis which are synthesized only later in outgrowth (12). The proteins degraded in this process are a group of small, acid-soluble spore proteins (SASPs) which are synthesized only during sporulation under transcriptional control (12). The exact number of SASPs in any one *Bacillus* species is not known, but seven different SASPs have been purified from *B. megaterium* spores and a number of minor SASPs have also been identified in this species (2, 12); other species also have multiple SASPs (9, 17). Although multiple SASPs exist in any one species, some of the different SASPs are very similar both immunologically and in primary sequence (12). Thus, two of the three major *B. megaterium* SASPs, SASPs A and C, are 85% identical in primary sequence, with the third major SASP, SASP B, more distantly related to the other two (12). *B. cereus* and *B. subtilis* spores also contain SASPs which are clearly homologous to *B. megaterium* SASPs A and C or B (9, 17). However, limited primary sequence analysis at the amino termini of the *B. cereus* and *B. subtilis* SASPs led to the suggestion that these proteins might be evolving extremely rapidly (17).

Despite the extensive information on SASPs which has accumulated over the past 10 years, a number of major questions have remained unanswered. These include (i) the number of different SASPs there are and what the function of this multiplicity of similar proteins is, (ii) whether SASPs evolve rapidly and if so by what mechanism, and (iii) what the molecular details of the regulation of transcription of these sporulation specific genes are. In an attempt to answer these questions, we have undertaken the cloning and analysis of SASP genes from various *Bacillus* species. In this communication we review our progress in this project.

EXPERIMENTAL PROCEDURES

Organisms used and isolation of DNA. The sources of all *Bacillus* and *E. coli* strains used have been described previously (1, 2, 6, 17; M. J. Connors and P. Setlow, submitted for publication; E. R. Fliss and P. Setlow, submitted for publication) with the exception of the λgt10 host *E. coli* POP-138, which was obtained from R. Davis (Stanford University, Stanford, Calif.). The sources of plasmids, M13 phage, and λ Charon 4A have also been described previously (1, 2; Connors and Setlow, submitted for publication; Fliss and Setlow, submitted for publication); λgt10 was also obtained from R. Davis.

Cloning methods and DNA sequences. The cloning of *Bacillus* DNA fragments followed published procedures: *Pvu*II fragments were cloned in pAD284 (1), and *Eco*RI fragments were cloned in pBR325 (2), λgt10, or λ Charon 4A (1). *Pvu*II-*Eco*RI fragments were cloned in λgt10 after *Eco*RI linkers were added to the *Bacillus* DNA fragments. The first SASP gene cloned (C) was detected immunologically by measuring SASP C production in the expression vector pAD284 (1). All subsequent SASP genes have been identified by their hybridization with other cloned SASP genes (usually the SASP C gene) under nonrestrictive hybridization conditions (2; Connors and Setlow, submitted for publication). The hybridizing region of all DNA fragments cloned was identified by restriction mapping and Southern blots, subcloned into various phage M13 derivatives, and subjected to DNA sequence analysis by the chain termination procedure (2; Fliss and Setlow, submitted for publication).

Northern blot analysis. *B. megaterium* was grown in supplemental nutrient broth (14), and *B. subtilis* was grown in 2× SG (17). RNA was isolated from cells in various stages of growth as described previously (4). RNA samples were treated with glyoxal, run on a 2% agarose gel, and transferred to nitrocellulose paper. The paper was baked and then hybridized to an SASP DNA probe under restrictive conditions as described by Thomas (16). Sizes of mRNAs were determined by reference to glyoxal-denatured *Hae*III fragments of φX174 DNA.

TABLE 1. *B. megaterium* SASP genes cloned[a]

SASP gene	Restriction fragment carrying gene	Cloning vector	Reference
C[b,c]	PvuII-PvuII (5)[d]	pAD284	1
	EcoRI-EcoRI (12)	λ Charon 4A	Loshon and Setlow, unpublished data
A[b]	PvuII-EcoRI (0.7)	λgt10	Loshon and Setlow, unpublished data
C-1[c]	EcoRI-EcoRI (4.9)	pBR325	2
C-2[c]	EcoRI-EcoRI (4.4)	pBR325	2
C-3[b,c]	EcoRI-EcoRI (3.2)	pBR325	2
C-4[e]	PvuII-EcoRI (0.5)	λgt10	Loshon and Setlow, unpublished data
C-5[e]	PvuII-EcoRI (0.6)	λgt10	Loshon and Setlow, unpublished data

[a] SASP genes were cloned as described in Experimental Procedures. Their identity as SASP genes has been confirmed by DNA sequence analysis.
[b] A protein derived from this SASP gene has been identified in spores.
[c] An mRNA from this SASP gene has been identified in sporulating cells.
[d] Size in kilobases.
[e] mRNA production from these genes has not yet been tested.

RESULTS

Cloning of the initial *B. megaterium* SASP gene. The strategy for the identification of the initial *B. megaterium* SASP gene was to detect SASP expression in *E. coli* immunologically by using antisera against SASPs A and C (1). However, since SASP genes are sporulation specific, they may be expressed poorly if at all in *E. coli*. Consequently, the cloning vector used (pAD284) contained an external promoter (the λ pL) and provided relief of termination of transcription beginning at pL by supplying λ N function (1). With the use of this cloning vector the SASP C gene was cloned (Table 1), and it was shown that SASP C gene expression in *E. coli* did indeed require both an external promoter and antitermination (Table 2). Strikingly, this initial clone, although 5 kilobases in length with the C gene located in the center of the fragment, contained

TABLE 2. Expression of SASP genes cloned in *E. coli*[a]

Conditions	Level of particular SASP (ng/ml of culture)	
	SASP C	SASP C-3
External promoter off	<4	<5
External promoter on, but transcription termination not suppressed	<100	<10
External promoter on, and transcription termination suppressed	9,500	1,720

[a] Values taken from data given in references 1 and 2.

no other SASP gene! Subsequently, the SASP C gene has been cloned on a 12-kilobase *Eco*RI fragment which again contains no other SASP gene (Table 1).

Cloning of additional *B. megaterium* SASP genes. Despite the success in cloning the SASP C gene by measuring SASP C expression in *E. coli*, no other SASP genes were identified in this way even though repeated attempts were made (2). Consequently, detection of other cloned SASP genes has relied on hybridization with a previously cloned SASP gene (usually the SASP C gene) under nonrestrictive hybridization conditions. By use of this technique six different *B. megaterium* DNA fragments have been cloned, all of which carry SASP genes, as shown by DNA sequence analysis (6a; Fliss and Setlow, submitted for publication; Table 1). One of these genes, the SASP C-3 gene, codes for a protein (termed SASP C-3) when in an *E. coli* expression vector (2). As was the case with the SASP C gene, SASP C-3 gene expression in *E. coli* required both an external promoter and supression of termination of transcription (Table 2). SASP C-3 has been purified and is distinct from previously isolated *B. megaterium* SASP, but is immunologically related to SASPs A and C. SASP C-3 also comigrates with a minor spore SASP which is immunologically related to SASPs A and C (2).

Of the remaining genes, the SASP A, C-4, and C-5 genes have not yet been tested in an expression vector, but the SASP C-1 and C-2 genes code for no detectable protein in an *E. coli* expression vector (2). The reasons for this are not clear, since no apparent defects in the genes were revealed by DNA sequence analysis. How-

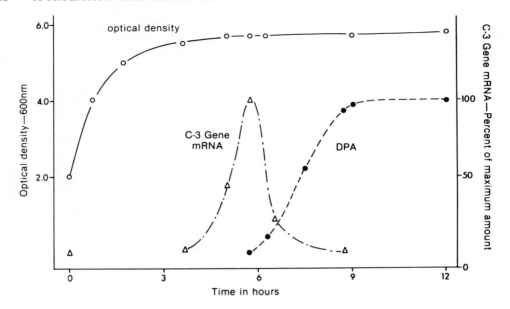

FIG. 1. Levels of SASP C-3 mRNA during growth and sporulation of *B. megaterium*. Cells were grown and sporulated, samples were taken, and RNA was extracted, run on gels, transferred to nitrocellulose, and hybridized to a SASP C-3 gene probe as described elsewhere (Fliss and Setlow, submitted for publication). The hybridizing band was cut out and counted in a toluene-based scintillation fluid. Dipicolinic acid was also analyzed in samples taken in parallel with those for RNA analysis.

ever, it may be that the gene products are extremely unstable in *E. coli*—too unstable to be detected. Indeed, whereas SASP C is relatively stable in *E. coli* ($t_{1/2} \sim 120$ min), SASP C-3 is rather unstable ($t_{1/2} \sim 4$ min) (Fliss and Setlow, in press).

Expression of different SASP genes during sporulation. Previous work using in vitro translation of total RNA isolated at various times during growth and sporulation has shown that both SASP A and C mRNAs appear in parallel midway in sporulation during the period of SASP synthesis (4). However, despite their identification as SASP genes by DNA sequence analysis, there was no firm proof that the SASP C-1 through C-5 genes were expressed during sporulation. SASP C-4 and C-5 gene expression has not yet been tested, but Northern blot analysis of RNAs isolated at various times in sporulation has shown that the SASP C-3 gene (Fig. 1), SASP C gene (data not shown), and SASP C-1 or C-2 gene, or both (Fliss and Setlow, submitted for publication), are expressed in parallel during sporulation. The uncertainty about SASP C-1 and C-2 gene expression is due to their being almost completely identical in DNA sequence; thus, a DNA probe from one gene would be expected to hybridize to an mRNA from the other gene (Fliss and Setlow, submitted for publication). This Northern blot analysis also gave the size of the various SASP gene mRNAs

(Table 3) and indicated that the level of expression of the SASP C-1, C-2, and C-3 genes was significant relative to the SASP C gene (Fliss and Setlow, submitted for publication).

TABLE 3. Properties of various SASP genes[a]

Gene	Length of mRNA bases[b]	Free energy of formation of mRNA-rRNA initiation complex
B. megaterium		
SASP C	390 (380)	−11.8
SASP C-1	290 (286)	−25.4
SASP C-2	290 (285)	−25.4
SASP C-3	330 (332)	−16.4
B. subtilis		
SASP 1	290 (277)	−14.4

[a] mRNA lengths were measured by hybridizations of specific SASP gene probes to Northern blots containing maximum levels of SASP gene mRNAs. The free energy values (kcal) were calculated as described elsewhere (Connor and Setlow, submitted for publication; Fliss and Setlow, submitted for publication).

[b] Values in parentheses are those measured by S1 nuclease mapping (SASP C gene) or calculated by using the designated transcription start point (Fig. 3) and assuming that transcription stops five bases downstream from the end of the dyad symmetry region following the gene's coding sequence (Fliss and Setlow, submitted for publication).

AMINO ACID SEQUENCES

SASP

A: ↓
      NH₂-MANTNKLVAPGSAAAIDQMKYEIASEFGVNLGPEATARANGSVGGEITKRLVQMAEQQLGGK-COOH

                                        ↓
C: NH₂-MANYQNASNRNSSNKLVAPGAQAAIDQMKFEIASEFGVNLGPDATARANGSVGGEITKRLVQLAEQNLGGKY-COOH

                                    ↓
C-1:   NH₂-MANNNSSNNNELLVYGAEQAIDQMKYEIASEFGVNLGADTTARANGSVGGEITKPLVQLAEQQLGGGRF-COOH

                                  ↓
C-2:   NH₂-MANNNKSSNNNELLVYGAEQAIDQMKYEIASEFGVNLGADTTARANGSVGGEITKRLVQLAEQQLGGGRSKTTL-COOH

                              ↓
C-3:   NH₂-MARTNKLLTPGVEQFLDQYKYEIAQEFGVTLGSDTAARSNGSVGGEITKRLVQQAQAHLSGSTQK-COOH

                            ↓
C-4:   EQAIDQMKYEIASEFGVNLGADTTARANGSVGGEITKRLVQLAEQQLGGGRF-COOH

C-5:   LGADTTARANGSVGGEITKRLVQLAEQQLGGGRF-COOH

                                    ↓
Conserved   N L^V_L G   I$_L$DQ^M_YK^Y_FEIA EFGV LG ^D_E AR NGSVGGEITKRLVQ A   L G
Residues

FIG. 2. Comparison of amino acid sequences of *B. megaterium* SASP genes. Amino acid sequences of the SASP C, C-1, C-2, and C-3 genes are from Fliss and Setlow (6; in press; submitted for publication). Data for the SASP A, C-4, and C-5 genes are from C. A. Loshon, E. Fliss, and P. Setlow (unpublished data). The sequences of the SASP C-4 and C-5 genes have not yet been completely determined. Sequences have been aligned to give maximum homology, and residues which are identical in all seven SASPs are given as conserved residues; where positions are occupied by one of two similar amino acid residues, both residues are given as conserved.

DNA sequence of SASP genes. Whereas the identity of the SASP C gene was established by its coding for a known SASP, other SASP genes were isolated initially as DNA fragments which hybridized well to a SASP C gene probe. These hybridizing fragments were subsequently shown to be SASP genes by DNA sequence analysis which showed that their DNA sequence coded for a small protein which was extremely homologous to a known SASP. To date we have completely sequenced four SASP genes (C, C-1, C-2, and C-3) (6; Fliss and Setlow, in press; Fliss and Setlow, submitted for publication) and have obtained partial sequence data on the SASP A, C-4, and C-5 genes (Fig. 2). Examination of all of these data allows the following conclusions. (i) SASP genes appear monocistronic, as all have transcription termination sites just downstream from the translation stop codon, and many have apparent transcription start sites upstream as well. (ii) When the SASP primary sequence is known, the amino acid sequence predicted from the gene sequence is identical, with the exception of the amino-terminal methionine predicted in the gene sequence which is presumably removed from the protein posttranslationally. (ii) All SASP genes examined have a strong ribosome binding site sequence just prior to the translation start point. The strength of the interaction of this ribosome bind-

ing site in the mRNA with the 3' end of *B. megaterium* 16S rRNA is generally stronger than that for formation of the analogous initiation complex in *E. coli* (Table 3; 11). However, there is a significant range of stabilities for this initiation complex, with the weakest one yet determined being that with the SASP C mRNA which codes for the protein which is the most abundant of this group (SASPs C, C-1, C-2, and C-3; Table 3; 12). (iv) A large amount of the SASP primary sequence is conserved. Comparison of all SASP sequences indicates that among the seven SASP genes analyzed to date about one-half of all amino acid residues are conserved exactly in all seven SASPs, with an additional five residues being homologous in all seven proteins (Fig. 2). The greatest differences among different SASPs are in the amino- and carboxyl-terminal regions, which vary considerably in both length and sequence. Possibly most striking is that the primary sequences of four of the SASP genes (C-1, C-2, C-4, and C-5) are almost completely identical (Fig. 2), with the coding sequences for SASP C-1 and C-2 genes differing in only five nucleotides (Fliss and Setlow, submitted for publication).

Although the sequence conservation between different SASPs is striking, the reason for the sequence conservation is only partially clear. One region of SASP sequence does have a

UPSTREAM FLANKING SEQUENCE

B. megaterium

| | | -50 | -40 | -30 | -20 | -10 | +1 |

SASP-C gene: $[(AT)_{40}(GC)_6]$CCCTTTCGCCATAATA<u>C</u><u>TAGTAAC</u>AACCGAGCACAACAGCTC<u>GCAAAC</u>ATTGAGTT

SASP-C-3 gene: $[(AT)_{27}(GC)_4]$GCCCAGCTTCT<u>TTGAAAG</u>CTGGGCTTTTTTATAC<u>TTTCTT</u><u>GCAAAC</u>CATTACCA   +1

SASP-C-1 gene: $[(AT)_{17}(GC)_3]$CCAC<u>GAATA</u>CATTTCTTCCACAATA<u>GGAAAC</u>TTAAGAAA   +1

SASP-C-2 gene: $[(AT)_{17}(GC)_3]$CCAT<u>GTAT</u>ATACTTCTTGGCCCCTA<u>GGAAAC</u>TTAAGAAA   +1

B. subtilis
SASP-1 gene: $[(AT)_{23}(GC)_3]$GGCGTGTATAAATTA<u>AAATAAT</u>CTCTCCATAATATGATT<u>CAAAC</u>AAGCTTGT   +1

FIG. 3. DNA sequences upstream from transcription start sites of SASP genes. The data are from Connors and Setlow (submitted for publication) and Fliss and Setlow (6; submitted for publication). The nucleotide at the transcription start site is labeled +1; putative −10 and −35 regions are double and single underlined, respectively, and the composition of adenine-plus-thymine-rich regions is given in brackets. The transcription start point precedes the translation start point by 108 bases for the SASP C gene, 20 bases for the SASP C-1 gene, 20 bases for the SASP C-2 gene, 79 bases for the SASP C-3 gene, and 29 bases for the B. subtilis SASP 1 gene.

known function, that of providing a recognition and cleavage site for the sequence-specific spore protease (12, 13), and as predicted from earlier work, the protease cleavage site (arrow, Fig. 2) is in a region of conserved sequence (3, 13). However, the function of the 14-residue conserved sequence toward the carboxy terminus is less clear. This sequence is not found in B. megaterium SASP B, which is also rapidly cleaved by the spore protease (13), so it does not appear necessary for protease cleavage. Elucidation of its function will require further work.

Promoter regions of B. megaterium SASP genes. Previous work suggested that the transcriptional regulation of SASP genes is via some form of positive control (7). Given the work from several laboratories indicating that sporulation-specific genes may have unique promoter regions and unique RNA polymerase σ factors to recognize them (8, 10), it has been of obvious interest to determine the transcription start points for SASP genes. To date, the SASP C gene's transcription start point has been determined by S1 nuclease mapping (6), and the start points for a number of other B. megaterium SASP genes have been determined (i) by homology with that for the SASP C gene, (ii) from the location of transcription stop sites downstream from SASP gene's coding sequences, and (iii) from the size of the SASP gene's mRNA. The transcription start points are given in Fig. 3, and putative −10 and −35 regions are double and single underlined, respectively; all show significant homology. All transcription start points also have large adenine-plus-thymine-rich regions 40 to 55 bases upstream. Although the −10 and −35 regions of all of these

genes have significant homologies, they are not particularly similar to the analogous regions of sporulation-specific genes.

Cloning and analysis of SASP genes from B. subtilis and B. cereus. In addition to SASP genes from B. megaterium, we have also cloned and analyzed a number of SASP genes from other Bacillus species. Again, these were isolated by hybridization with the previously cloned B. megaterium SASP C gene, since a DNA probe from the latter hybridizes well to a number of restriction fragments of both B. subtilis (Connors and Setlow, submitted for publication) and B. cereus (data not shown) DNA. There are at least four such hybridizing fragments in B. subtilis DNA (Connors and Setlow, submitted for publication) and five in B. cereus DNA (B. Setlow and P. Setlow, unpublished data). To date, four hybridizing B. subtilis fragments (termed SASP genes 1 through 4) have been cloned, as have two hybridizing B. cereus fragments. DNA sequence analysis on one B. subtilis fragment is complete, and partial data are available on the two B. cereus fragments (Fig. 4). These data show that all three fragments contain SASP genes and that the predicted SASP sequences are quite similar to that of the B. megaterium SASPs (Fig. 4). Although the amino-and carboxy-terminal sequences of SASPs from different species differ, the great majority of the residues conserved in B. megaterium SASPs are also conserved in the related SASPs of B. cereus and B. subtilis (Fig. 4).

Despite the extensive sequence information already available on these B. cereus and B. subtilis SASPs, it is not yet clear which spore SASPs these genes represent. To date, only the

AMINO ACID SEQUENCES

SASP-gene

B. megaterium: NH$_2$(X)$_{4-13}$N-L^V_L--G---AIDQMKYEIASEFGVNLG-D-TARANGSVGGEITKRLVQ-AEQ-LGG(X)$_{1-7}$-COOH
conserved   fl y f   q   t e a   s   qa   s

B. cereus-1:   -ALDQMKYEIAQEFGVQLG-D-TARSNGSVGGEITKRLVA-AEQ-LGG(X)$_3$-COOH

B. cereus-2:   -ALDQMKYEIAQEFGVQLG-D-TARSNGSVGGEITKRLVA-AEQ-LGG(X)$_4$-COOH

B. subtilis-1: NH$_2$(X)$_{12}$N-LL------AIEQMKLEIASEFGVQLG-E-TSRANGSVGGEITKRLVR-AQQ-MGG(X)$_3$-COOH

FIG. 4. Comparison of the conserved amino acid residues in *B. megaterium* SASPs with the sequence of *B. cereus* and *B. subtilis* SASPs. Conserved residues in all *B. megaterium* SASPs are taken from Fig. 2 and are in capital letters; residues in lowercase letters are present in only one of seven *B. megaterium* SASPs. Data for *B. subtilis* SASP 1 are from Connors and Setlow (submitted for publication). The data for the *B. cereus* SASPs are still incomplete and are from B. Setlow, E. Fliss, and P. Setlow (unpublished data). Sequences have been aligned to give maximum homology, and residues conserved in both *B. megaterium* and a particular *B. cereus* or *B. subtilis* SASP are underlined.

amino-terminal 15 residues of a *B. cereus* SASP immunologically related to *B. megaterium* SASP C have been determined (17), and the analogous *B. subtilis* SASP has not yet been isolated in pure form (9). Consequently, we have as yet no proof that the *B. cereus* SASP 1 and 2 genes are expressed. However, hybridization of Northern blots of *B. subtilis* RNAs has shown that the *B. subtilis* SASP 1 gene is expressed midway in sporulation (Connors and Setlow, submitted for publication) and produces an mRNA of 290 bases (Table 3). When the transcription start site for this SASP 1 gene was determined as described above for the *B. megaterium* SASP genes, it showed good homology in its −10 and −35 regions with those in *B. megaterium* SASP genes and also contained a large upstream adenine-plus-thymine region (Fig. 3).

DISCUSSION

The SASP genes described in this report represent the first extended, divergent, multigene family described in a procaryote. There are at least seven SASP genes in *B. megaterium* related to the SASP C gene. *B. cereus* and *B. subtilis* also have a number of similar SASP genes, although the exact number is not yet known. This family of SASP genes may be only one of many such SASP gene families in *Bacillus* species, since a number of SASPs immunologically distinct from SASPs A and C have been identified in *B. cereus*, *B. megaterium*, and *B. subtilis* (9, 12, 17). Although no gene for these other SASPs has yet been cloned, it appears likely that the situation for these bacterial spore storage protein genes can be as complex as the extensive multigene families which code for the storage proteins of plant seeds (15).

Strikingly, no restriction fragment carrying a SASP gene isolated to date contains more than one SASP gene. This suggests that even very similar SASP genes are not tightly clustered on the chromosome. Although mapping genes on the *B. megaterium* chromosome is difficult, there are now a number of techniques for mapping cloned *B. subtilis* genes for which there is no obvious selection (5). Consequently, with cloned *B. subtilis* genes in hand, their mapping becomes a feasible project, and the mapping of the SASP 1 gene is now in progress.

Initial amino acid sequence analysis of SASPs from *B. cereus* and *B. subtilis* suggested that the SASP genes may have evolved rapidly between species (17). This suggestion now appears to be incorrect as there is extensive conservation of SASP primary sequence across species. Indeed, the *B. subtilis* SASP 1 is more like *B. megaterium* SASP C than is *B. megaterium* SASP C-3 (1; Connors and Setlow, submitted for publication; Fig. 2 and 4). The reason for the initial incorrect suggestion is that all SASPs so far examined show significant heterogeneity in their amino-terminal regions, i.e., those sequences on which the initial suggestion of rapid evolution was made (17). Indeed, we must now explain the reason for the conservation of so much of the SASP primary sequence. Although the function of one conserved region of SASP sequence is the recognition and cleavage site for the spore protease, this does not explain the extensive sequence conservation. Indeed, this extensive sequence conservation suggests that SASPs may have some specific function in spores other than simply an amino acid storage role. Other functions for SASPs have been suggested (i.e., spore UV light resistance [12]), but they have never been proved and remain conjectural at this time.

A number of other questions also remain concerning this sporulation-specific multigene family. (i) What are the details of the transcriptional regulation of the SASP gene family (or families)? All SASP genes so far examined appear to have similarities in their promoter regions, but there is as yet no real knowledge of the factors involved in regulating their transcription. (ii) How have SASP genes evolved and become scattered on the chromosome? (iii) Why are there so many SASP genes, all of which appear to be expressed, but at very different levels? (iv) What is the function of SASPs (if any) beyond their role as amino acid storage proteins? Although the answer to all four of these questions is at present "we don't know," it now seems possible to tackle these questions and obtain some concrete answers.

SUMMARY

The proteins degraded during bacterial spore germination are a group of SASPs which are synthesized only during sporulation under transcriptional control. There are at least nine different SASPs in *B. megaterium* spores at various levels, and some of these (SASPs A and C) are very similar in amino acid sequence. The SASPs are the products of an extended multigene family, of which seven distinct genes coding for proteins similar to SASPs A and C have been cloned from *B. megaterium*. *B. cereus* and *B. subtilis* also contain similar multigene SASP families, of which six have been cloned. Analysis of a number of these genes has shown (i) that they are expressed in parallel during sporulation, (ii) that they are monocistronic and not tightly clustered on the chromosome, (iii) that over half of the residues in all the SASPs coded for by these genes are conserved, and (iv) that all have similar promoter sequences.

This work was supported by grants from the Army Research Office and the National Institutes of Health and an equipment grant from the Army Research Office.

The continued excellent technical assistance of Susan Goldrick is gratefully appreciated.

LITERATURE CITED

1. **Curiel-Quesada, E., B. Setlow, and P. Setlow.** 1983. The cloning of the gene for C-protein, a low molecular weight spore specific protein from *Bacillus megaterium*. Proc. Natl. Acad. Sci. U.S.A. **80:**3250–3254.

2. **Curiel-Quesada, E., and P. Setlow.** 1984. Cloning of a new low-molecular-weight spore-specific protein gene from *Bacillus megaterium*. J. Bacteriol. **157:**751–757.

3. **Dignam, S. S., and P. Setlow.** 1980. *Bacillus megaterium* spore protease: action of the enzyme on peptides containing the amino acid sequence cleaved by the enzyme *in vivo*. J. Biol. Chem. **255:**8408–8412.

4. **Dignam, S. S., and P. Setlow.** 1980. *In vivo* and *in vitro* synthesis of the spore-specific proteins A and C of *Bacillus megaterium*. J. Biol. Chem. **255:**8417–8423.

5. **Ferrari, F. A., A. Nguyen, D. Long, and J. A. Hoch.** 1983. Construction and properties of an integrable plasmid for *Bacillus subtilis*. J. Bacteriol. **154:**1513–1515.

6. **Fliss, E. R., and P. Setlow.** 1984. The complete nucleotide sequence and the start sites for transcription and translation of the *Bacillus megaterium* protein C gene. J. Bacteriol. **158:**809–813.

6a. **Fliss, E. R., and P. Setlow.** 1984. *Bacillus megaterium* spore protein C-3: nucleotide sequence of its gene and the amino acid sequence at its spore protease cleavage site. Gene **30:**167–170.

7. **Goldrick, S., and P. Setlow.** 1983. Expression of a *Bacillus megaterium* sporulation-specific gene during sporulation of *Bacillus subtilis*. J. Bacteriol. **155:**1459–1462.

8. **Johnson, W. C., C. P. Moran, Jr., and R. Losick.** 1983. Two RNA polymerase sigma factors from *Bacillus subtilis* discriminate between overlapping promoters for a developmentally regulated gene. Nature (London) **302:**800–804.

9. **Johnson, W. C., and D. J. Tipper.** 1981. Acid-soluble spore proteins of *Bacillus subtilis*. J. Bacteriol. **146:**972–982.

10. **Moran, C. P., Jr., N. Lang, S. F. J. LeGrice, G. Lee, M. Stephens, A. L. Sonenshein, J. Pero, and R. Losick.** 1982. Nucleotide sequences that signal the initiation of transcription and translation in *Bacillus subtilis*. Mol. Gen. Genet. **186:**339–346.

11. **Murray, C. L., and J. C. Robinowitz.** 1982. Species specific translation: characterization of *B. subtilis* ribosome binding sites, p. 271–285. *In* A. T. Ganesan, S. Chang, and J. A. Hoch (ed.), Molecular cloning and gene regulation in *Bacilli*. Academic Press, Inc., New York.

12. **Setlow, P.** 1981. Biochemistry of bacterial forespore development and spore germination, p. 13–28. *In* H. S. Levinson, A. L. Sonenshein, and D. J. Tipper (ed.), Sporulation and germination. American Society for Microbiology, Washington, D.C.

13. **Setlow, P., C. Gerard, and J. Ozols.** 1980. The amino acid sequence specificity of a protease from spores of *Bacillus megaterium*. J. Biol. Chem. **255:**3624–3628.

14. **Setlow, P., and A. Kornberg.** 1969. Biochemical studies of bacterial sporulation and germination. XVII. Sulfhydryl and disulfide levels in dormancy and germination. J. Bacteriol. **100:**1155–1160.

15. **Sorenson, J. C.** 1983. The structure and expression of nuclear genes in higher plants. Adv. Genet. **22:**109–144.

16. **Thomas, P. S.** 1980. Hybridization of denatured RNA and small DNA fragments transferred to nitrocellulose. Proc. Natl. Acad. Sci. U.S.A. **77:**5201–5205.

17. **Yuan, K., W. C. Johnson, D. J. Tipper, and P. Setlow.** 1981. Comparison of various properties of low-molecular-weight proteins from dormant spores of several *Bacillus* species. J. Bacteriol. **146:**965–971.

Cloning of *gerJ* and Other Genes from the *spoIIA-tyrA* Region of *Bacillus subtilis*

R. J. WARBURG,[1] M. P. DAVIS,[1] I. MAHLER,[1] D. J. TIPPER,[2] AND H. O. HALVORSON[1]

Rosenstiel Center, Brandeis University, Waltham, Massachusetts 02254,[1] and University of Massachusetts, Worcester, Massachusetts 01650[2]

GerJ mutants of *Bacillus subtilis*, like the wild-type strain, form phase-bright spores, but are altered in their germination response. GerJ spores respond to the same germinants as wild-type spores, but unlike the latter, which become phase dark during germination, GerJ spores reach only a phase-gray stage. The mutants are also delayed in the aquisition of resistances to heat and organic chemicals during sporulation (10) and produce spores much more sensitive than wild type to heating at 90°C (11). These properties suggest a defect in the cortex structure of the mutant spores, perhaps caused by some altered enzyme involved in cortical synthesis. Our aim is to isolate and characterize the gene(s) involved and to study its regulation. Since GerJ$^+$ cannot be selected, we are attempting to clone *gerJ* by first isolating closely linked auxotrophic or sporulation markers. GerJ$^+$ can then be detected in transformation assays by complementation.

The region between *spoIIA* and *tyrA*, which encompasses approximately 40 kilobases (kb) of DNA, includes *gerJ* and several *spo* loci (Fig. 1). The closest known markers to *gerJ* are *aroC* and *ser-22*, the nearest to *spoIIA* and *spoVA* is *lys-1*, and the nearest to *spoIVA* is *aroB*. We attempted, unsuccessfully, to find clones from existing *Bacillus subtilis* banks (1, 3, 8) which could transform *lys-1*, *aroC*, or *ser-22* strains to prototrophy. After construction of a new cosmid library (4), we were able to isolate an unstable clone, pRC12, containing DNA from the *spoIIA-tyrA* region. pRC12-5, a subclone of pRC12 containing only *spoVA*, was used as a probe for an existing *B. subtilis* λ bank. We describe here the use of pRC118, isolated from a shotgun bank in pHV33 (R. J. Warburg, I. Mahler, D. J. Tipper, and H. O. Halvorson, Gene, in press) and containing DNA with the *aroC* and *ser-22* genes, to probe banks of *B. subtilis* DNA. Finally, we describe the the use of the transposon Tn917 system developed by Youngman et al. (12) for cloning DNA from the region of the *B. subtilis* genome around the transposon inserts (P. J. Youngman, J. B. Perkins, and R. Losick, Mol. Gen. Genet., in press). A library of Tn917 insertions was used to isolate further stable clones containing *aroC* and *ser*.

EXPERIMENTAL PROCEDURES

The construction of a library of Tn917 insertions is described by K. Sandman, P. Youngman, and R. Warburg (manuscript in preparation). Insertion results in the MLS^r phenotype (resistance to macrolide, lincosamide, and streptogramin B antibiotics: erythromycin, 1 µg/ml; lincomycin, 25 µg/ml). To isolate insertions near *aroC*, Aro$^+$ MLS^r colonies were selected in strain 5098 (*aroC*) after transformation with chromosomal DNA from the transposon library. One of these was termed TnB.

Strain 5098 was also transformed with the library DNA to give MLS^r colonies from which spores were prepared on Difco sporulation medium plates. These spores were incubated in Penassay broth for 30 min at 37°C and then for 30 min at 70°C. After the spores were washed in Penassay broth, this process was repeated three more times. Finally, the surviving spores were plated on Difco sporulation medium plates, and their tetrazolium (6) reaction was scored. Inserts, termed Tn6 and Tn7, into *gerJ* gave tetrazolium white colonies, compared with the red of wild-type colonies, although other insertions giving tetrazolium white colonies were also isolated.

Kathleen Sandman generously provided us with strains containing transposons inserted into *spo* genes near *gerJ*. These were also isolated from the transposon library.

RESULTS AND DISCUSSION

Cosmid library. A recombinant cosmid clone containing a 15-kb *Bam*HI insert of *B. subtilis* DNA was isolated from our cosmid library by screening for complementation of *lys-1* (4). The cosmid clone and a plasmid clone, pRC12, constructed by transfer of the 15-kb fragment to pHV33 (7), could transform to prototrophy several markers in the *spoIIA-tyrA* region of the *B. subtilis* chromosome. Both recombinant clones were unstable in *Escherichia coli* and *B. subtilis*. Selection for insert function in *B. subtilis* resulted in loss of the antibiotic resistance vector function, and selection for vector antibiotic resistance resulted in deletions of the cloned fragment (4).

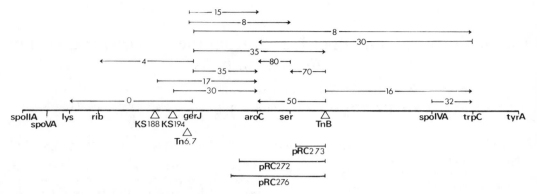

FIG. 1. Transposon insertions and clones in the *spoIIA-tyrA* region. Thick line, chromosomal markers; triangles under thick line, Tn*917* insertions. KS188 and 194 were isolated by Kathleen Sandman; TnB, Tn6 and Tn7 are described in Experimental Procedures. Upper lines represent percent cotransformation of a nonselected marker (start of arrow) with a selected marker (end of arrow). Lower lines are clones isolated from the TnB insert by using plasmids pTV20 and pTV21. (Not drawn to scale.)

A Southern analysis using labeled pRC12 DNA suggested that multiple deletions had probably occurred during isolation of the original clone (data not shown). A spontaneously derived subclone, pRC12-5, carrying a 3-kb insert, was isolated from pRC12 transformants of *B. subtilis* under chloramphenicol resistance selection. This plasmid had lost the ability to transform Ger⁺ but could still correct *lys-1* and *spoVA89*. Labeled pRC12-5 DNA was used as a probe for the Charon 4A library (1). From five positive plaques, lysates were prepared and tested for biological activity as described by Ferrari et al. (1). All five lysates transformed *lys-1* to Lys⁺, and three of the lysates also contained the *rib* marker. However, no transforming activity for *gerJ* was detectable.

Plasmid shotgun. A stable recombinant clone was isolated from the *spoIIA-tyrA* region by use of shotgun cloning (Warburg et al., in press). The clone, pRC118, contains 14 kb of inserted DNA and has transforming activity only for *aroC* and *ser*. Restriction enzyme fragments of pRC118 were used to construct subclones which contained only *aroC* or both *aroC* and *ser* on 2.5-kb fragments. These are termed pRC181 and pRC144, respectively. A Southern analysis using these subclones to probe genomic *B. subtilis* DNA had two possible interpretations: either deletions had occurred during the construction of the original clone, or DNA from several chromosomal regions had been ligated together. Probing of our cosmid library and a library we constructed using λBF101 (5) and a partial *Sau*3A *B. subtilis* DNA digest (molecular weight, 15 to 20 kb) revealed clones with DNA homologous to that of pRC118. The DNA from two of these clones, pRC24 and λIIIC4, had *B. subtilis* inserts of approximately 15 kb each but had no

transforming activity for any markers in the *spoIIA-tyrA* region. By using the mapping plasmid described by Haldenwang et al. (2), these inserts were located in the *glnA* (pRC24) and *glyB* (λIIIC4) regions of the *B. subtilis* chromosome. Presumably, this result reflects the presence of DNA from these regions in the original clone.

Thus, although we have indeed isolated DNA from the *aroC* region, near to *gerJ*, we have not been able to use any of these as probes for libraries to isolate *gerJ* itself. Since Youngman et al. (in press) have been successful in isolating contiguous fragments of DNA from regions next to Tn*917* insertions, we attempted to use this system to walk to *gerJ* from transposon insertions near *aroC*.

Transposon cloning. A series of Tn*917* insertions in the *aroC* region were isolated by transformation of strain 5098 (*aroC gerJ⁺*) to Aro⁺ as described in Experimental Procedures. One of these, TnB, was cotransformed 70% with *ser-22* and 50% with *aroC*. Its location with respect to *gerJ*, with which it was 35% cotransformed, was defined by three-point analysis (see Fig. 1), so that we would know in which direction to walk to *gerJ*. TnB was transformed with linearized pTV20 and pTV21 (plasmids containing pBR322 sequences and a chloramphenicol resistance [Cmʳ] gene inserted into the transposon sequences; Youngman et al., in press) to give Cmʳ colonies. One colony from each cross was purified, and the chromosomal DNA was isolated. This chromosomal DNA has pBR322 sequences inserted into the original Tn*917* site and can be used to isolate *B. subtilis* DNA from either side of that insertion (Youngman et al., in press). These DNAs were partially cut with *Hin*dIII and ligated to themselves after dilution to 1 μg/ml;

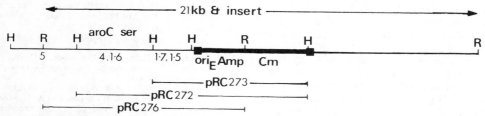

FIG. 2. Derivation of clones from TnB insertion. Thick line, pTV21 insertion into TnB (represented as boxes on either side of pTV21). Restriction enzyme sites: H, HindIII; R, EcoRI. Cm, Amp, and ori$_E$ represent the approximate sites of genes encoding chloramphenicol resistance, ampicillin resistance, and the origin of replication for *E. coli*, respectively. Approximate sizes, in kb, of HindIII fragments are given under the central line, which represents genomic DNA. pRC276 was derived as shown; pRC272 and pRC273 were derived from a pTV20 insertion into TnB, where the pBR322 insertion is inverted, but the HindIII site is located in TnB. The approximate locations of the *aroC* and *ser* genes are shown.

0.1 µg of each DNA was used to transform *E. coli* HB101 cells to ampicillin resistance (Ap^r). Plasmid DNA was isolated from 25 colonies, analyzed for HindIII fragment sizes, and tested for *aroC* or *ser* activity. The results are summarized in Fig. 1 and 2. pRC273, which was derived from the DNA with pTV20 inserted into TnB, contained HindIII fragments of 8, 1.7, and 1.5 kb, but had no *aroC* or *ser* transforming activity. pRC272 contained two additional HindIII fragments of 4.0 and 1.6 kb and carried transforming activity for both *aroC* and *ser* but not *gerJ*. Southern analysis of pRC272 against genomic DNA showed homology to HindIII fragments of 1.5, 1.6, 1.7, and 4 kb as expected and to a single 21-kb EcoRI fragment (see Fig. 3). Note that the 8-kb HindIII fragment, presumably consisting mostly of DNA derived from the TnB and pTV20 inserts, should also contain *B. subtilis* DNA from the TnB insert to the nearest HindIII site. Since a unique HindIII fragment was not seen, the TnB insert is either in a very small HindIII fragment or in one of a size similar to the other HindIII fragments. It is therefore apparent that clones isolated from TnB consisted of contiguous genomic DNA. The large size of the homologous EcoRI fragment probably explains the lack of *aroC7* and *ser* activity in the λ Charon 4A bank, which was constructed with EcoRI-digested DNA, since 21 kb probably exceeds the maximum insert size. We thus used EcoRI treatment of the TnB chromosomal DNA with the pTV21 insert to isolate clones as above but containing part of this EcoRI fragment (since the EcoRI and HindIII sites are on opposite sides of the Ap^r gene in these plasmids). pRC276 consists of a single 15-kb EcoRI fragment with *aroC* and *ser* activity but not *gerJ* activity (see Fig. 2). Southern analysis demonstrated homology to the same HindIII bands as pRC272, except that the 1.5-kb fragment is missing and there is an additional band of 5 kb (Fig. 3). As expected, only a single EcoRI fragment of 21 kb

had homology to pRC276 (Fig. 3). It appears that pRC276 contains a small deletion.

We now intend to use partial EcoRI digestion of TnB chromosomal DNA containing the pTV21 insert to walk to *gerJ*. We shall also screen for *gerJ* in clones obtained from other transposon inserts (KS188 and KS194), isolated by Kathleen Sandman, which appear to be closer to *gerJ* (Fig. 1). These inserts give rise to a SpoII or a SpoIV phenotype, respectively, in the strains into which they are transformed. Their close proximity to *gerJ* and the potential cortical

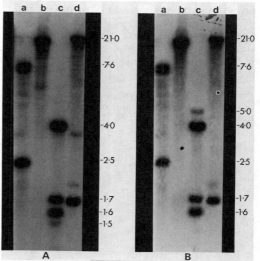

FIG. 3. Southern analysis of clones. *B. subtilis* genomic DNA was treated with excess restriction enzyme, and the fragments were separated on a 0.8% agarose gel. After blotting to nitrocellulose paper, the DNA was sequentially probed with ^{32}P-labeled (A) pRC272 or (B) pRC276. Lane a, BglII-treated DNA; b, EcoRI-treated DNA; c, HindIII-treated DNA; and d, PstI-treated DNA. The numbers on the right represent fragment sizes in kb, estimated from λ-HindIII-treated DNA molecular weight markers.

deficiency of the mutant blocked at stage IV of sporulation may be indicative of a cluster of genes in this area involved with cortical synthesis.

We have been successful in isolating transposon inserts (Tn6 and Tn7) in the *gerJ* gene itself, as described in Experimental Procedures. These inserts give rise to spores with germination responses similar to those of *gerJ* mutants, the spores becoming phase gray after incubation in germinants. The spores were also sensitive to heating at 90°C, their *D* values being approximately 6 min, compared with 45 min for wild-type spores. Unlike *gerJ* mutant spores, they were also sensitive to heating at 80°C. The location of the insertions was determined by a three-factor cross with *ser-22* and *aroC7*, to which they were both cotransformed 8 and 15%, respectively (Fig. 1). These inserts can now be used directly to isolate the control region of *gerJ* and any nearby related genes. Similarly, these clones can be used as probes for appropriate libraries to isolate an intact *gerJ* gene.

SUMMARY

Two clones, containing DNA in the *spoIIA-tyrA* region of the *B. subtilis* chromosome, were used as probes for libraries of *B. subtilis* DNA. We were unable to find any clones containing DNA with transforming activity for *gerJ*, a gene involved with spore germination. Using a transposon system, we were able to isolate transposon insertions near to *gerJ* and to clone genes from this region. These clones appear to be stable, and the transposon system allows us to walk towards *gerJ*. Inserts in *gerJ* have also been obtained.

We are grateful to Kathleen Sandman for the gift of strains containing Tn917 inserts near to *aroC*, and for the Tn library. We thank Kyle Parent for technical help.

The work was supported by Public Health Service grants AI-18904 to H.O.H. and AI-10806 to D.J.T. from the National Institute of Allergy and Infectious Diseases.

LITERATURE CITED

1. Ferrari, E., D. J. Henner, and J. A. Hoch. 1981. Isolation of *Bacillus subtilis* genes from a Charon 4A library. J. Bacteriol. **146:**430–432.
2. Haldenwang, W. G., C. D. B. Banner, J. F. Ollington, R. Losick, J. A. Hoch, M. B. O'Connor, and A. L. Sonenshein. 1980. Mapping a cloned gene under sporulation control by insertion of a drug resistance marker into the *Bacillus subtilis* chromosome. J. Bacteriol. **142:**90–98.
3. Hutchison, K. W., and H. O. Halvorson. 1980. Cloning of randomly sheared fragments from a φ105 lysogen of *Bacillus subtilis*. Gene **8:**267–280.
4. Mahler, I., R. Warburg, D. J. Tipper, and H. O. Halvorson. 1984. Cloning of an unstable *spoIIA-tyr*A fragment from *Bacillus subtilis*. J. Gen. Microbiol. **130:**411–421.
5. Maniatis, T., E. F. Fritsch, and J. Sambrook. 1982. Molecular cloning—a laboratory manual, p. 44. Cold Spring Harbor Laboratory, Cold Spring Harbor, N.Y.
6. Moir, A., E. Lafferty, and D. A. Smith. 1979. Genetic analysis of spore germination mutants of *Bacillus subtilis* 168: the correlation of phenotype with map location. J. Gen. Microbiol. **111:**165–180.
7. Primrose, S. B., and S. D. Ehrlich. 1981. Isolation of plasmid deletion mutants and study of their instability. Plasmid **6:**193–201.
8. Rapoport, G., A. Klier, A. Billault, F. Fargette, and R. Dedonder. 1979. Construction of a colony bank of *E. coli* containing hybrid plasmids representative of the *Bacillus subtilis* 168 genome. Expression of functions harbored by the recombinant plasmids in *B. subtilis*. Mol. Gen. Genet. **176:**239–245.
9. Roberts, T. M., S. L. Swanberg, A. Poteete, G. Reidel, and K. Bachman. 1980. A plasmid cloning vehicle allowing a positive selection for inserted fragments. Gene **12:**123–127.
10. Warburg, R. J. 1981. Defective sporulation of a spore germination mutant of *Bacillus subtilis* 168, p. 98–100. *In* H. S. Levinson, A. L. Sonenshein, and D. J. Tipper (ed.), Sporulation and germination. American Society for Microbiology, Washington, D.C.
11. Warburg, R. J., and A. Moir. 1981. Properties of a mutant of *Bacillus subtilis* 168 in which spore germination is blocked at a late stage. J. Gen. Microbiol. **124:**243–253.
12. Youngman, P. J., J. B. Perkins, and R. Losick. 1983. Genetic transposition and insertional mutagenesis in *Bacillus subtilis* with *Streptococcus faecalis* transposon Tn917. Proc. Natl. Acad. Sci. U.S.A. **80:**2305–2309.

Characterization of the Gene for Glucose Dehydrogenase and Flanking Genes of *Bacillus subtilis*

BRENDA URATANI, KEITH A. LAMPEL, ROBERT H. LIPSKY, AND ERNST FREESE

Laboratory of Molecular Biology, National Institute of Neurological and Communicative Disorders and Stroke, Bethesda, Maryland 20205

During the developmental process leading to spore production, numerous proteins are synthesized. These are used for the sequential biochemical and morphological changes which become components of the heat-resistant spore. Glucose dehydrogenase [GlcDH; β-D-glucose: NAD(P)-1-oxidoreductase, EC 1.1.1.47] is a developmentally regulated enzyme whose control is particularly interesting because GlcDH is made only in the forespore (3). Because GlcDH can be easily assayed, the effects of genetic manipulations can be easily studied.

Our laboratory has isolated the gene (*gdh*) coding for GlcDH from a λ Charon 4A phage library of *Bacillus subtilis* DNA (17). A 4.0-kilobase (kb) *Eco*RI DNA fragment was subcloned into the plasmid pBR322; *Escherichia coli* transformants containing the resulting plasmid, pEF1, synthesize enzymatically active GlcDH. This paper characterizes this 4.0-kb DNA fragment and shows that it contains *gdh* and two other genes, at least one of which is needed for sporulation.

EXPERIMENTAL PROCEDURES

Bacterial strains, plasmids, phages, and media. The *B. subtilis* strains used were 60001 (*trpC2*) and 60015 (*metC2 trpC2*). The *E. coli* strains used were JM101 and JM103 (9) for DNA sequencing and DH1 and CSR603 (12) for maxicell experiments. Bacteriophage M13mp8 and M13mp10 were used to subclone fragments for DNA sequence analysis (9). Plasmids used were pEF1 (17) and related derivatives and pE194 *cop6* (18). *E. coli* strains were grown in YT medium (10); *B. subtilis* strains were grown in nutrient sporulation medium (16).

Northern and Southern hybridizations. The Northern hybridization was performed as described by Silverman (14); the Southern hybridization was carried out by the method of Maniatis (8). The DNA probes were nick translated with [α-^{32}P]ATP by the procedures described by Bethesda Research Laboratories (BRL, Gaithersburg, Md.).

DNA sequencing. The DNA sequences were determined by the dideoxynucleotide chain termination method with the use of a 15-mer primer (5'-AGTCACGACGTTGTA-3') from Bethesda Research Laboratories (13).

In vitro transcription. Purified *B. subtilis* RNA polymerases used in the in vitro transcription assays (5) were gifts of R. Doi (Eσ^{55} and Eσ^{37}) and W. Haldenwang (Eσ^{29}).

Maxicells. Expression of *B. subtilis* genes was monitored in the *E. coli* maxicell strain CSR603 by the method of Sancar (12).

S1 mapping. The 500-base pair (bp) *Pvu*I fragment B and 700-bp *Pvu*I fragment C were purified and end labeled with [γ-^{32}P]ATP. S1 nuclease mapping was performed by the method of Gilman and Chamberlin (4).

RESULTS

DNA sequence of the *gdh* region. A partial restriction map of the 4.0-kb *Eco*RI DNA fragment of *B. subtilis* containing the *gdh* gene is shown in Fig. 1. The restriction endonuclease *Pvu*I digests the 4.0-kb DNA into four fragments (labeled A, B, C, and D) which nearly span the entire DNA. GlcDH was purified to homogeneity by R. F. Ramaley, and the sequence of the first 49 amino acids has been determined by S. Rudikoff (manuscript in preparation). In an attempt to locate the *gdh* gene, G. R. Chaudhry from our laboratory made several deletions of the *Pvu*I fragments within the 4.0-kb *Eco*RI fragment. He found that, after deleting *Pvu*I fragments A and B, the remaining 2.0-kb *B. subtilis* DNA (*Pvu*I fragments C and D) contained the genetic information encoding *gdh*.

The 2.0-kb *Xba*I fragment, located within pEF1 (see Fig. 1), was isolated and either subcloned directly into M13mp10 or digested with *Taq*I or *Hpa*II and then cloned into the *Acc*I site of M13mp8. The sequence of the entire 2.0-kb *Xba*I fragment was determined by the dideoxynucleotide chain termination reaction and is schematized in Fig. 2. The structural gene for GlcDH was located by matching the known NH$_2$-terminal amino acid sequence of GlcDH with the nucleotide sequence. The coding region for the structural gene of *gdh* is 864 bp, which corresponds to a protein of 31,000 molecular weight, the size of the GlcDH subunit. Upstream of the NH$_2$ terminus of *gdh* is a strong Shine-Dalgarno sequence (Fig. 2). No promoter-

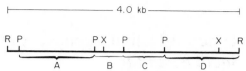

FIG. 1. Partial restriction map of the 4.0-kb *B. subtilis* DNA which is contained in plasmid pEF1 and includes the *gdh* gene. Cleavage sites for *Eco*RI (R), *Pvu*I (P), and *Xba*I (X) are marked. Restriction endonuclease *Pvu*I divides the DNA into four fragments: A, B, C, and D.

TABLE 1. In vitro transcription with the use of *B. subtilis* RNA polymerases Eσ^{55}, Eσ^{37}, and Eσ^{29a}

fragment	Sigma factor		
	55	37	29
Xba I	not detected	1.6 kb	1.6 kb
A	not detected	1.2 kb	1.2 kb

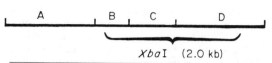

[a] The sizes of transcripts are shown in kb.

like sequence known to be recognized by *B. subtilis* RNA polymerases is located in the vicinity of this Shine-Dalgarno sequence. However, there is another open reading frame which begins in *Pvu*I fragment B and ends 19 bp upstream of the NH$_2$ terminus of *gdh*. This reading frame is again preceded by a Shine-Dalgarno sequence (Fig. 2).

Localization of transcriptional start site by S1 mapping. Labeled *Pvu*I fragments B and C were hybridized to total RNA from t$_5$ cells and subjected to S1 nuclease digestion. The *Pvu*I fragment C was completely protected against S1 nuclease, whereas *Pvu*I fragment B revealed a transcriptional start site located approximately 90 bp in front of the putative translational start site of the first open reading frame.

In vitro transcription with *B. subtilis* RNA polymerases. *Pvu*I fragment A and the *Xba*I fragment were used as templates for in vitro transcription experiments. As shown in Table 1, no transcripts were observed with the major vegetative RNA polymerase holoenzyme Eσ^{55} (now referred to as Eσ^{43}; R. H. Doi, M. Gitt, L.-F. Wang, C. W. Price, and F. Kawamura, this volume). However, when Eσ^{37} and Eσ^{29} were used, the *Xba*I fragment encoded for a transcript of 1.6 to 1.7 kb, and a 1.2-kb transcript was synthesized from the *Pvu*I fragment A. Thus, the two DNA fragments each contain promoters which are recognized by sigma factors that are primarily used during differentiation. As a control, the Eσ^{55} holoenzyme functioned normally when *B. subtilis* phage φ29 DNA was used as template.

Genes encoded on the 4.0-kb *B. subtilis* DNA fragment are developmentally regulated. To de-

termine whether GlcDH synthesis was under transcriptional or translational control, total cellular RNA was isolated from strain 60015 during vegetative growth and at t$_5$. The RNA was fractionated on an agarose-formaldehyde gel, blotted onto nitrocellulose filter paper, and hybridized separately against labeled *Pvu*I fragments A, B, C, and D. The results of one experiment are shown in Fig. 3A. None of the DNA probes hybridized with RNA from vegetative cells. An RNA corresponding to a size of 1.6 to 1.7 kb was detected with each of the four probes when they were hybridized with t$_5$ total RNA (Fig. 3A). Therefore, at least two RNA transcripts of nearly identical size are synthesized from the 4.0-kb DNA fragment. To determine at what time the *gdh* structural gene was transcribed, another Northern hybridization experiment was performed. Total RNAs from vegetative cells and from t$_1$ through t$_5$ cells were hybridized against an *Hpa*II fragment, which lies within the *gdh* gene. As shown in Fig. 3B, no hybridization was detected in cells harvested during vegetative growth up to t$_2$. From t$_3$ through t$_5$, one transcript of 1.6 to 1.7 kb was seen. This shows that *gdh* is transcribed at about the same time at which active GlcDH can be observed (17), i.e., that GlcDH expression is controlled at the transcriptional level.

Proteins synthesized by *E. coli* maxicells. To determine how many proteins are synthesized

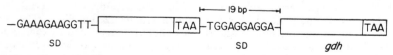

FIG. 2. Schematic representation of the DNA sequences of the 2.0-kb *Xba*I fragment. Boxes represent open reading frames, SD represents the two Shine-Dalgarno sequences, and TAA is the translational stop codon. The distance between the two open reading frames is 19 bp. The sequence on the left may be the equivalent of the "−10" region of vegetative or the usual sporulation promoters (6). No equivalent of the "−35" region was found.

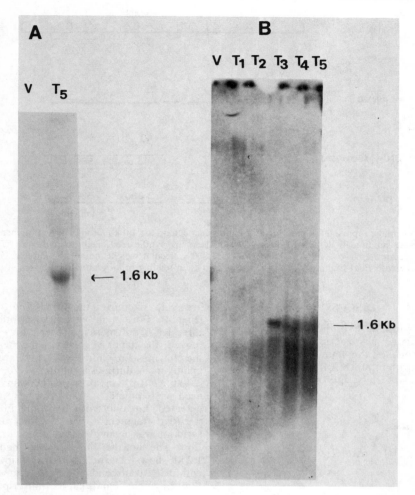

FIG. 3. Northern hybridization of DNA fragments within the 4.0-kb *B. subtilis* DNA with cellular RNA isolated from *B. subtilis* during vegetative growth and t_1 through t_5. (A) Total RNA from vegetative and t_5 cells hybridized with ^{32}P-labeled *Pvu*I fragment A. (B) Total RNA from vegetative and t_1 through t_5 cells hybridized with a ^{32}P-labeled internal fragment (*Hpa*II fragment) within the structural gene for GlcDH.

from the 4.0-kb *Eco*RI DNA fragment, maxicell experiments were performed with UV-irradiated *E. coli* CSR603 cells that carried plasmids containing either the entire 4.0-kb fragment or subfragments. As schematized in Fig. 4, a 62,000-molecular-weight protein was synthesized from a region spanning *Pvu*I fragments A and B; two 31,000-molecular-weight proteins were synthesized from within *Pvu*I fragments B, C, and D.

Construction and characterization of a deletion mutant of the *Xba*I region. To determine whether the 2.0-kb *Xba*I region has any physiological role in sporulation, we constructed a mutant that is deleted for this region in the chromosome. The construction of the plasmid used for creating this mutant is outlined in Fig. 5. pEF1 was digested with *Xba*I and ligated with pE194 *cop6* linearized with *Xba*I. The recombinant plasmid

(pIZA) contained pE194 *cop6*, which lacked *gdh* and its upstream DNA. It was then linearized with *Hin*dIII and transformed into *B. subtilis* *recE⁺* strain 60001. Erythromycin-resistant transformants (drug marker of pE194) were selected. A chromosomal deletion would result from a double crossover at the site of homology in the chromosome. Using Southern hybridization procedures, we confirmed (data not shown) that the deletion mutant has pE194 sequences replacing those within the *gdh* region of the chromosome. Table 2 shows the enzyme activity and sporulation frequency of the mutant compared with that of the parental strain (60001) and that of the transformant containing autonomously replicating pIZA. The deletion mutant exhibited no GlcDH activity and was defective in sporulation.

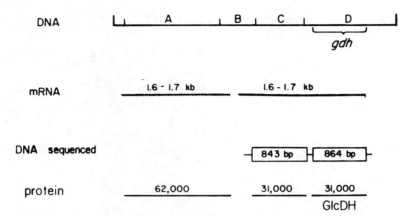

FIG. 4. Schematic representation of the information encoded on the 4.0-kb *Eco*RI DNA. Fragments A, B, C, and D are those subdivided by *Pvu*I (Fig. 1). Two mRNAs are synthesized (each about 1.6 to 1.7 kb) which encode for three proteins, one of 62,000 and two of 31,000 molecular weight; one of the latter is GlcDH. The DNA sequence with *Pvu*I fragments C and D shows two open reading frames (shown by boxes).

DISCUSSION

In the Northern hybridization experiments described above, a transcript of 1.6 kb appears when any one of the four *Pvu*I fragments is hybridized with total RNA from cells at t_5. This implies that two transcripts of about 1.6 kb each

FIG. 5. Construction of the pIZA plasmid. pEF1 was digested with *Xba*I and ligated with pE194, also digested with *Xba*I. pE194 was inserted into the *Xba*I site of pEF1, replacing *gdh* and flanking DNA.

are synthesized from the 4.0-kb *Eco*RI fragment (Fig. 4). For an in vitro transcription assay, three RNA polymerase holoenzymes were employed: the major vegetative enzyme $E\sigma^{55}$ and the enzymes $E\sigma^{37}$ and $E\sigma^{29}$, which are predominantly made during sporulation (2, 6). When the 2.0-kb *Xba*I fragment, which contains *gdh*, was used as template, a transcript of 1.6 kb was detected, but only with $E\sigma^{37}$ and $E\sigma^{29}$. When the *Pvu*I fragment A was similarly used, transcription was again observed only with $E\sigma^{37}$ and $E\sigma^{29}$. The fact that this transcript had a size of 1.2 kb shows that the 1.6-kb transcript observed with pEF1 transcends the *Pvu*I fragment A into *Pvu*I fragment B and that its transcription starts near the *Eco*RI site of fragment A.

TABLE 2. Glucose dehydrogenase activities and sporulation frequency of a deletion mutant[a]

Strain	GlcDH activity (nmol/min per mg of protein)	Sporulation frequency (spores/ml)
60001 (parent)	45.2	3×10^7
60001 (pIZA)	58.9	3×10^7
60001 deletion mutant	0.45	80

[a] Cells grown in nutrient sporulation medium. At t_5 (5 h after the end of exponential growth), a sample of each culture (50 ml) was centrifuged, and the pellet was suspended in 2 ml of 50 mM imidazole-HCl, pH 6.5, containing 20% glycerol, and broken by two passages through a French press cell. The extracts were centrifuged for 20 min at $100,000 \times g$, and the supernatants were assayed for GlcDH activity. Samples were taken at t_8 to determine the titer of heat-resistant spores by subjecting the cells to 75°C for 15 min and plating on tryptose blood agar base plates.

The expression of the 4.0-kb region in *E. coli* maxicells showed that three proteins, of molecular weights 62,000, 31,000, and 31,000, are synthesized. These data agree with the two open reading frames seen in the DNA sequence of the *Xba*I fragment (Fig. 1 and 4). Each reading frame is preceded by a Shine-Dalgarno sequence (Fig. 4). The *gdh* gene is within the second open reading frame.

The results of S1 nuclease mapping with the use of the *Pvu*I fragment B showed a transcriptional start site approximately 90 bp before the putative translation start (ATG) of the open reading frame. The total length from this translational start site to the final translational stop codon in the *gdh* gene is 1,726 bp. This result compares favorably with the 1.6- to 1.7-kb transcript measured in the Northern hybridization and the in vitro transcription experiments described above. This long mRNA of 1.6 to 1.7 kb agrees with the sequence data which do not show any consensus promoter sequences (7) for either vegetative genes or specific sporulation genes between the two open reading frames. Only upstream from the transcriptional start site identified by S1 mapping are sequences similar to the ``−10'' consensus sequence (Fig. 2). However, there are no sequences similar to the ``−35'' consensus sequence.

All evidence shows that the *gdh* gene is part of a polycistronic operon containing two structural genes. Other evidence of polycistronic transcription in *B. subtilis* includes the operons which contain *spoIIA* (11), *dnaE* and *rpoD* (Doi et al., this volume), and *gnt* (Y. Fujita, J.-I. Nihashi, and T. Fujita, this volume).

A chromosomal deletion of the 2.0-kb *Xba*I fragment containing this operon produced a strain unable to sporulate and to synthesize GlcDH activity. Strauss (15) showed that a strain containing a *gdh* point mutation could still sporulate. Therefore, the protein encoded for by the first open reading frame and presumably coordinately expressed with GlcDH is apparently needed for sporulation. Previous sporulation mutations have not been found near the *mtl* gene next to which the *gdh* operon is located (1). It will be interesting to determine at which stage this mutation blocks sporulation and whether the gene product is, like GlcDH, made only in the forespore.

SUMMARY

We have characterized a 4.0-kb *Eco*RI *B. subtilis* DNA fragment which encodes for two 1.6- to 1.7-kb mRNAs, both of which are under sporulation control. One mRNA encodes for a protein of 62,000 molecular weight whereas the other mRNA encodes for two 31,000-molecular-weight proteins, one of which is GlcDH. The latter two proteins compose a polycistronic operon whose DNA sequence and transcriptional start site have been determined.

A plasmid vector, pIZA, was constructed that enabled us to delete the polycistronic operon from the *B. subtilis* chromosome. The resulting mutant can no longer sporulate. Because sporulation does not require any active GlcDH (15), it does require the product of the first gene of this operon.

We thank R. Doi and W. Haldenwang for their generous gift of purified RNA polymerases. We also are indebted to E. Galliers for her assistance in preparing the illustrations.

LITERATURE CITED

1. Chaudhry, G. R., Y. S. Halpern, C. Saunders, N. Vasantha, B. J. Schmidt, and E. Freese. 1984. Mapping of the glucose dehydrogenase gene in *Bacillus subtilis*. J. Bacteriol. 160:607–611.
2. Doi, R. H., T. Kudo, and C. Dickel. 1981. RNA polymerase forms in vegetative and sporulating cells of *Bacillus subtilis*, p. 219–223. *In* H. S. Levinson, A. L. Sonenshein, and D. J. Tipper (ed.), Sporulation and germination. American Society for Microbiology, Washington, D.C.
3. Fujita, Y., R. Ramaley, and E. Freese. 1977. Location and properties of glucose dehydrogenase in sporulating cells and spores of *Bacillus subtilis*. J. Bacteriol. 132:282–293.
4. Gilman, M. Z., and J. J. Chamberlin. 1983. Developmental and genetic regulation of *Bacillus subtilis* genes transcribed by σ^{28}-RNA polymerase. Cell 35:285–293.
5. Goldfarb, D. S., S. L. Wong, T. Kudo, and R. H. Doi. 1983. A temporally regulated promoter from *Bacillus subtilis* is transcribed only by an RNA polymerase with a 37,000 dalton sigma factor. Mol. Gen. Genet. 191:319–325.
6. Losick, R. 1981. Sigma factors, stage 0 genes, and sporulation, p. 48–56. *In* H. S. Levinson, A. L. Sonenshein, and D. J. Tipper (ed.), Sporulation and germination. American Society for Microbiology, Washington, D.C.
7. Losick, R., and J. Pero. 1981. Cascades of sigma factors. Cell 25:582–584.
8. Maniatis, T., E. F. Fritsch, and J. Sambrook. 1982. Molecular cloning. A laboratory manual, p. 383. Cold Spring Harbor Laboratory, Cold Spring Harbor, N.Y.
9. Messing, J., and J. Viera. 1982. A new pair of M13 vectors for selecting either DNA strand of double digest restriction fragments. Gene 19:269–276.
10. Miller, J. 1972. Experiments in molecular genetics, p. 433. Cold Spring Harbor Laboratory, Cold Spring Harbor, New York.
11. Piggot, P. J., C. A. M. Curtis, and H. DeLencastre. 1984. Use of integrational plasmid vectors to demonstrate the polycistronic nature of a transcriptional unit (spoIIA) required for sporulation of *Bacillus subtilis*. J. Gen. Microbiol. 130:2123–2136.
12. Sancar, A., A. M. Hack, and W. D. Rupp. 1979. Simple method for identification of plasmid-coded proteins. J. Bacteriol. 137:692–695.
13. Sanger, F., S. Nicklen, and A. R. Coulson. 1977. DNA sequencing with chain terminating inhibitors. Proc. Natl. Acad. Sci. U.S.A. 74:5463–5467.
14. Silverman, S. J., M. Rose, D. Botstein, and G. R. Fink. 1982. Regulation of *HIS4-lacZ* fusions in *Saccharomyces cerevisiae*. Mol. Cell. Biol. 2:1212–1219.
15. Strauss, N. 1983. Role of glucose dehydrogenase in germination of *Bacillus subtilis* spores. FEMS Microbiol.

Lett. **20**:379–384.

16. **Vasantha, N., and E. Freese.** 1980. Enzyme changes during *Bacillus subtilis* sporulation caused by deprivation of guanine nucleotides. J. Bacteriol. **144**:1119–1125.

17. **Vasantha, N., B. Uratani, R. F. Ramaley, and E. Freese.** 1983. Isolation of a developmental gene of *Bacillus subtilis*

and its expression in *Escherichia coli*. Proc. Natl. Acad. Sci. U.S.A. **83**:785–789.

18. **Weisblum, B., M. Y. Graham, T. Gryczan, and D. Dubnau.** 1979. Plasmid copy number control: isolation and characterization of high-copy number mutants of plasmid pE194. J. Bacteriol. **131**:635–643.

Regulation of the *Bacillus spo0H* Gene

I. SMITH,[1] E. DUBNAU,[1] J. WEIR,[2] J. OPPENHEIM,[2] N. RAMAKRISHNA,[1] AND K. CABANE[1]

Department of Microbiology, The Public Health Research Institute of the City of New York, Inc.,[1] and Department of Microbiology, New York University School of Medicine,[2] New York, New York 10016

Bacilli, when exposed to certain environmental stresses, stop vegetative growth and begin a series of developmental changes which culminate in the appearance of dormant, resistant spores. The initiation of the sporulation process in *Bacillus subtilis* is dependent upon the proper functioning of nine genes, the *spo0* loci. It has been postulated that the *spo0* gene products are involved in the regulatory pathways responsible for the induction of sporulation-specific gene expression under conditions of nutrient limitation (8). The developmentally related expression of at least one gene, *spoVG*, requires the participation of several *spo0* loci (10, 17). In recent years several laboratories have isolated and studied some of the *spo0* genes in an attempt to understand the control of their expression and the function of their gene products in the sporulation process (2–5, 7, 13). We have cloned the *spo0F* and *spo0G* genes of *B. subtilis*, but since the major focus of our laboratory has been on the *spo0H* locus, this communication will concentrate on this gene.

EXPERIMENTAL PROCEDURES

All technical procedures were described previously (3, 11, 16). Other methods are described in the figure legends.

RESULTS

Physical characterization of the *Bacillus spo0H* gene. We previously described the cloning of a 1.2-kilobase (kb) chromosomal fragment from *B. licheniformis* which specifically complements all known *spo0H* mutations in *B. subtilis* (3). RNA complementary to the cloned fragment is transcribed during vegetative growth and sporulation. A protein with an apparent molecular weight of 27,000 (27K protein) is coded for by the 1.2-kb DNA segment in minicells. Deletions within the insert abolish the *spo0H*⁺ complementing activity of the cloned fragment and also eliminate the 27K protein.

The 1,228-base pair *B. licheniformis* fragment has now been sequenced (11). There is one open reading frame consisting of 168 codons, which corresponds to a 22K protein. It is preceded by a ribosomal binding sequence of AGGAAGG at bases 297 to 306 and uses an initiation codon of UUG at base 309. The deletions discussed above, which eliminate both the *spo0H*⁺ complement-

ing activity and the 27K minicell polypeptide, lie within the 168-codon open reading frame which terminates at base 813 (Fig. 1). In vivo and in vitro transcription mapping studies indicate the presence of one RNA species with a molecular size of approximately 1 kb which is coded for by the 1.2-kb insert. The 5′ transcription initiation site is at base 79, and the 3′ termination region is in the area of base 1,138. The promoter for this gene has the sequence 5′-TATAAT-3′ 10 bases upstream from the RNA start site at base 79, and the −35 sequence is 5′-TTGACG-3′. This promoter is transcribed, in vitro, by the *Escherichia coli* RNA polymerase and the major *B. subtilis* vegetative holoenzyme, containing sigma 43, and not by the polymerase with sigma 37.

Since the *B. licheniformis* clone specifically complemented *spo0H* mutations of *B. subtilis*, we assumed that we had cloned the homologous gene from *B. licheniformis*. However, the lack of sufficient DNA-DNA homology prevented the integration of the insert into the *B. subtilis* chromosome, and thus we were unable to show conclusively that we had isolated the *spo0H* gene and not some other gene which could suppress the *spo0H75* mutation. We have recently cloned a 2.8-kb fragment from the *B. subtilis* chromosome which complements all *spo0H* mutations in *trans* (16). Genetic studies, in which this cloned fragment was inserted into the *B. subtilis* chromosome, showed that the clone carries sequences derived from the *spo0H* gene. An internal segment of the 2.8-kb fragment, lying within the transcribed region of the *B. subtilis spo0H* gene, shows limited DNA-DNA homology with the 1.2-kb *B. licheniformis spo0H* complementing clone. This homology proves that the *B. licheniformis* clone does contain the *spo0H* gene. The results of the above section are summarized in Fig. 1.

Though the *spo0H* gene has now been completely characterized physically, the two major questions concerning it have not yet been answered. We still do not know how the expression of the *spo0H* gene is regulated, and we have no idea of the function of its gene product, the 22K protein.

Construction and integration of a *spo0H-lacZ* translational fusion. Extremely low levels of the *spo0H* protein are made in vivo, as two-dimensional gels of radioactive cell extracts containing

B. licheniformis

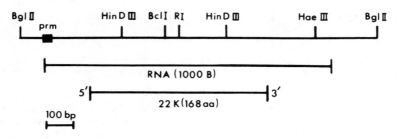

B. subtilis

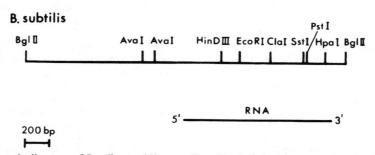

FIG. 1. Schematic diagrams of *Bacillus spo0H* genes. Top: The *B. licheniformis spo0H* gene. The position of the RNA initiation and termination sites was determined by in vivo and in vitro transcriptional mapping studies (11). The open reading frame was deduced from the DNA sequence. Removal of the internal *Hin*dIII fragment results in a loss of *spo0H+* complementing activity and in disappearance of a 27K protein in minicells and the loss of a 25K protein found in whole cells, which reacts with antibody prepared against *spo0H* protein. Bottom: The *B. subtilis spo0H* gene. Deletions around the *Hin*dIII site of this clone cause loss of *spo0H+* complementing activity and map in the chromosomal *spo0H* gene. The location of RNA synthesized on the clone was determined by in vivo transcriptional mapping experiments. The *Hin*dIII-*Pst*I fragment shows homology with the 1.2-kb *B. licheniformis* clone (16).

the cloned *spo0H* gene in multicopy plasmids do not show the 27K protein. To study expression of *spo0H*, we have constructed a translational *spo0H-lacZ* gene fusion. The fusion was effected by joining the plasmid pIS9a, carrying the *B. licheniformis spo0H* gene in the vector pBD97 (3), to pSK10Δ6, a pBR322 derivative containing the structural gene for β-galactosidase (17), at their unique *Eco*RI sites (Fig. 2). The *Eco*RI site of pIS9a lies within the *spo0H* open reading frame, 250 base pairs downstream from the putative translational start site, and is known from the nucleotide sequence to be in phase with the *Eco*RI site at the eighth codon of the *lacZ* reading frame in pSK10Δ6. The resulting chimeric plasmid, pIS52b, can replicate in *E. coli*, expressing ampicillin resistance, and in *B. subtilis*, where chloramphenicol resistance is expressed. An enzymatically active β-galactosidase hybrid protein is produced at high levels in both organisms containing pIS52b. High-resolution transcriptional mapping studies have also shown that the *spo0H* mRNA is initiated at the same nucleotide in *B. licheniformis* and in *B.*

subtilis and *E. coli* strains carrying pIS52b (11). Strains containing pIS52a, a fusion plasmid in which pIS9a and pSK10Δ6 are joined at their *Eco*RI sites, but in the reverse orientation, produce background levels of β-galactosidase in both *E. coli* and *B. subtilis*.

Since chromosomal genes cloned on multicopy plasmids frequently do not exhibit normal patterns of control (17), we proceeded to integrate the *spo0H-lacZ* fusion into the *B. subtilis* chromosome, so that its expression could be studied at the level of a single gene copy. However, no genetic homology exists in any part of the fusion plasmid which would be sufficient for integration into the *B. subtilis* chromosome. Although, as discussed above, DNA-DNA hybridization occurs between the *spo0H* genes of *B. subtilis* and *B. licheniformis*, there is insufficient homology to provide for detectable genetic recombination and integration at the *spo0H* locus.

The transposonlike activity of plasmid pE194 (6) provided us with a way to achieve this end, however. Plasmid pBD97, the parent plasmid of pIS52b, contains the temperature-sensitive replication region of pE194. The fusion plasmids pIS52a and pIS52b were transformed into the *B. subtilis* strain h, which contains the entire pE194 genome integrated into the *B. subtilis* chromosome between *cysB* and *hisA* (6). Transformants were selected for chloramphenicol resistance at 37°C, and these were purified by several single-colony passages at 51°C on chloramphenicol agar. The resulting colonies were chloramphenicol resistant at 51°C and were erythromycin susceptible, indicating that the incoming plasmids had integrated into the chromosome and had replaced the resident pE194 sequences. DNA prepared from these recombinants was capable of transferring the chloramphenicol resistance marker to recombination-proficient but not *recE* strains of *B. subtilis*, and there was no detectable free plasmid in the DNA preparations. The chloramphenicol resistance marker in these recombinants was also linked to *cysB* and *hisA*, though the transductional linkages observed were lower than those seen with originally integrated pE194. Southern hybridization experiments have also shown that the *spo0H-lacZ* fusions have integrated in a single copy.

In contrast to *B. subtilis* strains carrying free pIS52b, and as expected for genes integrated into the chromosome by a replacement, not a Campbell-type event, the chloramphenicol resistance and β-galactosidase activity are completely stable. This is true even when strains carrying the integrated pIS52 plasmids are grown in the absence of chloramphenicol.

We have also integrated pIS9a, with the intact *spo0H* gene, into the chromosome of *B. subtilis* which contains the integrated pE194. In this single-copy configuration, the integrated *B. licheniformis spo0H* gene can restore a Spo+ phenotype to *spo0H* strains. This provides further evidence for the integration of the incoming plasmid without gross alteration and indicates that the *B. licheniformis* gene can function normally as a single copy in *B. subtilis*.

Analysis of *spo0H*-driven β-galactosidase activity. The integration of the *spo0H-lacZ* fused gene now enabled us to study the regulation of the *spo0H* gene during growth, under conditions of catabolite repression, and in different genetic backgrounds, using β-galactosidase activity as an assay for the *spo0H* protein. Strains IS253, containing the integrated pIS52a (in which the *spo0H* N terminus is in the wrong orientation with respect to the *lacZ* gene), and IS254, with integrated pIS52b, were grown in Schaeffer's nutrient sporulation medium (12), in the presence and absence of 0.5% glucose. Extremely low levels of β-galactosidase activity were detected in IS253, approximately 0.10 to 0.15 Miller units (9) during early vegetative growth

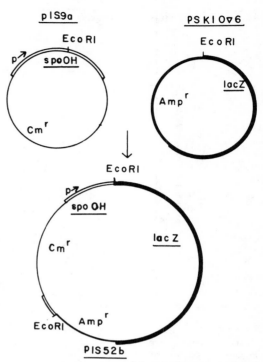

FIG. 2. Construction of a *spo0H-lacZ* translational fusion. pIS9a, composed of the *spo0H* gene in the vector pBD97, and pSK10Δ6, a pBR322 derivative carrying the β-galactosidase coding region of *lacZ*, were restricted at their unique *Eco*RI sites, ligated with T4 DNA ligase, and used to transform a *B. subtilis* recipient to Cm^r and Lac$^+$. pIS52b is the name given to the chimeric plasmid isolated from one such transformant. In pIS52b, the 5′ half of the *spo0H* gene is inserted upstream of the *lacZ* sequences in the correct orientation. pIS52a refers to an analogous chimeric plasmid carrying the pIS9a moiety in the opposite orientation with respect to *lacZ*.

reaching 0.35 to 0.5 Miller units at t_0 to t_1 (Fig. 3). (In these experiments, t_0 is defined as the end of logarithmic growth, and t_1, t_2, etc., occur 1, 2, etc., h after t_0.) At later stages (t_2 to t_3) these values can reach 1 to 1.5 Miller units. This low-level activity, which was completely repressed by glucose, was also observed in IS250, an isogenic *B. subtilis* strain carrying the original transposed pE194, as well as in other *B. subtilis* strains not carrying integrated plasmids. Much higher levels of β-galactosidase activity, dependent on the *spo0H* gene, were observed in IS254, at all stages of growth (Fig. 3). Enzyme levels, in the absence of glucose, were approximately 2 to 4 Miller units in early vegetative growth, showed a peak of activity at t_0 (approximately 5 to 8 Miller units), and showed a decline at later times. Interestingly, the presence of glucose resulted in the same general pattern of β-

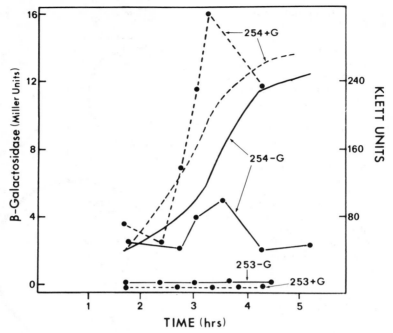

FIG. 3. Time course of *spo0H-lacZ* fusion activity. IS253 containing the integrated pIS52a and IS254 with integrated pIS52b were grown in Schaeffer's nutrient sporulation medium (12) with and without 0.5% glucose. Samples were removed at different times during growth, and β-galactosidase activity was assayed (9). The lines without points indicate the optical density of the cultures; the lines with points reflect β-galactosidase activity in Miller units (9). For the sake of clarity the optical density of the IS253 cultures is omitted. They were essentially identical to the IS254 samples. In all cases broken lines refer to cultures with glucose; solid lines, to cultures without glucose.

galactosidase activity during growth, but with significantly higher levels.

To see whether *spo0H* gene expression, as measured by the *spo0H*-driven β-galactosidase activity, was dependent on the presence of other *spo* genes, the integrated pIS52b sequences were transferred to a series of isogenic strains with mutations in the *spo0A*, *spo0B*, *spo0F*, *spo0H*, or *spo0IIIA* genes, selecting for the chloramphenicol resistance marker of the integrated pIS52b. The resulting recombinants were grown in liquid nutrient sporulation medium, and β-galactosidase activity was measured at different times during growth. As illustrated in Fig. 4, the *spo0H* gene was expressed in all genetic backgrounds tested. The levels of β-galactosidase activity in *spo0B*, *spo0F*, and *spoIIIA* strains were similar to those of Spo+ cells, whereas larger amounts of enzyme activity were observed in *spo0A* and *spo0H* strains.

Characterization of the *spo0H* gene product. To aid us in characterizing the *spo0H* protein, in vivo, and in elucidating its role in sporulation, we have developed an immunological assay for the *spo0H* gene product. The approach taken was essentially as described by Shuman et al. (14) for the preparation of antibodies against an

uncharacterized *E. coli* protein. A chimeric gene is constructed which directs the synthesis of a hybrid protein comprising both *spo0H* protein and β-galactosidase sequences. The hybrid protein is then purified on the basis of its β-galactosidase properties and used to elicit antibodies in rabbits. The antiserum generated against the hybrid protein can be used to detect the wild-type *spo0H* gene product in whole cells. For these studies we used pIS52b, described above (Fig. 2).

pIS52b carries sequences sufficient to encode a chimeric protein of about 145,000 daltons, composed of the first 83 amino acids of the *spo0H* gene product linked to the eighth amino acid of β-galactosidase. Because it is one of the largest proteins made by *E. coli* and is expressed at high levels in the pIS52b construct, the hybrid protein can readily be detected by direct staining of electrophoretically separated extracts of an *E. coli lac* strain carrying pIS52b (Fig. 5). As expected, the hybrid protein migrates slightly more slowly than the wild-type β-galactosidase. The same strain carrying pIS52a, a fusion plasmid in which pIS9a is inserted into the *Eco*RI site of pSK10Δ6 in the opposite orientation, does not encode a protein of this size.

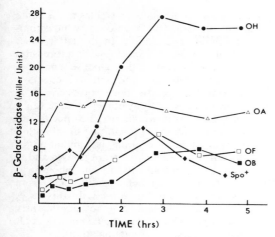

FIG. 4. Effect of *spo* mutations on *spo0H-lacZ* activity. Isogenic *B. subtilis* strains containing integrated pIS52b and different *spo* mutations were grown in Schaeffer's nutrient sporulation medium, and β-galactosidase was assayed at different times. For the sake of clarity, the optical density values for the cultures have been eliminated. In the different cultures, t_0 (200 Klett units) occurred at about 2 h. A strain with the *spoIIIA* mutation gave results similar to those obtained with the *spo*+ and *spoOB* and *spoOF* strains. The following *spo* mutations were used in these studies: *spo0H12*, *spo0B136*, *spo0F221*, *spo0H75*, and *spoIIIA7*.

E. coli was chosen as the source of the pIS52b-encoded fusion protein to be used to elicit antibodies specific to the *spo0H* protein, as this would minimize potential cross-reactivity due to antibodies that might be generated to contaminants in the hybrid protein preparation. Such antibodies are not likely to react strongly with *Bacillus* proteins. The scheme developed for the purification of large quantities of the *spo0H-lacZ* fusion protein took advantage of the tendencies of this protein to aggregate during processing of cell extracts. Cells from an overnight culture of the *E. coli lac* deletion strain MC1060 containing pIS52b were harvested and disrupted by passage through a French pressure cell. The cell lysate was centrifuged at 20,000 × *g* for 1 h to pellet unbroken cells and membrane debris. The supernatant fraction, possessing nearly all of the β-galactosidase activity, was subsequently subjected to centrifugation at 40,000 × *g* for 3 h. Under these conditions, 40 to 60% of the fusion protein was pelleted. Nonidet P-40, EDTA, high salt, and RNase were included in the extract prior to the 40,000 × *g* centrifugation to increase the solubility of other cellular proteins while retaining the relative insolubility of the hybrid.

Once pelleted, the *spo0H-lacZ* fusion protein could be completely solubilized only in the presence of sodium dodecyl sulfate (SDS). As shown in Fig. 6, lane 2, the hybrid protein is a major component of the SDS-solubilized 40,000 × *g* pellet fraction. The proteins of this fraction were applied to a Sephacryl S-300 molecular sieve column to further purify the hybrid protein (Fig. 6). The leading edges of the peak elution fractions containing the fusion protein were pooled and concentrated. The yield of the partially purified *spo0H-lacZ* fusion protein after Sephacryl chromatography was typically 2 to 3 mg per liter of overnight culture.

New Zealand White rabbits were immunized with the hybrid protein preparation to elicit antibodies against the N-terminal half of the *spo0H* protein. Once sufficient quantities of the serum were obtained and shown to react with purified β-galactosidase, the antiserum was used to probe extracts of *B. subtilis* containing the cloned *B. licheniformis spo0H* gene (Fig. 7). The antiserum was found to react strongly with two proteins, of approximately 25K and 60K, present in an extract prepared from a pIS9a-containing strain. The 25K protein band was not detected in extracts from strains carrying the parent vector, pBD97, or pIS26. pIS26 contains a deletion which disrupts the *spo0H* open reading frame

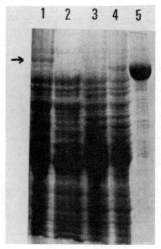

FIG. 5. Identification of the *spo0H-lacZ* fusion protein in *E. coli*. Lysates of the *E. coli lac* strain HB101 carrying pIS52b (lane 1) or pIS52a (lane 2) and the *E. coli lac*+ strain HB94 uninduced (lane 3) or induced with 0.2 mM isopropyl-β-D-thiogalactopyranoside (lane 4) were prepared and analyzed by electrophoresis on a 10% polyacrylamide-SDS gel. The Coomassie brilliant blue-stained gel is shown. An arrow indicates the position of the *spo0H-lacZ* fusion protein encoded by pIS52b, but not pIS52a. Purified β-galactosidase (14 μg, Sigma Chemical Co.) was run in lane 5 as a marker for the native enzyme. As expected, it comigrates with the isopropyl-β-D-thiogalactopyranoside-induced β-galactosidase of HB94.

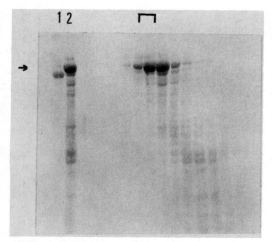

FIG. 6. Purification of the *spo0H-lacZ* fusion protein by Sephacryl S-300 chromatography. The SDS-solubilized 40,000 × *g* pellet fraction prepared from a cell lysate of strain MC1060 carrying pIS52b (lane 2) was applied to a Sephacryl S-300 column. The separation was carried out at room temperature in 50 mM Tris hydrochloride (pH 7.5)–0.1 mM NaCl–50 mM β-mercaptoethanol–0.1 mM phenylmethylsulfonyl fluoride. Every third fraction eluted from the column was analyzed by gel electrophoresis. Fractions from the leading edge of the fusion protein elution peak (indicated by bracket) were pooled and used for immunizations. The position of the fusion protein band is indicated by an arrow. Purified β-galactosidase was run in lane 1 as a size marker.

(Fig. 1) and results in the production of a severely truncated protein, which would not have been detected in this experiment. The 60K protein is also found in extracts from the pBD97- and pIS26-containing strains, although at lower levels than in the strain carrying pIS9a. It is not known whether this band represents a protein reacting with *spo0H* protein-specific antibodies or with antibodies elicited against contaminants in the hybrid protein preparation. No other strongly reacting proteins were observed in these extracts.

DISCUSSION

To study the regulation of the *B. licheniformis* *spo0H* gene and to characterize its gene product, we have constructed a translational fusion between the NH$_2$-terminal portion of the *spo0H* 22K protein and the *lacZ* protein. We have shown, with the fused gene integrated into the *B. subtilis* chromosome, that the *spo0H* gene is expressed throughout growth and early in sporulation and shows a peak of activity at the end of stationary growth, at t_0 to t_1. This temporal expression of the *spo0H* gene is observed in Schaeffer's nutrient sporulation medium, in

which cells will sporulate, as well as in the same medium containing 0.5% glucose, which completely inhibits sporulation. In the latter case, much higher levels of *spo0H* gene expression are observed. Whether this glucose enhancement of *spo0H*-driven *lacZ* activity reflects a specific effect on *spo0H* gene expression or is the result of a general physiological state (e.g., growth regulation in which enzyme specific activity is proportional to growth rate or inhibition of sporulation and resulting accumulation of various effector molecules) is not known and is currently being investigated in our laboratory. The *spo0H-lacZ* fusion behaves differently from fusions between *spo0B* and *lacZ*. In this case, enzyme values do not change significantly throughout growth and glucose has no effect (J. Szulmajster, personal communication).

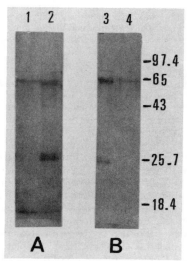

FIG. 7. Immunological detection of the *spo0H* gene product in *B. subtilis* extracts. Extracts were prepared from late-vegetative-phase cultures of *B. subtilis* *spo0H75* strains carrying pBD97 (lane 1), pIS9a (lanes 2 and 3), or pIS26 (the *spo0H* deletion derivative of pIS9a, lane 4) by incubation with lysozyme and boiling with SDS. The protein extracts were separated by electrophoresis through a 12.5% polyacrylamide-SDS gel and transferred to nitrocellulose (15). The blot was incubated with a 1:1,000 (A) or a 1:5,000 (B) dilution of the anti-hybrid protein antiserum. Bound antibody was detected by subsequent treatment with alkaline phosphatase-conjugated goat anti-rabbit immunoglobulin G antibodies. The reduction of tetrazolium salt to diformazan by the hydrogen ions released in the formation of indigo by the reaction of phosphatase on indoxyl phosphate served as a colorimetric indicator for specifically bound antibody (1). Blots A and B represent two separate Western blot experiments. In each case roughly equivalent amounts of total protein were applied to each lane of the gel. Commercially prepared protein size standards were run adjacent to the samples of blot B.

Significantly higher than wild-type *spo0H*-driven β-galactosidase activity is also observed throughout growth when *spo0H* and *spo0A* mutations are present in strains carrying the integrated *spo0H-lacZ* fusion. As with the glucose-enhanced *spo0H*-driven β-galactosidase activity described above, the significance of the elevated *spo0H* gene expression in *spo0A* and *spo0H* strains is not known. The tempting speculations that there is a specific interaction between the *spo0A* gene product and expression of the *spo0H* gene and also that the *spo0H* gene is itself autogenously regulated are also being investigated.

The fact that *spo0B* and *spo0F* mutations do not affect the levels of *spo0H* gene expression is interesting in view of the observation that these genes may act on the *spo0A* gene product at a posttranslational stage, in a process that can be bypassed by several suppressor mutations which map in the *spo0A* open reading frame (J. Hoch, personal communication). Our results suggest that the *spo0A* gene product, unmodified by the *spo0B* or *spo0F* gene function and unable to function in the initiation of sporulation, can still affect *spo0H* gene expression.

In the course of our studies on *spo0H-lacZ* fusions we have observed an endogenous, low-level β-galactosidaselike activity which increases late in growth and is glucose repressible. This activity can interfere with the determination of β-galactosidase in low-activity *lacZ* fusions and must be considered when mutagenizing *B. subtilis* containing integrated *lacZ* fused genes. In the latter case, elevations in β-galactosidase activity frequently result from "UP" mutations in the endogenous β-galactosidase gene, not at the integrated *lacZ* fusion locus (P. Zuber and R. Losick, personal communication; E. Dubnau and I. Smith, unpublished data).

The *spo0H-lacZ* fusion has enabled us to prepare antibodies which can be used to assay and purify the *spo0H* protein. Using these antibodies, we have detected a 25K protein in extracts from the *B. subtilis spo0H75* strain carrying pIS9a which appears to be the product of the *B. licheniformis spo0H* gene. Its presence in cell extracts correlates with the presence of an intact copy of the *spo0H* clone. Its size is approximately the same as that determined for the minicell-encoded *spo0H* gene product.

We are currently using the antiserum elicited from the *spo0H-lacZ* fusion protein to characterize the *B. licheniformis spo0H* gene product and to aid in its quantification during growth in *B. licheniformis*. In addition, we are determining the reactivity of our antiserum with the *B. subtilis spo0H* protein. Since the *B. licheniformis spo0H* gene shows homology to the *B. subtilis spo0H* clone, it is likely that their protein products are antigenically related, as well.

Although we still cannot give complete answers to the questions of what regulates *spo0H* and what the function of its gene product is, we are confident that the use of the *spo0H-lacZ* fusion will give significant information in the near future.

SUMMARY

A *B. licheniformis spo0H-lacZ* translation fusion has been constructed and integrated into the *B. subtilis* chromosome. In liquid medium which supports sporulation, *spo0H*-driven β-galactosidase activity is present in vegetative growth, reaches a maximum at t_0 to t_1, and declines thereafter. The presence of glucose in liquid media, while preventing sporulation, stimulates the expression of the *spo0H* gene 5- to 10-fold. Mutations in the *spo0B*, *spo0F*, or *spoIIIA* loci have no effect on *spo0H* expression, whereas *spo0A* and *spo0H* alterations result in significantly higher *spo0H*-dependent β-galactosidase activity. A hybrid *spo0H*-β-galactosidase protein has been used to elicit antibodies to the *spo0H* protein, which in turn have been used to detect the *B. licheniformis spo0H* gene product in cell extracts.

We thank D. Dubnau for bacterial strains and plasmids and helpful advice, and N. Gaur and M. Lewandoski for valuable discussions.

This work was supported by Public Health Service grant GM-19683 from the National Institute of General Medical Sciences awarded to I.S. J.W. is supported by Public Health Service training grant 5T32-AI-01780, awarded to the Department of Microbiology, New York University School of Medicine, by the National Institute of Allergy and Infectious Diseases.

LITERATURE CITED

1. **Blake, M. S., K. H. Johnston, G. J. Russell-Jones, and E. C. Gotschlich.** 1984. A rapid, sensitive method for detection of alkaline phosphatase-conjugated antibody on Western blots. Anal. Biochem. **136:**175–179.
2. **Bonamy, C., and J. Szulmajster.** 1982. Cloning and expression of *Bacillus subtilis* spore genes. Mol. Gen. Genet. **188:**202–210.
3. **Dubnau, E., N. Ramakrishna, K. Cabane, and I. Smith.** 1981. Cloning of an early sporulation gene in *Bacillus subtilis*. J. Bacteriol. **147:**622–632.
4. **Ferrari, F., D. Lang, E. Ferrari, and J. A. Hoch.** 1982. Molecular cloning of the *spo0B* sporulation locus in bacteriophage lambda. J. Bacteriol. **152:**809–814.
5. **Hirochika, H., Y. Kobayashi, F. Kawamura, and H. Saito.** 1981. Cloning of sporulation gene *spo0B* of *Bacillus subtilis* and its genetic and biochemical analysis. J. Bacteriol. **146:**494–505.
6. **Hofemeister, J., M. Israeli-Reches, and D. Dubnau.** 1983. Integration of plasmid pE194 at multiple sites on the *Bacillus subtilis* chromosome. Mol. Gen. Genet. **189:**58–68.
7. **Ikeuchi, T., J. Kudoh, and K. Kurahashi.** 1983. Cloning of sporulation genes *spo0A* and *spo0C* of *Bacillus subtilis* into ρ11 temperate bacteriophage. J. Bacteriol. **154:** 988–991.

8. **Losick, R., and J. Pero.** 1981. Cascades of sigma factors. Cell **25:**582–584.
9. **Miller, J. H.** 1972. Experiments in molecular genetics. Cold Spring Harbor Laboratory, Cold Spring Harbor, N.Y.
10. **Ollington, J. F., W. G. Haldenwang, T. V. Huynh, and R. Losick.** 1981. Developmentally regulated transcription in a cloned segment of the *Bacillus subtilis* chromosome. J. Bacteriol. **147:**432–442.
11. **Ramakrishna, N., E. Dubnau, and I. Smith.** 1984. The complete DNA sequence and regulatory regions of the *Bacillus licheniformis spo0H* gene. Nucleic Acids Res. **12:**1779–1790.
12. **Schaeffer, P., J. Millet, and J. P. Aubert.** 1965. Catabolic repression of bacterial sporulation. Proc. Natl. Acad. Sci. U.S.A. **54:**704–711.

13. **Shimotsu, H., F. Kawamura, Y. Kobayashi, and H. Saito.** 1983. Early sporulation gene *spo0F*: nucleotide sequence and analysis of gene product. Proc. Natl. Acad. Sci. U.S.A. **80:**658–662.
14. **Shuman, H. A., T. J. Silhavy, and J. R. Beckwith.** 1980. Labeling of proteins with β-galactosidase by gene fusions. J. Biol. Chem. **255:**168–174.
15. **Towbin, H., T. Staehlin, and J. Gordon.** 1979. Electrophoretic transfer of protein from polyacrylamide gels to nitrocellulose sheets: procedure and some applications. Proc. Natl. Acad. Sci. U.S.A. **76:**4350–4354.
16. **Weir, J., E. Dubnau, N. Ramakrishna, and I. Smith.** 1984. *Bacillus subtilis spo0H* gene. J. Bacteriol. **157:**405–412.
17. **Zuber, P., and R. Losick.** 1983. Use of a *lacZ* fusion to study the role of the *spo0* genes of *Bacillus subtilis* in developmental regulation. Cell **35:**275–283.

Cloning, Amplification, and Characterization of Sporulation Genes, Especially *spoIIG*, of *Bacillus subtilis*

YASUO KOBAYASHI AND HIROYUKI ANAGUCHI†

Department of Applied Biochemistry, Hiroshima University, Fukuyama 720, Japan

Sporulation of *Bacillus subtilis* consists of seven distinct morphological stages. Development of spores is regulated by about 40 sporulation genes randomly scattered on the chromosome (15, 29). Sequential expression of these sporulation genes is controlled at the levels of both transcription and translation (35). However, little is known of the details of the regulation of these genes. To elucidate the details of these regulatory mechanisms, we have cloned several sporulation genes by the prophage transformation method (23). This paper describes the cloning, amplification, and characterization of these sporulation genes, especially *spoIIG*.

EXPERIMENTAL PROCEDURES

B. subtilis strains, phages, and the prophage transformation method have been described previously (3, 16, 22). Construction and selection of recombinant plasmids and deletion plasmids were carried out as described previously (3). Transformation of bacterial cells with *B. subtilis* DNA and plasmid DNA was performed as described by Anaguchi et al. (2).

DNA sequence analysis was performed by the method of Sanger et al. (32). M13 manipulations were performed as described by Messing et al. (26). *Escherichia coli* JM105 and phages M13mp10 and M13mp11 (27) were obtained from the Amersham-Japan Co., Ltd. The polyacrylamide-urea gel electrophoresis system described by Sanger and Coulson (31) was used. Sequence analysis was performed in part by a computer.

Restriction endonucleases were purchased from Takara Shuzo Co., Ltd., Boehringer-Mannheim GmbH, and Nippon Gene Co., Ltd. *E. coli* DNA polymerase I (Klenow fragment) was purchased from Amersham-Japan Co., Ltd.

RESULTS AND DISCUSSION

Cloning of sporulation genes. Using the prophage transformation method (23), we constructed seven specialized transducing phages: φ105*spo0C* (T. Yamamoto and Y. Kobayashi,

unpublished data), φ105*spo0B* (16, 17), φ105*spo0F* (17, 22), φ105*spoIIC* (2), φ105*spoIIG* (3), φ105*spoIVC* (M. Fujita and Y. Kobayashi, Mol. Gen. Genet., in press), and φ105*spoVE* (36). These phages carry *B. subtilis* sporulation genes *spo0C* (and *spo0A*), *spo0B*, *spo0F*, *spoIIC*, *spoIIG*, *spoIVC*, and *spoVE*, respectively, and could transduce their respective asporogenous strains, which formed almost no spores (fewer than 10 spores per ml), to Spo⁺ at a frequency of 10^7 to 10^8 Spo⁺ transductants per ml. The sporulation efficiency of the asporogenous strains lysogenic for the respective transducing phages was normal (more than 10^8 spores per ml), indicating that the cloned *spo*⁺ genes function normally in the lysogen and that these *spo* mutations are recessive to the wild-type allele.

Analysis of the specialized transducing phage genomes by restriction endonuclease cleavage and transformation with various restriction fragments showed that sporulation genes *spo0B*, *spo0F*, *spoIIC*, *spoIIG*, *spoIVC*, and *spoVE* reside, respectively, on 1.4-, 1.3-, 2.7-, 6.2-, 2.4-, and 1.7-megadalton (MDa) *Eco*RI fragments derived from the *B. subtilis* chromosome. These fragments are inserted in or replaced one or more of the *Eco*RI-G, -I, and -E fragments of φ105 DNA (Fig. 1; 1). The sporulation gene *spo0C* (and *spo0A*) is present on a 5.3-MDa *Hin*dIII fragment derived from the *B. subtilis* chromosome (18; Yamamoto and Kobayashi, unpublished data).

To analyze the restriction fragments carrying various *spo* genes in detail, each *Eco*RI fragment was subcloned into the unique *Eco*RI site of the multicopy plasmid pUB110 (13); the *spo0C* *Hin*dIII fragment was subcloned in the *Hin*dIII site of plasmid pBD12 (14). The recombinant plasmids obtained were designated pUBSB, pUBSF, pSIIC, pSIIG, pSIVC, and pSVE. These plasmids carry the *spo0B*, *spo0F*, *spoIIC*, *spoIIG*, *spoIVC*, and *spoVE* genes, respectively. A recombinant plasmid having *spo0C*⁺ (and *spo0A*⁺) transforming activity could not be obtained, suggesting that the recombinant plasmid carrying the 5.3-MDa *spo0C* fragment cannot be maintained stably in the host cells or that the plasmid severely disturbs the growth of the recipient cells. A similar phenomenon was observed with the *spoIIG* fragment (6.2 MDa;

---

† On leave from the Department of Fermentation Technology, Hiroshima University, Higashihiroshima 724, Japan.

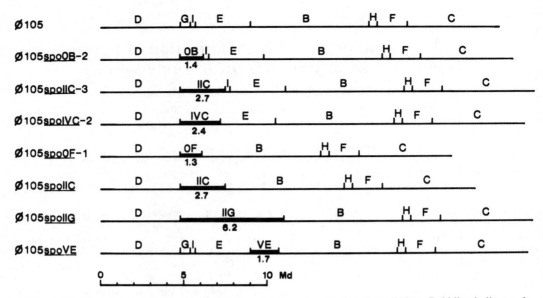

FIG. 1. *Eco*RI restriction maps of φ105 and its specialized transducing phage DNAs. Bold line indicates the cloned fragments derived from the *B. subtilis* chromosome. Figures indicate the sizes of the fragment in megadaltons (Md). The scale (in Md) is shown at the bottom.

see reference 3). We tried to construct a recombinant plasmid with pUB110 (3.0 MDa) and the 6.2-MDa *spoIIG* *Eco*RI fragment, but did not succeed. When the 6.2-MDa fragment was digested with *Bam*HI to a smaller fragment (2.25 MDa) which contained the *spoIIG*⁺ transforming activity, a recombinant plasmid, pSIIG, was obtained (3). However, with the *spo0C* fragment, digestion with *Bcl*I to a smaller fragment (1.5 MDa) which still had *spo0C*⁺ transforming activity and insertion into the *Bcl*I site of plasmid pBD9 (20; 5.4 MDa) did not give the appropriate recombinant plasmid. Although kanamycin-resistant, erythromycin-susceptible transformants were obtained, their recombinant plasmids carry a fragment smaller than the 1.5-MDa fragment.

Effect of the recombinant plasmids on sporulation. These six recombinant plasmids were introduced into the wild-type strain (*spo*⁺ *recE4*) and the respective asporogenous strains (*spo recE4*), and the effect of the recombinant plasmids on sporulation was examined. Table 1 shows that the recombinant plasmids can be classified into three types according to their effect on sporulation. The type I plasmid is pUBSB which does not inhibit sporulation of the wild-type strain and restores sporulation of strain UOT0436 (*spo0B136 recE4*) completely. The type II plasmids are pSIIC and pSIVC. They slightly inhibit the sporulation of the wild-type strain (10^7 spores per ml), and restore sporulation of strains 4305 (*spoIIC298 recE4*) and HU1214 (*spoIVC133 recE4*) to the level of 10^7 spores per ml. The type III plasmids are pUBSF,

pSIIG, and pSVE. Introduction of these plasmids into the wild-type strain does not inhibit vegetative growth, but inhibits sporulation drastically to the level of 10^3 to 10^4 spores per ml. Restoration of sporulation of the asporogenous strains UOT0294 (*spo0F221 recE4*), HU1002 (*spoIIG41 recE4*), and 5202 (*spoVE85 recE4*) is observed, but only to a level of 10^3 spores per ml. As mentioned earlier, when the specialized transducing phage φ105*spo0F*, φ105*spoIIG*, or φ105*spoVE* is lysogenized in the respective asporogenous mutant, the lysogen sporulates normally (10^8 spores per ml). This fact suggests that the inhibition of sporulation by type III plasmids may be due to the amplification of the *spo*⁺ fragment.

There are several possible explanations for the inhibition of sporulation by the recombinant plasmids: (i) excess production of the sporulation gene product, (ii) excess production of an unknown gene product which may exist on the cloned fragment, or (iii) titration of a regulatory protein or RNA polymerase, or both, necessary for the expression of other sporulation genes. Losick and his colleagues proposed, on the basis of analysis of deletion mutations, that amplification of the *spoVG* gene titrates a sporulation-specific regulatory protein that binds at or near the region of transcription (4). To clarify the cause of the inhibition of sporulation by the type III plasmids, detailed analysis of plasmid pSIIG was carried out in the following experiments.

Location of the *spoIIG* gene on pSIIG. To determine the location of the *spoIIG* gene and the inhibitory region on pSIIG, deletion derivatives of pSIIG were constructed by using restriction endonucleases *Pst*I, *Hin*dIII, and *Bcl*I (Fig. 2). Transformation experiments with pSIIG and its deletion plasmids were carried out with strain HU1001 (*spoIIG41*) as a recipient strain. pSIIGΔP16 and pSIIGΔH7 showed *spoIIG*⁺ transforming activity, but pSIIGΔH2 did not (Fig. 2). These results suggest that the *spoIIG41* mutation is located on the *Hin*dIII-A fragment within the *Pst*I-A fragment.

The sporulation ability of asporogenous strain HU1002 (*spoIIG41 recE4*) harboring pSIIG or its deletion plasmids is also shown in Fig. 2. Since the sporulation of strain HU1002 is restored only by pSIIG, pSIIGΔP16, or pSIIGΔB1, the *spoIIG*⁺ gene must be located in the region between the second *Pst*I site from the left (*Bam*HI) end and the *Bcl*I site located in the middle of the 2.25-MDa *Eco*RI-*Bam*HI fragment. Figure 2 also shows the inhibition of the sporulation of the wild-type strain 4309 (*spo*⁺ *recE4*) by pSIIG or its deletion plasmids. Since the inhibition was not observed in cells harboring pSIIGΔP9 or pSIIGΔB1, this suggests that the inhibitory region is between the middle *Bcl*I site and the first *Pst*I site from the right (*Eco*RI)

TABLE 1. Effect of the recombinant plasmid on the sporulation of wild-type and asporogenous strains of *B. subtilis*[a]

Plasmid	Sporulation (spores/ml) of:	
	spo⁺ strain[b]	*spo* strain[c]
None	10⁸	<10
pUB110	10⁸	<10
pUBSB(*spo0B*⁺)	10⁸	10⁸
pUBSF(*spo0F*⁺)	10³	ND
pSIIC(*spoIIC*⁺)	10⁷	10⁷
pSIIG(*spoIIG*⁺)	10³	10⁴
pSIVC(*spoIVC*⁺)	10⁷	10⁷
pSVE(*spoVE*⁺)	10⁴	10³

[a] Cells were cultivated for 24 h at 37°C in 5 ml of Schaeffer sporulation medium (33) supplemented with 5 μg of kanamycin per ml (in strains without plasmid, kanamycin was omitted). Heat-resistant spores were counted by heating the cells for 10 min at 80°C and plating them on NB plates (8 g of Difco nutrient broth, 15 g of agar, and 1,000 ml of water).
[b] Strain 4309 (*metB5 nonB1 recE4*) was used as a host strain for plasmids pSIIC, pSIIG, and pSVE. Strain UOT0277 (*hisA1 metB5 nonB1 recE4*) was used for pUBSB and pUBSF, and strain HU1018 (*trpC2 metB51 nonB1 recE4*) was used for pSIVC.
[c] Asporogenous strains UOT0436 (*pheA1 recE4 spo0B136*), 4305 (*metB5 nonB1 recE4 spoIIC298*), HU1002 (*metB5 nonB1 recE4 spoIIG41*), HU1214 (*trpC2 metB51 nonB1 recE4 spoIVC133*), and 5202 (*metB5 nonB1 recE4 spoVE85*) were used as host strains for plasmids pUBSB, pSIIC, pSIIG, pSIVC, and pSVE, respectively. ND, Not determined.

end. However, plasmid pSIIGΔH3, which possesses this region, exhibits only slight inhibition. This is contradictory to the above conclusion and will be discussed later.

Failure to detect the *spoIIG* gene product. Maxicell and minicell methods, and an in vitro transcription-translation coupled system (M. Okamoto et al., Agr. Biol. Chem., in press) were used in attempts to identify the *spoIIG* gene product, but without success. Since the products of other sporulation genes (*spo0F* [34], *spo0B* [Y. Kobayashi et al., *in* A. T. Ganesan and J. A. Hoch, ed., *Genetics and Biotechnology of Bacilli*, in press], and *spoIIC* [Kobayashi et al., in press]) were detected by these methods, the failure to detect the *spoIIG* gene product may reflect the very low expression of the gene at the level of transcription or translation. J. Szulmajster et al. (Abstracts of the U.S.-France Bacillus Meeting, Paris, 1984) also were unable to detect the *spoIIG* gene product by the maxicell method.

Nucleotide sequence analysis. The results shown in Fig. 2 indicate that the *spoIIG* gene is located on the left half of the 2.25-MDa *Bam*HI-*Eco*RI fragment (between the middle *Bcl*I site and the second *Pst*I site from the left [*Bam*HI] end). Using the *Bam*HI-*Pst*I fragment obtained by the partial digestion of the 2.25-MDa fragment (see Fig. 3), we determined the sequence of the DNA covering the *spoIIG* gene. The sequencing strategy is presented in Fig. 3. The nucleotide sequence of DNA fragments was determined by Sanger's dideoxynucleotide method (32) with the use of the single-stranded bacteriophage cloning vectors M13mp10 and M13mp11 (27). The 2,028-base-pair nucleotide sequence obtained was analyzed by a computer, and possible open reading frames were determined. Only one large continuous reading frame (ORF 2) which fits the possible location of the *spoIIG* gene and *spoIIG41* locus was found in reading frame no. 4. This frame consists of 283 codons, as shown in Fig. 4. However, there is no Shine-Dalgarno sequence in front of the first initiation codon ATG (538 to 540), although there is a Shine-Dalgarno sequence in front of the third ATG codon (670 to 672) which has ΔG of −15.0 kcal (ca. −62.8 kJ)/mol. Therefore, we suggest that the *spoIIG* gene starts at the third ATG codon, ends at the 283rd codon (TAA), and consists of 239 amino acid residues (Fig. 4). To confirm this assumption, further analyses, such as the determination of the N-terminal amino acid sequence of purified *spoIIG* gene product, will be necessary. In front of ORF 2, there is another large open reading frame (ORF 1). The termination codon TAG (654 to 656) of ORF 1 is located just in front of the ATG codon of the *spoIIG* gene (670 to 672) and shares two

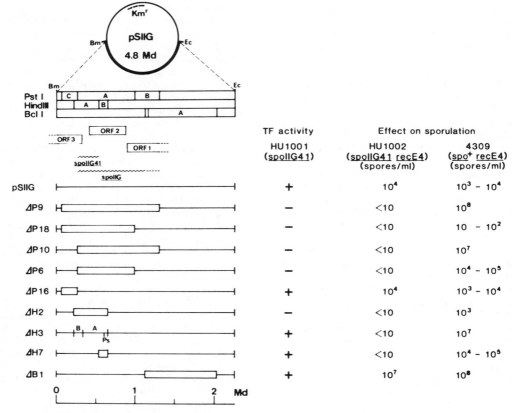

FIG. 2. Structure of pSIIG and its deletion plasmids, their *spo*⁺ transforming activity, and their effect on sporulation. pSIIG is a recombinant plasmid carrying the 2.25-MDa *Eco*RI-*Bam*HI fragment in the region between the *Eco*RI and *Bam*HI sites of pUB110. Endonuclease restriction sites for *Eco*RI, *Bam*HI, and *Pst*I are indicated as Ec, Bm, and Ps, respectively. For construction of deletion plasmids, pSIIG DNA (1 μg) was partially digested with *Pst*I or *Hin*dIII, or completely digested with *Bcl*I. After religation of the digested DNA, the ligated DNA was used to transform *B. subtilis* to kanamycin resistance. Plasmid DNAs from the transformants were analyzed by the respective restriction enzymes. The boxes indicate the deletion regions of the pSIIG derivatives. Wavy lines indicate the possible location of the *spoIIG41* marker and the *spoIIG* gene, and the dashed line shows the inhibitory region. ORF indicates the open reading frame determined by nucleotide sequence analysis (see later section of this paper). Transforming activity of these plasmids in transforming strain HU1001 (*spoIIG41*) to *spo*⁺ is shown by +. Effects of the recombinant plasmids on the sporulation of wild-type strain 4309 (*spo*⁺ *recE4*) and asporogenous strain HU1002 (*spoIIG41 recE4*) were determined by counting heat-resistant spores as described in footnote *a* of Table 1.

bases, A and G at 655 and 656, with the Shine-Dalgarno sequence of the *spoIIG* gene (Fig. 4).

There is no consensus promoter sequence in the upstream region of the *spoIIG* gene, except a promoterlike sequence at 547 to 552, TTGACG, and one at 568 to 573, AAAAAT. However, these sequences are separated by only 15 base pairs and may not be utilized by sigma-43 RNA polymerase, since the preferred spacing between "−10" and "−35" sequences for sigma-43 RNA polymerase is 17 or 18 base pairs (28). Therefore, it is likely that the *spoIIG* gene is transcibed by the read through from the preceding gene. After the *spoIIG* gene, there is no simple terminator structure (30) having a GC-

rich segment of dyad symmetry centered about 20 nucleotides before the stop site, which is followed by a region encoding four to eight uridine residues located near the 3' terminus of the transcript. However, there is a stem-and-loop structure having a ΔG of −28.6 kcal (ca. −119.7 kJ)/mol (Fig. 5). We do not know whether this structure functions as a terminator.

It is interesting to note that another large open reading frame (ORF 3) follows the *spoIIG* gene. This open reading frame also does not have a consensus promoter sequence for sigma-43 RNA polymerase. Therefore, it is possible that the transcription of this gene depends on the read through from the upstream gene. These results

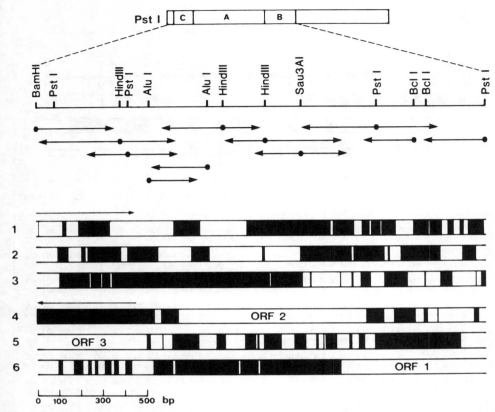

FIG. 3. Restriction sites and sequencing strategy of the cloned DNA fragment. Upper part: the restriction sites used in the sequencing experiment. Arrows show the direction and extent of sequencing in each experiment. Lower part: 1 to 3 indicate three reading frames from the *Bam*HI end to the *Pst*I end; 4 to 6 indicate three reading frames from the *Pst*I end to the *Bam*HI end. Open boxes are open reading frames (ORF 1, 2, and 3), and closed boxes are noncoding regions.

suggest that the *spoIIG* gene consists of a polycistronic operon whose transcription starts at the far upstream region of the *Pst*I end. To confirm this assumption, the upstream region of the *Pst*I end must be analyzed.

However, since the distance between the 3' end of the *spoIIG* gene and the initiation codon of the succeeding gene is long (113 base pairs), it is possible that an unidentified promoter is present in front of the succeeding gene. The initiation codon GTG (1,529 to 1,531) is preceded by a Shine-Dalgarno sequence which has a ΔG of −14.4 kcal (ca. −60.3 kJ)/mol and exists in the stem of the terminatorlike structure of the *spoIIG* gene (Fig. 5). Similar stem-and-loop structures which sequester Shine-Dalgarno sequences are present in the chloramphenicol-inducible *cat* gene of plasmids pC194 (6) and pL703 (10). Therefore, it is possible that this structure has a function in regulating the expression of the succeeding gene. We also found a short open reading frame (1,497 to 1,601) span-

ning the terminatorlike structure and the succeeding gene. This frame consists of 34 amino acid residues and is preceded by a Shine-Dalgarno sequence (ΔG, −12.8 kcal [ca. −53.6 kJ]/mol) existing in the stem of the terminatorlike structure (Fig. 5). The initiation codon TTG (1,497 to 1,499) exists in the loop of the terminatorlike structure, and its C terminus exists in the N-terminal region of the succeeding gene. We do not know whether this sequence has a special regulatory role and is translated into protein.

Possible function of the *spoIIG* gene product. As mentioned earlier, we could not detect the *spoIIG* gene product by the maxicell method, the minicell method, or an in vitro transcription-translation coupled system. However, the nucleotide sequence of the *spoIIG* gene indicates that the molecular weight of the *spoIIG* gene product is 27,592, consisting of 239 amino acid residues. The amino acid composition of the *spoIIG* protein is shown in Table 2. The protein is relatively

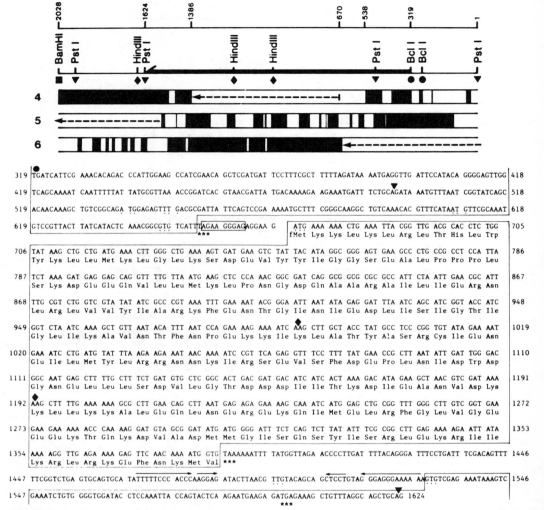

FIG. 4. Nucleotide sequence of the *spoIIG* gene. In general, the sequence was determined on both strands, and where it was determined on one strand, the sequence reactions were repeated at least twice. Dotted arrows indicate the direction of the *spoIIG* gene from N terminus to C terminus and that of the preceding and succeeding genes. Nucleotide base positions are numbered from 5' to 3' in the coding strand, starting with number one at the right end (*Pst*I site); the Shine-Dalgarno sequence (655 to 663) is boxed. Sequences exhibiting dyad symmetry are indicated by arrows. The sequence of nucleotides 1 to 318 and 1,625 to 2,028 will be presented elsewhere.

low in small aliphatic and hydroxylated amino acids and high in basic and hydrophobic residues. The protein contains 26 lysine and 15 arginine residues, and the calculated isoelectric point is about 9.0, indicating that the protein is basic. Table 2 also shows the amino acid composition of the *rpoD* gene product of *E. coli* (5).

It is interesting to note that a part of the amino acid sequence of the *spoIIG* protein is homologous to that of the sigma factor (*rpoD* gene product) of *E. coli* RNA polymerase (5). About 50% homology (28 of 59) was observed between amino acids 64 to 122 of the *spoIIG* protein and

amino acids 381 to 439 of the sigma factor (Fig. 6). In addition, nine amino acids belong to similar groups. This region of the sigma factor seems to be functionally very important, since the *htpR* gene product, which is a new sigmalike factor of *E. coli* and is involved in the induction of heat-shock genes (12), has an amino acid sequence homologous to this region of the sigma factor (24). Furthermore, the amino acid sequence of sigma-43 of *B. subtilis* is homologous with this region of the *E. coli* sigma factor (M. Gitt et al., Abstract of the Ninth International Spores Conference, Asilomar, Calif., 1984).

These results suggest that the *spoIIG* protein might have a sigmalike function. It is interesting, however, that the amino acid sequence of the *spoIIG* protein has almost no homology with that of sigma-gp28 (7), which is a gene product of bateriophage SPO1 and functions as a sigma-like factor to recognize the promoters of the middle genes of SPO1.

Several minor sigmalike factors, sigma-28, sigma-29, sigma-32, and sigma-37, are known in *B. subtilis* (9, 21, 25) in addition to a major sigma factor, sigma-43. We cannot compare the amino acid sequence of the *spoIIG* protein with that of these minor sigmalike factors, since the amino acid sequences of these minor sigmalike factors have not yet been determined. If the *spoIIG* protein is one of these sigmalike factors, it is unlikely that sigma-28 is the *spoIIG* protein, since transcription from the promoters recognized by sigma-28 is rapidly shut off after the first hour of sporulation (11), whereas the transcription of the *spoIIG* gene itself seems to occur during early stages of sporulation (Szulmajster et al., Abstract of the U.S.-France Bacillus Meeting, Paris, 1984) and the *spoIIG* product itself seems to function at an early stage of sporulation (37). The calculated molecular weights of *E. coli* sigma factor and related proteins, such as *nus* gene product (19) and sigma-43 (Gitt et al., Abstract of the Ninth International Spores Conference, Asilomar, Calif., 1984) are 78.1, 78.8, and 78.2%, respectively, of that estimated by sodium dodecyl sulfate-polyacrylamide gel electrophoresis. If this anomaly is caused by an unusually high amount of charged residues in these proteins, the expected molecular weight of the *spoIIG* protein by sodium dodecyl sulfate-polyacrylamide gel electrophoresis may be different than calculated from the nucleic acid sequence.

Possible cause of the inhibition of sporulation by pSIIG. Plasmid pSIIGΔB1 restores the sporulation of strain HU1002 (*spoIIG41 recE4*) almost completely (10^7 spores per ml), suggesting that the pSIIGΔB1 produces *spoIIG* gene product. Furthermore, pSIIGΔB1 does not inhibit the sporulation of strain 4309 (*spo⁺ recE4*) (see Fig. 2). On the other hand, pSIIGΔP6 lacks the structural gene for *spoIIG*, but inhibits the sporulation of strain 4309. These results suggest that the inhibition of sporulation of the *spo⁺* strain by pSIIG and its deletion plasmids is not caused by the *spoIIG* gene product. When the effects of the pSIIGΔP9 and pSIIGΔP18 on sporulation of the *spo⁺* strain are compared, pSIIGΔP9 does not inhibit sporulation, but pSIIGΔP18 severely inhibits sporulation. Similarly, when the effects of pSIIGΔP10 and pSIIGΔP6 on sporulation are compared, pSIIGΔP10 shows only slight inhibition, whereas pSIIGΔP6 shows strong inhibi-

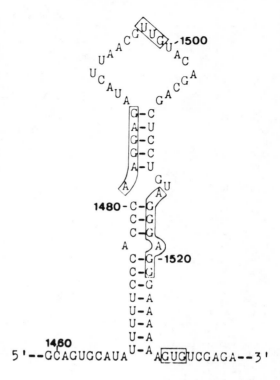

$$\Delta G = -28.6 \text{ kcal/mol}$$

FIG. 5. Stem-and-loop structure found between the *spoIIG* gene and the succeeding gene. Shine-Dalgarno sequences and initiation codons are boxed.

tion. These results suggest that the *Pst*I-B region within the *Bcl*I-A fragment (Fig. 2) has a close relation to inhibition of sporulation.

This region contains the 5'-flanking region of the *spoIIG* gene and a part of the preceding gene (ORF 1). Therefore, the inhibition of sporulation may be caused either by the upstream region of the *spoIIG* gene, as in the case of the *spo0F* gene (H. Yoshikawa et al., this volume) and *spoVG* (4), or by the product of the preceding gene. We could not find any significant signals, such as dyad symmetry and AT-rich regions, in the upstream region of the *spoIIG* gene. On the other hand, comparison of the inhibition of sporulation by pSIIGΔP18 and pSIIGΔP6 shows that inhibition by pSIIGΔP18 is more severe than by pSIIGΔP6, although both plasmids possess the *Pst*I-B region. If this difference in inhibition is due to the difference in the length of the *Bam*HI-end region, the cause of the sporulation inhibition might be the fusion protein synthesized from the N-terminal region of the preceding gene and the *Bam*HI-end region. These observations suggest that the inhibition of sporulation by pSIIG is more likely caused by

TABLE 2. Amino acid composition of *spoIIG* and *ropD* gene products[a]

Amino acid		From sequence data				Avg protein (8) (mol%)
		spoIIG		*rpoD* (5)		
		Residues	Mol%	Residues	Mol%	
Ala	A	11	4.6	49	8.0	8.6
Asx	B (D+N)	(29)	(12.1)	(74)	(12.1)	(9.8)
Cys	C	1	0.4	3	0.5	2.9
Asp	D	15	6.3	55	9.0	5.5
Glu	E	23	9.6	70	11.4	6.0
Phe	F	5	2.1	15	2.5	3.6
Gly	G	12	5.0	24	3.9	8.4
His	H	1	0.4	9	1.5	2.0
Ile	I	23	9.6	43	7.0	4.5
Lys	K	26	10.9	34	5.6	6.6
Leu	L	34	14.2	54	8.8	7.4
Met	M	8	3.3	25	4.1	1.7
Asn	N	14	5.9	19	3.1	4.3
Pro	P	6	2.5	19	3.1	5.2
Gln	Q	6	2.5	30	4.9	3.9
Arg	R	15	6.3	46	7.5	4.9
Ser	S	11	4.6	29	4.7	7.0
Thr	T	8	3.3	38	6.2	6.1
Val	V	11	4.6	34	5.6	6.6
Trp	W	2	0.8	4	0.7	1.3
Tyr	Y	7	2.9	13	2.1	3.4
Glx	Z (E+Q)	(29)	(12.1)	(100)	(16.3)	(9.9)
Small aliphatic (A+G)			9.6		11.9	16.9
Hydroxyl (S+T)			7.9		10.9	13.1
Acidic (D+E)			15.9		20.4	11.6
Acidic+acid amide (B+Z)			24.3		28.4	19.8
Basic (K+R+H)			17.6		14.5	13.5
Hydrophobic (L+V+I+M)			31.8		25.4	20.2
Aromatic (F+Y+W)			5.9		5.2	8.3
Charged (D+E+K+R+H)			33.5		34.9	25.1

[a] Other characteristics of the *spoIIG* and *rpoD* gene products, respectively, were as follows: residues, 239 and 613; unmodified molecular weight, 27,592 and 70,249; mean residue molecular weight, 115.4 and 114.6; and isoelectric point, 9.0 (calculated) and 4.4 to 5.1.

the preceding gene product than by the 5'-flanking region of the *spoIIG* gene. This will be clarified by deletion or insertional inactivation of the preceding gene.

When the effects of pSIIGΔH2 and pSIIGΔH3 on sporulation are compared, pSIIGΔH3 shows only slight inhibition, although it has the *Pst*I-B region. Since the only difference between pSIIGΔH2 and pSIIGΔH3 is the presence of the inverted *Hin*dIII-A and -B fragments in pSIIGΔH3, the cause of the inhibition may be in the orientation of this fragment. At present, we do not have any explanation for this result, but if an unidentified promoter of the succeeding gene (ORF 3) is present in this fragment, then in pSIIGΔH3 the transcription from this promoter (from left to right) might affect the production of the preceding gene (ORF 1) product. As discussed earlier, if the inhibition of sporulation is caused by the preceding gene product, the decrease in this product will reduce the inhibitory effect of pSIIGΔH3 on sporulation.

If the *spoIIG* gene and the preceding gene make up a polycistronic operon, it is likely that the function of the preceding gene product may be closely related to that of the *spoIIG* gene product, as observed in *E. coli* (*rpoD* and *dnaG*) and *B. subtilis* (*rpoD* and *dnaE*; Gitt et al., Abstract of the Ninth International Spores Conference, Asilomar, Calif., 1984) operons. Obviously, it is of great interest to learn which genes are present in this operon. Determination of the nucleotide sequence of the preceding and succeeding genes is now in progress in our laboratory.

SUMMARY

Recombinant plasmids harboring sporulation genes *spo0B*, *spo0F*, *spoIIC*, *spoIIG*, *spoIVC*, and *spoVE* of *B. subtilis* have been constructed and can be classified into three types according to their effect on sporulation of wild-type and asporogenous strains. One of the recombinant plasmids, pSIIG, was studied in detail.

```
spoIIG   64  GluArgAsnLeuArgLeuValValTyrIleAlaArgLysPheGluAsnThrGlyIleAsn  83
             =   = = = = = =  *   = = * = *    =    = = * *
rpoD    381  GluAlaAsnLeuArgLeuValIleSerIleAlaLysLysTyrThrAsnArgGlyLeuGln 400
               =  =     =  *  = =   = *    =        = =
htpR     55  LeuSerHisLeuArgPheValValHisIleAlaArgAsnTyrAlaGlyTyrGlyLeuPro  74
```

```
spoIIG   84  IleGluAspLeuIleSerIleGlyThrIleGlyLeuIleLysAlaValAsnThrPheAsn 103
                = = =    =   = = = *  = = =              =
rpoD    401  PheLeuAspLeuIleGlnGluGlyAsnIleGlyLeuMetLysAlaValAspLysPheGlu 420
                = = =  = = = = =  = = =  =  = = = =    *   = =
htpR     75  GlnAlaAspLeuIleGlnGluGlyAsnIleGlyLeuMetLysAlaValArgArgPheAsn  94
```

```
spoIIG  104  ProGluLysLysIleLysLeuAlaThrTyrAlaSerArgCysIleGluAsnGluIle    122
                *        =       = = *     =    =   *   =
rpoD    421  TyrArgArgGlyTyrLysPheSerThrTyrAlaThrTrpTrpIleArgGlnAlaIle   439
                = =  =  *    * * =  = = =   =     = = *     =
htpR     95  ProGluValGlyValArgLeuValSerPheAlaValHisTrpIleLysAlaGluIle   113
```

FIG. 6. Homology between the *spoIIG*, *htpR*, and *rpoD* proteins. The amino acids of the *spoIIG* and *htpR* proteins that are homologous with those of the *rpoD* protein are marked with equal signs (=). Amino acids of the two proteins that belong to groups of similar chemical nature are marked with asterisks (*). The numbers indicate the amino acid residues from the N terminus.

Recombinant plasmid pSIIG harboring sporulation gene *spoIIG* strongly inhibits sporulation of the wild-type strain of *B. subtilis*. Analysis using plasmids containing deletions of pSIIG showed that the inhibition is not due to the *spoIIG* gene product, but is due to either the upstream region of the *spoIIG* gene or the gene product of the preceding gene.

Nucleotide sequence analysis of the *spoIIG* fragment shows an open reading frame of 239 amino acid residues (calculated molecular weight, 27,592). A part of the amino acid sequence of the *spoIIG* gene product is homologous to that of sigma factors of *E. coli* and *B. subtilis* and to that of the *htpR* gene product which is a sigmalike factor newly found in *E. coli*. This result suggests that the *spoIIG* gene product may be a protein with a sigmalike function.

We thank J. Hoch, M. Young, and the *Bacillus* Genetic Stock Center for providing us with bacterial strains. We also thank A. Ishihama, H. Saito, F. Kawamura, and H. Shimotsu for useful suggestion and stimulating discussions.

This work was supported by grant 59214016 from the Ministry of Education of Japan.

LITERATURE CITED

1. **Anaguchi, H., S. Fukui, and Y. Kobayashi.** 1984. Revised restriction maps of *Bacillus subtilis* bacteriophage φ105 DNA. J. Bacteriol. **159:**1080–1082.
2. **Anaguchi, H., S. Fukui, H. Shimotsu, F. Kawamura, H. Saito, and Y. Kobayashi.** 1984. Cloning of sporulation gene *spoIIC* in *Bacillus subtilis*. J. Gen. Microbiol. **130:**757–760.
3. **Ayaki, H., and Y. Kobayahi.** 1984. Cloning of sporulation gene *spoIIG* in *Bacillus subtilis*. J. Bacteriol. **158:**507–512.
4. **Banner, C. D. B., C. P. Moran, Jr., and R. Losick.** 1983. Deletion analysis of a complex promoter for a developmentally regulated gene from *Bacillus subtilis*. J. Mol. Biol. **168:**351–365.
5. **Burton, E., R. R. Burgess, J. Lin, D. Moore, S. Holder, and C. A. Gross.** 1981. The nucleotide sequence of the cloned *rpoD* gene for the RNA polymerase sigma subunit from *E. coli* K12. Nucleic Acids Res. **9:**2889–2903.
6. **Byeon, W.-H., and B. Weisblum.** 1984. Post-transcriptional regulation of chloramphenicol acetyl transferase. J. Bacteriol. **158:**543–550.
7. **Costanzo, M., and J. Pero.** 1983. Structure of a *Bacillus subtilis* bacteriophage SPO1 gene encoding a RNA polymerase σ factor. Proc. Natl. Acad. Sci. U.S.A. **80:**1236–1240.
8. **Dayhoff, M. O., L. T. Hunt, and S. Hurst-Calderone.** 1978. Composition of proteins, p. 363–369. *In* M. O. Dayhoff (ed.), Atlas of protein sequence and structure, vol. 5, suppl. 3. Maryland National Biomedical Research Foundation, Silver Spring.
9. **Doi, R. H.** 1982. Multiple RNA polymerase holoenzymes exert transcriptional specificity in *Bacillus subtilis*. Arch. Biochem. Biophys. **214:**772–781.
10. **Duvall, E. J., D. M. Williams, P. S. Lovett, C. Rudolph, N. Vasantha, and M. Guyer.** 1983. Chloramphenicol-inducible gene expression in *Bacillus subtilis*. Gene **24:**171–177.
11. **Gilman, M. Z., and M. J. Chamberlin.** 1983. Developmental and genetic regulation of *Bacillus subtilis* genes transcribed by σ28-RNA polymerase. Cell **35:**285–293.
12. **Grossman, A. D., J. W. Erickson, and C. A. Gross.** 1984. The *htpR* gene product of *E. coli* is a sigma factor for heat shock promoters. Cell **38:**383–390.
13. **Gryczan, T. J., S. Contente, and D. Dubnau.** 1978. Characterization of *Staphylococcus aureus* plasmids introduced by transformation into *Bacillus subtilis*. J. Bacteriol. **134:**318–329.
14. **Gryczan, T., A. G. Shivakumar, and D. Dubnau.** 1980. Characterization of chimeric plasmid cloning vehicles in *Bacillus subtilis*. J. Bacteriol. **141:**246–253.
15. **Henner, D. J., and J. A. Hoch.** 1980. The *Bacillus subtilis* chromosome. Microbiol. Rev. **44:**57–82.
16. **Hirochika, H., Y. Kobayashi, F. Kawamura, and H. Saito.** 1981. Cloning of sporulation gene *spo0B* of *Bacillus subtilis* and its genetic and biochemical analysis. J. Bacteriol. **146:**494–505.
17. **Hirochika, H., Y. Kobayashi, F. Kawamura, and H. Saito.** 1982. Construction and characterization of φ105 special-

ized tranducing phages carrying sporulation genes *spo0B* and *spo0F* of *Bacillus subtilis*. J. Gen. Appl. Microbiol. **28**:225–229.

18. **Ikeuchi, T., J. Kudoh, and K. Kurahashi.** 1983. Cloning of sporulation genes *spo0A* and *spo0C* of *Bacillus subtilis* onto ρ11 temperate bacteriophage. J. Bacteriol. **154**:988–991.

19. **Ishii, S., M. Ihara, T. Maekawa, Y. Nakamura, H. Uchida, and F. Imamoto.** 1984. The nucleotide sequence of the cloned *nusA* gene and its flanking region of *Escherichia coli*. Nucleic Acids Res. **12**:3333–3342.

20. **Jalanko, A., I. Palva, and H. Söderlund.** 1981. Restriction maps of plasmids pUB110 and pBD9. Gene **14**:325–328.

21. **Johnson, W. C., C. P. Moran, Jr., and R. Losick.** 1983. Two RNA polymerase sigma factors from *Bacillus subtilis* discriminate between overlapping promoters for a developmentally regulated gene. Nature (London) **302**:800–804.

22. **Kawamura, F., H. Saito, H. Hirochika, and Y. Kobayashi.** 1980. Cloning of sporulation gene, *spo0F*, in *Bacillus subtilis* with ρ11 phage vector. J. Gen. Appl. Microbiol. **26**:345–355.

23. **Kawamura, F., H. Saito, and Y. Ikeda.** 1979. A method for construction of specialized transducing phage ρ11 of *Bacillus subtilis*. Gene **5**:87–91.

24. **Landick, R., V. Vaughn, E. T. Lau, R. A. VanBogelen, J. W. Erickson, and F. C. Neidhardt.** 1984. Nucleotide sequence of the heat shock regulatory gene of *E. coli* suggests its protein product may be a transcriptional factor. Cell **38**:175–182.

25. **Losick, R., and J. Pero.** 1981. Cascades of sigma factors. Cell **25**:582–584.

26. **Messing, J., R. Crea, and P. H. Seeburg.** 1981. A system for shotgun DNA sequencing. Nucleic Acids Res. **9**:309–321.

27. **Messing, J., and J. Vieira.** 1982. A new pair of M13 vectors for selecting either DNA strand of double-digest restriction fragments. Gene **19**:269–276.

28. **Moran, C. P., Jr., N. Lang, S. F. J. LeGrice, G. Lee, M. Stephens, A. L. Sonenshein, J. Pero, and R. Losick.** 1982. Nucleotide sequences that signal the initiation of transcription and translation in *Bacillus subtilis*. Mol. Gen. Genet. **186**:339–346.

29. **Piggot, P. J., and J. G. Coote.** 1976. Genetic aspects of bacterial endospore formation. Bacteriol. Rev. **40**:908–962.

30. **Platt, T., and D. G. Bear.** 1983. Role of RNA polymerase, ρ factor, and ribosomes in transcription termination, p. 123–161. *In* J. Beckwith, J. Davies, and J. A. Gallant (ed.), Gene function in prokaryotes. Cold Spring Harbor Laboratory, Cold Spring Harbor, N.Y.

31. **Sanger, F., and A. R. Coulson.** 1978. Use of thin acrylamide gels for DNA sequencing. FEBS Lett. **87**:107–110.

32. **Sanger, F., S. Nicklen, and A. R. Coulson.** 1977. DNA sequencing with chain-terminating inhibitors. Proc. Natl. Acad. Sci. U.S.A. **74**:5463–5467.

33. **Schaeffer, P., H. Ionesco, A. Ryter, and G. Balassa.** 1965. La Sporulation de *Bacillus subtilis*: ètude génétique et physiologique. Colloq. Intern. C.N.R.S. **124**:553–563.

34. **Shimotsu, H., F. Kawamura, Y. Kobayashi, and H. Saito.** 1983. Early sporulation gene *spo0F*: nucleotide sequence and analysis of gene product. Proc. Natl. Acad. Sci. U.S.A. **80**:658–662.

35. **Sonenshein, A. L., and K. M. Campbell.** 1978. Control of gene expression during sporulation, p. 179–192. *In* G. Chambliss and J. C. Vary (ed.), Spores VII. American Society for Microbiology, Washington, D.C.

36. **Yamada, H., H. Anaguchi, and Y. Kobayashi.** 1983. Cloning of the sporulation gene *spoVE* in *Bacillus subtilis*. J. Gen. Appl. Microbiol. **29**:477–486.

37. **Young, M.** 1976. Use of temperature-sensitive mutants to study gene expression during sporulation in *Bacillus subtilis*. J. Bacteriol. **126**:928–936.

Characterization of Cloned Genes and Their Regulatory Signals

Expression of Cloned Protease Genes in *Bacillus subtilis*

DENNIS J. HENNER,[1] MARIA YANG,[1] LOUISE BAND,[1] AND JAMES A. WELLS[2]

Departments of Molecular Biology[1] and Biocatalysis,[2] Genentech, Inc., South San Francisco, California 94080

The bacilli produce a number of commercially important enzymes. These include amylases, penicillinases, and proteases (6, 20). The proteases produced by bacilli are of two major types. One is a serine protease which is inhibitable by phenylmethylsulfonyl fluoride, has an alkaline pH optimum, and is called alkaline protease or subtilisin (11, 13). Although subtilisin is genetically unrelated to the pancreatic serine proteases, X-ray crystallographic studies indicate that it independently evolved a similar active site structure (11). Detailed stuctural and kinetic information on subtilisin makes it an excellent candidate for enzyme engineering by site-directed mutagenesis. The other protease is a metalloenzyme which is inhibitable by EDTA, has a neutral pH optimum, and is called neutral protease (14). Neutral protease from *Bacillus subtilis*, although not structurally well characterized, shares similar specificity and metal-binding properties with thermolysin (14).

Many species of bacilli produce these two types of proteases, and these enzymes show considerable similarities among the different species. The role of these proteases in the physiology of the bacilli has been the subject of much speculation. Both proteases are produced primarily during stationary phase, and strains that are deficient in the sporulation process often show much reduced levels of protease production (8, 19). We thought that the isolation of the genes for these proteases would be useful for a number of purposes. First, we could presumably place the genes for these proteases on high-copy plasmids and overexpress their products. This method should lead to hyperproducing strains more quickly than traditional methods of mutagenesis and screening. Second, the availability of the isolated genes would allow one to make directed changes in the sequence of the gene, leading to an elegant understanding of structure-function relationships. Such studies might be expected to lead to the creation of enzymes with improved properties. Third, the regulation of these genes could be studied at the molecular level, leading to an understanding of their apparent relationship with the sporulation process. Fourth, the isolated genes could be used to create in vitro mutations which could then be transferred back into the chromosome. Such mutations could be used to precisely determine the role(s) of these proteases in cell growth and physiology.

We have previously reported the isolation of the *B. amyloliquefaciens* subtilisin gene and the *B. subtilis* subtilisin and neutral protease genes (22, 26, 29). These isolated genes have been used for a number of studies, as outlined above (22, 29; J. A. Wells and D. A. Estell, personal communication; E. Ferrari, S. M. A. Howard, and J. A. Hoch, this volume). This report details some of our approaches to optimizing the yields of protease production by using cloned genes in *B. subtilis*. To increase subtilisin expression, we replaced the natural promoter with *B. amyloliquefaciens* amylase promoter. Site-directed mutagenesis was then used to make changes in the amylase promoter to further increase subtilisin expression. To increase neutral protease expression, we used the promoter region from a neutral protease hyperproducing strain of *B. subtilis* to express the neutral protease gene on a high-copy plasmid.

EXPERIMENTAL PROCEDURES

Bacterial strains and plasmids. *Escherichia coli* MM294 (F⁻ *supE44 endA1 thi-1 hsdR4*) was used for construction of the pBR322 and pUC derivatives, and *E. coli* JM101 [Δ(*lac-pro*) *supE thi* F′ *traD36 proAB lacIq Z*ΔM15] was used as a host for M13 (1, 15). M13 derivatives mp10 and mp11 were used for sequencing and site-directed mutagenesis (16). *B. subtilis* BG2036 (*aprE*Δ684 *nprE*Δ522) was developed in this laboratory (29). *B. subtilis* GSY264 (*mtr-264*) was obtained from the Bacillus Genetic Stock Center (BGSC 1A72). *B. subtilis* NT02 (*purB6 metB5 trpB3* Strʳ *nprR2 nprE02*) and NT17 (*purB6 metB5 trpB3* Strʳ *nprR2 nprE17*) were kindly provided by K. Yamane (23). Transformation, selection of recombinants, and construction of plasmids were done as previously described (29). Plasmids pBR322, pBS7, pBS42, pS4.5, pNPR10, and pUC13 have been previously described (2, 4, 16, 26, 29, 30). Screening of colonies containing recombinant plasmids was accomplished by hybridization techiques after transfer of colonies to nitrocellulose (7).

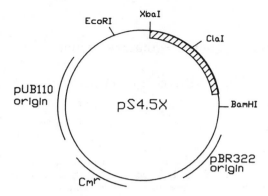

FIG. 1. Diagram of plasmid pS4.5X. Plasmid pS4.5X is derived from plasmid pS4 (26). The unique *Eco*RI site is 95 bp upstream from the initiation codon and was placed there by Bal-31 deletion and ligation of an *Eco*RI linker as described (26). The hatched double lines indicate the subtilisin structural gene; the direction of transcription is clockwise. The antibiotic resistance gene and origins of replication of the plasmid are indicated; however, they are not to scale. The *Xba*I site was introduced by mutagenesis in M13. Indicated below is the 32-bp synthetic fragment used for the mutagenesis and the wild-type sequence that it replaced. Differences are underlined, and the ribosome binding site (RBS) and initiation codon (met) are indicated.

```
        <  RBS  >              met
GAAAAAAGGAGGAGG TCTAGA ATG AGA GGC AA
GAAAAAAGGAGGAGGT AA AGA G TG AGA GGC AA
```

*Xba*I primer
wild-type sequence

For analysis of proteins in culture supernatants, cultures were grown in LB medium containing 10 μg of chloramphenicol or 10 μg of neomycin per ml for 16 h at 37°C. Cells were pelleted by centrifugation and the supernatant was adjusted to 3 mM phenylmethylsulfonyl fluoride and 25 mM EDTA. An equal volume of 20% trichloroacetic acid was added, and the protein was precipitated at 0°C for 40 min. The precipitate was pelleted by centrifugation, washed with acetone, and lyophilized. The precipitate was dissolved in sodium dodecyl sulfate sample buffer, heated at 90°C for 2 min, and electrophoresed on a 10% sodium dodecyl sulfate-polyacrylamide gel (12). Each sample was equivalent to 50 μl of the original culture supernatant.

Reagents. Restriction enzymes, polynucleotide kinase, Bal-31, and the Klenow fragment of DNA polymerase I were purchased from commercial sources and used as directed by the suppliers. Subtilisin activity was assayed by using the synthetic substrate succinyl[-L-ala-L-ala-L-pro-L-phe-] *p*-nitroanilide (Vega) as described (26). Oligonucleotides were provided by

the Genentech Organic Synthesis Group and were synthesized by the phosphotriester method (5). Molecular-weight standards for sodium dodecyl sulfate-polyacrylamide gel electrophoresis were purchased from Bio-Rad Laboratories.

Mutagenesis. Site-directed mutagenesis was performed essentially as previously described (31). Briefly, an oligonucleotide primer containing the desired sequence changes was annealed to an M13 clone containing an appropriate restriction fragment with the area to be mutagenized. The primer was extended with the Klenow fragment of polymerase I and treated with T4 DNA ligase. The reaction mixture was used to transform *E. coli* JM101. Plaques were transferred to nitrocellulose and screened by hybridization with the use of oligonucleotide primer (3). The primer was labeled for hybridization screening by phosphorylation with the use of [α-³²P]ATP (~5,000 Ci/mmol, New England Nuclear Corp.). Washing conditions were determined empirically for each probe. Positive clones were confirmed by sequencing by the dideoxy chain termination method (21).

RESULTS AND DISCUSSION

Subtilisin

Introduction of an *Xba*I site preceding the subtilisin gene. To facilitate changes in the regulatory signals of the subtilisin gene, an *Xba*I site was created immediately preceding the initiation codon of the structural gene for the *B. amyloliquefaciens* subtilisin. This *Xba*I site enclosed the promoter and ribosome binding site region of the subtilisin gene between two restriction sites, *Eco*RI and *Xba*I, which were unique to the subtilisin expression plasmid, pS4.5X (Fig. 1). The introduction of this *Xba*I site also resulted in the change of the initiation codon for subtilisin from GTG to ATG. The creation of this

TABLE 1. Subtilisin production

Promoter	Features[a]	Subtilisin (μg/ml)[b]		
		15 h	24 h	39 h
pS4	Natural subtilisin	75	135	123
pS4.5X	*Xba*I-ATG subtilisin	68	103	77
No. 1	Amylase	133	232	182
No. 2	Amylase − 6 bp	10	24	23
No. 3	−10 Amylase	119	189	157
No. 4	−10 Amylase − 6 bp	10	20	20
No. 5	−35, −10 Amylase	NT[c]	NT[c]	NT[c]
No. 6	Short amylase	71	86	73
pBS42		1.3	2.3	2.6

[a] Exact changes are detailed in Fig. 2.
[b] Cultures were grown for the indicated times at 37°C in LB + 12.5 μg of chloramphenicol per ml in 20-mL volumes. Each assay value is the average of duplicate determinations.
[c] Not tested.

```
                    EcoRI                              ClaI
                    GAATTCT CATGTTTGAC AGCTTATCAT CGATTGTTTG AGAAAAGAAG
AAGACCATAA AAATACCTTG TCTGTCATCA GACAGGGTAT TTTTTATGCT GTCCAGACTG TCCGCTGTGT AAAAATAAGG

        < -35>                   < -10 >                       ↓         ↓              met
AATAAAGGGG GGTTGTTATT ATTTTACTGA TATGTAAAAT ATAATTTGTA TAAGAAAATG AGAGGGAGAG GAAACATG

GAATTC        AC              C  T              ------                       TCTAGATG

< #6 >      < #5 >       < #'s 3,4,5 >      < #'s 2,4 >              < #'s 1-6 >
```

FIG. 2. Sequence of the amylase promoter derivatives. The amylase promoter was isolated from *B. amyloliquefaciens* ATCC 23844, and its sequence is shown from the *Cla*I site to the initiation codon. The sequence from the *Eco*RI site to the *Cla*I site is derived from pBR322. The positions of the probable −35 and −10 recognition sites are indicated. Underlined residues were changed in the various derivatives; the changes are indicated underneath by the replacing nucleotide or by a dash, indicating a deletion. Below the indicated changes are the derivatives that have those particular nucleotide changes. The sequence of the −10 region was changed by mutagenesis in M13 with the use of a primer of the sequence 5′-AAATTATATTATAGATATCAGTAA and the wild-type sequence as the starting template. The sequence of the −35 region was changed by M13 mutagenesis with the use of a primer of the sequence 5′-GTAAAATAATGTCAACCCCCCT and derivative number 3 as the starting template. An *Xba*I site was introduced preceding the initiation codon for derivatives 1–5 in the following manner. An appropriate DNA fragment containing the amylase promoter derivative was partially digested with *Mnl*I and then digested with *Eco*RI. *Mnl*I cuts twice within the promoter region, leaving blunt ends at the positions indicated by an arrow. These *Eco*RI-*Mnl*I fragments were then ligated to synthetic DNA fragments of the sequence

5′-AGAGGGAGAGGT

3′-TCTCCCTCTCCAGATC

and ligated into pUC13 which had been digested with *Eco*RI and *Xba*I. Since the *Mnl*I digest was partial, two derivatives were recovered for each ligation. Derivatives 1, 3, and 5 contained the entire sequence; 2 and 4 had a deletion of 6 bp. Derivative 6 was constructed entirely from synthetic DNA fragments which spanned the sequence from the *Eco*RI to the *Xba*I site.

*Xba*I site and change of initiation codons had no discernible effect on subtilisin production in *B. subtilis* (Table 1).

Derivatives of the *B. amyloliquefaciens* amylase promoter. Six derivatives of the *B. amyloliquefaciens* amylase promoter were created; the construction of each is described in Fig. 2. The first derivative is essentially the wild-type amylase promoter; the only difference is the presence of the *Xba*I site between the Shine-Dalgarno region and the initiation codon ATG. This *Xba*I site is constant in all the derivatives used. The next derivative (number 2) has a deletion of 6 base pairs (bp) between the Shine-Dalgarno region and ''−10'' region, which was an artifact of the creation of the *Xba*I site. Derivative number 3 differs from number 1 by two nucleotides. One is an A to T change which changes the sequence of the −10 region from AAATAT to TAATAT. The other is a G to C change, which introduces an *Eco*RV site between the −35 and −10 regions, allowing future changes in the −35 and −10 regions to be easily combined. Derivative number 4 combines the mutations of numbers 3 and 2. Derivative number 5 differs from number 3 by two nucleotides, which change the −35 region to a consensus TTGACA. In derivative number 6 approximately 100 bp of the region upstream of the −35

region was removed; otherwise, it is the same as the wild-type derivative number 1.

Use of the amylase promoter derivatives to express subtilisin. Each of the amylase promoter derivatives described above was isolated, ligated into the large *Eco*RI-*Xba*I fragment of plasmid pS4.5X, and transformed into strain BG2036; colonies were selected for chloramphenicol resistance on plates containing skim milk. The majority of the transformants had haloes, indicating the production of subtilisin. The only exception was derivative number 5. It is not possible to introduce derivative number 5 into pS4.5X. Few transformants were recovered, and none of those had haloes. Upon examination of the plasmids, none had inserted promoters.

Each of the derivatives was examined for the amount of subtilisin produced. The results presented in Table 1 show that these promoter changes can affect the production of subtilisin. The wild-type amylase promoter, derivative number 1, shows a twofold increase in subtilisin production compared with the natural subtilisin promoter. Derivative number 3, which has an altered −10 region, also produces almost twice as much subtilisin as the natural subtilisin promoter. The deletion of 6 bp between the −10 region and ribosome binding site in derivatives number 2 and number 4 leads in each case to a

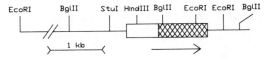

FIG. 3. Structure of the neutral protease gene. This diagram is a composite of sequence data, DNA blot hybridization, and restriction analysis of cloned DNA fragments (29). Double lines indicate the coding region for the neutral protease gene; filled double lines indicate the coding region for the mature protein. The arrow indicates the direction of transcription.

large reduction (8- to 10-fold) in the subtilisin levels when compared with the equivalent promoters without the 6-bp deletion. The results in Table 1 also show that the sequences upstream from the −35 region, as in derivative number 6, can affect production of subtilisin, leading to a 35% decrease in production as compared with the wild-type amylase promoter, derivative number 1.

The use of the amylase promoter derivatives to express subtilisin shows that changes in the regulatory region can lead to rapid increases in production levels of a desired product. The rationale for using the amylase promoter for the site-directed mutagenesis studies rather than the subtilisin promoter is that the amylase promoter appears to have a promoter recognition site recognized by the σ^{55} form of RNA polymerase. As indicated in Fig. 2, the amylase promoter has the sequence TTGTTA followed by 18 bp and by AAATAT, which shows reasonable similarity to the consensus σ^{55} promoter, i.e., TTGACA-17 bp-TATAAT (18). Which form of RNA polymerase recognizes the B. amyloliquefaciens subtilisin promoter is unclear. The B. subtilis subtilisin is recognized by the σ^{37} form of RNA polymerase (28); however, the B. amyloliquefaciens promoter has no clear similarity to the B. subtilis promoter or other known σ^{37} promoters (17, 25). Also, since the consensus σ^{55} recognition site is better defined than the σ^{37} recognition site, the bases targeted in the site-directed mutagenesis were more obvious. We believe that further changes in the promoter which make it conform more closely to the consensus sequence could lead to further increases in subtilisin expression. However, these data also suggest that some limits on the strength of the promoter might exist. The most likely explanation for the inability to place derivative number 5 in front of the subtilisin gene is that the changes made the promoter so strong that there were deleterious effects on either cell growth or plasmid replication. We have been unable to utilize derivative number 5 in a number of other constructions, suggesting that this effect was not specific to the subtilisin gene (L. Band and D. J. Henner, unpublished data).

The large negative effects on subtilisin expression of the 6-bp deletion in derivatives number 2 and number 4 were unexpected. We do not know whether the effects are due to decreased mRNA levels, possibly because of decreased transcription or increased mRNA turnover, or are due to decreased translation. The decreased expression levels in derivative number 6, which had the sequences upstream from −35 removed, were not entirely unexpected, as similar results have been reported in both B. subtilis and E. coli (9, 17).

Neutral Protease

The neutral protease gene. The neutral protease gene of B. subtilis was cloned and sequenced, as described in a previous report (29). A diagram of the gene structure and the relevant restriction sites is shown in Fig. 3. Like the subtilisin gene, there appears to be a "pro" sequence preceding the mature structural gene. However, in this case the pro sequence is approximately 220 amino acids, rather than the 75 amino acids seen in subtilisin. Our previous report indicated that placing the neutral protease gene on a high-copy plasmid led to an approximately 30-fold increase in the level of neutral protease expression (29).

Determination of the promoter region of neutral protease. To determine the location of the promoter for the neutral protease gene, a series of plasmids were constructed which progressively deleted the region upstream from the translation initiation site. Figure 4 outlines the strategy used to construct these deletion plasmids. Briefly, a series of deletions were created with Bal-31 starting at the StuI site (converted to BamHI in pNPR9), which lies 290 bp upstream from the translation initiation site. The promoter regions from a number of these derivatives were isolated and used to replace the promoter region of the neutral protease expression plasmid pNPR9 (the construction of pNPR9 is outlined in Fig. 6). These deletion plasmids were transformed into strain BG2036, a strain which lacks any detectable proteases, and scored for the presence of a halo on milk plates. Derivative 12 contained the largest insert which had lost protease activity, and derivative 19 had the smallest insert that appeared to retain full protease activity. Derivative 14 retained intermediate levels of protease activity. The inserts of derivatives 12, 14, and 19 were sequenced and are indicated in Fig. 5. The sequence analysis localizes the promoter region to within about 10 bp upstream from derivative 12.

Isolation of the nprR2 allele. A regulatory gene specific for hyperproduction of neutral protease was introduced into B. subtilis from B. natto IAM 1212 (24). This regulatory gene, designated

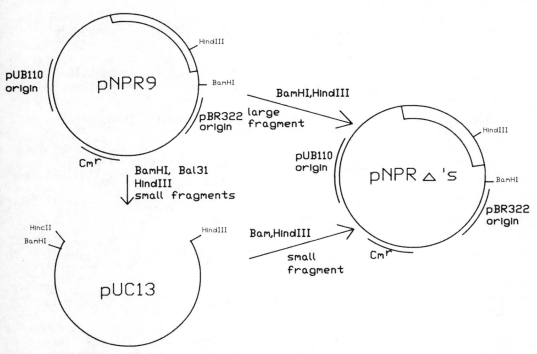

FIG. 4. Construction of promoter deletions. Plasmid pNPR9 (see Fig. 6 for pNPR9) was digested with *Bam*HI, digested with Bal-31 for 1 to 5 min, treated with the Klenow fragment of DNA polymerase I to repair the blunt ends, and then digested with *Hin*dIII. The DNA was electrophoresed on an acrylamide gel, and fragments in the size range of 300 to 400 bp were electroeluted and ligated into plasmid pUC13 which had been digested with *Hin*dIII and *Hin*cII. The size range of the inserts was determined by digestion with *Bam*HI and *Hin*dIII. *Bam*HI-*Hin*dIII inserts of various sizes were then isolated from the pUC derivatives, ligated into pNPR9 which had been digested with *Bam*HI and *Hin*dIII, and transformed into *B. subtilis* BG2036. Transcription of the neutral protease gene, indicated by double lines, is counterclockwise.

nprR2, was mapped and shown to be tightly linked to the structural gene for neutral protease (23). Since the most likely explanation for the increased neutral protease expression was changes in the promoter region, we isolated and characterized that region from a strain carrying the *nprR2* allele. Chromosomal DNA was isolated from strains NT02 and NT17, both of which carried the *nprR2* allele (23). DNA blot hybridizations performed with a variety of restriction enzymes and the previously cloned neutral protease gene as a probe showed that the restriction patterns of strains NT02 and NT17 were similar to that of strain GSY264 (Fig. 3). NT02 DNA was digested with *Eco*RI and *Hin*dIII and ligated into plasmid pBR322 which had been digested with *Eco*RI and *Hin*dIII. Recombinants were screened by colony hybridization techniques with a fragment of the wild-type gene used as a probe. Several positive colonies were identified; upon characterization they had the restriction map expected from previous blot hybridization studies. This plasmid carrying the neutral protease promoter region from strain

NT02 was designated pNPRRH1(NT02). The region from the *Hin*dIII site to the *Stu*I site, spanning the initiation codon and the promoter region, was sequenced. Figure 5 compares the sequence of the promoter regions of the neutral protease genes isolated from GSY264 and NT02.

Expression of neutral protease on plasmid vectors. To compare the promoters isolated from GSY264 and NT02, the neutral protease gene was placed on plasmid vectors under the control of each of the promoters. Figure 6 outlines the constructions of these plasmids. pNPR12 and pNPRH20 carry the protease gene on a pUB110 vector. However, there appears to be a plasmid promoter on pUB110 which transcribes across the *Bam*HI site in the direction of transcription of the protease gene (27; L. Band and D. J. Henner, unpublished data). Such transcription activity would make comparisons of the two promoters on these plasmids difficult. Plasmids pNPR9 and pNPRH21 were constructed to avoid this problem. The Bal-31 deletion experiments described above used pNPR9 and showed that removal of the neutral protease promoter led to

```
                       StuI                                                50
           GSY264      AGGCCTCATTCGGTTAGACAGCGGACTTTTCAAAAAGTTTCAAGATGAAA
           NT02        AGGCCTCATTCGGTTAGACAGCGGACTTTTCAAAAAGTTTCAAGATGAAA

                                                                          100
           GSY264      CAAAAATATCTCATCTTCCCCTTGATATGTAAAAAACATAACTCTTGAAT
           NT02        CAAAAATATCTCATCTTCCCCTTGAT-------------------------

                                                                          150
           GSY264      GAACCACCACATGACACTTGACTCATCTTGATATTATTCAACAAAAACAA
           NT02        ------------------------------------------AAAAATAA

                                            19                 14
                                            |                  |        200
           GSY264      ACACAGGACAATACTATCAATTTTGTCTAGTTATGTTAGTTTTTGTTGAG
           NT02        ACACATTACAATACTATCAATTTTGTCTAGTTATGTTAGTTTTTGTTGAG

                            12
                            |                                          250
           GSY264      TATTCCAGAATGCTAGTTTAATATAACAATATAAAGTTTTCAGTATTTTC
           NT02        TATTCCAGAATGCTGGTTTTTATATAACAATATAAAGTTTTCAGTATTTTC

                                         MET                            300
           GSY264      AAAAAGGGGGATTTATTGTGGGTTTAGGTAAGAAATTGTCTGTTCGTGTC
           NT02        AAAAAGGGGGATTTATTGTGGGTTTAGGTAAGAAATTGTCTGTTCGTGTC

                                                                        350
           GSY264      GCTGCTTCGTTTATGAGTTTATCAATCAGCCTGCCAGGTGTTCAGGCTGC
           NT02        GCTGTAATGTTTATGAGTTTATCAATCAGCCTGCCAGGTGTTCAGGCTGC

                                                                        400
           GSY264      TGAAGGTCATCAGCTTAAAGAGAATCAAACAAATTTCCTCTCCAAAAAGC
           NT02        TGAAGGTCATCAGCTTAAAGAGAATCAAACAAATTTCCTCTCCAAAAATG

                                                                        450
           GSY264      CGATTGCGCAATCAGAACTCTCTGCACCAAATGACAAGGCTGTCAAGCAG
           NT02        CGATTGCGCAATCAGAACTCTCTGCACCAAATGACAAGGCTGTCAAGCAG

                                                                        500
           GSY264      TTTTTGAAAAAGAACAGCAACATTTTTAAAGGTGACCCTTCCAAAAGCGT
           NT02        TTTTTGAAAAAGAACAGCAACATTTTTAAAGGTGACCCTTCCAAAAGCGT

                       HINDIII
           GSY264      GAAGCTT
           NT02        GAAGCTT
```

FIG. 5. Comparison of the neutral protease promoter regions isolated from strains GSY264 and NT02. The upper lines are the sequence from GSY264; the lower lines are from NT02. (– – – –) Gap created to maximize the homology. Mismatches between the two sequences are underlined. The endpoints of the Bal-31 deletions (12, 14, and 19) discussed in the text are indicated, as is the GTG initiation codon (met).

loss of protease activity. This confirmed that the protease gene is transcriptionally isolated on these constructions.

Each of the plasmids was transformed into strain BG2036, and the amount of neutral protease was determined by halo size on skim milk plates and by analysis of the proteins in the culture supernatants. In each case the plasmids that carried the promoter derived from strain NT02 showed a larger halo than the companion plasmid derived from GSY264 (data not shown). Analysis of the supernatants from the cultures is shown in Fig. 7. Each of the cultures shows a prominent band at a molecular weight of approximately 43,000 which is not present in the control culture (lane 5, Fig. 7). The amount of each band appears to agree qualitatively with the plate assay results; the band is more prominent in the lanes of the cultures carrying the plasmid with the promoter derived from strain NT02 (lanes 3 and 7, Fig. 7). The gel analysis also shows that under these conditions the neutral protease is the most abundant protein in the supernatant, indicating significant production levels.

The neutral protease promoter isolated from a hyperproducing strain of *B. subtilis* showed increased production when used to express the neutral protease gene on a plasmid. However, the levels of protease appeared to be only three to five times as great as when the wild-type promoter was used. The *nprR2* allele, as originally described, showed 30 to 50 times more protease production than the wild-type allele

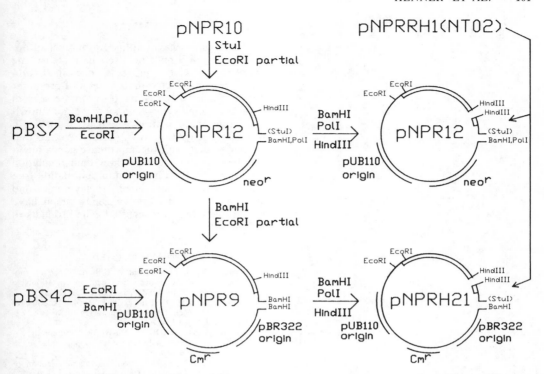

FIG. 6. Construction of neutral protease expression vectors. Plasmid pNPRRH1(NT02) carries the promoter region of the neutral protease gene isolated from strain NT02, as described in the text. Each of the ligation mixtures diagrammed was transformed into *B. subtilis* BG2036 and selected for transformants on LB milk plates containing 5 μg of chloramphenicol per ml. The presence of the origins of replication and antibiotic resistance markers are indicated; however, they are not necessarily to scale. The direction of transcription of the neutral protease gene, indicated by double lines, is counterclockwise.

(24). Several possibilities exist to explain this difference. One simple explanation is that, by placing the promoters on high-copy plasmids, transcription is no longer a major limiting factor in expression. Another possibility is that the *nprR2* allele has lost a regulatory site, which is the reason for its increased productivity. It is possible that the wild-type allele, when placed on a high-copy plasmid, is no longer affected by the postulated regulatory element as a result of titration. A third possibility is that the phenotype of the *nprR2* allele is not due solely to the promoter region, or at least not the region we have used in these expression studies.

The comparison of the two promoters shows several interesting features. The most prominent is the large 60-bp deletion from the promoter isolated from strain NT02. This deletion is 50 bp upstream from the region indicated in the Bal-31 experiments as critical for expression, which presumably define the promoter. The Bal-31 deletions 12 and 14 bracket 10 bp which seem to be necessary for expression. Contained within that 10 bp is the sequence TTGAGT which matches well with the consensus −35 sequence TTGACA for a σ^{55} promoter (18). Seventeen bp

downstream of the putative −35 sequence is the sequence TAATAT, which matches the consensus −10 sequence, TATAAT, and spacing, 17 or 18, for a σ^{55} promoter. Previous work has also shown that this gene is expressed in an *E. coli* transcription-translation reaction, suggesting that it might be recognized by a σ^{55}-type RNA polymerase (29). However, there is also reasonable homology to a σ^{37} promoter in this region. The sequence AGTTTAATAT at 215 is similar to the consensus −10 region of GG-ATT-TTT, and the two base changes in the NT02 promoter increase the similarity to GGTTTTATAT (10, 17, 25). There is also a possible −35 region at 199, AGTATTCCA, compared with a consensus sequence of AG--TT-A (25). An attractive hypothesis for the increased expression of the *nprR2* allele is the increased homology to the σ^{37} recognition sequence. However, in vitro studies to actually define the form of RNA polymerase are needed to interpret these data. It is interesting to note that there has been one promoter in *B. subtilis* which has been shown to use both σ^{37} and σ^{55} forms of polymerase (25).

More studies are needed to further characterize this interesting promoter. In vitro transcrip-

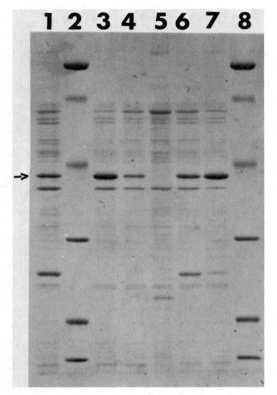

FIG. 7. Sodium dodecyl sulfate-polyacrylamide gel electrophoresis of culture supernatants. Cultures were grown and supernatants were prepared for electrophoresis as described in Materials and Methods. Lane 1, BG2036(pNPR10); lane 2, molecular-weight standards; lane 3, BG2036(pNPRH20); lane 4, BG2036(pNPR12); lane 5, BG2036; lane 6, BG2036(pNPR9); lane 7, BG2036(pNPRH21); lane 8, molecular-weight standards. Molecular-weight standards are 93,000, 66,000, 43,000, 31,000, 22,000, and 14,000.

tion studies with the different forms of RNA polymerase could be used to determine which form transcribes this promoter and also to map the start point of transcription. Preliminary results have shown that the neutral protease gene is regulated to various degrees by various *spo0* mutations (Ferrari and Henner, unpublished data). If the neutral protease gene is transcribed by a σ^{55} promoter, it would appear to be regulated in a different manner than other previously characterized genes under sporulation control. The exact causes of the *nprR2* phenotype also need to be determined. The two base pair changes in the *nprR2* promoter could be separated from the 60-bp deletion to determine which of these changes is responsible for the hyperproduction phenotype. Such studies would best be done in single-copy systems to avoid possible artifacts due to high-copy plasmids.

SUMMARY

These studies have outlined a number of approaches which may be used to increase the expression of cloned protease genes in *B. subtilis*. The site-directed mutagenesis of promoter regions and comparisons of promoters from hyperproducing genes also increase our knowledge about these features which are important in promoter recognition by *B. subtilis*. We feel that further studies with these protease genes might also help us to understand how their regulation interacts with the regulation of sporulation (see Ferrari, Howard and Hoch, this volume) and also with the many known genes which have pleiotropic effects on secreted gene products in *B. subtilis*.

LITERATURE CITED

1. **Backman, K., M. Ptashne, and W. Gilbert.** 1976. Construction of plasmids carrying the cI gene of bacteriophage lambda. Proc. Natl. Acad. Sci. U.S.A. **73**:4174–4178.
2. **Band, L., and D. J. Henner.** 1984. Bacillus subtilis requires a "stringent" Shine-Dalgarno region for gene expression. DNA **3**:17–21.
3. **Benton, W. D., and R. W. Davis.** 1977. Screening λgt recombinant clones by hybridization to single plaques *in situ*. Science **196**:180–182.
4. **Bolivar, F.** 1978. Construction and characterization of new cloning vehicles. III. Derivatives of plasmid pBR322 carrying unique *Eco*RI sites for selection of *Eco*RI generated recombinant DNA molecules. Gene **4**:121–136.
5. **Crea, R., and T. Horn.** 1980. Synthesis of oligonucleotides on cellulose by a phosphotriester method. Nucleic Acids Res. **8**:2331–2348.
6. **Glenn, A. R.** 1976. Production of extracellular proteins by bacteria. Annu. Rev. Microbiol. **30**:41–62.
7. **Grunstein, M., and D. S. Hogness.** 1975. Colony hybridization: a method for the isolation of cloned DNAs that contain a specific gene. Proc. Natl. Acad. Sci. U.S.A. **72**:3961–3965.
8. **Hoch, J. A.** 1976. Genetics of bacterial sporulation. Adv. Genet. **18**:69–98.
9. **Horn, G. T., and R. D. Wells.** 1981. The leftward promoter of bacteriophage lambda. Structure, biological activity and influence of adjacent regions. J. Biol. Chem. **256**:2003–2009.
10. **Johnson, W. C., C. P. Moran, Jr., and R. Losick.** 1983. Two RNA polymerase sigma factors from Bacillus subtilis discriminate between overlapping promoters for a developmentally regulated gene. Nature (London) **302**:800–804.
11. **Krant, J.** 1977. Serine proteases: structure and mechanism of catalysis. Annu. Rev. Biochem. **46**:331–358.
12. **Maizel, J. V., Jr.** 1971. Polyacrylamide gel electrophoresis of viral proteins. Methods Virol. **5**:179–246.
13. **Markland, F. S., and E. L. Smith.** 1971. Subtilisins: primary structure, chemical and physical properties, p. 561–608. *In* P. D. Boyer (ed.), The enzymes, vol. 3. Academic Press, Inc., New York.
14. **Matsubara, H., and J. Feder.** 1971. Other bacterial, mold, and yeast proteases, p. 721–795. *In* P. D. Boyer (ed.), The enzymes, vol. 3. Academic Press, Inc., New York.
15. **Messing, J.** 1979. A multipurpose cloning system based on the single-stranded DNA bacteriophage M13. Recombinant DNA Technical Bulletin, NIH Publication No. 79–99, **2**:43–48.
16. **Messing, J.** 1983. New M13 vectors for cloning. Methods Enzymol. **101**:20–78.
17. **Moran, C. P., Jr., N. Lang, C. D. B. Banner, W. G. Haldenwang, and R. Losick.** 1981. Promoter for a devel-

opmentally regulated gene in *Bacillus subtilis*. Cell **25**:783–791.

18. **Moran, C. P., Jr., N. Lang, S. F. J. LeGrice, G. Lee, M. Stephens, A. L. Sonenshein, J. Pero, and R. Losick.** 1982. Nucleotide sequences that signal the initiation of transcription and translation in *Bacillus subtilis*. Mol. Gen. Genet. **186**:339–346.

19. **Piggot, P. J., and J. G. Coote.** 1976. Genetic aspects of bacterial endospore formation. Bacteriol. Rev. **40**:908–962.

20. **Priest, F. G.** 1977. Extracellular enzyme synthesis in the genus *Bacillus*. Bacteriol. Rev. **41**:711–753.

21. **Sanger, F., S. Nicklen, and A. R. Coulson.** 1977. DNA sequencing with chain-terminating inhibitors. Proc. Natl. Acad. Sci. U.S.A. **74**:5463–5467.

22. **Stahl, M. L., and E. Ferrari.** 1984. Replacement of the *Bacillus subtilis* subtilisin structural gene with an in vitro-derived deletion mutant. J. Bacteriol. **158**:411–418.

23. **Uehara, H., K. Yamane, and B. Maruo.** 1979. Thermosensitive, extracellular neutral proteases in *Bacillus subtilis*: isolation, characterization, and genetics. J. Bacteriol. **139**:583–590.

24. **Uehara, H., Y. Yoneda, K. Yamane, and B. Maruo.** 1974. Regulation of neutral protease productivity in *Bacillus subtilis*: transformation of high protease productivity. J. Bacteriol. **119**:82–91.

25. **Wang, P.-Z., and R. H. Doi.** 1984. Overlapping promoters transcribed by Bacillus subtilis σ^{55} and σ^{37} RNA polymerase holoenzymes during growth and stationary phases. J. Biol. Chem. **259**:8619–8625.

26. **Wells, J. A., E. Ferrari, D. J. Henner, D. A. Estell, and E. Y. Chen.** 1983. Cloning, sequencing, and secretion of *Bacillus amyloliquifaciens* subtilisin in *Bacillus subtilis*. Nucleic Acids Res. **11**:7911–7925.

27. **Williams, D. M., E. J. Duvall, and P. S. Lovett.** 1981. Cloning restriction fragments that promote expression of a gene in *Bacillus subtilis*. J. Bacteriol. **146**:1162–1165.

28. **Wong, S.-L., C. W. Price, D. S. Goldfarb, and R. H. Doi.** 1984. The subtilisin E gene of *Bacillus subtilis* is transcribed from a σ^{37} promoter *in vivo*. Proc. Natl. Acad. Sci. U.S.A. **81**:1184–1188.

29. **Yang, M. Y., E. Ferrari, and D. J. Henner.** 1984. Cloning of the neutral protease gene of *Bacillus subtilis* and the use of the cloned gene to create an in vitro-derived deletion mutation. J. Bacteriol. **160**:15–21.

30. **Yansura, D. G., and D. J. Henner.** 1984. Use of the *Escherichia coli* lac repressor and operator to control gene expression in *Bacillus subtilis*. Proc. Natl. Acad. Sci. U.S.A. **81**:439–443.

31. **Zoller, M. J., and M. Smith.** 1982. Oligonucleotide-directed mutagenesis using M13-derived vectors: an efficient and general procedure for the production of point mutations in any fragment of DNA. Nucleic Acids Res. **10**:6487–6500.

Identification of a Positive Retroregulator That Functions in *Escherichia coli* and *Bacillus subtilis*

HING C. WONG AND SHING CHANG

Department of Microbial Genetics, Cetus Corporation, Emeryville, California 94608

It has been reported that mRNA from the crystal protein (*cry*) gene of *Bacillus thuringiensis* has a longer half-life than other mRNAs from genes expressed during sporulation (7). It has been speculated that the high stability of the *cry* gene mRNA is a contributing factor in the massive synthesis of the crystal protein during sporulation. Although little is known about the mechanism(s) involved in mRNA stability in bacteria, it has been postulated that the stability of an mRNA is correlated with the extent of its secondary structure. For instance, it is known that a stable procaryotic mRNA such as that of the lipoprotein (*lpp*) gene of *Escherichia coli* (4) can form an extensive secondary structure. However, analysis of the nucleotide sequence of the cloned *cry* gene from *B. thuringiensis* revealed that the predicted secondary structure of crystal protein mRNA is significantly less extensive than that of the lipoprotein mRNA of *E. coli*. Therefore, it is unlikely that the extraordinary stability of the crystal protein mRNA is the result of its extensive secondary structure.

From the studies on the regulation of coliphage λ integrase (*int*) gene, an unusual mechanism of regulation was discovered. The cis-acting element *sib*, located distal to the *int* gene, negatively regulates the expression of the *int* gene (1, 9). The *int* gene in λ is transcribed early after infection from the P_L promoter and later from the P_I promoter. Transcription from the P_I promoter generates a transcript which is terminated at a site (t_I) located 277 nucleotides beyond the *int* gene. The *sib* region overlaps with the terminator site t_I (11, 12). On the other hand, as a result of the antitermination effect of the N protein of λ, RNA polymerase initiating at the P_L promoter transcribes through the t_I termination signal into the *b* region of λ DNA. It was demonstrated that the *int* gene is expressed efficiently from the transcript terminated at the t_I, but is expressed poorly from the read-through transcripts initiated at the P_L promoter (1, 2). Court and co-workers showed that the difference in expression of these two transcripts exists because the mRNA that reads through t_I undergoes an RNase III-mediated processing event within the *sib* (t_I) region located ~260 nucleotides beyond *int*. The RNase III-processed mRNA is rapidly degraded, resulting in

very little *int* gene expression (2, 11, 12). This type of regulation of a target gene by an element distal to it, i.e., *sib*-controlled *int* expression, has been termed retroregulation (9).

We report here the finding of a positive retroregulatory mechanism for the control of *cry* gene expression. Earlier work on the structural analysis of the cloned *cry* gene (14) led one of us (H. C. Wong) to speculate that the sequence overlapping the transcriptional terminator of the *cry* gene may positively regulate the *cry* gene expression. It was observed that the insertion of a Tn5 transposon at the 3' end of the cloned *cry* gene resulted in a significant reduction of *cry* protein production in *E. coli*. Although detailed nucleotide sequence analysis to precisely locate the insertion site of this Tn5 transposon is lacking, immunoblotting and physical mapping data suggest that the insertion site is most likely located just beyond the coding region of the *cry* gene. We have recently employed molecular cloning techniques to investigate the role of the 3'-end noncoding region of the *cry* gene in gene expression. The data presented here indicate that the sequence containing the transcriptional terminator of the *cry* gene functions as a positive retroregulator in controlling gene expression in both *B. subtilis* and *E. coli*.

RESULTS

Structure and molecular cloning of the transcriptional terminator region of the *cry* gene. The restriction map of the recombinant plasmid pES1 bearing the cloned *cry* gene from *B. thuringiensis* subsp. *kurstaki* HD-1 is shown in Fig. 1. The transcriptional initiation and termination sites of the gene in *E. coli* and *B. thuringiensis* have been determined (14; H. E. Schnepf, H. C. Wong, and H. R. Whiteley, submitted for publication). Transcription of the *cry* gene originates at the site about 350 base pairs (bp) to the left of the fourth *Eco*RI site and terminates about 360 bp to the right of the third *Pvu*II site shown in Fig. 1.

The nucleotide sequence of the *Pvu*II-*Nde*I restriction fragment carrying the transcriptional terminator of *cry* is also shown in Fig. 1. There is an inverted repeat sequence located approximately 40 bp upstream of the *Nde*I site. The transcript made from this region can potentially

(A)

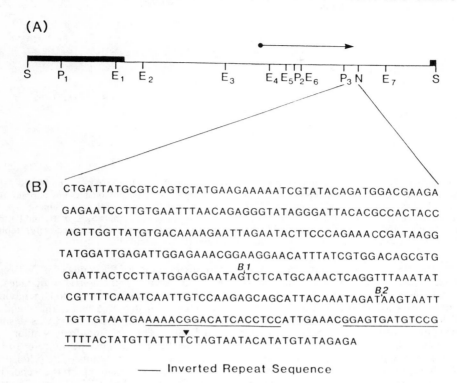

(B) CTGATTATGCGTCAGTCTATGAAGAAAAATCGTATACAGATGGACGAAGA

GAGAATCCTTGTGAATTTAACAGAGGGTATAGGGATTACACGCCACTACC

AGTTGGTTATGTGACAAAAGAATTAGAATACTTCCCAGAAACCGATAAGG

TATGGATTGAGATTGGAGAAACGGAAGGAACATTTATCGTGGACAGCGTG
                                                    B1
GAATTACTCCTTATGGAGGAATAGTCTCATGCAAACTCAGGTTTAAATAT
                                              B2
CGTTTTCAAATCAATTGTCCAAGAGCAGCATTACAAATAGATAAGTAATT

TGTTGTAATGAAAAACGGACATCACCTCCATTGAAACGGAGTGATGTCCG
          ▼
TTTTACTATGTTATTTTCTAGTAATACATATGTATAGAGA

———— Inverted Repeat Sequence

▼   Transcriptional Stop Site

FIG. 1. Structure of the *cry* gene. (A) Physical map of the *cry* gene in plasmid pES1 (14). The horizontal arrow indicates the coding region and the transcriptional direction of the *cry* gene. The thick lines represent the pBR322-derived sequences. (B) DNA sequence of the distal portion of the *cry* gene. The symbol ▼ represents the transcriptional stop site. The inverted repeat sequences are underlined. *B1* and *B2* indicate locations where new *Bgl*II sites were introduced by oligodeoxyribonucleotide-directed site-specific mutagenesis.

form a stem-and-loop structure. This putative stem-and-loop structure (Fig. 2) is relatively high in G+C content, with a predicted ΔG value of −30.4 kcal (ca. −127.3 kJ) as calculated by the rules of Tinoco et al. (13).

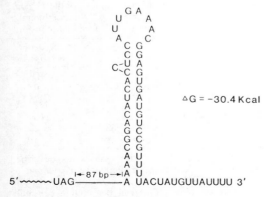

FIG. 2. Potential secondary structure at the 3' end of the *cry* mRNA.

To facilitate the construction of transcriptional fusions between the *cry* terminator-containing fragment and heterologous genes, the *Pvu*II-*Nde*I fragment was subcloned into the *Sma*I site of the bacteriophage M13mp9 in the orientation such that the *Pvu*II site is adjacent to the *Bam*HI site in the polylinker sequence of the phage genome. The resulting recombinant phage was designated M13mp9NP3.

Transcriptional fusion between the *penP* gene and the *cry* terminator fragment. The recombinant plasmid pSYC667 carrying the penicillinase (*penP*) gene from *B. licheniformis* was employed for this study (Fig. 3). Nucleotide sequence data revealed the presence of a unique *Bcl*I restriction site between the *penP* coding region and the presumptive transcription termination site of the gene (5).

To construct a transcriptional fusion of the *penP* and the *cry* terminator-containing fragment, we replaced the *Bcl*I-*Nru*I fragment on pSYC667 with the *Bam*HI-*Eco*RI fragment carrying the *cry* terminator from the phage M13mp9NP3 replicative form DNA (Fig. 3).

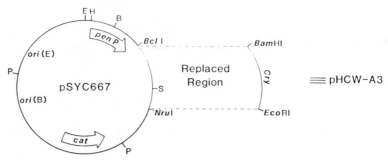

FIG. 3. Construction of a transcriptional fusion of the *penP* and the *cry* transcriptional terminator. Plasmid pSYC667 is constructed from plasmid pSYC660 (3) by eliminating the *Bam*HI site located downstream from the *penP* gene. The locations of the coding regions and the transcriptional directions of the penicillinase (*penP*) gene and the chloramphenicol acetytransferase (*cat*) gene are marked by arrows. Symbols S, E, B, H, and P represent the sites recognized by restriction enzymes *Sal*I, *Eco*RI, *Bam*HI, *Hind*III, and *Pvu*II, respectively. Replication origins (*ori*) for *E. coli* and *B. subtilis* are identified by (E) and (B), respectively.

This generated the bifunctional plasmid pHCW-A3, which replicates in both *E. coli* and *B. subtilis* and confers resistance to chloramphenicol on these hosts. As shown in Table 1, the *E. coli* and *B. subtilis* strains containing the fusion plasmid synthesized 2.6- to 5.3-fold-greater amounts of penicillinase than did the corresponding strains carrying the parental plasmid pSYC667. Since the copy numbers of pHCW-A3 and pSYC667 in either *E. coli* or *B. subtilis* hosts are very similar, based on the measurements of enzyme activity for the chloramphenicol acetyltransferase encoded by the *cat* gene on the same plasmid (data not shown), we concluded that the *cry* terminator-containing fragment positively regulates the expression of the penicillinase gene located immediately upstream. The *cry*-derived sequence in pHCW-A3 replaced the putative transcriptional terminator of *penP*. Therefore, the *cry*-derived sequence is most likely cotranscribed with the *penP* gene, and the transcript terminates at the site identical to that in the *cry* mRNA.

Transcriptional fusions of the human IL-2 cDNA and the *cry* terminator fragment. We also constructed transcriptional fusions of the cDNA encoding the human interleukin-2 (IL-2) protein and the *cry* terminator fragment to further examine the specificity of the enhancing effect of the terminator fragment on heterologous gene expression. A restriction map of the recombinant plasmid pLW1 bearing the cDNA of human IL-2 gene is shown in Fig. 4. The IL-2 sequence was modified so that its expression is under control of the *E. coli trp* promoter and translational initiation signal on the plasmid (8).

The unique *Stu*I site located just beyond the coding region of IL-2 on pLW1 was used to insert the terminator-carrying *Eco*RI-*Bam*HI fragment from M13mp9NP3. Ligation of the repaired, blunt-ended fragment carrying the *cry*

terminator with the *Stu*I ends of pLW1 regenerates the *Eco*RI and *Bam*HI sequences. One orientation of insertion resulted in having the *Bam*HI site located proximal to the 3' end of the IL-2 sequence, and the plasmid was designated pHCW701. (This orientation is similar to that found in the *cry* mRNA, and we refer to it here as the "native" orientation.) The other orientation of insertion resulted in the recombinant plasmid designated pHCW702 (see Fig. 4).

As shown in Table 2, *E. coli* strains containing either pHCW701 or pHCW702 produce higher levels (4.6- to 7-fold) of IL-2 than the strain carrying pLW1. This was confirmed independently by directly measuring the IL-2 protein in the cell extracts prepared from these strains (data not shown). Since the IL-2 gene is derived from a eucaryotic cell line, this suggests that the enhancing activity associated with the terminator fragment has a broad range for its target genes. Furthermore, the enhancing activity of the terminator fragment was observed to be

TABLE 1. Synthesis of penicillinase in the *E. coli* and *B. subtilis* strains carrying *penP* recombinant plasmids with or without the *cry* terminator fusion

Host	Plasmid	Sp act of penicillinase[a]
E. coli CS412		ND[b]
	pSYC667	1,055
	pHCW-A3	2,762
B. subtilis 1A510		ND
	pSYC667	3,280
	pHCW-A3	17,510

[a] Penicillinase activity was assayed as described (10). Specific activity of penicillinase is defined as nanomoles of PADAC (pyridine-2-azo-p-dimethylaniline chromophore) hydrolyzed per minute per milligram of protein at 25°C.

[b] ND, Not detectable.

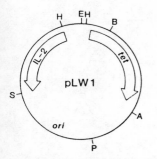 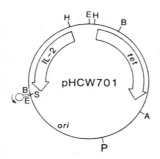 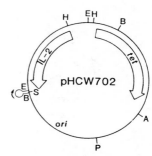

FIG. 4. *E. coli* plasmids containing the transcriptional fusions of the human IL-2 cDNA sequence and the *cry* transcriptional terminator. The direction of the arrows drawn above the terminator region indicates the "native" orientation of the terminator; the "native" orientation is the one similar to that found in the *cry* mRNA. Restriction sites recognized by *Ava*I, *Bam*HI, *Eco*RI, *Hin*dIII, *Pvu*II, and *Stu*I are presented as A, B, E, H, P, and S, respectively. The locations of the tetracycline resistance (*tet*) gene, the *trp* promoter-controlled IL-2 sequence (IL-2), and the replication origin (*ori*) are shown.

independent of its orientation in the plasmid with respect to the direction of transcription of the upstream target gene.

The enhancement of IL-2 expression of *B. subtilis* host by the terminator-containing fragment was also tested. A bifunctional plasmid derived from pLP1201 (6) was employed for the work. The *B. subtilis* promoter P_{156}, recognized by the vegetative RNA polymerase, was isolated from phage SP82. This promoter sequence equipped with a synthetic ribosome-binding-site sequence (detailed construction of this will be published elsewhere) was used to express the IL-2 gene with or without the terminator fragment in *B. subtilis*. Plasmid pHCW300 thus constructed contains the cloned IL-2 gene under the control of the P_{156} promoter. The *cry*-derived terminator fragment was inserted into pHCW300 at the *Stu*I site in its native orientation to generate plasmid pHCW301. The structure of these plasmids is shown in Fig. 5. Biological activity of IL-2 in cell extracts prepared from the strains harboring these plasmids was analyzed, and the results are summarized in Table 2. Again, we observed a higher level of expression of the IL-2 gene from the strain

carrying plasmid pHCW301, which contains the *cry* terminator, than from the strain bearing the parent plasmid pHCW300. Thus, the enhancing activity of the *cry* terminator-containing fragment is again observed in both the gram-negative *E. coli* and the gram-positive *B. subtilis*.

Deletion mapping of the retroregulator. Deletion studies were carried out to define more precisely the sequence within the *cry* terminator region responsible for the enhancing activity. Oligodeoxyribonucleotide-directed site-specific mutagenesis (15) was employed separately to introduce two *Bgl*II restriction sites at the locations 22 and 80 bp upstream from the inverted repeat sequence in the *cry* terminator fragment (Fig. 1). These shortened terminator-containing fragments can be excised from the respective, modified M13mp9NP3 phage genomes by *Bgl*II-*Eco*RI digestion. They were cloned into pSYC667 at the *Bcl*I-*Nru*I site by procedures similar to those employed for the construction of pHCW-A3 and generated plasmids pHCW-A4 and pHCW-A5. Data on the analyses of *B. subtilis* as well as *E. coli* strains harboring these plasmids for their ability to express the cloned *penP* gene are presented in Table 3. It is clear that the shortened fragments still contain the regulatory function observed in the original fragment. Since the two newly created *Bgl*II sites are outside the *cry* coding sequence, these data demonstrate that the locus that confers the enhancing activity is located in the noncoding region of the *cry* gene, and it probably overlaps with the terminator of the *cry* gene.

DISCUSSION

In this communication, we describe the identification of a DNA sequence that enhances the expression of the genes positioned upstream, which are presumably cotranscribed, in *E. coli* and *B. subtilis*. This is the first example demon-

TABLE 2. Enhancing effect of the *cry* gene terminator-containing fragment on the synthesis of IL-2 protein[a] in *E. coli* and *B. subtilis*

Host	Plasmid	Sp act of IL-2[b]
E. coli CS412	pLW1	2×10^5
	pHCW701	1.4×10^6
	pHCW702	9.2×10^5
B. subtilis 1A510	pHCW300	1×10^4
	pHCW301	3×10^4

[a] IL-2 was assayed as previously described (8).
[b] Units of activity per 150 μg of cellular proteins.

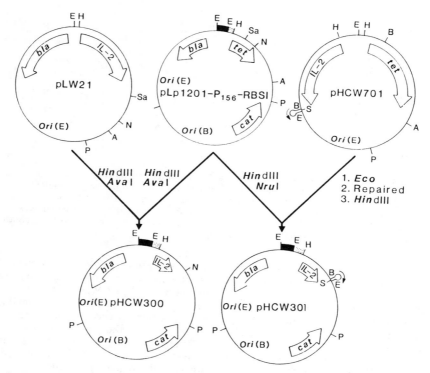

FIG. 5. Constructions of the IL-2 expression plasmids pHCW300 and pHCW301. The constructions of pLW21 (13a) and pLP1201 (6) have been described previously. The solid and dot-filled boxes represent the promoter P$_{156}$ and the synthetic ribosome-binding-site sequence, respectively. Symbols A, B, E, H, N, P, and S represent the sites recognized by the restriction enzymes *Ava*I, *Bam*HI, *Eco*RI, *Hin*dIII, *Nru*I, *Pvu*II, and *Stu*I, respectively. The β-lactamase (*bla*) gene is marked. Other symbols used are the same as in the previous figures.

strating that bacterial genes can be regulated positively by a retroregulator. The enhancement activity appears to be nonspecific; it is independent of the promoters used and of the origin of the target genes. Furthermore, it operates in both *E. coli* and *B. subtilis*.

At this point, we can only speculate on the possible mechanism(s) that contributes to the enhancing effect noted here for the *cry* transcriptional terminator fragment. One possibility is that the terminator fragment regulates gene expression by increasing its mRNA stability. This could be the result of: (i) the formation of the stable stem-and-loop structure at the 3' end of its mRNA to protect the fusion mRNA against exonuclease degradation or (ii) the recognition by an RNA-binding protein(s) of a specific nucleotide sequence(s) located within the *cry* terminator region that could make the end of the fusion mRNA physically inaccessible to exonuclease(s).

There is a potential ribosome binding site in the terminator fragment which is followed, at an appropriate interval, by an ATG and four additional codons (Fig. 2). Initially, we speculated

that this sequence might cause the stalling of ribosomes at the end of the mRNA that could in turn protect the end of the mRNA. However, our deletion mapping data conclusively ruled out this possibility.

If the enhancing activity of the terminator fragment is indeed sequence dependent, the essential sequence must be located in the in-

TABLE 3. Synthesis of penicillinase in *E. coli* and *B. subtilis* strains carrying the *penP-cry* fusion plasmids[a]

Host strain	Plasmid	Length of *cry*-derived fragments (bp)	Sp act of penicillinase[a]
E. coli CS412	pHCW-A3	380	2,762
	pHCW-A4	158	2,631
	pHCW-A5	78	2,861
B. subtilis 1A510	pHCW-A3	380	17,510
	pHCW-A4	158	18,600
	pHCW-A5	78	17,822

[a] The assay of penicillinase and the definition of specific activity are described in Table 1.

verted repeat regions. This conclusion was drawn on the basis of our observation that the enhancement activity of the terminator fragment is independent of its orientation with respect to the target gene. When present in either one of the two possible orientations, the only identical sequence between the two respective transcripts is the inverted repeats involved in base pairing to form the stem in the mRNA.

We propose that the enhancement effect of the *cry*-derived terminator fragment is a direct result of increased mRNA stability. This increased stability is probably contributed by the incorporation of the stem-and-loop structure in the mRNA as shown in Fig. 2. It is probably not important for the transcript to be terminated immediately following this structure. In fact, when the terminator fragment is inserted in the opposite orientation (as opposed to its "native" orientation), there is no polyuridine track following the stem-and-loop structure in the corresponding mRNA. Therefore, transcription could proceed beyond this region and terminate at a distal site. Whereas the terminator function may be orientation dependent, the enhancing function is not. Thus, the loci conferring the enhancement effect and the termination function physically overlap each other, and they probably function independently.

SUMMARY

Data presented here show that the transcriptional terminator fragment of the *cry* gene from *B. thuringiensis* subsp. *kurstaki* HD-1 contains a locus that enhances the expression of upstream genes in both *E. coli* and *B. subtilis*. The enhancing activity is observed when the terminator fragment is located at the distal end of the gene; it is independent of the orientation of the terminator fragment relative to the transcription direction of the target gene(s). Deletion analysis has localized the locus of this positive retroregulator within a 78-bp fragment; it includes the inverted repeat and the 22-bp upstream and the 13-bp downstream flanking sequences.

LITERATURE CITED

1. **Guarneros, G., and J. M. Galindo.** 1979. The regulation of integrative recombination by the *b*2 region and the *c*II gene of bacteriophage λ. Virology **95**:119–126.

2. **Guarneros, G., C. Montanez, T. Hernandez, and D. Court.** 1982. Post-transcriptional control of bacteriophage λ *int* gene expression from a site distal to the gene. Proc. Natl. Acad. Sci. U.S.A. **79**:238–242.

3. **Hayashi, S., S. Y. Chang, S. Chang, and H. C. Wu.** 1984. Modification and processing of *Bacillus licheniformis* prepenicillinase in *Escherichia coli*. Fate of mutant penicillinase lacking lipoprotein modification site. J. Biol. Chem. **259**:10448–10454.

4. **Nakamura, K., R. M. Pirtle, I. M. Pirtle, K. Takeishi, and M. Inouye.** 1979. Messenger ribonucleic acid of the lipoprotein of the *Escherichia coli* outer membrane. II. The complete nucleotide sequence. J. Biol. Chem. **225**:210–216.

5. **Neugebauer, K., R. Sprengel, and H. Schaller.** 1981. Penicillinase from *Bacillus licheniformis*: nucleotide sequence of the gene and implications for the biosynthesis of a secretory protein in a Gram-positive bacterium. Nucleic Acids Res. **9**:2577–2588.

6. **Ostroff, G. R., and J. Pene.** 1984. Molecular cloning with bifunctional plasmid vector in *Bacillus subtilis*. II. Transfer of sequences propagated in *Escherichia coli* to *B. subtilis*. Mol. Gen. Genet. **193**:306–311.

7. **Petit-Glatron, M. F., and G. Rapoport.** 1976. Translation of a stable mRNA fraction from sporulating cells of *Bacillus thuringiensis* in a cell-free system from *Escherichia coli*. Biochimie **58**:119–129.

8. **Rosenberg, S. A., E. A. Grimm, M. McGrogan, M. Doyle, E. Kawasaki, E. Koths, and D. F. Mark.** 1984. Biological activity of recombinant human interleukin-2 produced in *Escherichia coli*. Science **223**:1412–1415.

9. **Schindler, D., and H. Echols.** 1981. Retroregulation of the *int* gene of bacteriophage λ: control of translation completion. Proc. Natl. Acad. Sci. U.S.A. **78**:4475–4479.

10. **Schindler, P., and G. Huber.** 1980. Use of PADAC, a novel chromogenic β-lactamase substrate, for the detection of β-lactamase producing organisms and assay of β-lactamase inhibitors/inactivators, p. 169–176. *In* U. Brodbeck (ed.), Enzyme inhibitors. Verlag Chemie, Weinheim.

11. **Schmeissner, V., K. McKenney, M. Rosenberg, and D. Court.** 1984. Transcription terminator involved in the expression of the *int* gene of phage lambda. Gene **28**:343–350.

12. **Schmeissner, V., K. McKenney, M. Rosenberg, and D. Court.** 1984. Removal of a terminator structure by RNA processing regulates *int* gene expression. J. Mol. Biol. **176**:39–53.

13. **Tinoco, I., P. N. Borer, B. Dengler, M. D. Levine, O. C. Uhlenbeck, O. M. Corthers, and J. Gralla.** 1973. Improved estimation of secondary structure in ribonucleic acid. Nature (London) New Biol. **246**:40–41.

13a.**Wang, A., S. Lu, and D. F. Mark.** 1984. Site-specific mutagenesis of the human interleukin-2 gene: structure-function analysis of the cysteine residues. Science **224**:1431–1433.

14. **Wong, H. C., H. E. Schnepf, and H. R. Whiteley.** 1983. Transcriptional and translational start sites for the *Bacillus thuringiensis* crystal protein gene. J. Biol. Chem. **258**:1960–1967.

15. **Zoller, M. J., and M. Smith.** 1983. Oligonucleotide-directed mutagenesis of DNA fragments cloned in M13 vectors. Methods Enzymol. **100**:468–500.

Penicillinase Synthesis in *Bacillus licheniformis*: Nature of Induction

ANTHONY J. SALERNO, NANCY J. NICHOLLS, AND J. OLIVER LAMPEN

Waksman Institute of Microbiology, Rutgers, The State University of New Jersey, Piscataway, New Jersey 08854

One of the first systems selected for investigation of the regulation of enzyme production in microorganisms was the inducible production of penicillinase in *Bacillus cereus* and *Bacillus licheniformis*. Each of these organisms forms substantial quantities of a secreted enzyme that is highly active, relatively stable, and easily assayed. Induction has proved to be complex, however, and the nature of the process leading to enzyme formation is not yet clear (for reviews, see 4, 5, 12, 19). After addition of penicillin or other β-lactams to an exponential-phase culture, the rate of penicillinase formation increases for about 60 min (4). It then falls over the next hour, but remains above the uninduced level for several hours, even when the inducer is destroyed or removed early in the induction cycle. This extended synthesis prompted several investigators to postulate a long-lived *penP* mRNA (6, 18), but the apparent half-life, as determined by the decline in enzyme-forming activity after the addition of rifampin, is only 4 to 5 min (7). Induction is nonspecific in that a variety of non-β-lactam substances, for example, molybdate ions, carbodiimides, and a minor component of the bacitracin complex, have been reported to elicit penicillinase formation with kinetics similar to induction by a β-lactam (see 4). In addition, attempts to induce protoplasts have met with little success, although protoplasts prepared from mutants constitutive for penicillinase synthesis produce and secrete the enzyme efficiently (5, 9). Penicillinase production by protoplasts prepared from induced cells has been low and variable. Collins (4) suggested that removal of the cell wall interferes with the initial stages of induction but not with the later stages. Penicillin binds to a number of surface proteins in both of these bacilli, and it is assumed that such an interaction is an integral part of the induction process. The binding of penicillin by protoplasts and membrane preparations differs in both amount and distribution from that obtained with intact cells (2, 5, 9). This could account for the differences in inducibility between protoplasts and cells.

B. licheniformis 749 synthesizes a single penicillinase, whereas *B. cereus* 569 forms three distinct β-lactamases which appear to be coordinately regulated (4). Consequently, most of the available genetic information pertains to *B. licheniformis*. A number of structural and regulatory gene mutants have been isolated, and a linkage map has been constructed showing a regulatory gene, *penI*, 90% linked to the 3' segment of the structural gene (*penP*) corresponding to the C terminus of the protein (8, 22). Another regulatory gene, R1, is slightly linked, and a third, R2, is unlinked to the structural gene. *penI* is thought to have a negative regulatory function, and the production of a *trans*-acting repressor has been demonstrated by Imanaka et al. (11). On the basis of these findings, Collins (5) has put forward two models for induction of penicillinase. In one the inducer reacts with a specific membrane protein, thus altering the level of a cytoplasmic effector which is either a corepressor or an endogenous inducer. This effector finally combines with the repressor from *penI* to modulate penicillinase synthesis. The alternative model assumes that binding of inducer to the cell surface interferes with the biosynthesis or stability of the peptidoglycan. In either case "wall by-products" are formed which then modulate repression of *penP* expression. With the *penP* gene now sequenced (17) and specific probes available from subcloned fragments of *penP* and the adjacent regulatory stretches, we decided to reexamine the kinetics and characteristics of the formation of *penP* mRNA during induction in *B. licheniformis* 749.

EXPERIMENTAL PROCEDURES

Strains. *B. licheniformis* 749, inducible for penicillinase production, and the penicillinase-constitutive mutant 749/C (ATCC 25972) were obtained from M. R. Pollock (18).

RNA isolation. Cells were centrifuged, and the pellets were suspended in 10 mM Tris (pH 7.5), 1 mM EDTA, 4 mM pyrophosphate, 50 μg of chloramphenicol per ml, and 10 μg of actinomycin D per ml. The cells were then immersed in liquid N_2 and stored at −70°C. Cells were disrupted by adding lysozyme to 1 mg/ml immediately after thawing and were incubated at 30°C for 5 to 10 min, at which time 7 M guanidinium-HCl, 20 mM Tris (pH 7.5), 1 mM EDTA, and 1% Sarkosyl were added. RNA was recovered from the lysate by centrifugation through a

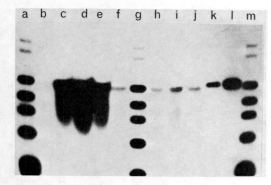

a b c d e f g h i j k l m

FIG. 1. Induction of *B. licheniformis* 749 *penP* mRNA. An overnight culture of *B. licheniformis* was diluted 1:50 into L broth and grown at 37°C. Cephalosporin C was added to 2 μg/ml when the log-phase cells reached a density of 0.13 mg (dry weight)/ml (0 h).Total RNA was prepared from cells either before addition of inducer (b) or 0.5 h (c), 1.0 h (d), 1.5 h (e), 2.0 h (f), 2.5 h (h), 3.0 h (i), and 3.5 h (j) after the addition. RNA samples (10 μg) were glyoxalated, electrophoresed on a 1.2% agarose gel, and hybridized to a 0.6-kb *Pst*I-*Cla*I *penP*-specific DNA probe after transfer to DBM paper. Markers (a, g, m) consisted of glyoxalated λ *Hind*III- and φX174 *Hae*III-restricted DNAs which were ^{32}P end labeled (10^5 cpm/lane); length (kb), 2.3, 2.0, 1.4, 1.1, 0.9, 0.6, and 0.3 to 0.1. (k) Glyoxalated 1.4-kb *Hind*III-*Bam*HI DNA fragment containing the *penP* structural gene, 0.15 ng. (l) Same as (k), 1.5 ng.

CsCl cushion as described (10), with minor modifications.

Northern analysis of *penP* mRNA. Nucleic acids were denatured by glyoxalation (16) and were separated on 1.2% agarose gels. Dimethyl sulfoxide was omitted from the glyoxalation reaction. Blotting of the gels onto diazobenzyloxymethyl (DBM) paper and hybridization with nick-translated DNA were essentially as described (1).

Preparation of probes. The plasmid pRWN1 contains the *penP* gene from *B. licheniformis* 9945A on a 4.2-kilobase (kb) *Eco*RI fragment derived from pTTE21 (11) and inserted into pRW4 at the *Eco*RI site. pRW4 is pBR322 with part of the *bla* gene deleted in the region which includes the *Pst*I site. A DNA fragment containing 83 bases that encode the signal sequence and 540 bases for the mature enzyme was removed from within *penP* by digestion of pRWN1 with *Pst*I and *Cla*I. This 0.6-kb *Pst*I-*Cla*I DNA fragment was used to probe *B. licheniformis* mRNA for *penP* specific sequences. A 1.65-kb region of DNA adjacent to, but not including, the 3′ end of the putative *penP* transcriptional terminator (15, 17) was isolated from pRWN1 by partial digestion with *Aha*III and *Eco*RI. This fragment was subcloned into pUC9 (23), which had been digested with *Sma*I and *Eco*RI, to give pRWS1.

The insert was removed from the multiple cloning site with *Bam*HI and *Eco*RI and was used to probe for transcripts derived from the region 3′ to *penP*. Probes were radiolabeled with [α-^{32}P]dCTP and [α-^{32}P]dGTP by nick translation as described (20).

RESULTS

Changes in level of *penP* mRNA during induction. Cultures of *B. licheniformis* 749 were induced with cephalosporin C under standard conditions and were sampled at intervals. The total RNA was extracted and fractionated by gel electrophoresis and transferred to DBM paper. Hybridization with ^{32}P-labeled probes prepared from subcloned DNA fragments containing *penP* or adjacent 3′ or 5′ fragments was used to detect and quantify the various mRNAs produced. The time course of a typical induction at 37°C is presented in Fig. 1 and 2. In an uninduced culture the level of *penP* mRNA per unit of total mRNA was low. After induction, the level rose to a maximum at 1 h and then fell sharply during the next hour, but remained above the uninduced level for several hours. When incubation was carried out at 30°C, the maximum level was not reached until 2 h, but the general shape of the curve was the same. At all times, the major *penP* mRNA had a size of 1,180 bases, as determined by S1 mapping, which is consistent with the 1,180-base mRNA obtained by McLaughlin et al. (15) from in vitro transcription of the *penP* gene. This mRNA species terminates 90 to 100 bases beyond the 3′ end of the structural gene and just beyond a typical transcription terminator stem loop (17).

The production of penicillinase lags behind the appearance of the *penP* mRNA. By 0.5 h after induction, the level of mRNA had reached nearly half maximum (Fig. 2) although only small amounts of penicillinase had yet been made. Nevertheless, the rate of penicillinase formation during the various time periods (shown by the shaded columns in Fig. 2) reflects the levels of *penP* mRNA. After the *penP* mRNA had fallen to severalfold the preinduction level, substantial amounts of penicillinase were still produced, but this is a result of the large amount of cell mass present, not of a high specific activity of synthesis.

B. licheniformis 749/C, constitutive for penicillinase synthesis, was grown under identical conditions but without inducer. The same major *penP* mRNA species was produced throughout and in amounts comparable to the maximum levels reached during induction of strain 749. Addition of cephalosporin C did not increase the level of *penP* mRNA formed.

Rate of turnover of *penP* mRNA. The half-life of the *penP* mRNA was measured by adding

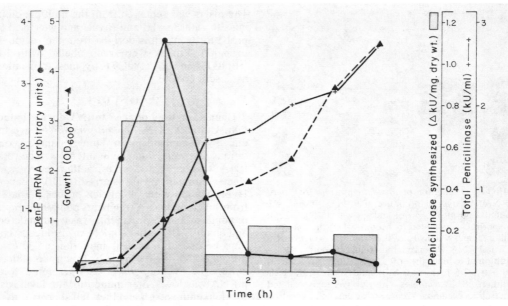

FIG. 2. Kinetics of *penP* mRNA and penicillinase formation. During the experiment described in Fig. 1, the relative levels of *penP* mRNA were determined by densitometry of the autoradiograms resulting from the Northern blot. Samples for total penicillinase were pipetted into tubes prechilled on dry ice. Penicillinase was measured as previously described (21). Growth was determined by diluting the cells 1:4 or 1:10 into 26 mM NaN$_3$ and measuring the opacity at 600 nm. The bars represent the net penicillinase synthesized during that period, normalized for the cell density.

rifampin (100 μg/ml) to inhibit RNA synthesis and following the subsequent decay in the amounts of *penP* mRNA present and in the synthesis of penicillinase. Addition of rifampin between 0.5 and 1.5 h after addition of inducer, a time when the level of *penP* mRNA is at its maximum, brought on a rapid fall in enzyme production and the *penP* mRNA decreased with a half-life of about 2 min (results not shown).

To determine whether the continued presence of inducer is needed for maintenance of *penP* mRNA formation or whether a long-lived stimulus has been produced, we induced a culture of strain 749 for 30 min, washed the cells thoroughly in "conditioned" medium (from a parallel uninduced culture), and suspended the cells in similar medium (Table 1). These cells continued to produce penicillinase for at least 30 to 60 min, although at about half the rate of a control culture treated in the same way but resuspended each time in the original culture fluid. A similar pattern was seen in the levels of *penP* mRNA. Although the manipulation and suspension of the induced cells in "conditioned" medium led to some decrease in mRNA and penicillinase synthesis, it is evident that a stable modification has occurred that continues to stimulate the formation of *penP* mRNA.

Formation of large mRNAs. When autoradiography of the RNA gels hybridized with the

*Pst*I-*Cla*I *penP* probe was continued for longer periods of time, two additional *penP* mRNA bands (2.9 and 3.4 kb) were detected (Fig. 3). The production and decay of these species have a time course similar to that of the major *penP* mRNA. Furthermore, induction with the pseudogratuitous inducer 2-(2'-carboxyphenyl)-benzoyl-6-aminopenicillanic acid (3) has given these bands in the same proportions.

It was possible to show definitively that these large mRNAs extend into the region 3' to the major *penP* mRNA. A probe was constructed that contains the region from *Aha*III to *Eco*RI, as shown in Fig. 4. This segment begins just downstream from the stem loop at which the major mRNA terminates. The probe hybridizes readily with the 2.9- and 3.4-kb mRNAs but not with the major band (Fig. 5). These longer bands are also present in the constitutive mutant 749/C in the same relative amounts. All three *penP* mRNAs appear to start at the *penP* promoter site (results not shown). Consequently, both large mRNAs extend into and through the 3' region that has been suggested as the site of the *penI* gene (13, 14, 22).

DISCUSSION

This work directly demonstrates that the overall regulation of *B. licheniformis* 749 penicillinase is primarily transcriptional in nature; any

TABLE 1. Formation of *penP* mRNA and penicillinase after removal of free inducer[a]

Treatment	Time after induction (min)	Penicillinase (total)		*penP* mRNA/ total mRNA (arbitrary units)
		U/ml	U/mg (dry wt) of cells	
Control	51	70	170	3.0
	81	440	590	10
	111	850	810	6.6
Washed	51	70	150	2.2
	81	340	380	3.1
	111	500	420	0.8

[a] An overnight culture of *B. licheniformis* 749 was diluted 1:50 into L broth and grown at 37°C. At a cell density of 0.17 mg (dry weight)/ml, the culture was split into four portions, two of which received 2 μg of cephalosporin C per ml. After 30 min, the cells were centrifuged at 5,000 × *g* for 5 min. The supernatant from one induced culture (washed) was discarded, and the pellet was suspended in the supernatant from one uninduced culture. The remaining cultures (control) were resuspended in their own supernatants. This washing step was repeated, leaving an induced control culture and the washed induced culture suspended in conditioned medium. Sampling and analysis for *penP* mRNA and penicillinase were carried out as described in the legend of Fig. 2.

regulatory roles for translation or processing of penicillinase are secondary. In addition, the short half-life for *penP* mRNA observed here confirms and extends the work of Davies (7), who reached the same conclusion on the basis of the kinetics of penicillinase production after inhibition of RNA synthesis with rifampin.

B. licheniformis 749/C produces penicillinase constitutively. One possibility is that the repressor protein known to be produced by *B. licheniformis* (11) has been inactivated by mutation. Our results support this argument in several ways. First, when fully induced wild-type strains are compared with *B. licheniformis* 749/C, similar levels of *penP* mRNA are seen. Furthermore, the sizes of *penP* mRNAs are the same (as are the physical maps of the DNA), indicating that both utilize the *penP* promoter. Finally, we have found that the DNA sequence of the *penP* promoter region is identical in strain 749 and in strain 749/C (unpublished data). From these findings we suggest that the constitutive phenotype in 749/C is not the result of a promoter/operator mutation but rather of a mutation in another regulatory element, possibly *penI*.

Our finding of larger *penP* mRNA in vivo is consistent with the previous report (22) that a mutation affecting an amino acid near the C terminus of penicillinase is 90% linked by trans-

formation to *penI*. On the basis of this linkage it has been suggested that *penP* and *penI* are at least part of an operon and may be transcribed on a polycistronic mRNA originating from the *penP* promoter (13, 14). To investigate this possibility, we have sequenced the DNA for about 0.5 kb downstream from the 3' end of the *penP* structural gene (unpublished data). The only sizable open reading frame in this region is shown schematically in Fig. 6. It extends 235 bases downstream from the termination codon for translation of *penP* and passes through the *penP* transcriptional termination signal. Initiation would occur 15 or 30 bases upstream from the penicillinase translation terminator, and the protein would be synthesized only if *B. licheniformis* 749 is an ochre suppressing strain. The smaller potential protein would contain 82 amino acid residues (Fig. 7). If such a protein is present in *B. licheniformis*, it is probably produced in small quantities. Less than 3% of the total *penP* mRNA extends past the *penP* transcriptional termination signal, and an ochre suppressor would be required. We do not know whether *B. licheniformis* is an ochre suppressing strain or whether the potential protein exists in *B. licheniformis*.

The elevated rate of enzyme and *penP* mRNA synthesis, even 3.5 h after addition of inducer,

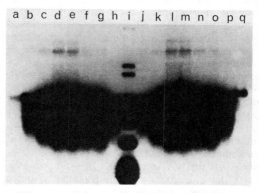

FIG. 3. Induction of high-molecular-weight *penP* mRNAs. An overnight culture of *B. licheniformis* was diluted 1:50 into L broth and grown at 30°C. When the cells reached 0.08 mg (dry weight)/ml, the culture was sampled (a) and then divided into two parts. One received 2-(2'carboxyphenyl)-benzoyl-6-aminopenicillanic acid (b–h) to a final concentration of 5 μM; cephalosporin C (j–p) was added to the other (2 μg/ml). Samples were taken at 0.5 h (b, j), 1.0 h (c, k), 1.5 h (d, l), 2.0 h (e, m), 2.5 h (f, n), 3.5 h (g, o) and 5.5 h (h, p). Total RNA was prepared from the samples, and 10 μg was analyzed as described in Fig. 1. (i) Glyoxalated markers (50,000 cpm) as in Fig. 1; the 4.3-kb λ HindIII DNA fragment can also be observed. (q) Glyoxalated 1.4-kb HindIII-BamHI fragment containing the *penP* structural gene, 1.5 ng.

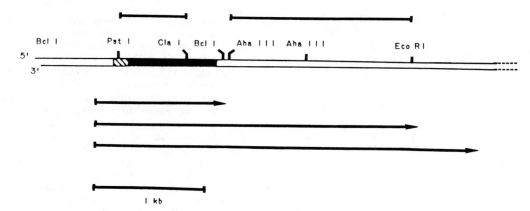

FIG. 4. *B. licheniformis* 749 *penP* transcripts. A partial map of the *penP* gene and adjacent regions. The solid segment is the mature protein, and the hatched area is the signal sequence. Significant restriction sites are indicated, along with the two major probes employed (top line). The major (small) transcript ends between the central *Bcl*I and *Aha*III sites. Beneath it are the two minor transcripts that run through the proposed *penI* region (14, 22).

suggests a long-lasting stimulus to *penP* expression. This is supported by the continued synthesis of penicillinase and *penP* mRNA in the absence of free inducer. The difference in magnitude of enzyme and mRNA synthesis between washed and control cells (Table 1) may reflect some damage to the cells during the washing process, as well as the possible continued presence of traces of cephalosporin C in the control but not the washed culture. It is unlikely that the long-lasting stimulus is due to entrapment of free inducer within the cell envelope. The delayed and relatively constant timing of maximum penicillinase synthesis, regardless of the inducer concentration (5), is inconsistent with such an explanation. Penicillin binds covalently to *B. licheniformis* cells, and this binding may alter the membrane in such a way as to produce the long-lasting stimulus. The cytotoxic inducer cephalosporin C and the pseudogratuitous inducer 2-(2'-carboxyphenyl)-benzoyl-6-aminopenicillanic acid yield very similar time courses for *penP* mRNA formation. This finding suggests that the toxicity of the β-lactam is not a critical feature in producing continued synthesis of *penP* mRNA.

In the DNA region 5' to the *penP* gene several mRNA transcripts are specifically induced by 2-(2'-carboxyphenyl)-benzoyl-6-aminopenicillanic acid or cephalosporin C. We are currently mapping the mRNAs and determining the kinetics of their formation. Characterization of these transcripts may aid us in identifying the stimulus to *penP* mRNA synthesis.

The studies presented here have enabled us to conclude that *penP* mRNA is made throughout the prolonged induction period and that the level of *penP* mRNA is a reasonable indicator of the rate of its synthesis by the cell. Determination of

the nature of the long-lasting stimulus must await further information on the molecular biology of penicillinase regulation.

SUMMARY

Hybridization with specific probes from the *B. licheniformis* 749 penicillinase structural gene,

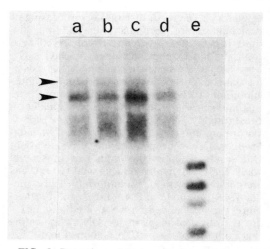

FIG. 5. Detection of high-molecular-weight *penP* mRNA by hybridization with a probe 3' to the *penP* gene. Samples of total RNA (7.5 μg) were prepared from *B. licheniformis* 749 cells after maximum induction of *penP* mRNA (a, b) and from log-phase cells of *B. licheniformis* 749/C (c, d). Cultures were grown, as described with previous figures, at 30°C (a, c) or 37°C (b, d). RNA was analyzed as described for Fig. 1 except that the probe was a subcloned 1.65-kb *Bam*HI-*Eco*RI DNA fragment from the region 3' to the major *penP* transcriptional terminator. (e) Glyoxalated markers (25,000 cpm) as in Fig. 1. Arrows indicate 2.9- and 3.4-kb bands.

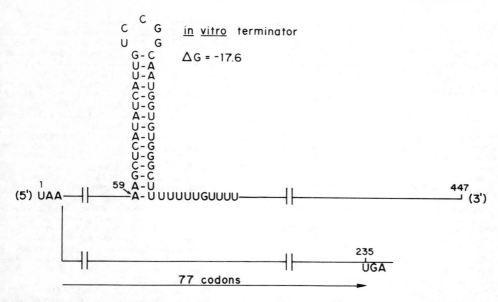

FIG. 6. Schematic structure of mRNA corresponding to the DNA region 3' to *penP*. The major translation terminator (nucleotides 1–3), the stem loop structure which serves as major transcription terminator, and the extent of the segment (to nucleotide 447) that has been sequenced (unpublished data) are indicated. Translation of the only sizable open frame (77 codons) depends on read through of the ochre terminator from an initiation site upstream.

penP, or the adjacent 3' and 5' regions has been used to identify and quantitate the specific mRNAs produced as a result of induction with cephalosporin C. The level of *penP* mRNA is low in uninduced cultures. On induction, it rises to a maximum at about 60 min and then declines, but remains above the uninduced level for several hours. Production of penicillinase follows the same sequence, and the rate of enzyme synthesis is roughly proportional to the level of *penP* mRNA. We conclude that regulation of penicillinase formation occurs primarily at the level of transcription.

The half-life of the *penP* mRNA was approximately 2 min, a result consistent with the 4 to 5 min previously calculated from the decay of

enzyme-forming capacity. Maintenance of free inducer was not required for continued *penP* mRNA synthesis; consequently, induction must produce a long-lasting stimulus for *penP* mRNA synthesis, perhaps a specific β-lactam–membrane protein adduct.

The major *penP* mRNA formed in vivo is the same size (ca. 1,180 bases) by S1 mapping as the in vitro product, but 2 to 3% of the in vivo product extends into and through the 3' region of the DNA that has been suggested as the site of the regulatory gene, *penI*. The only open frame in the 500-base pair 3' DNA region which has been sequenced to date begins either 15 or 30 base pairs upstream from the major *penP* translation terminator (UAA). Production of the

PENASE
→

(NH$_2$)M - N - G - K$^+$- X - K$^+$- H$^+$- C - D$^-$- E$^-$- S - V - K$^+$- T - S - S - Y - I - K$^+$-$\overset{20}{\text{S}}$ -

L - G - D$^-$- Q - A - H$^+$- I - I - V - R$^+$- Q - W - C - G - L - F - L - F - S -$\overset{40}{\text{I}}$ -

F - K$^+$- D$^-$- H$^+$- V - K$^+$- E$^-$- K$^+$- T - G - K$^+$- S - V - C - G - K$^+$- G - P - G -$\overset{60}{\text{F}}$ -

C - R$^+$- N - H$^+$- R$^+$- R$^+$- M - G - W - I - V - T - K$^+$- F - D$^-$- T - C - T - G -$\overset{80}{\text{G}}$ - A - T •

FIG. 7. Sequence of potential protein from region 3' to mature penicillinase. The arrow marks the end of the mature enzyme; X is the residue inserted by the putative ochre suppressor. The one-letter code for amino acids is used, with fixed charges indicated. The symbol • represents a terminator codon, either in frame (on the line) or in the second or third frame (one or two spaces below, respectively).

82 or 87 polypeptide would require read through by a suppressor.

Several mRNAs are produced from the region 5' to *penP*. Their precise location is being determined, as well as the kinetics of their appearance and decay.

LITERATURE CITED

1. **Alwine, J. C., D. J. Kemp, B. A. Parker, J. Reiser, J. Renart, G. R. Stark, and G. M. Wahl.** 1979. Detection of specific RNAs or specific fragments of DNA by fractionation in gels and transfer to diazobenzyloxymethyl paper. Methods Enzymol. **68:**220–242.
2. **Aten, R. F., and R. A. Day.** 1973. Penicillin-binding component of *Bacillus cereus.* J. Bacteriol. **14:**537–542.
3. **Bettinger, G. E., and J. O. Lampen.** 1970. Penicillinase (β-lactamase) induction in *Bacillus licheniformis* under pseudogratuitous conditions by 2-(2'-carboxyphenyl)-benzoyl-6-aminopenicillanic acid. J. Bacteriol. **104:**283–288.
4. **Collins, J. F.** 1971. The regulation of penicillinase synthesis in Gram-positive bacteria, p. 489–523. *In* H. J. Vogel (ed.), Metabolic pathways (Metabolic regulation), vol. 5. Academic Press, Inc., New York.
5. **Collins, J. F.** 1979. The *Bacillus licheniformis* β-lactamase system, p. 351–368. *In* J. M. T. Hamilton-Miller and J. T. Smith (ed.), Beta-lactamases. Academic Press, London.
6. **Csanyi, V., J. Gervai, I. Mile, and I. Ferencz.** 1971. Stability of penicillinase messenger RNA of *Bacillus cereus.* Biochim. Biophys. Acta **240:**420–428.
7. **Davies, J. W.** 1969. The stability and cell content of penicillinase messenger RNA in *Bacillus licheniformis.* J. Gen. Microbiol. **59:**171–184.
8. **Dubnau, D., and M. R. Pollock.** 1965. The genetics of *Bacillus licheniformis* penicillinase: a preliminary analysis from studies from mutation and inter-strain and intra-strain transformation. J. Gen. Microbiol. **41:**7–21.
9. **Duercksen, J. D.** 1964. Localization of the site of fixation of the inducer, penicillin, in *Bacillus cereus.* Biochim. Biophys. Acta **87:**123–140.
10. **Goodman, H. M., and R. J. MacDonald.** 1979. Cloning of hormone genes from a mixture of cDNA molecules.

Methods Enzymol. **68:**75–90.
11. **Imanaka, T., T. Tanaka, T. Tsunekawa, and S. Aiba.** 1981. Cloning of the genes for penicillinase, *penP* and *penI*, of *Bacillus licheniformis* in some vector plasmids and their expression in *Escherichia coli, Bacillus subtilis,* and *Bacillus licheniformis.* J. Bacteriol. **147:**776–786.
12. **Imsande, J.** 1978. Genetic regulation of penicillinase synthesis in gram-positive bacteria. Bacteriol. Rev. **42:**67–83.
13. **Kelly, L. E., and W. J. Brammar.** 1973. A frameshift mutation that elongates the penicillinase protein of *Bacillus licheniformis.* J. Mol. Biol. **80:**135–147.
14. **Kelly, L. E., and W. I. Brammar.** 1973. The polycistronic nature of the penicillinase structural and regulatory genes in *Bacillus licheniformis.* J. Mol. Biol. **80:**149–154.
15. **McLaughlin, J. R., S.-Y. Chang, and S. Chang.** 1982. Transcriptional analysis of the *Bacillus licheniformis penP* gene. Nucleic Acids Res. **10:**3905–3919.
16. **McMaster, G. K., and G. G. Carmichael.** 1977. Analysis of single- and double-stranded nucleic acids on polyacrylamide and agarose gels by using glyoxal and acridine orange. Proc. Natl. Acad. Sci. U.S.A. **74:**4835–4838.
17. **Neugebauer, K., R. Sprengel, and H. Schaller.** 1981. Penicillinase from *Bacillus licheniformis*: nucleotide sequence of the gene and implications for the biosynthesis of a secretory protein in a Gram-positive bacterium. Nucleic Acids Res. **9:**2577–2588.
18. **Pollock, M. R.** 1963. The differential effect of actinomycin D on the biosynthesis of enzymes in *Bacillus subtilis* and *Bacillus cereus.* Biochim. Biophys. Acta **76:**80–93.
19. **Pollock, M. R.** 1963. Exoenzymes, p. 121–178. *In* I. C. Gunsalus and R. Y. Stanier (ed.), The bacteria, vol. 4. Academic Press, Inc., New York.
20. **Rigby, P. W. J., M. Dieckmann, C. Rhodes, and P. Berg.** 1977. Labelling deoxyribonucleic acid to high specific activity *in vitro* by nick translation with DNA polymerase I. J. Mol. Biol. **113:**237–251.
21. **Sargent, M. G.** 1968. Rapid fixed-time assay for penicillinase. J. Bacteriol. **95:**1493–1494.
22. **Sherratt, D. J., and J. F. Collins.** 1973. Analysis by transformation of the penicillinase system of *Bacillus licheniformis.* J. Gen. Microbiol. **75:**217–230.
23. **Viera, J., and J. Messing.** 1982. The pUC plasmids, an M13mp7-derived system for insertion mutagenesis and sequencing with synthetic universal primers. Gene **19:**259–268.

Bacillus subtilis Secretion Vectors for Proteins and Oligopeptides Constructed from B. subtilis α-Amylase Genes

KUNIO YAMANE,[1] TERUAKI SHIROZA,[1] TAKASHI FURUSATO,[1] KOUJI NAKAMURA,[1] KIYOSHI NAKAZAWA,[1] KAZUO YANAGI,[2] MAKARI YAMASAKI,[3] AND GAKUZO TAMURA[3]

Institute of Biological Sciences[1] and Institute of Basic Medical Sciences,[2] University of Tsukuba, Sakura, Ibaraki, 305, and Department of Agricultural Chemistry, The University of Tokyo,[3] Bunkyo-ku, Tokyo, 113 Japan

Bacillus subtilis is an excellent host organism for genetic engineering because of its active protein secretion system. The use of this secretion system is a powerful tool for the production of proteins and peptides.

α-Amylase is one of the major extracellular enzymes of *B. subtilis*. We cloned the α-amylase structural gene, *amyE*[+], and its regulatory gene, *amyR2*, from the chromosomal DNA of an α-amylase–hyperproducing strain, *B. subtilis* NA64, using the temperate *B. subtilis* phage ρ11 and the plasmid pUB110. The plasmid constructed, pTUB4, has a 2.3-kilobase (kb) DNA insert (19). From DNA nucleotide sequence analysis of the *amyR2* and *amyE*[+] genes of pTUB4 (12, 23) and amino acid sequence analysis of the NH_2-terminal region of *B. subtilis* extracellular α-amylase (7), it was shown that the precursor molecule of α-amylase contains 41 amino acids in its putative signal peptide (signal sequence) and that one of the *Alu*I sites in the plasmid is coincident with the site of cleavage of the signal sequence which generates the extracellular enzyme. A cleavage at this *Alu*I site and at another *Alu*I site upstream from *amyE*[+] generates a 424-base pair (bp) DNA fragment. This fragment contains an inverted repeat structure rich in A and T (32/40), the potential RNA polymerase recognition site, a Pribnow box, a Shine-Dalgarno sequence, and, finally, the DNA sequence for the signal sequence which was made up of 123 bp, as schematically shown in Fig. 1. Figure 1 also shows the most stable base-pairing scheme in the promoter region of the *B. subtilis* NA64 α-amylase gene which has a free energy of −21.2 kcal/mol. The DNA nucleotide sequences of the promoter and signal sequence coding regions of the *B. subtilis* α-amylase gene were quite different from those of the *B. amyloliquefaciens* α-amylase gene (14, 20).

We used the 424-bp *Alu*I-*Alu*I fragment described above to construct a secretion vector for high-molecular-weight polypeptides and enzymes in *B. subtilis*. When the *Escherichia coli* β-lactamase gene or the mouse interferon-β gene was fused downstream from the 424-bp DNA fragment and transferred into *B. subtilis* cells,

the genes were expressed and active β-lactamase or mouse interferon-β was secreted into the culture medium (11; T. Shiroza, K. Nakazawa, N. Tashiro, K. Yamane, K. Yanagi, M. Yamasaki, G. Tamura, H. Saito, Y. Kawade, and T. Taniguchi, in press).

We also developed another secretion vector for oligopeptides in *B. subtilis*, using the α-amylase genes of pTUB4. When chemically synthesized human angiotensin I gene was inserted in the vector and the hybrid plasmid was transferred into *B. subtilis* cells, a human angiotensin I–α-amylase chimeric protein with amylase activity was secreted into the culture medium. The CNBr-cleaved product from the chimeric protein had angiotensin I activity indistinguishable from the activity of authentic human angiotensin I.

EXPERIMENTAL PROCEDURES

Bacterial strain and plasmids. *B. subtilis* 207-25 (m_{168}^- *hsrM recE4 amyE07 aroI906 leuA8 lys21*) is a derivative of *B. subtilis* Marburg 168. *B. subtilis* plasmid pTUB4 is pUB110 with a 2.3-kb insert containing the *amyR2* and *amyEm*[+] genes from the chromosomal DNA of *B. subtilis* NA64. *B. subtilis* plasmid pTUB101 had a 2.5-kb insert containing the *amyR2* and *amyEn*[+] genes derived from *B. natto* IAM1212 (21a). *E. coli* plasmid pMβ-3, kindly provided by T. Taniguchi (Osaka University, Osaka, Japan), had a 680-bp insert of mouse interferon-β cDNA (5).

Medium and culture conditions. The composition of LG medium and the growth conditions were as described previously (22).

Isolation of plasmid DNAs. Plasmid DNAs were prepared by the rapid alkaline lysis method of Birnboim and Doly (1) and were purified by CsCl equilibrium centrifugation in the presence of ethidium bromide. Large DNA fragments and plasmids were separated by agarose gel electrophoresis, and small fragments were separated by polyacrylamide gel electrophoresis. For preparative purposes, DNA fragments in agarose gels were electroeluted onto hydroxylapatite (18) or DNAs in polyacrylamide gels were extracted from the crushed and ground gels.

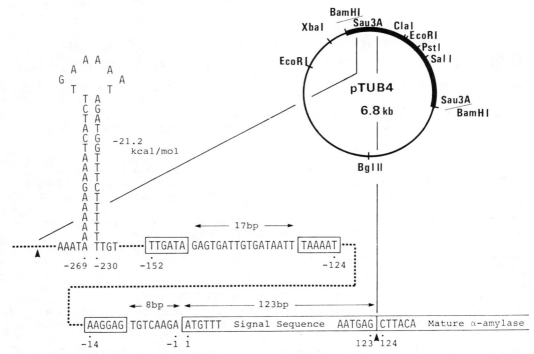

FIG. 1. Schematic drawing of a possible hairpin structure of the inverted repeat sequence, the potential RNA polymerase recognition site, the potential Pribnow box, a Shine-Dalgarno sequence, and the signal sequence coding region for secretion in the upstream and promoter region of the α-amylase structural gene of pTUB4. The arrowhead indicates the *Alu*I restriction site.

Transformation of *B. subtilis*. Protoplast transformation of *B. subtilis* was performed by the method of Chang and Cohen (2).

DNA nucleotide sequence determination. DNA nucleotide sequences were determined by the Maxam and Gilbert method (8) or by the dideoxy chain-termination method of Sanger et al. (16). For the latter method, DNA fragments were cloned into M13mp8 or M13mp9 DNA (9).

Western blot analysis. After electrophoresis in sodium dodecyl sulfate–10% polyacrylamide gels (6), mouse interferon-β was visualized by the Western blot method (21) with the use of sheep anti-mouse interferon-β serum, rabbit anti-sheep immunoglobulin G serum, and ^{125}I-labeled F(ab′)₂ fragment of donkey serum against rabbit immunoglobulin G.

Assay of antiviral activity. The antiviral activity of interferon-β molecules was measured on L-cells by the inhibition of the cytopathic effect caused by vesicular stomatitis virus (17).

Assay of angiotensin I activity. Angiotensin I activity was measured by radioimmunoassay with the use of ^{125}I-labeled angiotensin I and anti-angiotensin I serum (4).

Enzymes. Restriction enzymes, bacterial alkaline phosphatase, exonuclease Bal-31, T4 DNA ligase, and DNA polymerase I (Klenow fragment) were purchased from Takara Shuzo Co. Ltd., Kyoto, Japan, or from Bethesda Research Laboratories, Bethesda, Md. Each enzyme was used as specified by the manufacturer.

Chemicals. *Hin*dIII linker DNA, d(CAAG-CTTG), was purchased from Takara Shuzo Co. Ltd., and kanamycin was from Meiji Seika Co. Ltd., Tokyo, Japan. All other chemicals were of reagent grade.

RESULTS

Construction of a secretion vector from the *B. subtilis* α-amylase genes and secretion of active β-lactamase by *B. subtilis* cells. To test the utility of the 424-bp DNA fragment for secretion of proteins, the *E. coli* β-lactamase gene, provided by I. Palva, Finland, was fused downstream from the 424-bp fragment. The *E. coli* β-lactamase gene which was derived from pBR322 lacks the DNA region of the signal peptide and contains *Hin*dIII termini (15). The cleavage maps of the constructed plasmids pTUB226 and pTUB228 are derived as shown in Fig. 2. The orientation of the inserted DNA fragments relative to the vector DNA was different in pTUB226 and pTUB228. The *B. subtilis* 207-25 strains

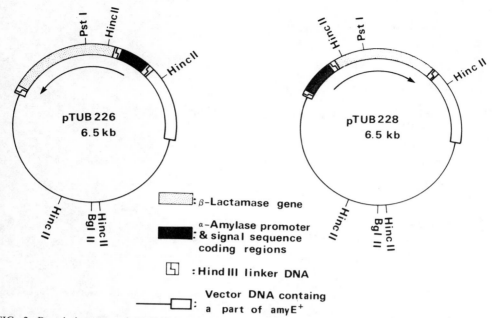

FIG. 2. Restriction maps of pTUB226 and pTUB228. Arrows indicate the possible orientation of transcription and translation in each plasmid.

carrying pTUB226 or pTUB228 secreted almost the same amount of β-lactamase activity into the culture media. These results indicated that active β-lactamase was expressed and secreted by the aid of the promoter and the signal sequence of the *B. subtilis* α-amylase gene. More than 95% of active β-lactamase synthesized was secreted into the culture medium from *B. subtilis* cells. The *Hind*III site located upstream from the 424-bp fragment in pTUB228 was removed, and an improved secretion vector, pTUB285, was constructed.

Insertion of the mouse interferon-β cDNA gene into pTUB285. Since the secretion vector constructed from the *B. subtilis* α-amylase gene was able to produce *E. coli* β-lactamase in *B. subtilis*, the mouse interferon-β gene was inserted in the secretion vector, pTUB285. Figure 3 shows the relevant structure of *E. coli* plasmid pMβ-3, containing the 680-bp mouse interferon-β cDNA insert at the *Pst*I site of pBR322 by the GC

tailing method. DNA nucleotide sequence analysis and prediction of the corresponding amino acid sequence revealed that the precursor protein of mouse interferon-β was composed of 182 amino acids. The first 21 amino acids were considered to be a signal peptide, and the remaining 161 amino acids were considered to be the mature interferon-β polypeptide (5). pMβ-3 was digested with both *Bam*HI and *Pst*I, and a 645-bp fragment, lacking a part of signal peptide, was isolated. This DNA fragment was treated with exonuclease Bal-31 to remove the signal peptide. Fragments approximately 550 bp in length were recovered from a 5% polyacrylamide gel, *Hind*III linkers were added, and the fragments were ligated with vector DNA from pTUB285.

Interferon-β–secreting transformants were identified by use of a new method for screening for secretory proteins. Two nitrocellulose filters were put on one agar plate, and kanamycin-re-

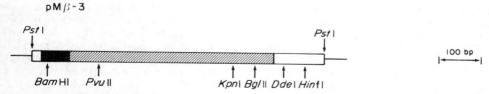

FIG. 3. Relevant structure of pMβ-3. The mouse interferon-β cDNA gene, which was cloned in pBR322 (——), is composed of 5′ and 3′ noncoding regions (open bar), signal peptide for mouse interferon-β (solid bar), and the DNA sequence for mature mouse interferon-β (shaded bar).

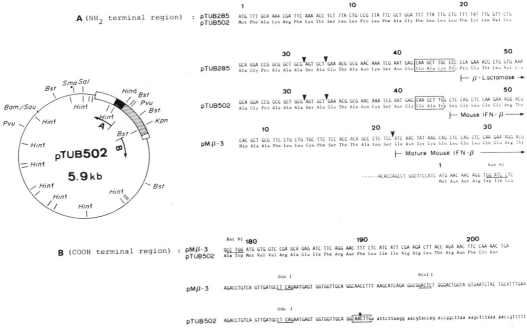

FIG. 4. Restriction map of pTUB502 expressing active mouse interferon-β and DNA nucleotide sequences at and near both junction regions of the secretion vector and mouse interferon-β cDNA. The arrows A and B indicate the direction and extension in the junction regions for the NH₂-terminal (upper) and the COOH-terminal (lower) regions of the mouse interferon-β gene, respectively. Lowercase letters represent the sequence of vector DNA originating from pUB110. Arrowheads show possible cleavage sites between signal peptide and mature protein.

sistant (Km^r) B. subtilis transformants were inoculated on the upper side of the first membrane filter. After overnight incubation, secretory proteins were adsorbed on the secondary filter. The first filter was used for colony hybridization analysis (3), and the second filter was used for immunoblot analysis to visualize the mouse interferon-β. Eighty-eight transformants secreting antigenically active proteins were selected from about 2,000 Km^r transformants, their production of antiviral activity was tested, and the transformant producing the highest antiviral activity was selected. The plasmid harbored in this transformant was designated as pTUB502. The physical map of pTUB502 was derived as shown in Fig. 4. To study the relationships between antiviral activity and amino acid sequence of the protein, the nucleotide sequence at and near the two junctions of the vector DNA and mouse interferon-β cDNA in pTUB502 was determined by the dideoxy chain-termination method (Fig. 4). The mouse interferon-β gene started from the codon at position 27, a Leu residue (codon position 6 in mature mouse interferon-β) in the NH₂-terminal region. The nucleotide sequences in the coding region of the COOH-terminal region and the

termination codon (TGA) were identical to those of pMβ-3. The vector DNA, shown by lowercase letters in Fig. 4, started 50 bp downstream from the termination codon. The sequence CAACTTG 43 to 49 bp downstream from the termination codon seemed to be derived from the HindIII linker by a single base deletion of G. The DNA nucleotide sequence in the mouse interferon-β structural gene of pTUB502 was assumed to be the same as that of pMβ-3 because their restriction maps were identical.

Since the precursor protein of β-lactamase from pTUB228 is thought to be cleaved between amino acids at positions 31 (Ala) and 32 (Ser) or between 33 (Ala) and 34 (Gln) (10), the mouse interferon-β expressed from pTUB502 will contain 8 or 10 amino acids of the B. subtilis α-amylase signal sequence, a tripeptide encoded in the linker DNA, and mature mouse interferon-β proper starting from position 6. It is noteworthy that the hybrid protein encoded in pTUB502 has the highest antiviral activity among the proteins in which the mouse interferon-β gene starts from codon positions −5, +1, and +3 of the mature mouse interferon-β.

Expression and secretion of biologically active mouse interferon-β by B. subtilis. The time course

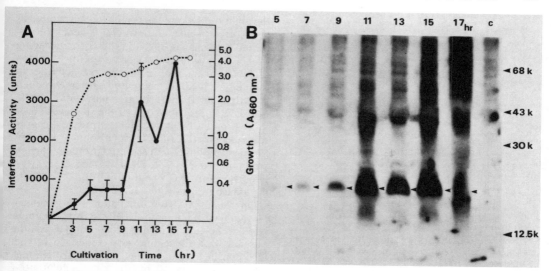

FIG. 5. Time course of the production of antiviral activity (A) and antigenic activity (B) in the supernatants of the *B. subtilis* strain carrying pTUB502. The strain was cultured in LG medium containing 10 μg of kanamycin per ml at 37°C for indicated periods. Cells were removed by centrifugation, and antiviral activity in the supernatants was assayed by the reduction of cytopathic activity. Symbols: ●, antiviral activity; ○, growth. Antigenic activity in the supernatants was assayed by Western blot analysis (B). Small arrows indicate the protein bands corresponding to mature interferon-β.

of the production of mouse interferon-β activities in the culture supernatants of the strain containing pTUB502 is shown in Fig. 5. The maximum antiviral activity, approximately 4×10^3 U/ml, was observed at 11 to 15 h after inoculation. However, the activity disappeared rapidly after further growth. Similar results were obtained by Western blot analysis of the culture supernatants (Fig. 5B). The protein band corresponding to mouse interferon-β became faint at 17 h of growth. Thus, the decrease in antiviral activity at 17 h might be caused by degradation of the mouse interferon-β molecule, probably by extracellular proteases of the host cells.

Construction of a secretion vector for oligopeptides. Yang et al. cloned the α-amylase gene of *B. subtilis*, using the *E. coli* vector systems (24). The DNA nucleotide sequence analysis indicated that the α-amylase was composed of 660 amino acids, including its signal sequence. Our previous experiments suggested that a shorter polypeptide, consisting of the first 477 amino acids encoded in the α-amylase gene, is capable of giving both α-amylase enzyme activity and immunological activity. A *Hin*dIII linker DNA was inserted in one of the *Ava*II sites (nucleotides 1,531 to 1,535) of pTUB4, and plasmid pTUB12 was constructed. The precursor of α-amylase from pTUB12 was deduced to consist of 520 amino acids by a shift of the reading frame downstream from the linker DNA. pTUB12 was used as the secretion vector for oligopeptides.

Production of human angiotensin I by *B. subtilis*. We selected human angiotensin I as a model because of its intrinsic biological interest and its sensitive radioimmunoassay. Human angiotensin I is a decapeptide whose amino acid sequence is Asp-Arg-Val-Tyr-Ile-His-Pro-Phe-His-Leu. The gene for human angiotensin I was chemically synthesized by the solid-phase method of phosphotriesters (13). The codon usage bias observed in the *B. subtilis* α-amylase gene was used in the design of the gene. The gene was equipped with *Hin*dIII termini, an ATG codon preceding the first Asp codon, and two stop codons (TGA and TAG) after the COOH-terminal Leu codon (Fig. 6). The gene was inserted into the *Hin*dIII site of pTUB12, and the plasmid constructed was transferred into *B. subtilis* cells. One thousand Km^r and amylase-positive transformants were tested by

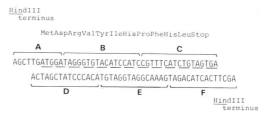

FIG. 6. Chemically synthesized human angiotensin I gene. Each bracket (A, B, C, D, E, and F) indicates the fragment synthesized.

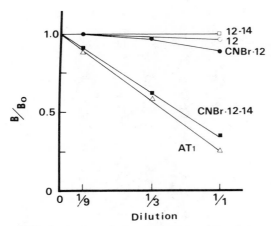

FIG. 7. Angiotensin I activity of the authentic human angiotensin I (AT$_1$, △), CNBr cleavage product of α-amylase from pTUB12-14 (CNBr · 12-14, ■), CNBr cleavage product of α-amylase from pTUB12 (CNBr · 12, ●), α-amylase from pTUB12-14 (12-14, □), and α-amylase from pTUB12 (12, ○).

colony hybridization (3), with ^{32}P-labeled human angiotensin I DNA as the probe. Thirty-four positive colonies were selected. Extracellular proteins having amylase activity from the 34 strains were partially purified and cleaved by CNBr, and the human angiotensin I activity of the cleaved product was assayed. One strain from which α-amylase showed the highest angiotensin I activity was selected. The plasmid harbored in this strain was designated as pTUB12-14. Nucleotide sequence analysis of the DNA fragment containing the gene for human angiotensin I revealed that the human angiotensin I gene in pTUB12-14 joined with pTUB12 in the correct reading frame from nucleotides 1,537 to 1,581.

The extracellular α-amylase from pTUB12-14 was purified by ammonium sulfate precipitation and chromatography (twice) on DEAE-cellulose. When the α-amylase was cleaved by CNBr, the cleavage product showed angiotensin I activity indistinguishable from the activity of the authentic human angiotensin I. Thus, α-amylase from pTUB12-14 was a chimeric protein of human angiotensin I and B. subtilis α-amylase (Fig. 7).

DISCUSSION

We constructed two vectors from the B. subtilis α-amylase gene useful for secretion of high-molecular-weight polypeptides and enzymes or for oligopeptides. The former vector was able to express and secrete active E. coli β-lactamase and mouse interferon-β into the culture medium from B. subtilis cells. Thus, the vector will have

the ability to secrete the proteins, irrespective of their origin, when the structural genes of exported proteins are fused in the correct reading frame with the α-amylase signals. However, the β-lactamase and mouse interferon-β activities in the culture medium decreased during extended growth (later than 15 h after inoculation). The decrease is probably due to the degradation of these proteins by extracellular proteases. A protease-deficient B. subtilis strain should be useful for hyperproduction of secreted proteins. Such a strain, improved by R. H. Doi's laboratory, was effective for the production of E. coli β-lactamase (unpublished data).

On the other hand, the human angiotensin I–α-amylase chimeric protein stably accumulated in the culture medium like α-amylase. It is possible that the amino acid sequence of the angiotensin I region in the chimeric protein has no site for cleavage by the extracellular proteases because of its low molecular weight. Further improvements in secretion vectors, host cells, and culture conditions are necessary to utilize the secretion ability of B. subtilis fully in the production of proteins derived from other organisms.

SUMMARY

B. subtilis secretion vector systems for the production of enzymes and high-molecular-weight polypeptides and for oligopeptides were constructed from the B. subtilis α-amylase gene. When an E. coli β-lactamase gene and a mouse interferon-β gene were fused downstream from the promoter and signal sequence of the α-amylase gene, the genes were expressed, and active β-lactamase and mouse interferon-β were secreted into the culture medium. A chemically synthesized human angiotensin I gene was inserted into another secretion vector which was constructed from the promoter and signal sequence coding regions and the major part of structural gene of the α-amylase. Human angiotensin I–α-amylase chimeric proteins having amylase activity were secreted into the culture medium. The CNBr cleavage product from one of the chimeric proteins showed angiotensin I activity indistinguishable from the activity of authentic human angiotensin I.

We thank H. Saito (The University of Tokyo), Y. Kawade (Kyoto University), and T. Taniguchi (Osaka University) for helpful discussions and for kindly supplying antiserum and plasmids. We are also grateful to I. Palva (Finland) for kindly supplying the E. coli β-lactamase gene and to S. Hirose and K. Murakami (Tsukuba University) for assay of angiotensin I.

This work was supported by a Grant-in-Aid for Special Distinguished Research from the Ministry of Education, Science and Culture of Japan.

LITERATURE CITED

1. **Birnboim, H. C., and J. Doly.** 1979. A rapid alkaline extraction procedure for screening recombinant plasmid DNA. Nucleic Acids Res. **7**:1513–1523.

2. **Chang, S., and S. N. Cohen.** 1979. High frequency transformation of *Bacillus subtilis* protoplasts by plasmid DNA. Mol. Gen. Genet. **168**:111–115.

3. **Grunstein, M., and D. Hogness.** 1975. Colony hybridization: a method for the isolation of cloned DNAs that contain a specific gene. Proc. Natl. Acad. Sci. U.S.A. **72**:3961–3965.

4. **Haber, E., T. Kaerner, L. B. Page, B. Kliman, and A. Purnode.** 1969. Application of radioimmunoassay for angiotensin I to the physiologic measurements of plasma renin activity in normal human subject. J. Clin. Endocrinol. **29**:1349–1355.

5. **Higashi, Y., Y. Sokawa, Y. Watanabe, Y. Kawade, S. Ohno, C. Takaoka, and T. Taniguchi.** 1983. Structure and expression of a cloned cDNA for mouse interferon-β. J. Biol. Chem. **258**:9522–9529.

6. **Laemmli, U. K.** 1970. Cleavage of structural proteins during the assembly of the head of bacteriophage T4. Nature (London) **227**:680–685.

7. **Mäntsälä, P., and H. Zalkin.** 1979. Membrane-bound and soluble extracellular α-amylase from *Bacillus subtilis*. J. Biol. Chem. **254**:8540–8547.

8. **Maxam, A. M., and W. Gilbert.** 1980. Sequencing end-labeled DNA with base-specific chemical cleavages. Methods. Enzymol. **65**:499–560.

9. **Messing, J., and J. Vieira.** 1982. A new pair of M13 vectors for selecting either DNA strand of double-digest restriction fragments. Gene **19**:269–276.

10. **Ohmura, K., K. Nakamura, H. Yamazaki, T. Shiroza, K. Yamane, Y. Jigami, H. Tanaka, K. Yoda, M. Yamasaki, and G. Tamura.** 1984. Length and structural effect of signal peptides derived from *Bacillus subtilis* α-amylase on secretion of *Escherichia coli* β-lactamase in *B. subtilis* cells. Nucleic Acids Res. **12**:5307–5319.

11. **Ohmura, K., T. Shiroza, K. Nakamura, A. Nakayama, K. Yamane, K. Yoda, M. Yamasaki, and G. Tamura.** 1984. A *Bacillus subtilis* secretion vector system derived from the *B. subtilis* α-amylase promoter and signal sequence region, and secretion of *Escherichia coli* β-lactamase by the vector system. J. Biochem. **95**:87–93.

12. **Ohmura, K., H. Yamazaki, Y. Takeichi, A. Nakayama, K. Otozai, K. Yamane, M. Yamasaki, and G. Tamura.** 1983. Nucleotide sequence of promoter and NH₂-terminal signal peptide region of *Bacillus subtilis* α-amylase gene cloned in pUB110. Biochem. Biophys. Res. Commun. **112**:678–683.

13. **Ohtsuka, E., M. Ikehara, and D. Söll.** 1982. Recent developments in the chemical synthesis of polynucleotides. Nucleic Acids Res. **10**:6553–6570.

14. **Palva, I., R. F. Petterson, N. Kalkkinen, P. Lehtovaara, M. Sarvas, H. Söderlund, K. Takkinen, and L. Kääriäinen.** 1981. Nucleotide sequence of the promoter and NH₂-terminal signal peptide region of α-amylase gene from *Bacillus amyloliquefaciens*. Gene **15**:43–51.

15. **Palva, I., M. Sarvas, P. Lehtovaara, M. Sibakov, and L. Kääriäinen.** 1982. Secretion of *Escherichia coli* β-lactamase from *Bacillus subtilis* by the aid of α-amylase signal sequence. Proc. Natl. Acad. Sci. U.S.A. **79**:5582–5586.

16. **Sanger, F., S. Nicklen, and A. R. Coulson.** 1977. DNA sequencing with chain-terminating inhibitors. Proc. Natl. Acad. Sci. U.S.A. **74**:5463–5467.

17. **Shimokata, K.** 1978. Studies on the pathogenicity of human-origin parainfluenza virus in the brain of mice. Microbiol. Immunol. **22**:535–543.

18. **Tabak, H. F., and R. A. Flavell.** 1978. A method for the recovery of DNA from agarose gels. Nucleic Acids Res. **5**:2321–2332.

19. **Takeichi, Y., K. Ohmura, A. Nakayama, K. Otozai, and K. Yamane.** 1983. Cloning of *Bacillus subtilis* α-amylase structural gene in plasmid pUB110. Agric. Biol. Chem. **47**:159–161.

20. **Takkinen, K., R. F. Petterson, N. Kalkkinen, I. Palva, H. Söderlund, and L. Kääriäinen.** 1983. Amino acid sequence of α-amylase from *Bacillus amyloliquefaciens* deduced from the nucleotide sequence of the cloned gene. J. Biol. Chem. **258**:1007–1013.

21. **Towbin, H., T. Staehelin, and J. Gordon.** 1979. Electrophoretic transfer of proteins from polyacrylamide gels to nitrocellulose sheets: procedure and application. Proc. Natl. Acad. Sci. U.S.A. **76**:4350–4354.

21a. **Yamane, K., Y. Hirata, T. Furusato, H. Yamazaki, and A. Nakayama.** 1984. Changes in the properties and molecular weights of *Bacillus subtilis* M-type and N-type α-amylase resulting from a spontaneous deletion. J. Biochem. (Tokyo) **96**:1849–1858.

22. **Yamane, K., Y. Takeichi, T. Masuda, F. Kawamura, and H. Saito.** 1982. Construction and physical map of a *Bacillus subtilis* specialized transducing phage ρ11 containing *B. subtilis lys⁺* gene. J. Gen. Appl. Microbiol. **28**:417–428.

23. **Yamazaki, H., K. Ohmura, A. Nakayama, Y. Takeichi, K. Otozai, M. Yamasaki, G. Tamura, and K. Yamane.** 1983. α-Amylase genes (*amyR2* and *amyE⁺*) from an α-amylase-hyperproducing *Bacillus subtilis* strain: molecular cloning and nucleotide sequences. J. Bacteriol. **156**:327–337.

24. **Yang, M., A. Galizzi, and D. Henner.** 1983. Nucleotide sequence of the α-amylase gene from *Bacillus subtilis*. Nucleic Acids Res. **11**:237–249.

Two Structural Genes for Alkaline Phosphatase in *Bacillus licheniformis* Which Require Different RNA Polymerase Holoenzymes for Transcription: a Possible Explanation for Complex Synthesis and Localization Data

F. MARION HULETT

Department of Biological Sciences, University of Illinois at Chicago, Chicago, Illinois 60680

The study of alkaline phosphatase (APase) from *Bacillus* species has been hindered by the resistance of the system to conventional genetic analysis. However, renewed interest in APase has been heightened because of available technology which allowed the cloning of the APase genes for use in (i) studying the molecular basis of regulation of vegetative and sporulation APase genes and (ii) constructing plasmid secretion vectors by fusing a structural gene of interest to the appropriate portion of the APase gene coding for a well-regulated secreted gene product.

Early studies suggested that there were two APase proteins: one was synthesized during vegetative growth and was repressible by P_i (37; C. Anagnostopoulos, Fed. Proc. **19**:48, 1960), and the other was produced during sporulation in the presence of inorganic phosphate concentrations which would repress the synthesis of the vegetative enzyme (4, 39, 40). Glenn (7) indeed suggested that there might be two structural genes for APase (one sporulation and one vegetative) but favored the hypothesis that the two enzyme activities were identical and the product of the same gene under distinct control elements in the vegetative and sporulating states. These authors took the latter position because there were no discernible chemical-physical differences in the APase enzyme from vegetative or sporulating cells.

Our interest in the alkaline phosphatase in *Bacillus* spp. stemmed from the fact that reports of APase localization varied from a membrane-bound species to a secreted species (3, 6, 16, 37, 44). We suspected that the membrane-bound species might be associated with the outer surface of the cytoplasmic membrane and be a precursor to the secreted form. This appeared to be an excellent system for studies on protein localization and secretion in procaryotes. We chose a strain of *B. licheniformis* on the basis of its overproduction of APase relative to other *Bacillus* species. Our histochemical (24) and lactoperoxidase-^{125}I (34) localization studies indicated that active APase dimers were peripherally associated with the inner surface of the cytoplasmic membrane. Growth conditions sig-

nificantly affect the distribution and the amount of APase synthesis in the organism. Conditions of growth are described below in which the organism secretes 99% of the APase synthesized. It is difficult to interpret these localization data (active dimer associated with inner leaflet of cytoplasmic membrane [24, 34] versus active dimer secreted [13, 32]) on the basis of any of the current models for protein secretion (1, 5, 38, 41, 42) or insertion of proteins into membranes (22, 30) if a single gene is responsible for these APase species.

We also, but for different reasons, were compelled to hypothesize either that there is a single structural gene which is under uniquely complex transcriptional and translational regulatory control or that there are multiple structural genes for APase which account for the synthesis of different species directed to different locations depending on the growth conditions and phase of growth (9, 32). Despite an extensive search by a number of groups, including ours (7, 8, 11, 21), no phosphatase mutant has yet been isolated which is the result of a structural gene mutation. (The loci of a number of phosphatase regulatory mutants have been mapped on the *B. subtilis* chromosome.) Since the APase gene(s) itself had proved resistant to conventional genetic analysis, it was clear that it would be necessary to clone the structural gene(s) for APase to test the above alternative hypothesis.

We describe below the current biochemical knowledge of APase species from *B. licheniformis* MC14 and the genes which are responsible for their synthesis.

METHODS

Bacterial strains and plasmids. The bacterial strains and plasmids used are given in Table 1. *E. coli* Xph90a, which contains deletion E15 within *phoA* (*phoA8*) was obtained from J. Beckwith via P. Berg.

Media. Growth media for *B. licheniformis* were described before (32). Inoculation and culture growth in a defined minimal salts (DMCo^{2+}) medium have been described previously (32). In labeling experiments, [^{35}S]methionine (10

TABLE 1. Bacterial strains and plasmids

Bacterial strain or plasmid	Genotype or phenotype	Origin
E. coli Xph90a	F⁻ lacZ624 phoAE15 proC⁺ phoR⁺ trp rpsL	J. Beckwith (18)
B. licheniformis MC14	phoA⁺	F. M. Hulett (16)
pMK2004	Ampʳ Tetʳ Kanʳ	M. Kahn (19)
pMH8	Ampˢ Tetʳ Kanʳ phoA⁺	This study
pMH81	Ampʳ Tetˢ Kanˢ phoA⁺	This study
pMH87	Kanʳ phoA⁺	This study

μCi/ml) and cobalt (0.1 mM) were added when the APase activity reached 0.2 U/ml, unless otherwise stated. Screening and selection were carried out on 1% Neopeptone plates (13). The indicator dye 5-bromo-4-chloro-3-indolylphosphate-p-toluidine (XP) was obtained from Sigma Chemical Co.

DNA hybridization. DNA-DNA hybridization was carried out by the method of Southern (31).

ExoIII-S1 deletion mapping. Deletion mapping was carried out using a procedure described by Roberts and Lauer (28).

Preparation of DPT-paper and Western blotting of protein from gel to DPT-paper. The procedures for preparation of DPT (diazophenylthioether)-paper and Western blotting have been outlined previously (43). The only modification was that iodinated goat anti-rabbit immunoglobulin G was substituted for iodinated protein A for labeling the primary antibody. The primary antibody was made to the salt-extractable, membrane-associated APase.

In vitro transcription. The methods described by Goldfarb et al. were used for in vitro transcription (10).

Fluorography techniques. The slab gels for fluorographic detection of ^{35}S-labeled proteins were prepared by the method of Bonner and Laskey (2).

RESULTS

A number of biochemical, physiological, and immunological studies concerning the synthesis and localization of APase (orthophosphoric monoester phosphohydrolase; EC 3.1.3.1) in B. subtilis and B. licheniformis have been reported. APase of B. licheniformis has been identified in a secreted from (32), a soluble cell-bound form released by lysozyme (19), a salt-extractable membrane endoprotein (16, 17, 24), and a detergent-extractable membrane ectoprotein (33). These various APase species are immunologically related. Although localization studies of

APase in B. subtilis are less extensive, seemingly contradictory reports of cell-bound (37, 44) or membrane-associated (inner leaflet) (6) and secreted (3) APase in the organism have been made.

APase is the major secreted protein. We reported that the addition of 0.1 mM Co^{2+} to a defined medium increased the concentration of assayable APase 35-fold (32) in B. licheniformis MC14. This increased synthesis is totally repressed by concentrations of inorganic phosphate above 0.075 mM. These studies showed that the maximum secretion (20%) of this activity into the medium occurred 4.5 h after the onset of APase production. More recently, we have observed that if the cobalt is added at the onset of APase synthesis (after 6.5 h of growth; 0.075 mM PO_4) rather than at the time of inoculation of the flask, 99% of the total APase produced is secreted into the medium. Coomassie blue-stained gels of the medium show that over 50% of the secreted protein is APase. (The APase activity remains in the supernatant after a $100,000 \times g$ centrifugation, indicating that it is a truly soluble enzyme.)

Fluorographs of sodium dodecyl sulfate (SDS) gels of secreted [^{35}S]methionine-labeled proteins from cultures grown with or without cobalt showed that significantly more (at least fourfold more) APase protein was secreted in the culture with cobalt. It does not appear that Co^{2+} is involved in a generalized secretion mechanism for proteins since the only protein which is secreted in increased amounts is APase. More likely, Co^{2+} influences some component of the regulatory mechanism for APase synthesis, in addition to its being required for enzyme activity (four cobalt atoms per APase dimer) (32). The effects of cobalt might be explained by conformational changes in the APase (nascent chain or inactive cytosol species) which could make it more or less subject to processing, or the cobalt could activate different processing proteases. However, since there are two or more structural genes for APase (see below), the different structural genes might differ in regulation of expression, with cobalt positively affecting expression of the gene for the secreted APase.

The secreted APase can be purified to homogeneity by subjecting medium in which the cells are grown to batch carboxymethyl Sephadex adsorption and elution, followed by carboxymethyl Sephadex chromatography with gradient elution. This procedure resulted in a 2.24-fold purification, with approximately 10% recovery of the initial activity. The specific activity of the enzyme was 2,100 U/mg (approximately 10 times higher than for other APase species). The secreted APase has a subunit molecular weight of 60,000 (SDS gels) and a native molecular weight of 120,000 (G-200 Sephadex gel filtration chromatography). Antibody to the medium APase can immunoprecipitate each of the other species of APase previously observed. Further evidence that the nonsecreted APase (Mg^{2+}-extractable membrane APase—peripherally associated with the inner surface of the cytoplasmic membrane [34]) is quite similar to the secreted APase was shown by limited proteolysis performed by the procedure of Randall et al. (27). Since fluorograms of *Staphylococcus aureus* V-8 protease-treated Mg^{2+}-extracted APase or secreted APase show no significant difference in the digestion pattern (Fig. 1), the active dimer located on the inner leaflet of the cytoplasmic membrane and the secreted APase dimer are quite similar proteins.

Inactive cytosol APase precursor. At no time does the cytosol fraction of the cells contain APase activity. However, the cytosol fraction of cells actively synthesizing and secreting APase (culture age between 7 and 8 h) does contain a protein which can be isolated by anti-APase immuno-bead adsorption (protein A-Sepharose) that has a molecular weight 3,000 heavier than the active APase species. Figure 2 shows an SDS gel which compares the migration of se-

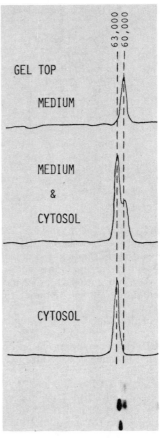

FIG. 2. SDS subunit molecular weight difference between the secreted APase and the anti-APase-immunoadsorbed cytosol protein. Top gel lane and top scan are of purified medium APase. Center lane and scan are a mix of immunoprecipitated cytosol protein and secreted APase. Bottom lane and scan are of immunoprecipitated cytosol protein alone.

creted APase with that of the inactive cytosol species.

Figure 3 summarizes the location and properties of the APase species in *B. licheniformis*. The predominant species at any time depends upon the growth conditions and the phase of growth of the culture. APase is formed as (i) an inactive soluble cytosol protein, (ii) a peripheral (salt-extractable) APase located primarily in the inner leaflet (34), (iii) an integral (detergent-extractable) APase located on the outer leaflet (33), and (iv) a soluble species secreted into the medium (above). (When cells are grown in 1% Neopeptone, a soluble dimer is formed but protoplast formation is required to release it from the cell-bound state [9].)

The precise precursor relationship of the inactive cytosol protein to any one or all of the active dimers is not totally clear. Pulse-chase

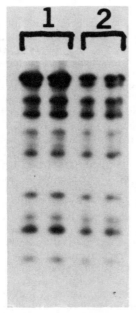

FIG. 1. Fluorograph of peptides from limited proteolysis of Mg^{2+}-extracted membrane APase and secreted APase electrophoretically resolved on SDS gels. (1) Two lanes containing identical digests of Mg^{2+}-extracted membrane APase. (2) Two lanes containing identical digests of secreted APase.

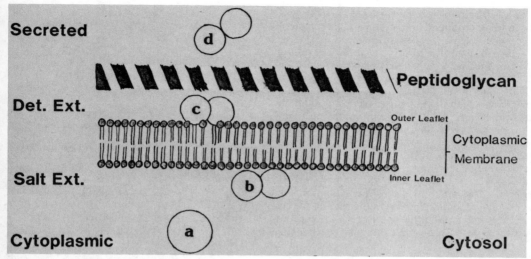

FIG. 3. Summary of APase species. (a) Inactive 63,000-molecular-weight anti-APase-immunoprecipitated cytosol species. (b) Membrane APase: active APase dimer peripherally associated with inner leaflet. (c) Membrane APase: active APase dimer integrally associated with outer leaflet. (d) Secreted APase: soluble dimer.

experiments comparing the whole-cell soluble APase fraction to the secreted fraction show that the 63,000-molecular-weight inactive cytosol species is processed to a 60,000-subunit-molecular-weight protein while both are cell bound. The cell-bound soluble APase protein (60,000-subunit-molecular-weight APase which has been secreted through the membrane but is retained by the cell wall [32]) is then chased into the medium (Fig. 4). The 60,000-subunit-molecular-weight membrane APase species do not chase (data not shown). It would not be surprising if the peripheral endomembrane APase (active dimer) was not a precursor to the secreted APase. However, there is precedent for integral ectoproteins being precursors to secreted proteins (penicillinase, *B. licheniformis* 744/c [14]).

Genes Responsible for APase Production in *B. licheniformis*

To examine the possibility that the multiple forms of APase in *B. licheniformis* and their subcellular membrane association or secretion might be the result of selective expression of multiple structural genes under different regulatory control, we cloned the structural gene(s) for APase.

Cloning and restriction mapping of the *phoA* gene. To clone the APase gene of *B. licheniformis* MC14, we constructed a *Pst*I chromosomal DNA fragment library containing 6,000 independently isolated clones. *B. licheniformis* MC14 chromosomal DNA was cut with *Pst*I and mixed with the vector pMK2004, which had been cut with *Pst*I and pretreated with calf intestinal APase. These DNA fragments were ligated and

used to transform *E. coli* Xph90a. Selection and screening were carried out by using 1% Neopeptone plates containing kanamycin and XP (15). In these experiments it was found that between 80 and 85% of colonies which were kanamycin and tetracycline resistant were ampicillin susceptible. Putative APase colonies turned blue on XP indicator plates. Approximately 1 in 3,000 colonies was blue. Plasmid DNA from the blue colonies was isolated and digested with *Pst*I. The sizes of the two DNA fragments observed corrreponded to linear pMK2004 (5.2 kilobases [kb]) and to an insertion fragment of 8.45 kb.

When the plasmid DNA was used to transform *E. coli* Xph90a, all transformed cells (kanamycin and tetracycline resistant, ampicillin susceptible) were blue on XP indicator plates. This plasmid, pMH8, is shown in Fig. 5 with restriction sites indicated. Deletion plasmids and subclones generated from pMH8 determined that the *Pvu*II-3 fragment was in the coding region of the gene (15). The *Eco*RI-2/*Xho*I-2 fragment of pMH8 was subcloned into the *Eco*RI-*Xho*I region of pMK2004 (resultant plasmid was pMH81, see Table 1). Xph90a transformed with pMH81 exhibited the blue-colored colony phenotype on indicator plates. Figure 6 shows the restriction map of pMH81. The *Pvu*II site (*Pvu*II-3 fragment) implicated in the coding region maps at 5.0 kb. The following deletions and subclones were used to map the coding region. The *Hin*dIII subclone (*Hin*dIII-3/*Hin*dIII-4 fragment) did not contain the complete gene; the *Bgl*II deletion (*Bgl*II-2/*Bgl*II-3 fragment) did not interrupt the gene. Therefore, the *Hin*dIII site at 4.1 kb must be in the coding region.

Tn5 insertional mutagenesis of pMH81 was used to determine the minimum coding region of the gene. pMH81 plasmids carrying Tn5 were used to transform strain Xph90a. The resulting transformed cells were selected and screened on Neopeptone-XP-kanamycin plates. The location of the Tn5 insertion was mapped in all of the plasmids which did not cause APase production in transformed *E. coli* Xph90a, indicating that the insertion was in the coding region of the gene. Of the 15 separate clones, all mapped between fragments *Xho*I-2 and *Pvu*II-3. The one which mapped farthest from the *Pvu*II-3 site was at 3.64 kb, as indicated on the map of pMH81 (Fig. 6). The minimum size of the coding region is calculated to be 1.3 kb, with the right terminus of the gene in pMH81 close to 3.64 kb and the left terminus of the gene containing the *Pvu*II-3 fragment.

The maximum coding region of the gene was established by constructing deletions approaching each terminus of the gene. Deletions were constructed approaching the right terminus by cutting pMH81 with *Xho*I, followed by nuclease(s) *Exo*III-S1 digestion and ligation (Fig. 6). Deletion plasmids were used to transform strain Xph90a. Plasmids from blue and white colonies were isolated, and the extent of the deletion in each was mapped. Colonies containing plasmids with deletions from fragment *Xho*I-2 to 3.6 kb (on pMH81) remained blue. Plasmids with larger deletions (from fragment *Xho*I-2), past the point (3.64 kb) at which Tn5 insertion was shown (above) to cause inactivation of the APase gene on pMH81, showed no APase production when used to transform Xph90a. This locates one terminus of the gene between 3.60 and 3.64 kb.

*Exo*III-S1 nuclease deletion mapping was also used to locate the terminus of the APase I gene in the *Pvu*II-3,4,5 region of pMH81 by digesting pMH81 with *Bgl*II followed by *Exo*III and S1 nucleases. Deletion plasmid DNA from blue and white colonies was isolated, and the extent of the deletion in each was mapped by digesting deletion plasmid with *Pvu*II nuclease. *Pvu*II-5 and *Pvu*II-4 sites could be deleted and the APase phenotype was retained (blue colonies). If the *Pvu*II-3 site was deleted, *E. coli* Xph90a transformed with such a plasmid made white colonies. Therefore, the *Pvu*II-4/*Pvu*II-3 fragment contains one end of the coding region of APase I (Fig. 5). The maximum coding region was thereby established on 1.7 kb.

Southern blot analysis of *B. licheniformis* genomic DNA and pMH8 plasmid DNA by use of a DNA fragment from within the coding region of the APase I gene. The 0.97-kb *Pvu*II-3/*Hind*III-3 fragment within the coding region of APase I on pMH81 (Fig. 6) was used as a probe for APase

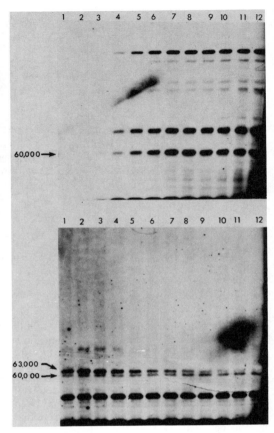

FIG. 4. Fluorogram of SDS gels showing processing and secretion of APase cytosol precursor. Cells were grown as described previously (32). [^{35}S]methionine was added at the same time cobalt was added (at 6.5 h of growth). After a 1-min pulse, excess unlabeled methionine was added. Samples were taken at 2-min intervals during a 22-min chase. Cells were separated from the medium by centrifugation (100,000 × *g*, 1 h). Cells were lysed, and the soluble fraction was separated from the membrane fraction (100,000 × *g*, 1 h). The bottom picture shows an SDS gel of the whole-cell soluble fractions; 1 through 12 correspond to 0 to 22 min after the chase was initiated. The top picture shows the labeled secreted proteins from the corresponding sample. Anti-APase immunoprecipitates the 63,000- and 60,000-molecular-weight proteins of the whole-cell soluble fraction and the 60,000-molecular-weight protein from the medium.

sequences in genomic DNA. Southern transfer analysis showed that *B. licheniformis* MC14 chromosomal DNA contained three regions of strong homology to DNA within the coding region of the APase I gene cloned on pMH81. One area of hybridization was located adjacent to the cloned APase I gene and was also cloned on pMH8.

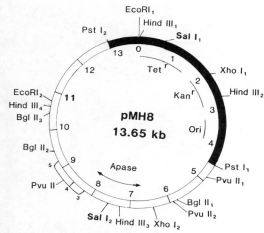

FIG. 5. Restriction map of pMH8 (13.65 kb). Vector DNA is represented by a heavy solid line. Insert DNA is represented by double lines. Fragments are referred to by the flanking restriction site numbers, e.g., $XhoI_2$-$PvuII_4$ identifies the region which includes the *phoA* gene (APase I).

Figure 7 shows a Southern blot of the restriction fragments of genomic DNA (lanes 2, 3, 4, 5) or pMH8 (lanes 1, 7, 8) separated on a 1% agarose gel and hybridized with this probe. When pMH8 was digested with *PstI* the 8.45-kb *B. licheniformis* DNA insert showed hybridization and the vector did not (lane 1). The *BglII* digest of the genomic DNA (lane 3) shows two fragments of strong hybridization. The top fragment is the same size (3.6 kb) as the *BglII* fragment from pMH8 (lane 7) which contains the APase I gene. The bottom fragment (lane 3) is the size of a second *BglII* fragment ($BglII$-2/$BglII$-3) of pMH8; however, this fragment does not show hybridization in the control *BglII* digest of pMH8 (lane 7). This region of hybridization has not been cloned.

In a *BglII*-*HindIII* double digest of the genomic DNA (Fig. 7, lane 4), the 3.6-kb fragment (top fragment of lane 3) is reduced to two smaller fragments, 2.0 kb and 1.6 kb. The 2.0-kb fragment is the size of the DNA fragment ($HindIII$-3/$BglII$-2 on pMH8) which should carry the DNA homologous to the entire DNA probe. That the 1.6-kb fragment ($BglII$-1/$HindIII$-3) also hybridizes (lane 4) indicates that another region of hybridization is quite close to the APase gene which we have characterized previously (15). Note that the *BglII* digest of pMH8 (lane 7) shows two fragments which hybridize. The top fragment (lane 7) does not contain any portion of the APase gene we have mapped, but does carry DNA adjacent to the 1.6-kb ($BglII$-1/$HindIII$-3) fragment. This shows that DNA on both sides of the $BglII$-1 fragment on pMH8 hybridizes to the

probe. There are not two areas of hybridization within the *XhoI*-RI digest of pMH8 because the two DNA fragments, one which contains APase I and one which contains the second region of homology identified above, comigrate at 4.15 kb (lane 8).

The Southern transfer data suggest the following. (i) There are at least two regions of strong hybridization with the internal probe on pMH8. (ii) They are located close together on the genomic DNA, and one (the APase I described above) and at least part of another is cloned on pMH8. (iii) Another area of hybridization (the second band of the *BglII* genomic digest [Fig. 7, lane 4]) is not cloned on pMH8 and suggests the possible presence of a third APase gene. This DNA has not been cloned.

pMH87, a second APase gene cloned on pMH8. DNA flanking the *BglII*-1 site on pMH8 was implicated as a second region of hybridization by the Southern hybridization. A *SalI* deletion of pMH8 (pMH87) was constructed (Fig. 8). This removes over half the coding region of the APase gene subcloned on pMH81 (APase I) and contains 1.1 kb of DNA which is also on pMH81. When this deletion plasmid was used to transform *E. coli* Xph90a and plated on Neopeptone-

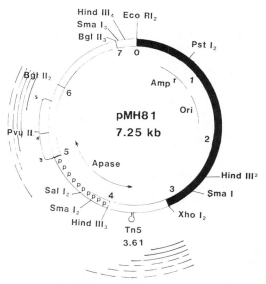

FIG. 6. Restriction map of pMH81 (7.25 kb). DNA fragments are referred to by the flanking restriction site numbers as in Fig. 1. The fragment of DNA used for the probe for Southern hybridization ($PvuII$-3 to $HindIII$-3) is indicated by pppppppppp. *ExoIII*-S1 deletions of pMH81: solid lines indicate deletion distance from $BglII$-2 or *XhoI* in pMH81 that retained the APase phenotype when the deletion plasmid was transformed into Xph90a; dashed lines indicate the length of DNA deleted which resulted in an APase-negative phenotype.

XP-kanamycin plates, the colonies turned blue in 1 day. This subclone is called pMH87. The possibility that the APase phenotype of *E. coli* Xph90a transformed with pMH87 was dependent on DNA previously assigned to the coding region of APase I was eliminated by constructing and analyzing deletion plasmids of pMH87 which retained none of the APase I gene. Such deletion plasmids of pMH87 complement *phoA*-negative *E. coli*.

Deletions of pMH87 were constructed by digestion of pMH87 with *Sal*I, followed by *Exo*III nuclease and S1 nuclease digestion as described above. Deletion plasmids were used to transform *E. coli* Xph90a. Colonies containing plasmids in which all of the DNA from the coding region of the APase gene on pMH81 (APase I) was deleted remained blue (Fig. 6). Further deletion (from *Sal*I-1) up to a point approximately 400 kb before the *Xho*I-2 fragment also showed APase (APase II) production when used to transform Xph90a. Deletion plasmids in which all DNA from fragment *Sal*I-2 to a point approximately 260 bases before fragment *Xho*I-2 or farther was removed showed no APase (APase II) production when used to transform *E. coli* Xph90a (Fig. 8). Therefore, one terminus of the APase II gene is mapped at approximately 6.1 to 6.3 kb on pMH87.

Both APase I and APase II code for 60,000-molecular-weight proteins which cross-react with anti-APase. We have previously shown (using Western transfer analysis) that pMH81 carries a gene which codes for a protein of 60,000 molecular weight (subunit molecular weight of *B. licheniformis* APase) and cross-reacts with anti-APase (15). Similar experiments with cell lysates of *E. coli* Xph90a cells carrying pMH87 showed that pMH87 also carried a gene (APase II) which codes for a 60,000-molecular-weight protein which cross-reacts to anti-APase.

Preliminary characterization of the promoter for APase I. Deletion mapping and in vitro transcription studies showed that the promoter for APase I lies between *Pvu*II-3 and *Pvu*II-4 on pMH81 (shown above) and that a σ^{55}-containing RNA polymerase holoenzyme is required for transcription. The *Bgl*II/*Xho*I-2 fragment of pMH81 (APase I) was used as template for in vitro transcription (Fig. 9). The σ^{55} RNA polymerase holoenzyme from *B. subtilis* produced a large transcript (lane B in Fig. 9). To better map the promoter and the direction of transcription, the *Bgl*II-2/*Xho*I-2 fragment was digested with *Hind*III or *Hind*III and *Pvu*II. The transcript was shortened by the *Hind*III digestion of the template (lane C), and a *Hind*III-*Pvu*II digestion resulted in a unique 150-nucleotide transcript (lane D). These data, coupled with the *Exo*III-S1 mapping, suggested that the promoter was lo-

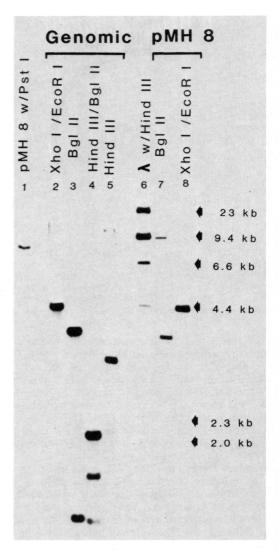

FIG. 7. Southern blot analysis of *B. licheniformis* genomic DNA and pMH8 plasmid DNA with the use of an internal probe isolated from pMH81 within the coding region of APase I. The probe, isolated from pMH81, corresponds to the *Hind*III-3/*Pvu*II-3 fragment of DNA on pMH8. Genomic DNA or pMH8 DNA was digested by restriction enzyme(s) and separated on a 0.7% agarose gel as follows. pMH8: lane 1, *Pst*I; lane 7, *Bgl*II; or lane 8, *Xho*I and *Eco*RI. Genomic DNA: lane 2, *Xho*I and *Eco*RI; lane 3, *Bgl*II; lane 4, *Bgl*II and *Hind*III; lane 5, *Hind*III. A *Hind*III digestion of bacteriophage lambda was used as a control in lane 6. After electrophoretic separation, the DNA fragments were transferred from the agarose to nitrocellulose paper by diffusion transfer. ^{32}P-labeled *Hind*III-*Pvu*II probe DNA and a ^{32}P-labeled *Hind*III digest of λ was hybridized to the DNA on the filter. Fragments which hybridized were visualized by autoradiography.

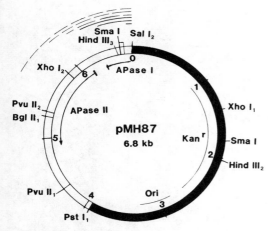

FIG. 8. Restriction map of pMH87. *Exo*III-S1 deletion mapping of pMH87. Solid lines indicate the deletion distance from fragment *Sal*I-2 in pMH87 that retained the APase phenotype when the constructed plasmid was transformed into Xph90a. Dashed lines indicate the length of DNA deleted which resulted in an APase-negative phenotype.

cated to the right end of the fragment about 150 bases before *Pvu*II-3 and is read rightward by a σ^{55} RNA polymerase holoenzyme.

Preliminary characterization of the promoter for APase II. Deletion mapping and in vitro transcription studies showed that the promoter for APase II lies between *Hin*dIII-4 and *Xho*I-2 approximately 300 base pairs (bp) from *Xho*I-2 and that it requires a σ^{37}-containing RNA polymerase holoenzyme for transcription. We described above *Exo*III-S1 nuclease deletion mapping which showed that deleting DNA from *Sal*I (on pMH87) to approximately 400 bp before *Xho*I-2 did not affect APase II activity (Fig. 8). Larger deletions which removed DNA from *Sal*I to approximately 260 bp before *Xho*I-2 or farther resulted in deletion plasmids which no longer expressed the *phoA* phenotype when transformed into *E. coli* Xph90a. This positioned one terminus of the gene between *Sal*I and *Xho*I-2, 400 to 260 bp from *Xho*I-2.

In vitro transcription studies were carried out with the *Sal*I-2/*Pst*I-1 fragment of pMH87. The σ^{55} holoenzyme of *B. subtilis* did not yield a transcript with this template (Fig. 10A, lane 2); σ^{37} holoenzyme of *B. subtilis* did. *Xho*I digestion of the fragment produced a transcript of approximately 300 bp when σ^{37} holoenzyme was used (lane 3). Transcription using the isolated *Xho*I-2/*Sal*I-2 fragment of pMH87 digested with *Sma*I produced the same 300-bp run-off fragment, showing that the run-off transcript (Fig. 10B, lane 1) was at the *Xho*I end of the fragment. These data, combined with the *Exo*III-S1 deletion mapping data, position the promoter approx-

imately 300 bp from the *Xho*I site and suggest that it is read counterclockwise by a σ^{37} RNA polymerase holoenzyme.

Thus, we have cloned two tandemly positioned APase genes from *B. licheniformis*, one which requires the major polymerase, σ^{55}-containing holoenzyme, and another which requires a minor polymerase, σ^{37}-containing holoenzyme, for transcription.

DISCUSSION

Our localization data show that there are active APase dimers inside the cell (peripherally associated with the cell membrane) as well as APase dimers which have been translocated across the membrane (integrally associated with

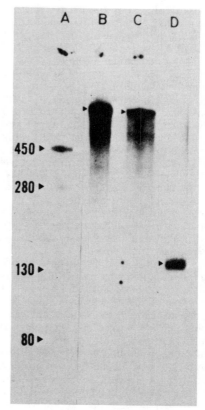

FIG. 9. Transcription mapping of the *Bgl*II-2/*Xho*I-2 fragment of pMH81 with σ^{55} containing *B. subtilis* RNA polymerase. Lane A, *Hin*dIII fragment of phage φ29 DNA transcribed by σ^{55} enzyme which produces transcripts of 450, 280, 130, and 80 bases for use as markers; lane B, intact *Bgl*II-2/*Xho*I-2 transcribed by σ^{55} enzyme; lane C, *Bgl*II-2/*Xho*I-2 fragment digested with *Hin*dIII and transcribed by σ^{55} enzyme; lane D, *Bgl*II-2/*Xho*I-2 fragment digested with *Hin*dIII and *Pvu*II and transcribed by σ^{55} enzyme.

the outer leaflet of the cytoplasmic membrane or secreted soluble). None of the species which we have studied is likely to be the APase associated with sporulation, considering that the studies were carried out on vegetative cells and the expression of the APase which occurs as the cells enter stationary growth (due to phosphate starvation only) is repressible by inorganic phosphate. The vegetative APase species are all very similar proteins except that the inactive cytosol protein is slightly larger. Similarly, the APases from vegetative and sporulating cells of *Bacillus* were indistinguishable by a large number of criteria (8).

Our data show that there are at least two structural genes for APase in *B. licheniformis* MC14. Hybridization studies of genomic DNA under stringent conditions (requirement of 84% homology for hybridization) suggest that there may be a third gene in this APase gene family. Construction of a double APase mutant strain of *B. licheniformis* (by using in vitro-generated deletion of APase I and APase II on pMH8) will allow us to determine whether APase I and APase II are the only expressed structural APase genes in *B. licheniformis* MC14.

Having cloned the APase genes, we can now generate structural gene mutants in *B. licheniformis* which retain only one APase gene to determine unambiguously the final destination of the gene product of each gene. It will be of interest to determine whether divergence of duplicated genes in *Bacillus* spp. might have resulted in altered regulation of the gene and location of the protein gene product.

It has been suggested that different genes account for the intracellular serine protease and the extracellular serine protease (subtilisin) of *B. subtilis* and that they arose via gene duplication and divergence (36). As in the case of *Bacillus* APase(s), mutations in the structural gene(s) for subtilisin could not be isolated. (In vitro-derived deletion mutations of subtilisin have recently been constructed [35].) Amino acid sequence homology (50%) was determined via Edman degradation of 50 amino acids of the NH$_2$ terminus of the secreted and nonsecreted proteases. Recently, the gene for the secreted protease has been cloned, sequenced, and shown to require a σ^{37} polymerase for transcription (43a). This is the only other *Bacillus* gene of known function thus far isolated which has a σ^{37}-requiring promoter. These authors suggest that the role of the σ^{37} promoter may include, in addition to expression of sporulation-specific genes (12, 23, 25, 26), expression of genes encoding extracellular enzymes as well as genes regulated by growth phase. We can now determine whether the gene requiring σ^{37}, APase II, is responsible for secreted APase species or a sporulation-specific

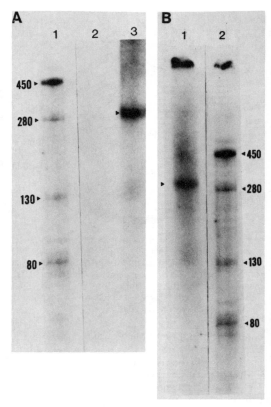

FIG. 10. Transcription mapping of the *Sal*I/*Pst*I-1 fragment of pMH87 with σ^{55}- and σ^{37}-containing *B. subtilis* RNA polymerase. (A) Lane 1, markers same as lane A in Fig. 9; lane 2, the *Sal*-2/*Pst*I-2 fragment intact and transcribed by σ^{55} enzyme; lane 3, the *Sal*I-2/*Pst*I-1 fragment digested by *Xho*I and transcribed by σ^{37} RNA polymerase holoenzyme. (B) Lane 1, the isolated *Sal*I-2/*Xho*I-2 fragment, digested with *Sma*I, transcribed by the σ^{37} RNA polymerase holoenzyme; lane 2, markers same as in lane 1 of (A).

APase of *B. licheniformis*, whereas the gene requiring σ^{55} makes the internalized APase.

SUMMARY

We have cloned two APase genes arranged in tandem on the chromosome of *B. licheniformis* MC14 (Fig. 11). Evidence that the genes code for APase includes enzymatic activity (ability to hydrolyze XP), protein subunit size, and antigenic cross-reactivity with anti-APase. In vitro transcription mapping of each promoter agrees well with the *Exo*III-S1 nuclease deletion mapping of the coding region of each gene. The finding that each gene has its own promoter and that they differ with respect to the sigma factor required for transcription suggests that the regulation for the two genes differs. Analysis of the regulation and location of the gene product of these two APase genes (and additional APase

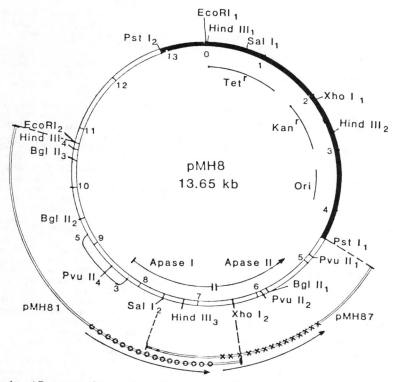

FIG. 11. Tandem APase genes from *B. licheniformis* MC14. APase I and APase II arranged in tandem on the chromosome. APase I terminates approximately 400 bp after fragment *Hin*dIII-3 (at about 6.9 on pMH8), and transcription initiation for APase II starts approximately 300 bp before fragment *Xho*I-2 at 6.7 on pMH8 (proceeding to the right). Therefore, the APase genes are separated by approximately 200 bp. APase I is transcribed by an RNA polymerase with a σ^{55} subunit (transcript starts 150 bp before fragment *Pvu*II-3 and reads to the right), whereas APase II is transcribed by RNA polymerase with a σ^{37} subunit. At the bottom of the figure, $\bigcirc\bigcirc\bigcirc$ indicates the coding region for APase I on pMH81, $\times\times\times$ indicates the coding region for APase II on pMH87, and the arrows indicate direction of transcription.

genes, if they exist) will be used to interpret our earlier data which show that, under different growth conditions and at different phases of growth, different amounts of APase are produced and localized in different cellular fractions.

LITERATURE CITED

1. **Blobel, G.** 1980. Intracellular protein topogenesis. Proc. Natl. Acad. Sci. U.S.A. **77:**1496–1500.
2. **Bonner, W. M., and R. A. Laskey.** 1974. A film detection method of tritium-labelled proteins and nucleic acids in polyacrylamide gels. Eur. J. Biochem. **46:**83–88.
3. **Cashel, M., and E. Freese.** 1964. Excretion of alkaline phosphatase by *Bacillus subtilis*. Biochem. Biophys. Res. Commun. **16:**541–544.
4. **Coote, J. G.** 1972. Sporulation in *Bacillus subtilis*. Characterization of oligosporogenous mutants and comparison of their phenotypes with those of asporogenous mutants. J. Gen. Microbiol. **71:**1–15.
5. **Emr, S. D., M. N. Hall, and T. J. Silhavy.** 1980. A mechanism of protein localization: the signal hypothesis and bacteria. J. Cell Biol. **86:**701–711.
6. **Ghosh, B. K., J. T. M. Wouters, and J. O. Lampen.** 1971.

Distribution of the sites of alkaline phosphatase(s) activity in vegetative cells of *Bacillus subtillis*. J. Bacteriol. **108:**928–937.
7. **Glenn, A. R.** 1975. Alkaline phosphatase mutants of *Bacillius subtilis*. Aust. J. Biol. Sci. **28:**323–330.
8. **Glenn, A. R., and J. Mandelstam.** 1971. Sporulation in *Bacillus subtilis* 168. Comparison of alkaline phosphatase from sporulating and vegetative cells. Biochem. J. **123:**129–138.
9. **Glynn, J. A., S. D. Schaffel, J. M. McNicholas, and F. M. Hulett.** 1977. Biochemical localization of alkaline phosphatase of *Bacillus licheniformis* as a function of culture age. J. Bacteriol. **129:**1010–1019.
10. **Goldfarb, D. S., S. L. Wong, T. Kudo, and R. H. Doi.** 1983. A temporally regulated promoter from *Bacillus subtilis* is transcribed only by an RNA polymerase with a 37,000 dalton sigma factor. Mol. Gen. Genet. **191:**319–325.
11. **Grant, W. D.** 1974. Sporulation in *Bacillus subtilis* 168. Control of synthesis of alkaline phosphatase. J. Gen. Microbiol. **82:**363–69.
12. **Haldenwang, W. G., and R. Losick.** 1980. Novel RNA polymerase σ factor from *Bacillus subtilis* (promoter recognition/sporulation genes/RNA nucleotidyltransferase). Proc. Natl. Acad. Sci. U.S.A. **77:**7000–7004.
13. **Hansa, J. G., M. Laporta, M. A. Kuna, R. Reimschuessel, and F. M. Hulett.** 1981. A soluble alkaline phosphatase from *Bacillus licheniformis* MC14: histochemical localiza-

tion, purification and characterization and comparison with the membrane-associated alkaline phosphatase. Biochim. Biophys. Acta 675:340–401.

14. **Hayashi, S., and H. C. Wu.** 1983. Biosynthesis of *Bacillus licheniformis* penicillinase in *Escherichia coli* and *Bacillus subtilis*. J. Bacteriol. 79:773–777.

15. **Hulett, F. M.** 1984. Cloning and characterization of the *Bacillus licheniformis* MC14 gene coding for alkaline phosphatase. J. Bacteriol. 158:978–982.

16. **Hulett, F. M., and L. L. Campbell.** 1971. Purification and properties of an alkaline phosphatase of *Bacillus licheniformis*. Biochemistry 10:1364–1370.

17. **Hulett, F. M., S. D. Schaffel, and L. L. Campbell.** 1976. Subunits of alkaline phosphatase of *Bacillus licheniformis*: chemical, physiochemical, and dissociation studies. J. Bacteriol. 128:651–657.

18. **Inouye, H., S. Michaelis, A. Wright, and J. Beckwith.** 1981. Cloning and distribution mapping of the alkaline phosphatase structural gene (*phoA*) of *Escherichia coli* and generation of deletion mutants in vitro. J. Bacteriol. 146:668–675.

19. **Kahn, M., R. Kolter, C. Thomas, D. Figurski, R. Meyer, E. Remaut, and D. Helinski.** 1979. Plasmid cloning vehicles derived from plasmids CalE1, F, R6K, and RK2. Methods Enzymol. 68:268–280.

20. **Laskey, R. A., and A. D. Mills.** 1975. Quantitative film detection of ^{3}H and ^{14}C in polyacrylamide gels by fluorography. Eur. J. Biochem. 56:335–341.

21. **Le Hégarat, J. C., and C. Anagnostopoulos.** 1973. Purification, subunit structure and properties of two repressible phosphohydrolases of *Bacillus subtilis*. Eur. J. Biochem. 39:525–539.

22. **Lodish, H. F., and J. E. Rothman.** 1979. The assembly of cell membranes: the two sides of a biological membrane differ in structure and function; studies of animal viruses and bacteria have helped to reveal how this asymmetry is preserved as the membrane grows. Sci. Am. 240:48–63.

23. **Losick, R., and J. Pero.** 1981. Cascades of sigma factors. Cell 25:582–584.

24. **McNicholas, J. M., and F. M. Hulett.** 1977. Electron microscope histochemical localization of alkaline phosphatase in *Bacillus licheniformis*. J. Bacteriol. 129:501–515.

25. **Moran, C. P., Jr., N. Lang, C. D. B. Banner, W. G. Haldenwang, and R. Losick.** 1981. Promoter for a developmentally regulated gene in *Bacillus subtilis*. Cell 25:783–791.

26. **Moran, C. P., Jr., N. Lang, and R. Losick.** 1981. Nucleotide sequence of a *Bacillus subtilis* promoter recognized by *Bacillus subtilis* RNA polymerase containing σ^{37}. Nucleic Acids Res. 9:5979–5990.

27. **Randall, L. L., S. J. S. Hardy, and L.-G. Josefsson.** 1978. Precursors of three exported proteins in *Escherichia coli*. Proc. Natl. Acad. Sci. U.S.A. 75:1209–1212.

28. **Roberts, T. M., and G. D. Lauer.** 1979. Maximizing gene expression on a plasmid using recombination *in vitro*. Methods Enzymol. 68:473–482.

29. **Schaffel, S., and F. M. Hulett.** 1978. Alkaline phosphatase from *Bacillus licheniformis* solubility dependent on magnesium, purification and characterization. Biochim. Biophys. Acta 526:457–467.

30. **Silhavy, J. T., S. A. Benson, and S. D. Erm.** 1983. Mechanisms of protein localization. Microbiol. Rev. 47:313–344.

31. **Southern, E. M.** 1975. Detection of specific sequences among DNA fragments separated by gel electrophoresis. J. Mol. Biol. 98:503–517.

32. **Spencer, D. B., C.-P. Chen, and F. M. Hulett.** 1981. Effect of cobalt on synthesis and activation of *Bacillus licheniformis* phosphatase. J. Bacteriol. 145:926–933.

33. **Spencer, D. B., J. G. Hansa, K. V. Stuckmann, and F. M. Hulett.** 1981. Membrane-associated alkaline phosphatase from *Bacillus licheniformis* that requires detergent for solubilization: lactoperoxidase ^{125}I localization and molecular weight determination. J. Bacteriol. 150:826–834.

34. **Spencer, D. B., and F. M. Hulett.** 1981. Lactoperoxidase-^{125}I localization of salt-extractable alkaline phosphatase on the cytoplasmic membrane of *Bacillus licheniformis*. J. Bacteriol. 145:934–945.

35. **Stahl, M. L., and E. Ferrari.** 1984. Replacement of the *Bacillus subtilis* subtilisin structural gene with an in vitro-derived deletion mutation. J. Bacteriol. 158:411–418.

36. **Strongin, A. Ya., L. S. Izotova, Z. T. Abramov, D. I. Gorodetsky, L. M. Ermakova, L. A. Baratova, L. P. Belyanova, and V. M. Stephanov.** 1978. Intracellular serine protease of *Bacillus subtilis*: sequence homology with extracellular subtilisins. J. Bacteriol. 133:1401–1411.

37. **Takeda, K., and A. Tsugita.** 1967. Phosphoesterases of *B. subtilis*. J. Biochem. 61:231–241.

38. **von Heijne, G.** 1980. Trans-membrane translocation of proteins: a detailed physics-chemical analysis. Eur. J. Biochem. 103:431–438.

39. **Waites, W. M., D. Kay, I. W. Dawes, D. A. Wood, S. C. Warren, and J. Mandelstam.** 1970. Sporulation in *Bacillus subtilis*. Correlation of biochemical events with morphological changes in asporogenous mutants. Biochem. J. 118:667–676.

40. **Warren, S. C.** 1968. Sporulation in *Bacillus subtilis*. Biochemical changes. Biochem. J. 109:811–818.

41. **Wickner, W.** 1979. Assembly of proteins into biological membranes: membrane trigger hypothesis. Annu. Rev. Biochem. 48:23–45.

42. **Wickner, W.** 1980. Assembly of proteins into membranes. Science 210:861–868.

43. **Wong, S. L., and R. H. Doi.** 1982. Peptide mapping of *Bacillus subtilis* RNA polymerase factors and core-associated polypeptides. J. Biol. Chem. 257:11932–11936.

43a. **Wong, S. L., C. W. Price, D. S. Goldfarb, and R. A. Doi.** 1984. The subtilisin F gene of *Bacillus subtilis* is transcribed from a σ^{37} promoter in vivo. Proc. Natl. Acad. Sci. U.S.A. 81:1184–1188.

44. **Wood, D., and H. Tristam.** 1970. Localization in the cell and extraction of alkaline phosphatase from *Bacillus subtilis*. J. Bacteriol. 104:1045–1051.

Transcription, Processing, and Expression of tRNA genes from *Bacillus subtilis*

BARBARA S. VOLD AND CHRISTOPHER J. GREEN

Biomedical Research Division, SRI International, Menlo Park, California 94025

Recent studies on tRNA genes in *Bacillus subtilis* indicate that certain aspects of tRNA gene organization differ between *B. subtilis* and *Escherichia coli* (10–12). In *E. coli*, tRNA genes are scattered around the genome, the largest group contains seven tRNA genes, genes within a group are often repeated, and all tRNA genes encode the sequence CCA at the 3' terminus (7). In *B. subtilis*, tRNA genes are highly clustered, the largest group contains 21 tRNA genes, there are no repeated genes within a cluster, and some tRNA genes encode the CCA terminus whereas others do not (B. Vold, Microbiol. Rev., in press). Such differences in gene organization may influence how processing of tRNA precursors is accomplished in these two organisms. However, further studies of *B. subtilis* are necessary before comparisons can be made.

Processing of RNA in procaryotes has recently been reviewed by Pace (8) and Apirion and Gegenheimer (1), although most of what is known comes from data on *E. coli*. The only enzyme which has been purified for processing tRNA genes from *B. subtilis* is RNase P (9). RNase P is the RNase in *B. subtilis* and in *E. coli* responsible for cutting at the 5' end of tRNA precursors. The holoenzyme is composed of two parts: an RNA component and a protein component. It has recently been demonstrated that the catalytic activity resides solely in the RNA component (5). In *E. coli*, the RNA component is called M1 RNA; in *B. subtilis*, the RNA component is called P-RNA and the protein component is called P-protein. So far the action of the *B. subtilis* RNase P and P-RNA has been described only on monomeric or dimeric transcripts from heterologous organisms (5, 6). How a large cluster of homologous tRNA genes is transcribed and processed by *B. subtilis* RNase P and P-RNA is, therefore, of interest. In this paper, the transcription and processing in vitro and the expression in vivo of a DNA segment from *B. subtilis* containing 21 tRNA genes is described. We also present a system for generating smaller transcripts containing one or six tRNA genes, which may be useful as a model system.

EXPERIMENTAL PROCEDURES

Transcription reaction. The transcription reactions were carried out essentially as described by Green et al. (4) and Zinn et al. (13). The system used was purchased from Promega Biotec, Madison, Wis., under the name Riboprobe. Transcription reactions using [α-^{32}P]CTP contained 40 mM Tris (pH 7.5), 6 mM MgCl$_2$, 2 mM spermidine, 10 mM dithiothreitol, 50 U of RNasin, 400 μM of all four ribonucleotide triphosphates, 50 μCi (0.065 nmol) of [α-^{32}P]CTP, 5 μg of template DNA, and 37 to 50 U of SP6 RNA polymerase. Reactions using [α-^{32}P]GTP substituted 400 μM CTP, UTP, and ATP plus 12 μM GTP and 50 μCi of [α-^{32}P]GTP. Transcription products labeled with [α-^{32}P]CTP were extracted with phenol and chloroform-isoamyl alcohol and were precipitated with ethanol. Transcription products labeled with [α-^{32}P]GTP were further purified over a column of Sephacryl S-300 in buffer containing 100 mM NaCl, 50 mM Tris (pH 7.4), 1 mM EDTA, and 0.1% sodium dodecyl sulfate.

Processing reaction with *B. subtilis* P-RNA and reconstituted RNase P. Unless otherwise noted, processing reactions contained 1 $\times$ 10^5 to 2 $\times$ 10^5 cpm of transcript, 24 ng of P-RNA, 1.2 M NH$_4$Cl, 0.25 M MgCl$_2$, and 50 mM Tris (pH 8.0), or transcript, 24 ng of P-RNA, 0.25 ng of P-protein, 60 mM MgCl$_2$, 100 mM NH$_4$Cl, and 50 mM Tris (pH 7.5) in 15 μl (N. Pace and T. Marsh, personal communication). Reactions were incubated at 37°C for 30 min unless otherwise noted. After incubation, reactions were diluted fivefold with water and precipitated with ethanol after the addition of 0.08 A_{260} unit of carrier tRNA. Precipitated reaction products were collected by centrifugation, dried, and dissolved in a loading buffer containing 8 M urea for analysis by polyacrylamide gel electrophoresis.

Electrophoresis system. Polyacrylamide gels, 5% cross-linked, were run in 8 M urea and buffer containing 10.8 g of Tris base, 5.5 g of boric acid, and 0.84 g of EDTA per liter. The 3% gels of 1 mm thickness were soaked in 10% methanol–10% acetic acid–2% glycerol before drying.

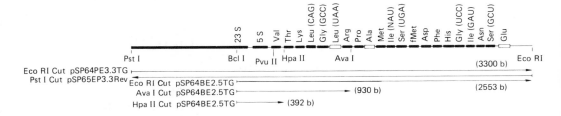

FIG. 1. Schematic representation of transcription products from 5P6 RNA polymerase in vitro system. Approximate length and direction of transcription are indicated by arrows. Length in bases is given in parentheses on the right. tRNA genes without the 3'-terminal CCA sequence encoded are designated as open boxes.

Processing reaction with E. coli M1 RNA and RNase P. E. coli processing reactions were carried out as described by Guerrier-Takada et al. (5). M1 RNA and E. coli RNase P were generously provided by C. Guerrier-Takada and S. Altman.

Plasmids. A 2.5-kilobase (kb) BclI-EcoRI and a 3.3-kb PstI-EcoRI DNA segment from B. subtilis containing 21 tRNA genes was cloned into the vectors pSP64 or pSP65 (Promega Biotec). The complete sequence of the DNA segment containing the tRNA genes from B. subtilis has been published (3). This DNA segment also contained the 3' end of the 23S rRNA gene and a 5S rRNA gene. The 3.3-kb piece was cloned in the appropriate direction to make a multimeric precursor tRNA pSP64PE3.3TG and, as a control, in the reversed direction, pSP65EP3.3Rev. EcoRI was used to linearize the plasmids containing tRNA genes in the correct orientation, and PstI was used to linearize the plasmids containing the reversed orientation. Figure 1 shows a schematic representation of the entire tRNA gene region; it also gives the direction of reading and the designation of the various constructions.

RESULTS AND DISCUSSION

Transcription reactions. Transcription reactions were carried out as described in Experimental Procedures. Results of transcription reactions using the full-length transcript cloned into pSP64, in the proper direction for precursor tRNAs, and into pSP65, in the opposite orientation, are shown in Fig. 2 (see Fig. 1 for template constructions). TG indicates the template read in the sense direction, and Rev indicates the template read in the reversed direction. The positions of molecular weight markers are consistent with the full-length transcripts being 3,300 bases long. Transcripts from reactions with all four ribonucleotide triphosphates in excess were usually full length. However, the Rev template generated a distinct partial transcript under these

conditions, possibly as a result of an RNA polymerase pause site.

Processing the large transcript in vitro with B. subtilis RNase P and P-RNA. Processing experi-

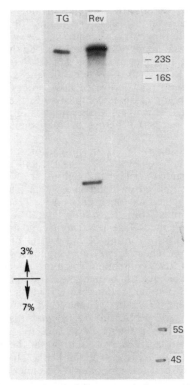

FIG. 2. Products of the transcription reaction analyzed by polyacrylamide gel electrophoresis. Templates designated EcoRI-cut pSP64PE3.3TG and PstI-cut pSP65EP3.3Rev were produced in the reaction using [α-^{32}P]CTP, extracted with phenol, precipitated with ethanol, and analyzed on a 3% polyacrylamide gel over a 7% lower gel. The 4S marker was ^{32}P-labeled pA$_3$tRNA^{fMet} (E. coli) with 81 bases. The 5S marker was ^{32}P-labeled 5S rRNA from Chlorobrium with 116 bases. Both labeled markers were provided from the laboratory of Norman Pace.

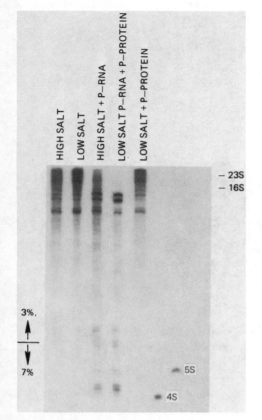

FIG. 3. Results of processing reaction on full-length transcript of precursor tRNAs. Processing reaction products were analyzed on a 3% polyacrylamide gel over a 7% lower gel. The initial transcript was produced by using the [α-^{32}P]GTP reaction. High- and low-salt conditions are described in Experimental Procedures. The sources of the 4 and 5S markers are given in Fig. 2; the 16 and 23S markers were unlabeled rRNAs. All reactions were incubated for 2 h at 37°C.

ments with the transcript containing 21 tRNA genes were done in collaboration with Norman Pace and Terry Marsh, National Jewish Hospital, Denver, Colo. It was of interest to us to see whether the P-RNA component could function as well as the P-RNA plus P-protein in processing the large transcript. Pre-tRNA transcript was incubated alone in high- or low-salt conditions and under conditions optimal for P-RNA by itself, P-RNA in conjunction with P-protein, or P-protein by itself. Reaction products were analyzed on a 3% polyacrylamide gel over a 7% gel to emphasize the larger products, as shown in Fig. 3, or on an 8% thin gel to emphasize the smaller products (Fig. 4). In no case was any processing or degradation detected when the transcript was incubated by itself or in the presence of P-protein without the P-RNA component. Figure 4 shows the pre-tRNA transcript

incubated for 20 min, 1 h, or 2 h, with either P-RNA alone or P-RNA plus P-protein. Products between the 4 and 5S markers are of a size consistent with products containing an unprocessed 3' sequence. The 4S marker used in this experiment was ^{32}P-labeled E. coli pA$_3$tRNA^{fMet} (5). These results show that P-RNA can act on this large transcript and give approximately the same cleavage pattern as P-RNA plus P-protein, giving further support to the concept that specificity lies completely in the RNA component of the RNase P.

A critical control experiment for the specificity of P-RNA and RNase P is shown in Fig. 5. In this experiment, the full cluster of 21 tRNA genes is transcribed in the reverse orientation, making an RNA complementary to the tRNA precursor. This transcript was then incubated under the same conditions as the pre-tRNA transcript. It was not cleaved by either P-RNA or P-RNA plus P-protein, even though this transcript must have a significant amount of base-paired hairpin loops.

Transcription and processing of transcripts containing one or six tRNA genes. Although the large transcript is useful for asking questions about the ability of processing enzymes to make comparative cuts, the generation of over 21 products makes analysis of the products difficult. In addition, we have not been successful in complete cleavage of the transcript to final products, and all of our reactions are characterized by intermediate reaction products. Therefore, we decided to produce transcripts containing only a few tRNA genes to make analysis of the products easier. To accomplish this, the segment diagrammed in Fig. 1 was cut with BclI and cloned into pSP64, still terminated at the 3' end by the EcoRI restriction site. Before adding it to the transcription reaction, this pSP64BE2.5TG construct was cleaved with either AvaI or HpaII restriction endonuclease. The AvaI-cut template generates a tRNA precursor containing all of the first six tRNAs and ends 12 bases before the 3' end of arginine tRNA. The HpaII-cut template generates a tRNA precursor containing all of the first tRNA (for valine), ending at the beginning of threonine tRNA.

An analysis of the transcription and processing products of the AvaI- and HpaII-cut templates is given in Fig. 6. The HpaII-cut template generates two transcripts of slightly different molecular weights, possibly as a result of premature termination because of some aspect of the secondary structure of the resulting transcripts. The next largest fragment in the lanes showing the migration of processing products resulting from the action of P-RNA or M1 RNA is approximately 285 bases long, estimated from migration distance on the gel. This band is also

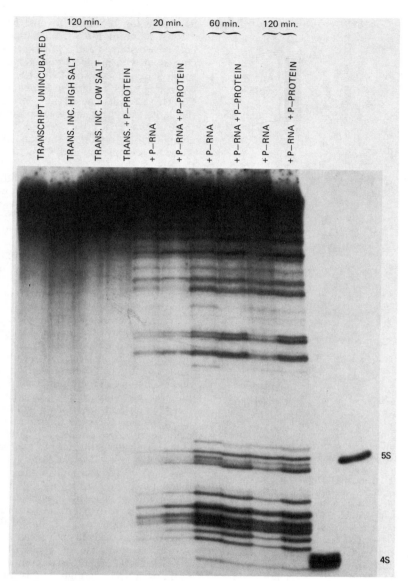

FIG. 4. Results of processing reaction on full-length transcript. Processing reaction products were analyzed on an 8% polyacrylamide gel. Products on this gel are the same as those described in Fig. 3 except that this gel shows a comparison of products incubated for various time periods.

found in the processing reactions of the AvaI-cut transcript and probably represents the 5′ end of the transcript which contains the rRNA segment (refer to Fig. 1). The sequence of this region shows that it should have 282 nucleotides.

Two double bands, presumably resulting from the double transcripts from the HpaII template, occur in the 4 to 5S region of the HpaII lanes. The expected length of a pre-tRNA from the 5′ terminus of tRNA^{Val} to the HpaII cut would be 110 bases. The HpaII cut is 2 bases into the

threonine tRNA gene. The 5S rRNA marker shown in Fig. 6 is 116 bases.

Expected molecular weights of processing reaction products of the AvaI-cut template would include the 282-base segment of rRNA genes and pre-tRNAs from 86 to 113 bases. Further characterization of these products will have to wait until we have determined conditions to achieve more complete processing and have determined the sequence of the reaction products.

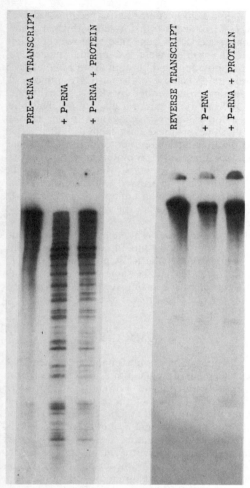

FIG. 5. Process reaction using precursor tRNAs or the reverse transcript as substrate. Reactions were incubated under conditions optimal for P-RNA or holoenzyme as described in Experimental Procedures. Products were analyzed on a 3% polyacrylamide gel over a 7% lower gel.

Comparison of processing by RNase P components from *B. subtilis* or *E. coli*. Two experiments were done in an attempt to compare processing by the RNase P components from *B. subtilis* and *E. coli*. Although these experiments are preliminary and the products have not been characterized, the data are shown here in Fig. 6 and 7 to demonstrate that all three of the transcripts described in this paper can be cleaved by the *E. coli* system. Some differences are noticeable in the products produced by the two systems, especially when bands in Fig. 7 are compared; however, these may be due to incomplete processing or slightly different efficiencies between the two systems. In the future, we intend

to investigate how RNase P from these two systems functions on a tRNA gene without the CCA terminus encoded. It has been observed that the lack of the CCA terminus in the *E. coli*

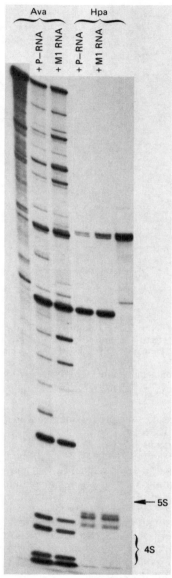

FIG. 6. Analysis of processing products from smaller transcripts containing one or six tRNA genes and comparison of action by P-RNA (*B. subtilis*) and M1 RNA (*E. coli*). Transcription reactions used pSP64BE2.5TG as a template after treatment with *Ava*I or *Hpa*II. Resulting transcripts contained 930 or 392 nucleotides, respectively. The 5S marker contained 116 bases; the 4S markers were unlabeled, unfractionated tRNAs from *B. subtilis*. Processing products were analyzed on a 48-cm-long 3% gel.

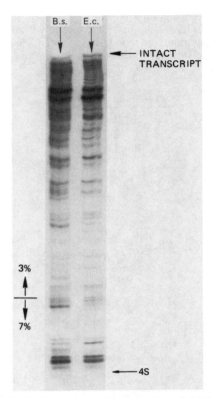

FIG. 7. Comparison of processing of full-length transcript with RNase P from either *B. subtilis* or *E. coli*. Substrate for the reaction was the transcript generated from pSP64PE3.3TG. Holoenzyme was used in both cases. Processing conditions are given in Experimental Procedures. The 4S marker contained 81 bases.

system makes the processing reaction much less efficient (2, 6).

Expression in vivo. When the gene segment containing the 21 tRNA genes was originally sequenced, a DNA sequence that may be used as a promoter for the sigma 55-containing RNA polymerase of *B. subtilis* was observed (3). To determine whether this promoter was a functional promoter, the DNA segment containing the end of the 23S rRNA, 5S rRNA, and 21 tRNA genes was cloned into pUC8. The tRNAs from *E. coli* cells containing pUC8 with or without the inserted genes were tested for levels of aminoacylation and for chromatographic properties on RPC-5. The inserted genes were oriented opposite to the β-galactosidase promoter of the pUC8 plasmid to test whether the putative internal promoter was active in vivo. The tRNAs for amino acid acceptor groups corresponding to the first six tRNA genes on the inserted segment, following the internal promoter, were the only tRNAs to show significantly elevated levels of amino acid acceptor activity compared with tRNAs from cells transformed with the plasmid lacking the tRNA gene insert. However, when column profiles were examined, species co-chromatographing differently from *E. coli* tRNA could be detected even from tRNA amino acid-accepting groups coded by tRNA genes farther away from the origin of transcription. This phenomenon is under investigation in our laboratory; however, these data seem to indicate that the putative internal promoter in the tRNA gene set functions as a promoter in vivo and that some of the *B. subtilis* tRNA genes can be transcribed and processed to tRNA species capable of accepting an amino acid in *E. coli*. In view of the fact that there are several RNA polymerase activities in *B. subtilis*, some of which occur at different stages of development, it is of interest that this internal promoter may be recognized by the RNA polymerase from *E. coli*, since this is the same type of promoter recognized by the major vegetative RNA polymerase from *B. subtilis*.

SUMMARY

The internal promoter found within the tRNA gene segment of a cluster of 21 tRNA genes from *B. subtilis* functions as a promoter in vivo in *E. coli*. At least some of these *B. subtilis* tRNA genes are transcribed and processed and can be detected as amino acid-accepting tRNAs.

A transcript containing these 21 tRNA genes can be produced in vitro which acts as a substrate for P-RNA. A transcript of this region cloned in the opposite direction does not act as a substrate for P-RNA or RNase P. The products of the processing reaction are the same with P-RNA alone or with P-RNA plus protein, indicating that the specificity of cleavage lies with the RNA component of RNase P and not with the protein moiety, supporting observations made by others using substrates containing a single tRNA gene. RNase P and M1 RNA from *E. coli* can also act to cleave the transcripts of *B. subtilis* tRNA genes; however, they may not produce exactly the same products.

A system has been presented for the production of transcripts of *B. subtilis* DNA containing one or six tRNA genes, and these may provide a valuable model system for exploring processing phenomena in *B. subtilis*.

This research was supported in part by Public Health Service research grant GM 29231 from the National Institute of General Medical Sciences.

We acknowledge the special help of Norman Pace and Terry Marsh in our experiments involving *B. subtilis* RNase P and P-RNA. We are also grateful to Cecilia Guerrier-Takada and Sidney Altman for the gift of *E. coli* RNase P and M1 RNA.

LITERATURE CITED

1. **Apirion, D., and P. Gegenheimer.** 1984. Molecular biology of RNA processing in prokaryotic cells, p. 35–62. *In* D. Apirion (ed.), Processing of RNA. CRC Press, Boca Raton, Fla.
2. **Fukada, K., and J. Abelson.** 1980. DNA sequence of a T4 transfer RNA gene cluster. J. Mol. Biol. **139:**377–391.
3. **Green, C., and B. Vold.** 1983. Sequence analysis of a cluster of twenty-one tRNA genes in *Bacillus subtilis.* Nucleic Acids Res. **11:**5763–5774.
4. **Green, M. R., T. Maniatis, and D. A. Melton.** 1983. Human beta-globin pre-mRNA synthesized in vitro is accurately spliced in Xenopus oocyte nuclei. Cell **32:**681–694.
5. **Guerrier-Takada, C., K. Gardiner, T. March, N. Pace, and S. Altman.** 1983. The RNA moiety of ribonuclease P is the catalytic subunit of the enzyme. Cell **35:**849–857.
6. **Guerrier-Takada, C., W. McClain, and S. Altman.** 1984. Cleavage of tRNA precursors by the RNA subunit of E. coli ribonuclease P (M1 RNA) is influenced by 3′-proximal CCA in the substrates. Cell **38:**219–224.
7. **Ozeki, H.** 1980. The organization of transfer RNA genes in *Escherichia coli,* p. 173–183. *In* S. Osawa, H. Ozeki, H. Uchida, and T. Yur (ed.), Genetics and evolution of RNA polymerase, tRNA and ribosomes. University of Tokyo Press, Tokyo.
8. **Pace, N. R.** 1984. Protein-polynucleotide recognition and the RNA processing nucleases in prokaryotes, p. 1–34. *In* D. Apirion (ed.), Processing of RNA. CRC Press, Boca Raton, Fla.
9. **Sogin, M. L., B. Pace, and N. R. Pace.** 1977. Partial purification and properties of a ribosomal RNA maturation endonuclease from *Bacillus subtilis.* J. Biol. Chem. **252:**1350–1357.
10. **Wawrousek, E. F., and J. N. Hansen.** 1983. Structure and organization of a cluster of six tRNA genes in the space between tandem ribosomal RNA gene sets in *Bacillus subtilis.* J. Biol. Chem. **258:**291–298.
11. **Wawrousek, E. F., N. Narashimhan, and J. N. Hansen.** 1984. Two large clusters with thirty-seven transfer RNA genes adjacent to ribosomal RNA gene sets in *Bacillus subtilis*: sequence and organization of trrnD and trrnE gene clusters. J. Biol. Chem. **259:**3694–3702.
12. **Yamada, Y., M. Ohki, and H. Ishikura.** 1983. The nucleotide sequence of *Bacillus subtilis* tRNA genes. Nucleic Acids Res. **11:**3037–3045.
13. **Zinn, K., D. DiMaio, and T. Maniatis.** 1983. Identification of two distinct regulatory regions adjacent to the human beta-interferon gene. Cell **34:**865–879.

Control of Developmental Transcription

Isolation of Genomic Segments from *Bacillus subtilis* that Contain Promoters for σ^{28} RNA Polymerase, and Studies on Regulation of Their Transcription in *B. subtilis* and *Escherichia coli*

M. J. CHAMBERLIN, J.-F. BRIAT, V. L. SINGER, J. S. GLENN, J. HELMANN, A. LEUNG, M. B. MOOREFIELD, AND M. Z. GILMAN

Department of Biochemistry, University of California, Berkeley, California 94720

The transcriptional machinery of *Bacillus subtilis* is complex. The predominant RNA polymerase (σ^{43} RNA polymerase) found in vegetatively growing cells has a subunit structure (1, 5) and promoter specificity (22) which are homologous with those of RNA polymerases of other eubacteria. However, the specificity of *B. subtilis* RNA polymerase is altered by interaction of the core polymerase with at least four minor sigma factors, leading to holoenzymes that recognize novel sets of promoter sites (14, 23). Some of these sites control transcription of genes that are specific to sporulation (8, 11, 15), which raises the question as to whether this multiplicity of transcriptional specificities is a unique feature of bacteria that undergo differentiation.

For some time we have been interested in the properties of a minor form of *B. subtilis* RNA polymerase controlled by a sigma factor of apparent M_r 28,000. This enzyme was first detected as a form of RNA polymerase which was unable to use normal bacterial promoters (10), but had a high degree of specificity for other promoter sites which are found primarily on *B. subtilis* DNA (23). These σ^{28}-specific promoters (P_{28}) are distinguished by unique DNA sequences at the -10 and -35 regions (6) that control promoter recognition (20). *B. subtilis* P_{28} promoters are used in vegetatively growing cells at the same start sites used by purified σ^{28} RNA polymerase in vitro. However, P_{28}-specific transcription in vivo is temporally regulated and also depends on the products of *B. subtilis* *spo0* genes of complementation groups A, B, E, and F (6). Hence, the use of P_{28} in cells depends on regulatory factors other than σ^{28}, and these factors are related to the control mechanisms for endospore formation.

Our major goals in the study of *B. subtilis* σ^{28} RNA polymerase and its regulation are (i) to identify the genes controlled by P_{28} and the role they play in cellular growth and (ii) to understand how the activity of RNA polymerases

such as σ^{28} RP is controlled temporally and by *spo0* products. At the same time we are interested in the more general question of what factors determine whether a set of genes (regulon) is controlled by a repressor-operator system, by positive control proteins such as the *Escherichia coli* CAP protein, or by a minor sigma factor.

EXPERIMENTAL PROCEDURES

Except where noted otherwise the experimental procedures used have been set forth in detail (6, 6a; J.-F. Briat, M. Gilman, and M. Chamberlin, J. Biol. Chem., in press).

The nomenclature used is that of Chamberlin (4). Sigma factors are designated by their apparent molecular weights, e.g., σ^{28}, except for *B. subtilis* σ^{43} (formerly σ^{55}; M. Gitt, L. F. Wang, and R. H. Doi, personal communication), *E. coli* σ^{70} (3), and *E. coli* σ^{32} (12), for which exact molecular weights are known from sequencing. P_{28} is the designation for a promoter utilized by σ^{28} RNA polymerase.

RESULTS AND DISCUSSION

Isolation and characterization of additional P_{28} from *B. subtilis*. *B. subtilis* σ^{28} RNA polymerase shows a high degree of promoter specificity and gives little synthesis, in vitro, from templates that lack P_{28} sequences (23). This selectivity allowed us previously to screen a plasmid library of *B. subtilis* genomic sequences using in vitro transcription, and gave two plasmids, pMG201 and pMG102, bearing the strain's σ^{28}-specific promoters P_{28-1} and P_{28-2}, respectively (6).

However, this was an inefficient way of screening for *B. subtilis* P_{28}, and we have since devised a more rapid procedure (6a) which takes advantage of the in vitro specificity of σ^{28} RNA polymerase to prepare a labeled RNA probe using *B. subtilis* DNA as template. To restrict transcription to P_{28}-promoter-proximal se-

TABLE 1. Directory of isolated *B. subtilis* DNA sequences that contain unique P_{28}

Promoter designa-tion	Form isolated	Relative promoter strength[a]	Smallest subclone[b]		Reference
			Size	Designa-tion	
P_{28-1}	Plasmid (pCD4136)	++	1.1 kb	pMG102[c]	6
P_{28-2}	Plasmid (pCD4322)	++	1.6 kb	pMG201[c]	6
P_{28-3}	Plasmid (pGR71-91)	±	0.64 kb	pMG301[c]	Gilman[d]
P_{28-4}	λ phage (1-62)	±	7.5 kb	1-62	6a
P_{28-5}	λ phage (1-9)	+	0.8 kb	pJG502	6a
P_{28-6}	λ phage (1-26)	++	2.2 kb	pMG601	6a
P_{28-7}	λ phage (1-56)	++	1.6 kb	pJG702	6a
P_{28-8}	λ phage (1-67)	+	1.1 kb	pJG802	6a
P_{28-9}	λ phage (2-12)	+	2.2 kb	pVS901	6a
P_{28-10}	λ phage (2-18)	+	1.2 kb	pMM1001	6a
P_{28-11}	λ phage (2-27)	±	9.1 kb	2-27	6a
P_{28-12}	λ phage (2-74)	±	12.25 kb	2-74	6a

[a] Relative promoter strength is measured by the amount of RNA synthesis in 10 min supported by a particular P_{28}-containing DNA in the presence of a fixed amount of purified σ^{28} RNA polymerase. Key: ++, 10 to 100 pmol; +, 3 to 10 pmol; ±, 1 to 3 pmol. Experimental conditions given by Gilman et al. (6a).

[b] Subclones are normally in pBR322 or in pUC18.

[c] P_{28} promoter regions in these plasmids have been sequenced (6; M. Gilman, Ph.D. thesis, University of California, Berkeley, 1983).

[d] Ph.D. thesis, 1983.

quences, the DNA template was first digested with restriction endonucleases *Hin*dIII and *Eco*RI.

Using this probe, we have screened a library of *B. subtilis* inserts in λgtWES, generously provided by James Hoch. Of 34 plaques which tested as positive in our initial screen of about 2,400 plaques, 13 phage strains bearing unique inserts were identified, 9 of which carried new P_{28} (6a). Identification of inserts and of P_{28} was based on three criteria: (i) restriction endonuclease analysis of insert sizes and structures; (ii) measurement of the activity of each template for in vitro transcription with purified σ^{28} RNA polymerase; (iii) Southern blot analysis of restriction digests of phage inserts using RNA transcribed in vitro with purified σ^{28} RNA polymerase.

The nine phage strains bearing unique, P_{28}-containing *B. subtilis* inserts have been analyzed further by subcloning and by fine-structure Southern blot analysis; our current information on these and other P_{28}-containing segments is summarized in Table 1. Detailed understanding of the P_{28} promoter sequences and of the regulatory patterns of each promoter will require sequencing and physiological studies, which are now in progress. However, we can make several generalizations at this point concerning *B. subtilis* P_{28}-containing fragments and their isolation.

(i) This method using in vitro-generated probes is highly effective in screening for promoters used by RNA polymerases that are reasonably specific in vitro. Of the 34 positive plaques mentioned above, 13 contained unique inserts,

and virtually all of the others were duplicates or contained the same insert in an opposite orientation.

(ii) Analysis of the frequency of isolation of P_{28} inserts (6a) suggests that there are probably only 20 to 30 such promoters in the entire *B. subtilis* genome. This is a relatively small number of promoters for a potential regulon. Of course, this does not include σ^{28} promoters that are very weak in vitro or that might require a positive control factor and hence do not work well in vitro with σ^{28} RNA polymerase alone.

(iii) There appears to be little or no clustering of P_{28} in the *B. subtilis* genome. We have yet to identify an insert bearing two or more P_{28} promoters, and it seems most likely that P_{28} promoters are distributed throughout the genome.

Regulation of σ^{28}-specific transcription; effect of the *crsA* mutation on transcription from P_{28} in *spo0* mutants. Transcription from *B. subtilis* promoters P_{28-1} and P_{28-2} is greatly reduced or eliminated in *spo0* mutants of complementation groups A, B, E, and F (6). Furthermore, P_{28} transcription is shut off rapidly after the onset of sporulation. Hence, although P_{28} transcription is not required during sporulation, P_{28} transcripts might code for gene products needed to trigger onset of sporulation.

However, a large number of different gene products are not synthesized in *spo* mutants (9, 18), and hence the effect on P_{28} transcription might be indirect and not reflect a role of P_{28} in sporulation. A more critical test of the relationship between P_{28} transcription and sporulation phenotype was carried out by following P_{28} transcription in second-site revertants (*rvt*) that restore sporulation proficiency to several groups of *spo0* mutants (19). In each case where the *rvt* mutant restored sporulation proficiency, P_{28} transcription was also restored (6).

To extend this correlation further, we have, in collaboration with Terry Leighton's laboratory, tested the effect of the *crsA47* mutation on P_{28-1} transcription in *spo0* mutants (14). The *crsA* mutation was originally isolated as an alteration that relieves catabolite sensitivity of sporulation (21); its ability to restore sporulation proficiency to *spo0* mutants is made more intriguing by the fact that C. W. Price and R. H. Doi (personal communication) have mapped the mutation to the *rpoD* gene which codes for the predominant *B. subtilis* sigma factor, σ^{43}.

We were surprised to find that, although *crsA* mutants restored sporulation proficiency to *spo0* mutants in complementation groups B, E, and F, there was little or no restoration of transcription from P_{28-1} as determined by S1 nuclease mapping by a 5′ P^{32}-labeled probe (Fig. 1). This would seem to break the invariant relationship we have previously seen between sporulation proficiency and transcription from P_{28} promoters (6). However, the result remains to be confirmed with other P_{28} promoters. In addition, it appears that there is a great deal of full-length protected probe in the *crsA spo0* mutants (Fig. 1). Hence, it is possible that the *crsA* mutation leads to activation of an upstream promoter that leads to transcription of the genes normally read from P_{28-1}. This would not be an unreasonable possibility given that the *crsA* mutation is in *rpoD* (Price and Doi, personal communication).

B. subtilis P_{28} promoters are utilized in *E. coli* by a form of *E. coli* RNA polymerase bearing the minor sigma factor *htpR* (*E. coli* σ^{32}). In the course of our studies on *B. subtilis* P_{28-1} and P_{28-2} we observed that DNA fragments bearing these promoters were able to facilitate transcription of downstream sequences such as *lacZ*, *galK*, or chloramphenicol acetyltransferase when cloned in plasmids in *E. coli*. This transcription could have originated from a separate promoter for the normal major *E. coli* RNA polymerase (σ^{70} RNA polymerase) carried on the *B. subtilis* insert. However, S1 nuclease mapping of the start site in *E. coli* on such a fragment bearing *B. subtilis* P_{28-2} shows it to be identical to that used by *B. subtilis* σ^{28} RNA polymerase in vitro and in vivo (Fig. 2; the experiment shown is the same one shown in Fig. 3B of Gilman and Chamberlin [6] except that lane F was deleted in that figure). The same result is obtained when the start site for *B. subtilis* P_{28-1} is mapped in *E. coli* (data not shown). Although the actual levels of transcription are low, there is no question that both P_{28} promoters are read in *E. coli* at the correct *B. subtilis* σ^{28} start site.

These results were surprising since these σ^{28}-specific promoters are not used detectably by the predominant *B. subtilis* (σ^{43}) or *E. coli* (σ^{70}) RNA polymerases (10, 23), a result which we have repeatedly confirmed.

We concluded that it was likely that *E. coli* contained an additional sigma factor or some novel transcriptional activator that allowed use of *B. subtilis* P_{28} by σ^{70} RNA polymerase. When the sequences of the *E. coli rpoD* promoter region which is activated by heat shock were obtained (Fig. 3), it became evident to us that they contained a region which closely resembles *B. subtilis* P_{28} promoters (7), but does not fit well to σ^{70}-specific promoter sequences (20). This

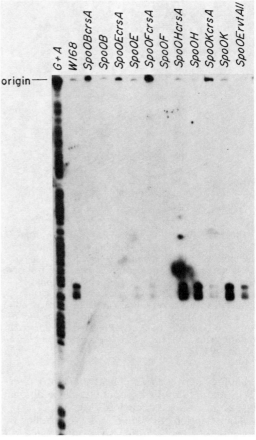

FIG. 1. S1 nuclease mapping of P_{28-1} transcripts in *B. subtilis* strains containing *spo0* and *spo0 crsA*. Mapping was carried out by the method of Gilman and Chamberlin (6), using a 5′ ^{32}P-labeled, single-stranded probe prepared from the *Hpa*II fragment of pMG102. Transcription from P_{28-1} gives two sizes of protected fragments (see W168 control, for example), probably as a result of the use of two different A residues at the start site; these are located 8 and 10 nucleotides to the right of the rightmost T in the −10 sequences (CCGATAT).

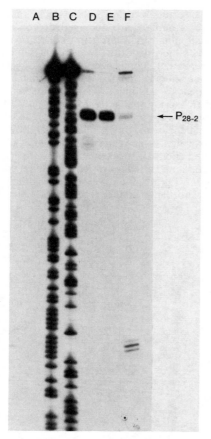

FIG. 2. S1 nuclease mapping of P_{28-2} transcripts in *B. subtilis* and in *E. coli*. Mapping was carried out by the method of Gilman and Chamberlin (6), using a 5′ ^{32}P-labeled, single-stranded probe prepared from the *Hpa*II fragment of pMG201. Track A, Control, no added RNA; track B, A+G cleavage markers; track C, T+C cleavage markers; track D, RNA transcribed in vitro from pMG201 by using purified *B. subtilis* σ^{28} RNA polymerase; track E, RNA isolated from *B. subtilis* W168 cells; track F, RNA isolated from *E. coli* bearing pMG201.

homology between *E. coli* heat shock promoters (P_{HS}) and *B. subtilis* P_{28} continues as additional P_{HS} sequences are compared (Fig. 3).

Transcription of P_{HS} in *E. coli* is controlled by a regulatory gene, *htpR* (16). Mutants in *htpR* fail to induce any of the genes in the *E. coli* heat shock regulon when subjected to heat or other stress. To test whether *htpR* also controls transcription from *B. subtilis* P_{28} in *E. coli*, we prepared *htpR*⁺ and *htpR* strains which carried a plasmid, pJH102T, which bears P_{28-1} (Briat et al., in press). The *htpR*⁺ strain initiates transcription at the normal *B. subtilis* start site on P_{28-1}, and there is no transcription from this region in the *htpR* mutant. Hence, the *htpR* regulatory

gene controls transcription of both *E. coli* P_{HS} and *B. subtilis* P_{28}. Surprisingly, however, there was no enhancement of P_{28} transcription in *E. coli* when cells were exposed to a 45°C heat shock (R. Foster, unpublished data).

This result, taken with the close sequence relationships between P_{HS} and P_{28-1}, led us to test whether the *B. subtilis* σ^{28} RNA polymerase could transcribe from an *E. coli* P_{HS}, that which controls *rpoD* (see Fig. 3 for sequence). When purified σ^{28} RNA polymerase is used, initiation of transcription of the cloned *rpoD* promoter region takes place at the same nucleotides found during heat shock in vivo (Briat et al., in press). Hence, *B. subtilis* σ^{28} RNA polymerase and the *E. coli* *htpR*-dependent activity share an overlapping promoter specificity.

How does the *E. coli* *htpR* protein activate transcription from P_{HS} and P_{28}? These properties fit with those expected of a minor sigma factor, a possibility made more likely by the demonstration (12) that the *htpR* gene sequence shares regions of homology with *rpoD*, the gene for *E. coli* σ^{70}. We found that, in vitro, *htpR*⁺-dependent initiation at the *rpoD* P_{HS} copurifies with *E. coli* RNA polymerase through DNA cellulose chromatography (2) and hence is determined by a tightly bound factor. However, before we could identify that factor as *htpR* protein, Grossman et al. (7a) independently reported studies that demonstrate convincingly that purified *htpR* protein can activate core polymerase to read heat shock promoters in vitro. Hence, *E. coli* *htpR* protein is a minor *E.*

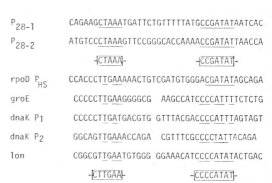

FIG. 3. Comparison of promoter sequences for *B. subtilis* σ^{28} RNA polymerase and *E. coli* heat shock genes. Promoter sequences for *B. subtilis* P_{28-1} and P_{28-2} are from Gilman et al. (7). Sequences for *E. coli* P_{HS} are taken from Neidhardt et al. (16). Boxed sequences are average or consensus sequences in the −10 and −35 regions of the promoters shown. The start sites for transcription on P_{28-1} and P_{28-2} are eight nucleotides to the right of the rightmost T in the -CCGATAT- box. The start sites for *rpoD* P_{HS} are 8 and 14 nucleotides to the right of the T in the -CGATAT- box (Briat et al., in press).

coli sigma factor, which they have designated *E. coli* σ^{32}.

What are the implications of this finding for those interested in gene regulation in *B. subtilis*? It is evident that the use of minor sigma factors as a regulatory mechanism is not restricted to *Bacillus* or to procaryotes that carry out complex developmental patterns, as has sometimes been supposed. Instead, use of minor sigma factors to control a regulon must now be viewed as only one of several alternative mechanisms that are available to all procaryotes. It will be interesting to learn whether different regulons controlled by minor sigma factors prove to have other features in common.

The genes for *B. subtilis* σ^{43} and *E. coli* σ^{70} are highly homologous (3; Gitt, Wang, and Doi, personal communication). It has been known for some time that polymerases bearing these factors share an overlapping promoter specificity (13, 22), which is defined by homologous sequences in the -35 and -10 regions upstream from the promoters they use. The overlapping promoter specificity of *B. subtilis* σ^{28}- and *E. coli* σ^{32}-directed RNA polymerases is also explained by the conserved sequences in the P_{28} and P_{HS} promoters, which are quite similar to each other and quite different from those used by the major RNA polymerases (Fig. 3). It will be interesting to compare the gene sequences for σ^{28} with those of *htpR* to determine whether the genes are homologous or have separately evolved an overlapping promoter specificity.

If *B. subtilis* σ^{28} and *E. coli* σ^{32} are related, it will reduce even further the apparent differences in the transcriptional machinery for these two organisms. One might hope to glean information about such relatedness by comparing the genes controlled by the two factors. The role of *htpR* in activation of *E. coli* P_{HS} is well defined (16). However, none of the *B. subtilis* genes controlled by P_{28} is known. We do know that neither *B. subtilis* P_{28-1} nor P_{28-2} is induced by elevated temperatures up to 49°C in *B. subtilis* (6); hence, these are probably not *B. subtilis* P_{HS}. However, there are at least 20 to 30 more *B. subtilis* P_{28} promoters that we have not yet studied, and there is some evidence that there may be essential *E. coli* promoters (P_{32}) read by σ^{32} that are not P_{HS} (16). In fact, we know already that *B. subtilis* P_{28-2} transcription in *E. coli* is not heat shock induced. Thus, it is still possible that *B. subtilis* P_{28} and *E. coli* P_{32} control homologous functions in their two respective cells, one of which involves the heat shock response and one of which does not.

The fact that transcription from P_{28} in *E. coli* is dependent on σ^{32}, but not on heat shock, has potential significance for those interested in regulation of transcription from promoters con-

trolled by minor sigma factors in *Bacillus*. In particular, we conclude that the sequences needed for σ^{32} recognition do not themselves lead to regulation by heat shock. Hence, the control of the heat shock regulon is probably determined by sequences other than the -10 and -35 conserved regions shown in Fig. 3. In addition, this result rules out models for such regulation in which heat shock simply controls the activity of σ^{32} in the cell. Thus, whereas σ^{32} controls the promoter specificity of the heat shock response, control over the activation or repression of the response is defined by another regulatory molecule.

The control of the heat shock regulon in *E. coli* is very reminiscent of what is known about control over transcription from P_{28} and P_{37} in *B. subtilis*. In both of the latter cases, the presence of the σ protein in the cells is apparently insufficient for transcription from the respective promoters (6, 14, 17). If the analogy with *htpR* regulation holds, this regulation may not be controlled by altering the activity of the sigma factor, but by other factors that may be promoter specific. It will be quite interesting to identify these factors and determine how they act.

We are grateful to Terry Leighton (University of California, Berkeley) and his associates for many helpful discussions.

This research was supported by Public Health Service research grant GM12010 from the National Institute of General Medical Sciences and by National Research Award Training Grant GM07232.

LITERATURE CITED

1. **Burgess, R. R.** 1976. Purification and physical properties of *E. coli* RNA polymerase, p. 69–100. *In* R. Losick and M. Chamberlin (ed.), RNA polymerase. Cold Spring Harbor Press, Cold Spring Harbor, N.Y.
2. **Burgess, R. R., and J. Jendrisak.** 1975. A procedure for the rapid, large-scale purification of *Escherichia coli* DNA-dependent RNA polymerase involving polymin P precipitation and DNA-cellulose chromatography. Biochemistry **14:**4634–4638.
3. **Burton, Z., R. R. Burgess, J. Lin, D. Moore, S. Holder, and C. Gross.** 1981. The nucleotide sequence of the cloned *rpoD* gene for the RNA polymerase sigma subunit from *E. coli*. Nucleic Acids Res. **9:**2889–2903.
4. **Chamberlin, M.** 1982. Bacterial DNA-dependent RNA polymerases, p. 87–108. *In* P. Boyer (ed.), The enzymes, vol. 15. Academic Press, Inc., New York.
5. **Fukuda, R., A. Ishihama, T. Saitoh, and M. Taketo.** 1977. Comparative studies of RNA polymerase subunits from various bacteria. Mol. Gen. Genet. **154:**135–144.
6. **Gilman, M., and M. Chamberlin.** 1983. Developmental and genetic regulation of *Bacillus subtilis* genes transcribed by σ^{28}-RNA polymerase. Cell **35:**285–293.
6a. **Gilman, M., J. Glenn, V. Singer, and M. Chamberlin.** 1984. Isolation of DNA sequences from *B. subtilis* that contain *in vitro* promoter sites for σ^{28}-RNA polymerase. Gene **32:**11–20.
7. **Gilman, M., J. Wiggs, and M. Chamberlin.** 1981. Nucleotide sequence of two *Bacillus subtilis* promoters used by *Bacillus subtilis* sigma-28 RNA polymerase. Nucleic Acids Res. **9:**5991–6000.
7a. **Grossman, A., J. W. Erickson, and C. A. Gross.** 1984. The *htpR* gene product of *E. coli* is a sigma factor for heat-

shock promoters. Cell **38**:383–390.

8. **Haldenwang, W., N. Lang, and R. Losick.** 1981. A sporulation-induced sigma-like regulatory protein from *B. subtilis*. Cell **23**:615–624.

9. **Hoch, J.** 1976. Genetics of bacterial sporulation. Adv. Genet. **18**:67–98.

10. **Jaehning, J., J. Wiggs, and M. Chamberlin.** 1979. Altered promoter selection by a novel form of *Bacillus subtilis* RNA polymerase. Proc. Natl. Acad. Sci. U.S.A. **76**:5470–5474.

11. **Johnson, W. C., C. P. Moran, and R. Losick.** 1983. Two RNA polymerase sigma factors from *Bacillus subtilis* discriminate between overlapping promoters for a developmentally regulated gene. Nature (London) **302**:800–804.

12. **Landick, R., V. Vaughn, E. Lau, R. Van Bogelen, J. Erickson, and F. C. Neidhardt.** 1984. Nucleotide sequence of the heat shock regulatory gene of *E. coli* suggests its protein product may be a transcription factor. Cell **38**:175–182.

13. **Lee, G., C. Talkington, and J. Pero.** 1980. Nucleotide sequence of a promoter recognized by *Bacillus subtilis* RNA polymerase. Mol. Gen. Genet. **180**:57–65.

14. **Losick, R., and J. Pero.** 1981. Cascade of sigma factors. Cell **25**:582–584.

15. **Moran, C., N. Lang, and R. Losick.** 1981. Nucleotide sequence of a *Bacillus subtilis* promoter recognized by *Bacillus subtilis* RNA polymerase containing σ^{37}. Nucleic Acids Res. **9**:5979–5990.

16. **Neidhardt, F. C., R. A. Van Bogelen, and V. Vaughn.** 1984. Genetics and regulation of heat-shock proteins. Annu. Rev. Genet. **18**:295–329.

17. **Ollington, J., W. Haldenwang, T. V. Huynh, and R. Losick.** 1981. Developmentally regulated transcription in a cloned segment of the *Bacillus subtilis* chromosome. J. Bacteriol. **147**:432–442.

18. **Piggot, P. J., and J. G. Coote.** 1976. Genetic aspects of bacterial endospore formation. Bacteriol. Rev. **40**: 908–962.

19. **Sharrock, R., S. Rubenstein, M. Chan, and T. Leighton.** 1984. Intergenic suppression of *spo0* phenotypes by the *Bacillus subtilis* mutation *rvtA*. Mol. Gen. Genet. **194**:260–264.

20. **Siebenlist, U., R. Simpson, and W. Gilbert.** 1980. *E. coli* RNA polymerase interacts homologously with two different promoters. Cell **20**:269–281.

21. **Sun, D., and I. Takahashi.** 1982. Genetic mapping of catabolite-resistant mutants of *Bacillus subtilis*. Can. J. Microbiol. **28**:1242–1251.

22. **Wiggs, J., J. Bush, and M. Chamberlin.** 1979. Utilization of promoter and terminator sites on bacteriophage T7 DNA by RNA polymerases from a variety of bacterial orders. Cell **16**:97–109.

23. **Wiggs, J., M. Gilman, and M. Chamberlin.** 1981. Heterogeneity of RNA polymerase in *B. subtilis*: evidence for an additional σ factor in vegetative cells. Proc. Natl. Acad. Sci. U.S.A. **78**:2762–2766.

Localizing the Site of *spo0*-Dependent Regulation in the *spoVG* Promoter of *Bacillus subtilis*

PETER ZUBER

The Biological Laboratories, Harvard University, Cambridge, Massachusetts 02138

Initiation of sporulation in *Bacillus subtilis* is dependent upon the products of at least eight *spo0* genes, a mutation in any one of which blocks sporulation at its earliest stage, stage 0 (10, 11, 18, 19). Mutations in *spo0* genes also cause a variety of changes in cell physiology unrelated to sporulation, such as acquisition of genetic competence and sensitivity to certain phages and antibiotics. I have been studying the transcription and regulation of a sporulation gene known as *spoVG* as a model system for investigating the mechanism of action of *spo0* gene products in early sporogenesis. The *spoVG* gene encodes a stable 400-base RNA whose synthesis is induced at the onset of sporulation and whose protein product participates in maturation of the spore cortex (16, 23). Synthesis of *spoVG* RNA is significantly reduced in strains which bear a mutation in any one of seven *spo0* genes. That induction of *spoVG* RNA synthesis is impaired in Spo0 mutants suggests that one or more of the *spo0* genes (or genes under their control) encode a transcription factor upon which the expression of *spoVG* and perhaps many other early induced sporulation genes is dependent. Another useful system for studying early sporulation control is the *apr* gene, which encodes the sporulation-associated alkaline protease of *B. subtilis* and whose transcription is under *spo0* control (6, 18, 25).

Previous studies of the *spoVG* gene have focused on the structure of its transcription initiation region and on the requirements for *spoVG* RNA synthesis in vitro and in vivo (1, 9, 10, 15). The *spoVG* transcription initiation region consists of two overlapping promoters, P1 and P2, which are separately utilized by the $E\sigma^{37}$ and $E\sigma^{32}$ forms of *B. subtilis* RNA polymerase, respectively. An upstream A-T-rich sequence strongly stimulates transcription from both the P1 and P2 start sites in vitro (Fig. 1). Even though accurate transcription of *spoVG* can be achieved in vitro simply by the action of $E\sigma^{32}$ and $E\sigma^{37}$, as yet unidentified transcription factors (*spo0* gene products) in addition to these holoenzyme forms appear to be required for efficient utilization of P1 and P2 in vivo; for example, both $E\sigma^{37}$ and $E\sigma^{32}$ are known to be present under conditions (vegetative growth and stationary phase in Spo0 mutants) in which *spoVG* is expressed at only low levels (W. C. Johnson and C. Binnie, unpublished data).

I have been attempting to study the way in which *spo0* gene products influence transcription of *spoVG*. The strategy I have adopted employs an in-frame gene fusion consisting of the promoter and translational initiation region of *spoVG* and the *Escherichia coli lacZ* gene (27). The fusion was constructed by inserting a fragment of *B. subtilis* DNA containing the promoter and amino-terminal coding end of *spoVG* upstream from a promoterless *lacZ* gene and then using Bal-31 nuclease to bring the two coding sequences into the same translational reading frame. The *spoVG-lacZ* fusion was introduced in single copy into *B. subtilis* and several of its Spo0 mutants by (i) Campbell integration with a fusion-bearing plasmid that recombines into the chromosome near the *spoVG* locus or (ii) transduction using an SPβ specialized transducing phage carrying the *spoVG-lacZ* fusion. When tested in several Spo0 mutants, expression of the *spoVG-lacZ* fusion was found to be regulated in a manner consistent with the earlier, direct studies of *spoVG* transcription (16). Spo⁺ cells of *B. subtilis* bearing the *spoVG-lacZ* fusion induced the synthesis of β-galactosidase under conditions which promote sporulation, whereas cells of Spo0 mutants synthesized only low levels of β-galactosidase under the same conditions, the reduction being more or less severe depending on the *spo0* locus. The *spo0*-dependent induction of *spoVG-lacZ* expression was found to be controlled entirely from within a 157-base-pair region of *spoVG* consisting of the upstream A-T-rich sequence, the overlapping promoters (P1 and P2), the ribosome binding site, and the first nine codons of the *spoVG* gene protein-coding sequence (27).

Here I exploit SPβ::*spoVG-lacZ* to survey the expression of *spoVG-lacZ* in a variety of Spo0 and SpoII mutants, to examine the effect of deletion mutations of the *spoVG* promoter region on its utilization in vivo, and to further localize, through the use of these deletions, the target of *spo0*-dependent regulation.

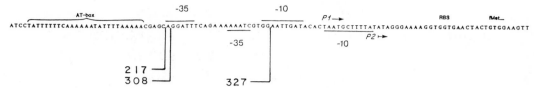

FIG. 1. Nucleotide sequence of the *spoVG* promoter region. The lines labeled 217, 308, and 327 indicate the 5' endpoints of each of the promoter deletion mutations described in the text.

EXPERIMENTAL PROCEDURES

Bacterial strains and plasmids. The bacterial strains bearing the *spoVG-lacZ* fusion are listed in Table 1. The Spo⁺ control strain used in all experiments is a derivative of JH642 (*trpC2 pheA1*, obtained from J. Hoch). The Spo0 strains (except for the *spo0HΔHind* mutant) were obtained from J. Hoch. The *spo0H* strain used in the analysis of *spoVG* promoter deletions was IS233 (*trpC2 pheA1 spo0HΔHind*) (24). The SpoII strains were obtained from the Bacillus Genetic Stock Center (Ohio State University). The *spoVG-lacZ* specialized transducing phage was derived from a thermoinducible form of the *B. subtilis* temperate phage SPβ (20, 26, 27). High-frequency transducing lysates were obtained by heat treatment of ZB239 (Table 1). The *lacZ* fusion plasmids were propagated in the *E. coli* strain MC1061 [*araD139* Δ(*ara-leu*)7697 Δ*lacX74f galU galK hsr hsm⁺ strA*] (2). Plasmids bearing the *spoVG-lacZ* fusion are derivatives of pZL205 (see below; 27) and were introduced into *B. subtilis* cells by transformation (7).

Determination of β-galactosidase specific activity. The *spoVG-lacZ B. subtilis* strains were grown in 50 ml of Difco sporulation medium (DSM; 22), and growth was monitored by measuring the absorbance at 595 nm. Cells from 1-ml samples were collected every hour and washed in cold 0.05 M Tris hydrochloride (pH 8.0) (3). The cell pellets were stored at −70°C. The cells were quickly thawed, suspended in 1 ml of Z buffer (14), and mixed with 10 µl of toluene. β-Galactosidase specific activity was determined by the method of Miller (14).

Construction of *spoVG* promoter deletions. The diagram in Fig. 2 describes the construction of plasmids carrying mutated *spoVG* promoters fused to *lacZ*. The shuttle plasmid pZL205 consists of the ampicillin resistance and replication functions of pBR322, the chloramphenicol resistance plasmid of *B. subtilis*, pBD64 (8), and the *spoVG-lacZ* fusion (27). pZL205 was cleaved at a unique *Hin*dIII site upstream from the *spoVG* promoter and was treated with Bal-31. The Bal-31-treated linear plasmid molecules were appended with *Hin*dIII linkers and recircularized by using T4 DNA ligase to yield a pZL205 derivative with a *spoVG* promoter deletion mutation. The *Hin*dIII-*Sal*I fragment containing the *spoVG* promoter mutant was purified and, by ligation, inserted into *Hin*dIII-*Sal*I-cleaved pZL205, thereby replacing the intact *spoVG* promoter *Hin*dIII-*Sal*I fragment of pZL205. The resulting plasmid, pZΔ324, carries the mutant *spoVG* promoter (fused to *lacZ*) and all sequences upstream from the *Hin*dIII site in pZL205 which were removed by Bal-31 treatment.

Plasmid pZΔ324 was modified to yield a plasmid which could recombine by a Campbell mechanism into the *B. subtilis* chromosome. This was accomplished by substituting the pBD64 plasmid *Eco*RI fragment of pZL205 with a 1.3-kilobase *Eco*RI fragment carrying a gene encoding chloramphenicol acetyltransferase (*cat-86*) (from pMI1082; M. Igo, unpublished data). The resulting plasmid, pZΔ327, when introduced into *B. subtilis* cells by transformation (selecting for chloramphenicol resistance) can recombine upstream from the *spoVG* gene (27). Other mutant *spoVG* promoters generated by

TABLE 1. *B. subtilis* strains used

Strain	Genotype
ZB239	*trpC2 pheA1 spoVG-lacZ (SPβc2Δ2::Tn917)*
ZB212	*trpC2 pheA1 spoVG-lacZ*
ZB233	*trpC2 pheA1 spoVG-lacZ (SPβc2Δ2::Tn917) spo0A12*
ZB259	*trpC2 pheA1 spoVG-lacZ (SPβc2Δ2::Tn917) spo0HΔHind*
ZB215	*trpC2 pheA1 spoVG-lacZ spo0B136*
ZB219	*trpC2 pheA1 spoVG-lacZ spo0C9V*
ZB227	*trpC2 pheA1 spoVG-lacZ spo0E11*
ZB228	*trpC2 pheA1 spoVG-lacZ spo0J93*
ZB294	*trpC2 lys-3 spoVG-lacZ (SPβc2Δ2::Tn917) spoIIA69*
ZB295	*trpC2 spoVG-lacZ (SPβc2Δ2::Tn917) spoIIB131*
ZB296	*trpC2 spoVG-lacZ (SPβc2Δ2::Tn917) spoIIC298*
ZB297	*trpC2 rpoB2(Rifʳ) spoVG-lacZ (SPβc2Δ2::Tn917) spoIID66*
ZB298	*trpC2 rpoB2(Rifʳ) spoVG-lacZ (SPβc2Δ2::Tn917) spoIIE61*
ZB299	*trpC2 spoVG-lacZ (SPβc2Δ2::Tn917) spoIIF96*
ZB300	*trpC2 tal-1 spoVG-lacZ (SPβc2Δ2::Tn917) spoIIG41*

Bal-31 treatment of pZL205 were inserted into HindIII-SalI-cleaved pZΔ327. All plasmids containing the promoter mutations were subsequently transferred by transformation into competent *B. subtilis* cells, with selection for chloramphenicol resistance. To determine the nucleotide sequence of the mutated promoters the HindIII-SalI fragments of each plasmid were cloned into M13mp8 (13) and sequenced by the method of Sanger et al. (21).

RESULTS

Figure 3 shows the time course of increase in the specific activity of β-galactosidase in a variety of *spoVG-lacZ*-bearing mutants. In confirmation and extension of previous work (16, 27), the induction of *spoVG* was found to be most severely impaired by the stage 0-blocked mutations *spo0A*, *spo0B*, and *spo0H*, with the most dramatic reduction in *lacZ* expression taking place in a strain which bears a deletion in the *spo0H* gene (Fig. 3A). This deletion mutation conferred a Lac⁻ (white) phenotype to *spoVG-lacZ*-bearing strains grown on DSM plates containing the chromogenic substrate X-gal (5-bromo-4-chloroindolyl galactoside) even after prolonged incubation. (In contrast, Spo⁺ bacteria containing *spoVG-lacZ* turn deep blue on these plates after overnight incubation.) Strains bearing *spo0E*, *spo0F*, *spoIIA*, and *spoIID* represented a second class of mutants in which only a partial reduction of *spoVG* expression could be detected (Fig. 3B). Finally, strains bearing *spo0J*, *spoIIB*, *spoIIC*, *spoIIE*, *spoIIF*, and *spoIIG* seemed to constitute a third class of mutants, as no impairment of *lacZ* expression could be detected in these strains and in some instances (*spo0J*, *spoIIC*, and *spoIIF*) even greater levels of β-galactosidase were produced than that observed in the Spo⁺ wild-type strain (Fig. 3C).

Effect of upstream deletion mutations on the expression of spoVG-lacZ. To define more precisely the sequences required for efficient transcription of *spoVG* in vivo, a series of deletions were generated in vitro that extended into the *spoVG* promoter region from the upstream direction. DNA was removed progressively from the HindIII site located just upstream from *spoVG* by Bal-31 treatment of HindIII-cut pZL205 DNA (see Fig. 2). The plasmids bearing the *spoVG* promoter deletions fused to *lacZ* were introduced by transformation into the Spo⁺ strain JH642. The endpoints of three such deletion mutations are shown in the nucleotide sequence of Fig. 1. One of the mutated promoters, *spoVG217*, had lost the A-T box and sequences up to the putative "−35" region of P1. A *spoVG217-lacZ*-containing strain was grown (in parallel with a *spoVG-lacZ* strain) in liquid

DSM, and 1-ml samples were assayed for β-galactosidase activity at hourly intervals from exponential-phase growth to t₅. Figure 4A shows the characteristic induction pattern of *spoVG-lacZ* in Spo⁺ cells and the pattern obtained from assays of cells harboring the deletion-mutated

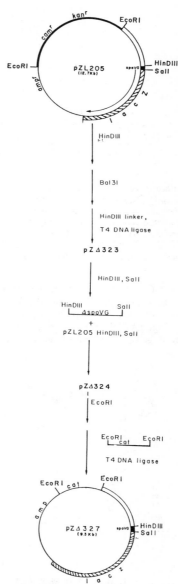

FIG. 2. Construction of the *spoVG* promoter deletion mutations and plasmids used for their transfer into *B. subtilis*. The figure shows the procedure used for generating deletions extending from the HindIII site into the *spoVG* promoter region and the construction of the plasmid used to insert the promoter deletions into the chromosome at the *spoVG* locus (see Experimental Procedures for details).

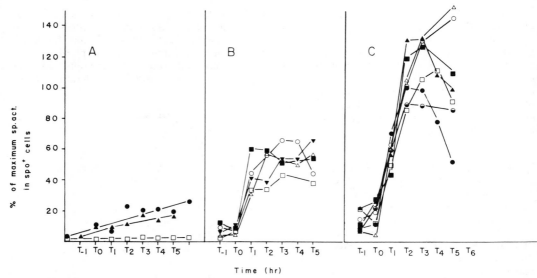

FIG. 3. Pattern of *spoVG-lacZ* expression in Spo0 and SpoII mutants. Various Spo0 and SpoII mutants bearing the *spoVG-lacZ* fusion were grown in liquid DSM. One-milliliter samples were collected at hourly intervals and assayed for β-galactosidase. Activity of β-galactosidase is reported as a percentage of the maximum specific activity detected in Spo⁺ cells. (A) ZB233 (*spo0A12*), ●; ZB215 (*spo0B136*), ▲; ZB259 (*spo0HΔHind*), □. (B) ZB219 (*spo0C9V*), ▼; ZB227 (*spo0E11*), ■; ZB228 (*spo0F221*), ○; ZB294 (*spoIIA69*), △; ZB297 (*spoIID66*), □. (C) ZB239 (Spo⁺), ●; ZB223 (*spo0J93*), ▲; ZB295 (*spoIIB131*), □; ZB296 (*spoIIC298*), ○; ZB298 (*spoIIE61*), △; ZB299 (*spoIIF96*), ■; ZB300 (*spoIIG41*), ◖.

fusion *spoVG217-lacZ*. Even though the *spoVG217* mutation reduced *spoVG-lacZ* expression to only about 2 to 3% of wild-type levels at t₂ (Fig. 4A), plotting the β-galactosidase activity on an expanded scale (Fig. 4B) reveals

that the *spoVG217-lacZ*-bearing strain had retained the normal pattern of induction. Thus, it appears that removal of the A-T box reduced the level of *spoVG-lacZ* expression but did not affect the timing of its induction.

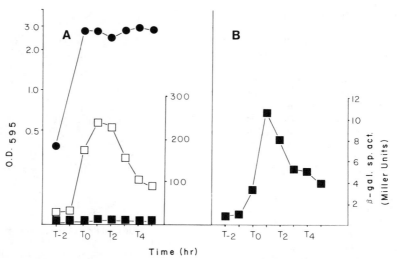

FIG. 4. Time course of β-galactosidase production in Spo⁺ cells bearing *spoVG-lacZ* or *spoVG217-lacZ*. β-Galactosidase specific activity was measured at hourly intervals beginning at t₂ to t₅. (A) Strains bearing the *spoVG-lacZ* (□) and *spoVG217-lacZ* (■) fusion plasmids integrated at the *spoVG* locus. Growth was monitored by measuring the absorbance at 595 nm (●). (B) Induction of *spoVG217-lacZ* from (A), plotted on an expanded scale.

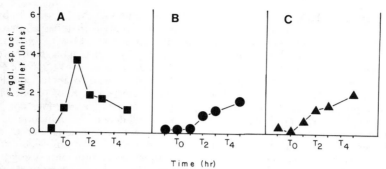

FIG. 5. Time course experiment showing the effects of promoter deletion mutations on the expression of *lacZ*. β-Galactosidase specific activity was measured at hourly intervals during exponential and stationary phase in DSM for strains harboring (A) *spoVG217-lacZ*, (B) *spoVG308-lacZ*, and (C) *spoVG327-lacZ* plasmids integrated near the *spoVG* locus of the Spo⁺ strain JH642. The *o*-nitrophenyl-β-D-galactopyranoside hydrolysis as indicated in (B) and (C) represents a background of promoter-independent β-galactosidase activity.

A second deletion mutation, *spoVG308*, lacked the A-T box, sequences up to the putative −35 region of P1, and one additional base, the adenine in the −35 region of P1. Because this adenine residue is conserved in other promoters utilized by $E\sigma^{37}$ (12, 25), I expected that removal of this base would damage P1 and partially impair *spoVG-lacZ* expression. A *spoVG308-lacZ* and a *spoVG217-lacZ* strain were grown in DSM, and β-galactosidase specific activity was determined during the exponential and stationary phases of growth. Again, the *spoVG217-lacZ* strain exhibited the characteristic pattern of low-level induction whereas the *spoVG308-lacZ* strain produced negligible levels of β-galactosidase and showed little (or at best delayed) induction of *lacZ* (Fig. 5B). The effect of *spoVG308* could be observed quite strikingly on DSM plates containing X-gal. Whereas a *spoVG217-lacZ* strain turned pale blue, the *spoVG308-lacZ* strain remained Lac⁻ (white) even after prolonged incubation. This was an unexpected result since the $E\sigma^{37}$ and $E\sigma^{32}$ forms of RNA polymerase holoenzyme can utilize P1 and P2 separately in vitro and it might have been anticipated that *spoVG308* would allow transcription to initiate from P2 (12). This result and other unpublished information suggest a more complex interaction between the two $E\sigma^{37}$ and $E\sigma^{32}$ RNA polymerase forms in their utilization of *spoVG* in vivo than has been possible to detect so far in vitro.

A third deletion mutation to be examined in this study, *spoVG327*, lacked sequences extending well into the P1 and P2 promoter region and was therefore lacking in promoter activity. In liquid DSM, a *spoVG327-lacZ*-bearing strain produced the same negligible level of β-galactosidase as did the *spoVG308-lacZ*-containing strain, and like *spoVG308*, the *spoVG327* mutant failed to induce *lacZ* (Fig. 5C) at the end of the exponential phase. The low level of β-

galactosidase to appear later in the stationary phase represents a background of *spoVG* promoter-independent β-galactosidase production.

spo0H-dependent induction of a deletion-mutated promoter. Given that a deletion mutation of the A-T box showed the normal temporal pattern of *spoVG* induction (albeit at a greatly reduced level), I wondered whether sequences located upstream of the P1 −35 region are required for *spo0H*-dependent promoter utilization. To investigate this, I determined the induction pattern of *spoVG217-lacZ* in a *spo0H* deletion mutant. The *spo0H* deletion mutant IS233 was transformed with chromosomal DNA from a *spoVG217-lacZ* strain, and the resulting *spo0H spoVG217-lacZ* transformant was grown in DSM and assayed for β-galactosidase production (Fig. 6). The Spo⁺, *spoVG217-lacZ* strain induced *lacZ* postexponentially (Fig. 6A), whereas the *spo0H* strain showed no significant induction of *lacZ* expression (Fig. 6B). Thus, the target of *spo0H*-dependent regulation must reside within a region consisting of the P1 and P2 promoters, the ribosome binding site, and the first eight codons of *spoVG*.

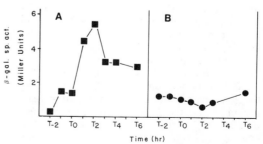

FIG. 6. *spo0*-dependent induction of *spoVG217-lacZ* expression in *spo0H⁺*-bearing cells (A) and *spo0HΔHind*-bearing cells (B).

DISCUSSION

My interest in the *spoVG* gene centers on the tight coupling of its induction to regulatory events that govern the initiation phase of sporulation (16, 23). I have shown that the transcription of the *spoVG-lacZ* gene fusion is switched on at the end of exponential growth, this induction being dependent to a greater or lesser extent (depending on the locus) on a wide variety of *spo0* and *spoII* genes. In confirmation and extension of earlier results, I have been able to distinguish three classes of Spo mutants on the basis of their ability to support *spoVG-lacZ* RNA synthesis. Mutations in *spo0J* and several *spoII* genes caused little or no impairment of *spoVG-lacZ*, and in certain instances (e.g., *spo0J*, *spoIIC*, *spoIIE*, and *spoIIF*) caused a somewhat greater accumulation of β-galactosidase than that observed in Spo⁺ cells. Mutations in *spo0C*, *spo0F*, *spoIIA*, and *spoIID*, on the other hand, caused a partial impairment of β-galactosidase synthesis. Finally, *spo0A*, *spo0B*, and *spo0H* mutations caused a rather substantial impairment of β-galactosidase synthesis. Particularly striking is the case of the *spo0H* deletion mutation, which caused an almost total block in *spoVG* expression. Possibly, the *spo0A*, *spo0B*, and *spo0H* genes are more directly involved in the *spoVG* transcriptional induction mechanism than are the genes in which mutations caused a more moderate impairment of expression. Despite the severity of the *spo0A*, *spo0B*, and *spo0H* mutation effects, it is interesting to recall that expression of *spoVG-lacZ* is *spo0* independent when the *spoVG-lacZ* fusion is propagated on a high-copy-number plasmid (27). By determining how transcription of *spoVG* escapes *spo0* dependency, we may gain important insights into the mechanism of *spoVG* induction. It will be interesting to see whether other genes whose expression is sensitive to *spo0* gene activity escape *spo0* dependence when amplified on a high-copy-number plasmid.

Previous deletion analysis to determine the requirements for in vitro transcription had shown that utilization of P1 and P2 by Eσ³⁷ and Eσ³² was strongly dependent upon the A-T box, which precedes the putative −35 and −10 regions of P1 and P2 (1). The use of the *spoVG-lacZ* fusion has enabled me to confirm the importance of the A-T box for *spoVG* expression in vivo. Thus, the deletion mutation *spoVG217* which removes the A-T box and sequences to the putative −35 region of P1 caused a greater than 40-fold reduction in *lacZ* expression. Nevertheless, strains bearing *spoVG217-lacZ* produce a low but measurable level of β-galactosidase, and the appearance of enzyme activity during postexponential growth parallels that observed for the intact parental promoter. An important observation of the current study is that the residual *lacZ*-inducing activity exhibited by the *spoVG217* deletion was strongly dependent on *spo0H* protein. Thus, the introduction of *spoVG217-lacZ* fusion into *spo0H* cells served to abolish the low-level induction observed in Spo⁺ cells. These results suggest that the A-T box does not play a direct role in *spo0*-dependent induction of *spoVG*, but simply serves as an enhancer of *spoVG* promoter utilization. The observed effects of A-T-rich sequences on promoter strength may be due to their DNA helix conformation, which permits more stable "nonspecific" interactions to be formed with RNA polymerase (4). The model in which the A-T box of the *spoVG* promoter acts as a nonspecific enhancer of promoter utilization could be further tested by placing it upstream of other *B. subtilis* promoters and assessing its effect on transcriptional activity by employing gene fusion technology or quantitative transcriptional mapping procedures (5).

From deletion analysis we can begin to discern functional domains in the *spoVG* promoter region. The A-T box upstream from P1 and P2 serves to enhance *spoVG* promoter utilization, and the region composed of the putative −35 and −10 sequences of P1 and P2 contains the sites recognized by RNA polymerase sigma factors σ³⁷ and σ³². The target of *spo0*-dependent control of *spoVG* transcription must be located downstream of the sequence deleted by the *spoVG217* mutation, either within the P1 and P2 sequence or further downstream in the vicinity of the ribosome binding site or the amino-terminal coding region. One could imagine several mechanisms of *spo0*-dependent induction of transcription. A positive activator (perhaps a *spo0* gene product) may bind near the *spoVG* promoter and enhance promoter strength by altering local DNA conformation or making direct protein-protein contact with RNA polymerase. Another model is the inactivation by a *spo0*-specified protein or small molecule with a repressor which is bound to the *spoVG* promoter region. An observation that may be pertinent to this discussion regards a region of dyad symmetry which exists in the *spoVG* promoter (Fig. 7). A similar structure can be detected in the promoter for the *sprE* gene (recently renamed *apr*) which encodes the alkaline protease of *B. subtilis*, another gene whose expression is under *spo0* control (25). This sequence is located just upstream of the −10 region of P1 and includes the putative −35 region of P1. In *E. coli*, several transcriptional regulatory proteins have been characterized which interact with regions of approximate twofold symmetry located near the site of RNA polymerase-promoter contact (17). Whether a similar mechanism may be operating

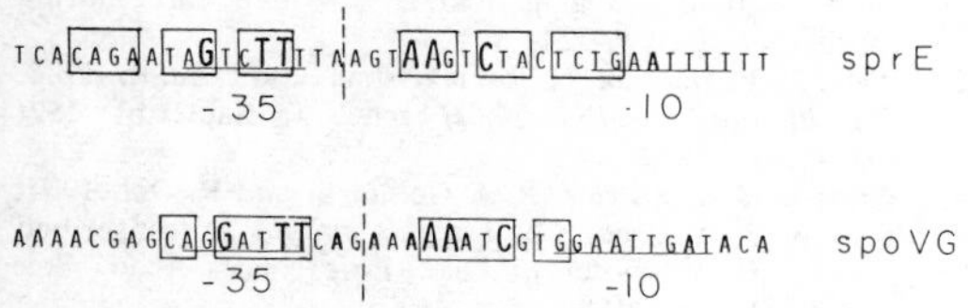

TCACAGAATAGTCTTTTAAGTAAGTCTACTCTGAATTTTTT   sprE
-35   -10

AAAACGAGCAGGATTTCAGAAAAAATCGTGGAATTGATACA   spoVG
-35   -10

FIG. 7. Regions of dyad symmetry in the *spoVG* promoter and the *sprE* promoter. The center of symmetry is indicated with a vertical broken line, and complementary sequences are shown in boxes. Large capital letters represent conserved nucleotides.

in the regulation of *spoVG* transcription is currently being investigated through in vitro point mutagenesis which may yield promoter mutations that alter the sensitivity of *spoVG* transcription to *spo0* control.

SUMMARY

A *spoVG-lacZ* gene fusion was used to examine the effects of various *spo0* and *spoII* mutations on the transcription of the *B. subtilis* sporulation gene *spoVG*. Spo0 and SpoII mutants could be placed into one of three classes according to their ability to support the expression of the *spoVG-lacZ* fusion. Strains carrying *spo0J*, *spoIIB*, *spoIIE*, *spoIIF*, and *spoIIG* mutations produced levels of β-galactosidase equivalent to or higher than that observed in Spo+ cells. Strains carrying *spo0C*, *spo0E*, *spo0F*, *spoIIA*, and *spoIID* mutations, on the other hand, were partially defective in their ability to induce *spoVG*-directed β-galactosidase synthesis. Finally, strains carrying *spo0A*, *spo0B*, and *spo0H* mutations represent a third class of mutants in which *spoVG-lacZ* expression was severely impaired. Especially striking is the case of a *spo0H* deletion mutation which caused almost a complete impairment of *spoVG-lacZ* expression. To further define the site within the *spoVG* promoter at which *spo0*-dependent control is exerted, a series of deletion mutations were constructed that extended into the *spoVG-lacZ* fusion from the upstream direction. One such deletion mutation removed the upstream A-T-rich sequence (A-T box), leaving the overlapping P1 and P2 promoters intact. The effect of this deletion was to greatly diminish, but not to entirely eliminate, the postexponential-phase induction of *spoVG-lacZ*. The residual *lacZ*-inducing activity of the deletion-mutated gene fusion was found to be strongly dependent upon *spo0H* product. These results suggest that the A-T box is an enhancer of P1 and P2 promoter utilization and that the site of *spo0H*-dependent regulation lies downstream from the A-T box, within a 95-base pair region consisting of the P1 and P2 sequences, the ribosome binding site, and the amino-terminal coding sequence of the *spoVG* gene.

I acknowledge I. Smith for the gift of strain IS233, and I thank R. Losick and M. Igo for comments on the manuscript. This work was carried out in the laboratory of R. Losick and was supported by Public Health Service grant GM 18568-12 from the National Institute of General Medical Sciences and by American Cancer Society postdoctoral fellowship PF-2066.

LITERATURE CITED

1. **Banner, C. D. B., C. P. Moran, Jr., and R. Losick.** 1983. Deletion analysis of a complex promoter for a developmentally regulated gene from *Bacillus subtilis*. J. Mol. Biol. **168:**351–365.
2. **Casadaban, M. J., and S. N. Cohen.** 1980. Analysis of gene control signals by DNA fusion and cloning in *Escherichia coli*. J. Mol. Biol. **138:**179–207.
3. **Donnelly, C. E., and A. L. Sonenshein.** 1982. Genetic fusion of *E. coli lacZ* gene to a *B. subtilis* promoter, p. 63–72. *In* A. T. Ganesan, S. Chang, and J. A. Hoch (ed.), Molecular cloning and gene regulation in bacilli. Academic Press, Inc., New York.
4. **Drew, H. R., and A. A. Travers.** 1984. DNA structural variations in the *E. coli tyrT* promoter. Cell **37:**491–502.
5. **Gilman, M., and M. Chamberlin.** 1983. Developmental and genetic regulation of *Bacillus subtilis* genes transcibed by σ^{28} RNA polymerase. Cell **35:**285–293.
6. **Goldfarb, D. S., S.-L. Wang, T. Kudo, and R. H. Doi.** 1983. A temporally regulated promoter from *Bacillus subtilis* is transcribed only by an RNA polymerase with 37,000 dalton sigma factor. Mol. Gen. Genet. **191:**319–325.
7. **Gryczan, T., and D. Dubnau.** 1978. Construction and properties of chimeric plasmids in *Bacillus subtilis*. Proc. Natl. Acad. Sci. U.S.A. **75:**1428–1432.
8. **Gryczan, T., A. G. Shivakumar, and D. Dubnau.** 1980. Characterization of chimeric plasmid cloning vehicles in *Bacillus subtilis*. J. Bacteriol. **141:**246–253.
9. **Haldenwang, W. G., and R. Losick.** 1979. A modified RNA polymerase transcribes a cloned gene under sporulation control in *Bacillus subtilis*. Nature (London) **282:**256–260.
10. **Hoch, J. A.** 1976. Genetics of bacterial sporulation, p. 67–98. *In* E. W. Caspari (ed.), Advances in genetics. Academic Press, Inc., New York.
11. **Hoch, J. A., M. Shiflett, J. Trowsdale, and S. M. A. Chen.** 1978. Stage 0 genes and their products, p. 127–130. *In* G. Chambliss and J. C. Vary (ed.), Spores VII. American Society for Micobiology, Washington, D.C.
12. **Johnson, W. C., C. P. Moran, Jr., and R. Losick.** 1983. Two RNA polymerase sigma factors from *Bacillus subtilis* discriminate between overlapping promoters for a developmentally regulated gene. Nature (London) **302:**800–804.
13. **Messing, J., and J. Viera.** 1982. A new pair of M13 vectors for selecting either DNA strand of double-digested restriction fragments. Gene **19:**269–276.
14. **Miller, J. H.** 1972. Experiments in molecular genetics. Cold Spring Harbor Laboratory, Cold Spring Harbor, N.Y.
15. **Moran, C. P., Jr., N. Lang, C. D. B. Banner, W. G. Haldenwang, and R. Losick.** 1981. Promoter for a developmentally regulated gene in *Bacillus subtilis*. Cell **25:**783–791.
16. **Ollington, J. F., W. G. Haldenwang, J. V. Huynh, and R. Losick.** 1978. Developmentally regulated transcription in a cloned segment of the *Bacillus subtilis* chromosome. J. Bacteriol. **147:**432–442.
17. **Pabo, C. O., and R. T. Sauer.** 1984. Protein-DNA recognition. Annu. Rev. Biochem. **53:**293–321.
18. **Piggot, P. J., and J. G. Coote.** 1976. Genetic aspects of bacterial endospore formation. Bacteriol. Rev. **40:**908–952.

19. **Piggot, P. J., A. Moir, and D. A. Smith.** 1981. Advances in the genetics of *Bacillus subtilis* differentiation, p. 29–39. *In* H. S. Levinson, A. L. Sonnenshein, and D. J. Tipper (ed.), Sporulation and germination. American Society for Microbiology, Washington, D.C.

20. **Rosenthal, R., P. A. Toye, R. Z. Korman, and S. A. Zahler.** 1979. The prophage of *SPβc2dcitK1*, a defective specialized transducing phage of *Bacillus subtilis*. Genetics **92**:721–739.

21. **Sanger, F., S. Nicklen, and A. R. Coulson.** 1977. DNA sequencing with chain-terminating inhibitors. Proc. Natl. Acad. Sci. U.S.A. **74**:5463–5467.

22. **Schaeffer, P., J. Millet, and J. Aubert.** 1965. Catabolite repression of bacterial sporulation. Proc. Natl. Acad. Sci. U.S.A. **54**:704–711.

23. **Segall, J., and R. Losick.** 1977. Cloned *Bacillus subtilis* DNA containing a gene that is activated early during sporulation. Cell **11**:751–761.

24. **Weir, J., E. Dubnau, N. Ramakrishna, and I. Smith.** 1984. The *Bacillus subtilis spo0H* gene. J. Bacteriol. **157**:405–412.

25. **Wong, S.-L., C. Price, P. J. Goldfarb, and R. Doi.** 1984. The subtilisin E gene of *Bacillus subtilis* is transcribed from a σ^{37} promoter *in vivo*. Proc. Natl. Acad. Sci. U.S.A. **81**:1184–1188.

26. **Zahler, S.A., R. Z. Korman, R. Rosenthal, and H. E. Hemphill.** 1977. *Bacillus subtilis* bacteriophage SPβ: localization of the phophage attachment site, and specialized transduction. J. Bacteriol. **129**:556–558.

27. **Zuber, P., and R. Losick.** 1983. Use of a *lacZ* fusion to study the role of *spo0* genes in developmental regulation in *Bacillus subtilis*. Cell **35**:275–283.

Major Sigma Factor, Sigma-43, of *Bacillus subtilis* RNA Polymerase and Interacting *spo0* Products Are Implicated in Catabolite Control of Sporulation

ROY H. DOI, MICHAEL GITT, LIN-FA WANG, CHESTER W. PRICE, AND FUJIO KAWAMURA

Department of Biochemistry and Biophysics, University of California, Davis, California 95616

The multiplicity of RNA polymerase sigma factors has been well documented in *Bacillus subtilis* (3, 14). The possible role of the minor sigma factors in regulating gene expression is expanding beyond sporulation control with the recent demonstration that the sigma-37 enzyme is involved in the expression of the alkaline serine protease or subtilisin gene (*aprA*) (28) as well as the *spoVG* gene (18). The role of the other minor sigma factors remains unknown. Sigma-28 enzyme appears to transcribe only a small number of genes during log phase (4). Sigma-32 enzyme transcribes two of the genes recognized by sigma-37 enzyme (9). Sigma-29 enzyme is a stationary-phase enzyme (6), but no specific gene has been identified which is controlled by sigma-29 enzyme. Since the sigma-43 (previously called sigma-55; M. Gitt, L.-F. Wang, and R. H. Doi, submitted for publication) enzyme is the major vegetative cell enzyme and is present at all stages (3), we began an investigation of the genetic properties of sigma-43 factor to understand its role during growth, during the initial stages of transition to the stationary phase, and during sporulation, as well as its relationship to other transcription factors.

The sigma-43 gene (*rpoD*) has been cloned, mapped physically and genetically (22), and sequenced completely (Gitt et al., submitted for publication). As with the *Escherichia coli* sigma factor (1), the molecular weight derived from the amino acid sequence (43,000) was much smaller than that derived from sodium dodecyl sulfate-polyacrylamide gel electrophoresis (55,000), and therefore we will now designate the major sigma factor of *B. subtilis* as sigma-43.

One of our major goals is to determine the functional relationship of sigma-43 to the minor sigma factors and to other transcriptional control systems. Since sporulation is one of a number of responses of the cell to environmental stress and many genes are derepressed during these responses, we are particularly interested in the relationship of sporulation and other control elements which function during early stationary and sporulation stages. Recent results from our laboratory concerning the role of sigma-43 in repression of sporulation by glucose and the unexpected interaction of sigma-43 and certain *spo0* gene products provide an insight into this complex problem.

Although catabolite repression is poorly understood in *B. subtilis*, glucose appears to exert a classical type of carbon catabolite repression on a number of genes in addition to sporulation genes (2, 17, 27). In this paper we show that the sigma-43 factor has a role in the catabolite repression of sporulation, since a mutation within the *rpoD* gene allows the cell to sporulate in the presence of glucose. Takahashi (27) and Sun and Takahashi (25) identified six *crs* loci which play a role in "catabolite-resistant sporulation." They have recently mapped *crsE* within the genes for the β-β' subunits of RNA polymerase core enzyme (26). Here, we show that the *crsA* mutation lies within the *rpoD* coding sequence. Thus, the RNA polymerase holoenzyme and probably other undefined gene products play important roles in catabolite repression.

The other factors that are significant during early sporulation are the *spo0* gene products which are suggested to control the initiation of spore formation (8). Previous studies showed that *spo0A* mutations, which are the earliest blocked developmental lesions (21), inhibit both the production of the extracellular alkaline series protease (subtilisin) and the formation of spores. By cloning (28) and deleting (10) part of the subtilisin gene (*aprA*) in the chromosome, we have been able to uncouple the production of subtilisin and the formation of spores; i.e., subtilisin-deficient strains can still sporulate (10). These data indicate that *spo0A* does not exclusively control sporulation and is required for expression of other genes not directly concerned with sporulation. Similar reasoning suggests that, since a sigma-37 promoter partly controls the expression of the subtilisin gene (5, 28), sigma-37 is not involved solely in controlling sporulation-specific events.

Although the product of *spo0A* is required for the expression of the subtilisin and sporulation genes (8), it is perhaps more of an umbrella control with broad regulatory effects rather than a spore-specific effect. Therefore, since the *spo0A* gene and, by implication, other *spo0* genes are not exclusively spore specific, *spo0* genes may play a more general role in detecting

```
M A D K Q T H E T E L T F D Q V K E Q L T E S G K K R G V L T V E E I        35

A E R M S S F E I E S D Q M D E Y Y E F L G E Q G V E L I S E N E E T        70

E D P N I Q Q L A K A E E E F D L N D L S V P P G V K I N D P V R M Y       105

L K E I G R V N L L S A K E E I A Y A Q K I E E G D E E S K R R L A E       140

A N L R L V V S I A K R Y V G R G M L F L D L I H E G N M G L M K A V       175

E K F D Y R K G Y K F S T Y A T W W I R Q A I T R A I A D Q A R T I R       210

I P V H M V E T I N K L I R V Q R Q L L Q D L G R E P T P E E I A E D       245

M D L T P E K V R E I L K I A Q E P V S L E T P I G E E D D S H L G D       280

F I E D Q E A T S [P] S D H A A Y E L L K E Q L E D V L D T L T D R E E     315

N V L R L R F G L D D G R T R T L E E V G K V F G V T R E R I R Q I E       350

A K A L R K L R H P S R S K R L K D F L E                                   371
```

FIG. 1. Predicted amino acid sequence of sigma-43, derived from the base sequence determined by Gitt et al. (submitted for publication). The boxed Pro residue (no. 290) was changed to Phe in *crsA1*, *crsA4*, and *crsA47* (Kawamura et al., submitted for publication).

signals for "stress" conditions of the cell. These conditions might include nutritional deprivation (7, 13), heat shock, irradiation, and other environmental insults.

AMINO ACID SEQUENCE OF THE SIGMA-43 GENE (*rpoD*) PRODUCT

We carried out a base sequence analysis of the cloned *rpoD* gene to determine the size and amino acid composition of the sigma-43 factor, to find the regulatory sequences which control its expression, and to elucidate the genetic organization of the *rpoD* region. The cloned fragment containing the complete *rpoD* gene (22) was mapped by restriction enzymes, and suitable restriction fragments were sequenced (Gitt et al., submitted for publication), either by the Maxam and Gilbert method (15) or by the M13 phage-dideoxy method (16, 23). The amino acid sequence deduced from the base sequence revealed a protein with 371 amino acids and a molecular weight of 42,959 (Fig. 1). The molecular weight of the mature protein is 42,828, since the N-terminal methionine is absent. In comparing the sigma-43 with the sigma-70 factor from *E. coli*, we observed an overall amino acid homology of about 50%. The highest degree of homology, about 80%, occurred between the C-terminal halves of the two sigma factors. We found a lesser degree of homology between the N-terminal regions, and sigma-43 lacked a 245-amino acid sequence found in sigma-70, roughly in the middle of the molecule (Gitt et al., submitted for publication).

Additional DNA sequencing studies (L. F. Wang, C. W. Price, and R. H. Doi, submitted for publication) and genetic studies (C. W. Price

and R. H. Doi, submitted for publication) have revealed the presence of the *dnaE* gene immediately upstream from the *rpoD* gene. The derived amino acid sequence of the *dnaE* gene product has a very high homology with the DNA primase (*dnaG*) of *E. coli*. Further, no typical *B. subtilis* promoter sequences were observed in the immediate 5'-proximal location of the *rpoD* gene. These results suggest the possibility of an operon structure similar to that found in the *E. coli* sigma operon (1), in which the promoter region is followed by *rpsU*, *dnaG*, and *rpoD*. Further studies of the area upstream from the *dnaE* gene should reveal the location of the promoter region and the genetic organization of the sigma operon in *B. subtilis*.

GENETIC LOCUS OF *rpoD* AND ITS NEIGHBORING GENES

The genetic locus of the *rpoD* gene was determined by use of pCP115, an integrative mapping plasmid (Price and Doi, submitted for publication). The *rpoD* gene is located at about 225 degrees on the *B. subtilis* map, between *aroD* and *lys* (Fig. 2). The order of the genes in this region is *aroD-dnaE-rpoD(crsA)-spo0G-lys*. The *dnaE* gene is immediately proximal to the *rpoD* gene.

RELATIONSHIP BETWEEN *rpoD* AND *crsA*

Since the map position of *rpoD* was inseparable from that observed for the *crsA* gene, the cloned *rpoD* gene was used in transformation and gene conversion studies to determine their relative map positions. The *crsA47* mutation was found to map within the *rpoD* gene (Price and Doi, submitted for publication), and the precise location of each *crsA* mutation was determined by gene conversion and molecular mapping studies (F. Kawamura, L.-F. Wang, C. W. Price, and R. H. Doi, submitted for publication). All three *crsA* mutations isolated independently by Takahashi (27) were found by base sequencing studies to have identical two-base changes at the same site (amino acid residue 290) in the *rpoD* gene in which a proline codon, CCT, has been mutated to a phenylalanine codon, TTT (Kawamura et al., submitted for publication). Since these *crs* mutants are able to sporulate in the presence of high concentrations of glucose, a condition which ordinarily represses sporulation, these results indicate that sigma-43 is involved, either directly or indirectly, in catabolic repression of sporulation.

RELATIONSHIP OF *rpoD* TO *spo0* MUTANTS

Results from several laboratories suggest that the major sigma-43 factor interacts with the *spo0A*, *spo0F*, and *spo0K* gene products. Sharrock et al. (24) reported that *crsA47*, which we have identified as an *rpoD* mutation, can sup-

TABLE 1. Relationship between sigma-43, *crs*, *spo0* mutants, and sporulation

Phenotype	*spo0F* genotype	Suppressor	Sporulation Low glucose	Sporulation High glucose
Crs	*spo0F+*	None	+	−
		rpoD47[a] (*crsA47*)	+	+
		sof-1[a]	+	+
Sup	*spo0F221*	None	−	−
		rpoD47 (*crsA47*)	+	+
		sof-1	+	+

[a] *rpoD47* and *sof-1* have Crs phenotypes. *sof-1* was identified as an allele of *spo0A* by Ferrari et al. (this volume) in collaboration with Kawamura and Saito (11). *crsA47*, which was isolated by Takahashi (27), was identified as an *rpoD* mutation by us.

press the *spo0F* and *spo0K* phenotypes, restoring sporulation ability. We have confirmed their results (Table 1). Also, the *sof-1* suppressor, originally isolated as a complete suppressor of the Spo⁻ phenotype of a *spo0F* deletion mutation (11), has been shown to be allelic with *spo0A* (F. A. Ferrari, K. Trach, D. LeCoq, and J. A. Hoch, this volume). The relationship between these events is summarized in Table 1. Thus, the identification of *crsA47* as a sigma mutation and these suppression results suggest a link between sigma-43, *spo0* gene products, and carbon catabolite repression in *B. subtilis*. Gilman and Chamberlin (4) and Zuber and Losick (29) have also shown that *spo0* mutations affect the transcriptional process. If, in fact, this integral relationship exists, it suggests that the *spo0* gene products are either direct or ancillary transcription factors which regulate the expression of stress-related genes.

REGULATION OF STATIONARY-PHASE EVENTS

Another approach to understanding regulation of gene expression at the stationary phase has included an analysis of promoters which are ordinarily expressed only after active growth has ceased. We have recently cloned (5) and characterized (28) a promoter which controls the expression of the subtilisin gene (*aprA*) of *B. subtilis*. The subtilisin gene promoter is recognized by the sigma-37 enzyme and is the first minor promoter-controlled gene whose product has been established. By partially deleting the *aprA* gene in vitro and reinserting it into the chromosome by the gene conversion technique, we have isolated a *B. subtilis* mutant lacking extracellular neutral and alkaline protease activity (10) which is still capable of sporulating at wild-type levels (Table 2). The significance of this observation is twofold. Since the *spo0A*

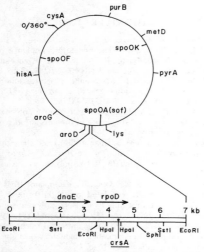

FIG. 2. Genetic locus of the sigma-43 gene, *rpoD* (Price and Doi, submitted for publication). The *rpoD* mutation maps between *aroD* and *lys* as shown on the abbreviated *B. subtilis* linkage map. The *crsA47* mutation restores sporulating ability to strains carrying lesions within the *spo0F* and *spo0K* genes shown within the map (24; this paper), and the *sof-1* allele of *spo0A* also suppresses *spo0F* mutations (11; Ferrari et al., this volume). The physical map of the *rpoD* region is modified from Price and Doi (submitted for publication) and Gitt et al. (submitted for publication). The arrows indicate the location and direction of transcription for the *dnaE* and *rpoD* genes. The *crsA* mutations have been molecularly mapped in *rpoD* (Kawamura et al., submitted for publication), as indicated by the dot within the *rpoD* coding sequence.

TABLE 2. Extracellular protease and sporulation frequency of protease-minus *B. subtilis* strains[a]

Strain	Activity of proteases (%)	Sporulation	
		Cells/ml	Spores/ml
DB101	100	5.3×10^8	4.1×10^8
DB102 (*nprE18*)	65	1.0×10^9	8.8×10^8
DB104 (*nprE18 aprAΔ3*)	2.6	1.0×10^9	7.9×10^8
DB105 (*nprE18 aprAΔ3*)	4.1	9.1×10^8	7.6×10^8

[a] From Kawamura and Doi (10).

gene product is required for the expression of both the *aprA* gene and certain sporulation genes (8) and these functions can be unlinked, it indicates that the *spo0A* control system regulates independent "pathways" of events, e.g., sporulation and synthesis of extracellular protease. It is likely that this umbrella control system also regulates other pathways which are turned on during nutritional or other stresses. Since *spo0A*, *spo0F*, and sigma-43 appear to interact and are related to catabolite repression by the suppression results (Table 1), the view emerges of a network of regulatory factors amenable to genetic and biochemical analyses.

The second important observation is that the minor sigma promoters are not exclusively involved in sporulation control. A sigma-37 promoter partly controls the subtilisin gene which is completely dispensable for sporulation. Although expression of two sigma-28 promoters is correlated with ability to sporulate and is perhaps regulated by *spo0A*, *spo0B*, *spo0E*, and *spo0F* (4), no direct evidence is available that sigma-28 promoter-controlled genes are involved in sporulation. Thus, a picture is emerging of the relationship between genes expressed by the major sigma-43 promoters and minor promoters. The sigma-43 enzyme definitely has a major role during growth and is implicated in events controlling sporulation initiation, but its other functions during stationary phase and sporulation are still uncertain. The sigma-28 enzyme appears to function only during growth (4). The sigma-37 enzyme definitely appears to increase in activity during the early stationary phase (12), and it may play a role in the derepression of catabolite-repressed genes (28) and the initiation of sporulation. Since sigma-29 enzyme is made only during the sporulation phase (6, 20) and is enriched in the forespore (19), it is likely to play a more central role in expressing sporulation genes.

The possibility that other sigma or sigmalike factors will be found seems high, since the mechanism of gene regulation in *B. subtilis* appears to entail a large number of transcriptional factors and multiple promoters which are transcribed at various growth stages by different holoenzyme forms (3, 14). It also seems likely that we will understand the role of the *spo0* genes and their relationship to the transcriptional machinery and catabolite repression in the near future.

SUMMARY

An interaction of the major sigma factor (sigma-43) of *B. subtilis*, catabolite repression, and the *spo0* genes is demonstrated by the suppression of the *spo0F* gene by a mutation (*crsA*) in *rpoD* that confers resistance to catabolite repression of sporulation (Crs phenotype). The *spo0F* mutations are also suppressed by the *spo0A* mutation, *sof-1*, which also has a Crs phenotype. Our results indicate that *spo0A* and sigma-37 do not control sporulation exclusively, since *aprA* expression is regulated by both those factors and yet an *aprA* deletion mutant which is deficient in subtilisin is still capable of sporulating normally.

This research was supported in part by National Science Foundation grant PCM-8218304 and Public Health Service grant GM-19673 from the National Institute of General Medical Sciences.

We thank I. Takahashi for the *crsA* mutants.

LITERATURE CITED

1. Burton, Z., R. R. Burgess, J. Lin, D. Moore, S. Holder, and C. A. Gross. 1981. The nucleotide sequence of the cloned *rpoD* gene for the RNA polymerase sigma subunit from *E. coli* K12. Nucleic Acids Res. 9:2889–2903.
2. Chasin, L. A., and B. Magasanik. 1968. Induction and repression of the histidine-degrading enzymes of *Bacillus subtilis*. J. Biol. Chem. 243:5165–5178.
3. Doi, R. H. 1982. Multiple RNA polymerase holoenzymes exert transcriptional specificity in *Bacillus subtilis*. Arch. Biochem. Biophys. 214:772–781.
4. Gilman, M. Z., and M. J. Chamberlin. 1983. Developmental and genetic regulation of the *Bacillus subtilis* genes transcribed by sigma-28 RNA polymerase. Cell 35:285–293.
5. Goldfarb, D. S., S.-L. Wong, T. Kudo, and R. H. Doi. 1983. A temporally regulated promoter from *Bacillus subtilis* is transcribed only by an RNA polymerase with a 37,000 dalton sigma factor. Mol. Gen. Genet. 191:319–325.
6. Haldenwang, W. G., N. Lang, and R. Losick. 1981. A sporulation-induced sigma-like regulatory protein from *Bacillus subtilis*. Cell 23:615–624.
7. Hoch, J. A., M. A. Shiflett, J. Trowsdale, and S. M. H. Chen. 1978. Stage 0 genes and their products, p. 127–130. *In* G. Chambliss and J. C. Vary (ed.), Spores VII. American Society for Microbiology, Washington, D.C.
8. Hoch, J. A., and J. Spizizen. 1969. Genetic control of some early events in sporulation of *Bacillus subtilis* 168, p. 112–120. *In* L. L. Campbell (ed.), Spores IV. American Society for Microbiology, Bethesda, Md.
9. Johnson, W. C., C. P. Moran, Jr., and R. Losick. 1983. Two RNA polymerase sigma factors from *Bacillus subtilis* discriminate between overlapping promoters for a developmentally regulated gene. Nature (London) 302:800–804.
10. Kawamura, F., and R. H. Doi. 1984. Construction of a *Bacillus subtilis* double mutant deficient in extracellular alkaline and neutral proteases. J. Bacteriol. 160:442–444.
11. Kawamura, F., and H. Saito. 1983. Isolation and mapping of a new suppressor mutation of an early sporulation gene *spo0F* mutation in *Bacillus subtilis*. Mol. Gen. Genet.

192:330–334.

12. **Losick, R.** 1981. Sigma factors, stage 0 genes, and sporulation, p. 48–56. *In* H. S. Levinson, A. L. Sonenshein, and D. J. Tipper (ed.), Sporulation and germination. American Society for Microbiology, Washington, D.C.

13. **Losick, R.** 1982. Sporulation genes and their regulation, p. 179–201. *In* D. A. Dubnau (ed.), The molecular biology of the bacilli, vol. 1, *Bacillus subtilis*. Academic Press, Inc., New York.

14. **Losick, R., and J. Pero.** 1981. Cascades of sigma factors. Cell **25:**582–584.

15. **Maxam, A. M., and W. Gilbert.** 1977. New method for sequencing DNA. Proc. Natl. Acad. Sci. U.S.A. **74:**560–564.

16. **Messing, J., R. Crea, and P. H. Seeburg.** 1981. A system for shotgun DNA sequencing. Nucleic Acids Res. **9:**309–321.

17. **Monod, J.** 1958. Recherches sur la croissance des cultures bacteriennes (thesis 1942), p. 145. Hermann, Paris.

18. **Moran, C. P., Jr., N. Lang, C. D. B. Banner, W. G. Haldenwang, and R. Losick.** 1981. Promoter for a developmentally regulated gene in *Bacillus subtilis*. Cell **25:**783–791.

19. **Nakayama, T., M. Irikura, Y. Kurogi, and H. Matsuo.** 1981. Purification and properties of RNA polymerases from mother cells and forespores of sporulating cells of *Bacillus subtilis*. J. Biochem. (Tokyo) **89:**1681–1691.

20. **Nakayama, T., V. Williamson, K. Burtis, and R. H. Doi.** 1978. Purification and properties of two RNA polymerases from sporulating cells of *Bacillus subtilis*. Eur. J. Biochem. **88:**155–164.

21. **Piggot, P. J., and J. G. Coote.** 1976. Genetic aspects of bacterial endospore formation. Bacteriol. Rev. **40:**908–962.

22. **Price, C. W., M. A. Gitt, and R. H. Doi.** 1983. Isolation and physical mapping of the gene encoding the major sigma factor of *Bacillus subtilis* RNA polymerase. Proc. Natl. Acad. Sci. U.S.A. **80:**4074–4078.

23. **Sanger, F., S. Nicklen, and A. R. Coulson.** 1977. DNA sequencing with chain-terminating inhibitors. Proc. Natl. Acad. Sci. U.S.A. **74:**5463–5467.

24. **Sharrock, R. A., S. Rubinstein, M. Chan, and T. Leighton.** 1984. Intergenic suppression of *spo0* phenotypes by the *Bacillus subtilis* mutation *rvtA*. Mol. Gen. Genet. **194:**260–264.

25. **Sun, D., and I. Takahashi.** 1982. Genetic mapping of catabolite-resistant mutants of *Bacillus subtilis*. Can. J. Microbiol. **28:**1242–1251.

26. **Sun, D., and I. Takahashi.** 1984. A catabolite-resistance mutation is localized in the *rpo* operon of *Bacillus subtilis*. Can. J. Microbiol. **30:**423–429.

27. **Takahashi, I.** 1979. Catabolite repression-resistant mutants of *Bacillus subtilis*. Can. J. Microbiol. **25:**1283–1287.

28. **Wong, S.-L., C. W. Price, D. S. Goldfarb, and R. H. Doi.** 1984. The subtilisin E gene of *Bacillus subtilis* is transcribed from a sigma-37 promoter *in vivo*. Proc. Natl. Acad. Sci. U.S.A. **81:**1184–1188.

29. **Zuber, P., and R. Losick.** 1983. Use of a *lacZ* fusion to study the role of the *spo0* genes of *Bacillus subtilis* in developmental regulation. Cell **35:**275–283.

P^{31}, a σ^{29}-Like RNA Polymerase Binding Protein of *Bacillus subtilis*

JANINE E. TREMPY, TERRY L. LaBELL, G. L. RAY, AND W. G. HALDENWANG

Department of Microbiology, University of Texas Health Science Center at San Antonio, San Antonio, Texas 78284

In an attempt to understand the mechanisms by which sporulation genes are controlled, numerous laboratories have isolated segments of the *Bacillus subtilis* chromosome that are transcribed at various times during growth and differentiation (13). Reconstruction of gene-specific transcription in vitro demonstrated that *B. subtilis* contains multiple forms of RNA polymerase and that some of these enzymes are clearly superior in transcribing sporulation genes (3–6, 10, 23). A hypothesis has developed that RNA polymerase modification could be a means of controlling sequential gene activation during sporulation (13). To date, five forms of RNA polymerase have been isolated from *B. subtilis*. The enzymes have a common core component ($\beta\beta'\alpha_2$), but differ in the template specificity determinant (sigma factor) that is associated with the core. Four of these sigma factors, σ^{55}, σ^{37}, σ^{32}, and σ^{28}, are present in holoenzymes that are isolated from vegetatively growing cells (6, 10, 17, 22), and the fifth, σ^{29}, has been found associated with a form of the enzyme routinely isolated only from sporulating cells (2, 4, 12); σ^{29} is thus an RNA polymerase specificity determinant that is not only likely to be involved in spore gene control but also likely to be controlled itself by sporulation-specific regulatory agents. To investigate the factors that influence the synthesis of σ^{29} protein without the need to purify it as a component of an RNA polymerase holoenzyme, we developed a σ^{29}-specific monoclonal antibody with which we analyzed crude cell extracts for the presence of σ^{29} protein. This analysis established that σ^{29} is absent from vegetatively growing cells and is abundant only in sporulating cells for a restricted period during differentiation (t$_2$ to t$_4$) (J. E. Trempy, J. M. Plummer, and W. G. Haldenwang, submitted for publication). On the basis of the narrowness of its period of abundance, we conclude that σ^{29} synthesis is highly regulated and that σ^{29} is likely to direct the transcription of only a subpopulation of genes that are expressed at a precise, intermediate stage of spore formation. An offshoot of our analysis was the discovery of a previously unrecognized 31,000-dalton protein (P^{31}) in extracts of sporulating *Bacillus* spp. which appears to be related to σ^{29}. P^{31} is recognized by the σ^{29}-specific monoclonal antibody that reacts with no other *B. subtilis* protein in crude cell extracts (including the other known sigma factors). In addition, P^{31} and σ^{29} yield similar peptide fragments when digested by *Staphylococcus aureus* V8 protease. P^{31} is also an RNA polymerase binding protein, and the elution pattern of RNA polymerase containing P^{31} (EP^{31}) from DNA-cellulose is identical to the elution profile of Eσ^{29}. The proteins differ, however, in that the ability of P^{31} to associate with RNA polymerase appears to be substantially weaker than that of σ^{29}. In addition, EP^{31} fails to transcribe known Eσ^{29} promoters in vitro. The inactivity of EP^{31} also extends to genes transcribed in vitro by Eσ^{55}, Eσ^{37}, and Eσ^{32}. The properties of P^{31} and σ^{29} suggest that they could be the products of the same structural gene, with P^{31} being a relatively inactive version of that product.

EXPERIMENTAL PROCEDURES

Bacterial strains and plasmids. The Spo$^-$ strains IS31 (*spoIIA26*), IS43 (*spoIIC298*), and IS35 (*spoIIE64*) were obtained from the Bacillus Stock Center (Ohio State University). The Spo$^+$ strain SMY was from R. Losick. The plasmids p63 and p213-1 have been described (15). The plasmids pGLR1 through pGLR5 consist of small segments of *B. subtilis* DNA (approximate size, 1.5 kilobases), identified as having σ^{29}-recognized in vitro promoters (G. L. Ray and W. G. Haldenwang, Abstr. Annu. Meet. Am. Soc. Microbiol. 1983, I129, p. 161) and cloned into pBR322.

Growth of bacteria and purification of RNA polymerase. *B. subtilis* vegetative cells were obtained by growing cells in Difco sporulation (DS) medium to one-half of the maximum cell density. Sporulating cells were harvested at the indicated times after completion of exponential growth in DS medium (16).

RNA polymerase was purified and in vitro transcriptions were performed as previously described (4).

Production of monoclonal antibody. BALB/c mice received a series of intraperitoneal injections of Eσ^{29} totaling 100 µg. Hybridomas were produced by the procedure of Oi and Herzenberg (14), using the non-immunoglobulin-secreting SP2/O-Ag$_{14}$ BALB/c myeloma cell line (18). Anti-σ^{29}–secreting hybridomas were detected by an enzyme-linked immunosorbent assay as described by Voller et al. (21). After single-cell cloning, antibody was routinely produced during overgrowth of hybridomas in tissue culture.

Analysis of crude extracts for P^{31} and σ^{29}. *B. subtilis* cells, washed and suspended in buffer I (17), were disrupted by passage (twice) through a French pressure cell (15,000 lb/in^2) and were clarified by centrifugation (10,000 × g for 30 min). After precipitation with ammonium sulfate (63% of saturation), the extract was dialyzed against buffer I. Extracts were fractionated by sodium dodecyl sulfate-polyacrylamide gel electrophoresis (SDS-PAGE) (11), transferred electrophoretically to nitrocellulose, and probed with antibody (20), with BLOTTO (9) as the blocking buffer. Bound antibody was visualized by using horseradish peroxidase-conjugated goat immunoglobulin G against mouse immunoglobulin (HyClone).

Peptide analysis of P^{31} and σ^{29}. Crude protein extracts (500 μg) prepared as described above of *spoIIE* cells that contained P^{31} (t$_2$ extracts) or vegetative-cell extracts (500 μg) that were supplemented with Eσ^{29} (2 μg) were incubated with *S. aureus* V8 protease (170 U) as described (7). The resulting peptide mixtures were separated on SDS polyacrylamide gels (15 to 20% acrylamide), electrophoretically transferred to nitrocellulose, and probed with monoclonal antibody; the antigen-antibody complexes were visualized with ^{125}I-labeled rabbit anti-mouse immunoglobulin and autoradiography.

"Dot blot" hybridizations to in vivo RNA. RNA was extracted from the indicated bacteria as described by Gilman and Chamberlin (3). Dot blot hybridizations were performed essentially as described by Thomas (19) except that pretreatment of the RNA was omitted. Two-microliter samples of RNA in 2× SSC (0.15 M NaCl plus 0.015 M sodium citrate) were spotted onto nitrocellulose and then processed as described (19). Hybridization probes were the indicated plasmids into which [^{32}P]dAMP had been incorporated by the nick-translation activity of DNA polymerase I.

RESULTS

Analysis of P^{31} and σ^{29} in cell extracts. A hybridoma cell line was isolated that secreted a murine immunoglobulin G which is highly specific for an antigenic site on σ^{29}. Figure 1 illustrates a typical analysis, in which crude cell extracts of *B. subtilis* SMY were size fractionated on SDS-polyacrylamide gels and either stained with Coomassie blue (Fig. 1A) or transferred to nitrocellulose and allowed to react with

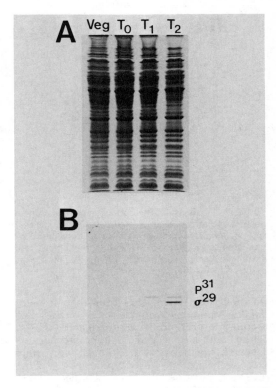

FIG. 1. Detection of σ^{29} and P^{31} in crude cell extracts. Crude protein extracts (100 μg) from SMY cells that were gathered during vegetative growth or harvested at hourly intervals during sporulation (T$_0$ represents the end of exponential growth; T$_1$, 1 h past T$_0$; etc.) were fractionated in duplicate by SDS-PAGE (12% acrylamide). One preparation was stained with Coomassie blue (A). The other was transferred to nitrocellulose, and allowed to react with monoclonal anti-σ^{29} antibody; bound antibody was visualized as described in Experimental Procedures (B).

the monoclonal antibody (Fig. 1B). Bound antibody was visualized by incubation with a horseradish peroxidase-conjugated goat anti-mouse antibody that catalyzes a colorimetric reaction. The monoclonal antibody was selectively bound to the nitrocellulose at positions which correspond to molecular weights of 31,000 (P^{31}) and 29,000 (σ^{29}). Proteins carrying the reactive epitope are absent from vegetatively growing cells. P^{31} was first detected at 1 h into sporulation (t$_1$), whereas σ^{29} accumulation did not become obvious until 1 h later (t$_2$) (Fig. 1B). We have observed this temporal appearance of P^{31} and σ^{29} not only in other strains of *B. subtilis* but also in different *Bacillus* species: proteins of 31,000 and 29,000 daltons have been detected in extracts prepared from cultures of sporulating *B. amyloliquefaciens*, *B. licheniformis*, *B. cereus*, and *B. natto* (J. Trempy, unpublished data).

FIG. 2. Analysis of P^{31} and σ^{29} accumulation in stage II mutants. Cultures of the indicated stage II Spo$^-$ *B. subtilis* mutant were grown in DS medium and harvested at 2 h after the end of the growth phase (t$_2$). Extracts were prepared, and 100-μg samples were analyzed as for Fig. 1.

Our consistent finding has been that in every *Bacillus* strain that accumulates σ^{29} or a σ^{29}-like protein, a protein of approximately 31,000 daltons with a σ^{29}-specific epitope precedes its appearance in the extracts. Although every strain that synthesizes σ^{29} also synthesizes P^{31}, the reverse is not true. We have identified a Spo$^-$ mutant (*spoIIE*) of *B. subtilis* which accumulates P^{31} but fails to make σ^{29}. The inability to synthesize σ^{29} while accumulating P^{31} is not a general characteristic of mutants blocked at this stage of development. As can be seen in Fig. 2, a *spoIIA* mutant failed to synthesize either of these proteins, whereas a *spoIIC* strain accumulated both. The discovery of a *B. subtilis* strain (*spoIIE62*) that accumulates P^{31} but fails to synthesize σ^{29} provided us with an organism in which the properties of P^{31} can be investigated without significant contamination by σ^{29}.

Crude t$_2$ extracts of *spoIIE* cells, partially digested with *S. aureus* V8 protease, fractionated by SDS-PAGE, and probed with monoclonal anti-σ^{29} antibody, should yield a pattern of peptide bands derived exclusively from P^{31}. This profile could be compared with that obtained from a vegetative-cell extract that is supplemented with highly purified Eσ^{29} prior to digestion. Although the peptide fragments visualized by this technique will exclude those which have lost the antigenic site recognized by the anti-σ^{29} antibody, the results should indicate whether substantial physical similarities exist between these proteins. Figure 3 illustrates that the *S. aureus* protease generates four major peptide bands from both P^{31} and σ^{29} which have identical mobilities in our gel system. We repeated this experiment with the protease chymotrypsin A$_4$ and again observed a distinct peptide

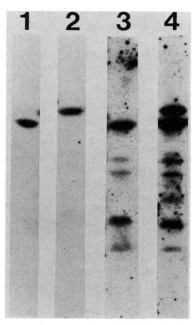

FIG. 3. Peptide analysis of P^{31} and σ^{29}. Crude cell extracts (500 μg) containing either P^{31} (*spoIIE*, t$_2$ extract) or σ^{29} (vegetative-cell extract supplemented with 2 μg of purified Eσ^{29}) were digested with *S. aureus* V8 protease (170 U per reaction), size fractionated by SDS-PAGE (15 to 20% acrylamide), transferred to nitrocellulose, and probed with monoclonal antibody. (1) Undigested σ^{29} extract; (2) undigested P^{31} extract; (3) protease-treated σ^{29} extract; (4) protease-treated P^{31} extract.

pattern that was conserved between P^{31} and σ^{29} digests (data not shown). These experiments demonstrate that P^{31} and σ^{29} have a closely related peptide substructure.

Purification of RNA polymerase containing P^{31} and σ^{29}. RNA polymerase containing σ^{29} can be readily purified from sporulating *B. subtilis* by a series of biochemical manipulations which includes a chromatography step on DNA-cellulose (2, 4). Elution, with a gradient of KCl, of an RNA polymerase preparation isolated from sporulating bacteria from such a column resolves the polymerase into two peaks of enzyme. The polymerase population that is eluted at high salt (approximately 0.8 M KCl) is highly enriched for Eσ^{29} (4). In view of the physical similarities between P^{31} and σ^{29}, we attempted to detect a possible EP^{31} peak by performing this chromatography step with polymerase preparations from cells that contain either P^{31} and σ^{29} (*spoIIC*) or only P^{31} (*spoIIE*). Figure 4A illustrates the protein composition of fractions that were sequentially eluted from the DNA-cellulose column. The proteins were size-fractionated on SDS-polyacrylamide gels and stained with Coomassie

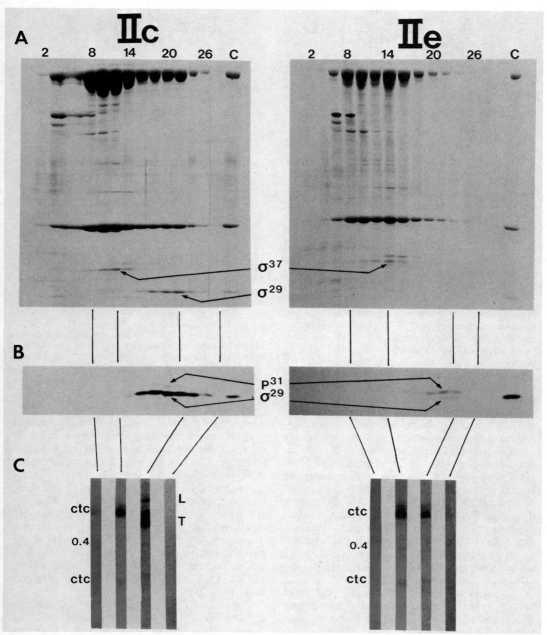

FIG. 4. Gradient elution of RNA polymerase from DNA-cellulose. Samples (20 g) of the indicated *B. subtilis* mutant strains were collected at the second hour of sporulation (t_2) in DS medium. Cells were disrupted, and polymerase was purified through phase extraction followed by gel filtration and applied to a calf thymus DNA-cellulose column (3 by 1 cm) as described (4). Polymerase was eluted with a 0.2 to 1.2 M linear gradient of KCl in 50 ml of buffer C (16). Fractions of 2 ml were collected, and 100-μl samples from the indicated fractions were resolved by SDS-PAGE (linear gradient 7 to 15% acrylamide) in duplicate. The preparations were either stained with Coomassie blue (A) or transferred to nitrocellulose and probed with anti-σ^{29} antibody as for Fig. 1 (B). The lane labeled C in the gel profiles contained 5 μg of purified Eσ^{29}. (C) RNA generated in vitro from plasmid p63 by 5-μl samples of the indicated fraction was hybridized to Southern strips containing *Hpa*II fragments of p63 DNA. Hybrids were visualized by autoradiography. The locations of p63 DNA fragments containing the σ^{37}-recognized transcription units *ctc* and *0.4* and the σ^{29}-recognized transcription units *L* and *T* are as indicated.

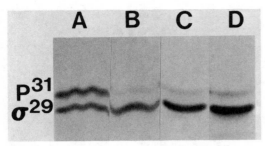

FIG. 5. Analysis of σ^{29} and P^{31} during RNA polymerase purification. Samples of an RNA polymerase preparation (*spoIIC* extract) equivalent to 100 µg of initial crude cell extract were taken at different stages during the purification and fractionated on SDS-PAGE (15% acrylamide). After transfer to nitrocellulose, the bound proteins were probed as for Fig. 1. (A) Fraction 1 enzyme (17) (extract after disruption of cells and high-speed centrifugation); (B) fraction 2 enzyme (17) (extract after phase partition); (C) ammonium sulfate-precipitated fraction 2 enzyme; (D) fraction 3 enzyme (17) (extract after gel filtration).

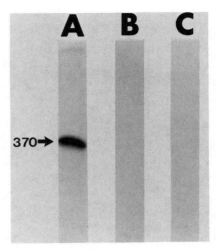

FIG. 6. Synthesis of an $E\sigma^{29}$ specific run-off RNA. ^{32}P-labeled RNA was synthesized in vitro from *Eco*RI-cleaved pGLR4 DNA and subjected to electrophoresis through an 8% polyacrylamide gel containing 7 M urea. RNA was visualized by autoradiography. (A) Transcribing enzyme: 5 µl of fraction 22 DNA-cellulose RNA polymerase of *spoIIC* strain (Fig. 3). Fraction 24 DNA-cellulose RNA polymerase from *spoIIE* strain was used as transcribing enzyme in lanes B (5 µl of enzyme) and C (25 µl of enzyme). The amount of *spoIIE* enzyme used in C had polydeoxyadenylate-thymidylate-transcribing activity equivalent to that of the 5 µl of *spoIIC* enzyme used in A.

blue. The σ^{29} protein is clearly visible in the Coomassie blue-stained *spoIIC* preparation but, as the antibody data predicted, is absent from the *spoIIE* extract. The σ^{37} protein is visible in both preparations. No obvious candidate for P^{31} was apparent among the proteins stained by this technique. If, however, similar gels are transferred to nitrocellulose and probed with anti-σ^{29} antibody, a very different picture emerges (Fig. 4B). This sensitive detection system clearly visualizes small amounts of P^{31} protein in the *spoIIC* extract. Larger amounts of P^{31} protein are detectable in the *spoIIE* extract, as well as a very small amount of a protein with a mobility similar to that of σ^{29}. The striking feature of this analysis is that the P^{31} molecules are eluting from the DNA-cellulose column at the exact position at which σ^{29} is being eluted. This elution behavior is seen with no other RNA polymerase holoenzyme; that is, each enzyme has a unique salt concentration at which it elutes from DNA-cellulose (2, 4, 6, 10, 17).

The purification steps that preceded the DNA-cellulose chromatography included several procedures which selectively enriched the preparation for RNA polymerase (4, 17). The discovery of P^{31} as a component of the RNA polymerase preparation at this stage of the purification strongly suggests that it is an RNA polymerase binding protein and that it confers on RNA polymerase the same chromatographic properties as does σ^{29}.

If the amount of P^{31} relative to σ^{29} in the crude extract of the *spoIIC* strain (Fig. 2) is compared with that of the purified polymerase (Fig. 4A), it is clear that a substantial amount of P^{31} was selectively lost during the purification. To deter-

mine where that loss might have occurred, portions of the polymerase preparation, at different stages of the purification, were analyzed by the "Western blot" technique (20) for the presence of P^{31} and σ^{29} (Fig. 5). Anti-σ^{29} antibody detects considerable P^{31} in the crude extract (Fig. 5A); however, after the first step that enriches for RNA polymerase (extraction from dextran into a polyethylene glycol phase), there is a dramatic decrease in P^{31} (Fig. 5B). A small amount of P^{31} persists with the RNA polymerase through a subsequent ammonium sulfate precipitation (Fig. 5C) and chromatography on a gel filtration column (Fig. 5D).

The loss of P^{31} from the polymerase fraction at the phase-partitioning step is apparently not due to degradation: we were able to detect significant amounts of P^{31} in the residual dextran phase after the polymerase had been extracted (data not shown).

We conclude that, although P^{31} is capable of associating with RNA polymerase, most P^{31} protein is either not bound to polymerase or loosely joined to it so that they are separated during purification of the RNA polymerase.

Comparison of the transcriptional activity of $E\sigma^{29}$ and EP^{31}. Chromatography on DNA-cellulose resolves forms of RNA polymerase that

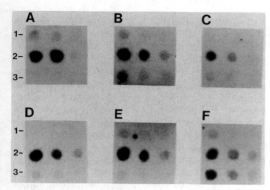

FIG. 7. Analysis of σ^{29}-directed RNAs in mutant *B. subtilis*. RNA was extracted from *B. subtilis* cultures that had been grown in DS medium and harvested at 5 h (t_5) after the end of exponential growth. Decreasing amounts of RNA (1.0, 0.33, and 0.11 µg) in 2 µl of 2× SSC were spotted horizontally across sheets of nitrocellulose. The nitrocellulose sheets were then processed, hybridized to a radioactive probe (10^6 cpm of [^{32}P]DNA), and washed as described by Thomas (19). Bound radioactive probe was visualized by autoradiography. RNA source: lane 1, *spoIIA*; lane 2, *spoIIC*; lane 3, *spoIIE*. DNA probes: A, p213-1; B, pGLR1; C, pGLR2; D, pGLR3; E, pGLR4; F, pGLR5.

contain σ^{29} or P^{31} from the bulk of the RNA polymerase peak (Fig. 4A and B). Analysis of Eσ^{29} that was purified in this fashion was used to demonstrate its novel template specificity (4). If the physical similarities between P^{31} and σ^{29} extend to conferring a similar specificity on RNA polymerase, it should be possible to detect this by examining the in vitro transcriptional activity of EP^{31} on cloned *B. subtilis* genes. RNA polymerase from fractions eluted off the DNA-cellulose column were used in an in vitro transcription system with p63 DNA as template. p63 is a recombinant plasmid which contains a number of *B. subtilis* genes that are transcribed in vitro by RNA polymerase with σ^{55}, σ^{37}, σ^{32}, or σ^{29}. The pattern of genes transcribed is unique to each of the transcribing enzymes. Figure 4C illustrates an autoradiograph of a Southern hybridization in which P^{32}-labeled RNA, transcribed in vitro by RNA polymerase from the indicated fractions, is annealed to restricted p63 DNA. The RNA polymerase from the *spoIIC* mutant has two peaks of recognizable promoter specificity: σ^{37} at fraction 11 (genes *ctc* and *0.4*) (6, 15) and σ^{29} at fraction 21 (genes *L* and *T*) (4; N. E. Lang-Unnasch, Ph.D. thesis, Harvard University, Cambridge, Mass., 1982). The RNA polymerase from the *spoIIE* mutant, in contrast, displays only the peak of σ^{37} activity at fraction 13. This experiment was performed with equal volumes of RNA polymerase from the indicated fractions. It is clear from the protein profiles in Fig. 4A that there is considerably more enzyme

in the σ^{29} region of the gradient from the *spoIIC* mutant relative to the *spoIIE* mutant. To minimize the possibility that the amount of enzyme used from the *spoIIE* fractions was too small to detect a potential P^{31}-directed transcription, we transcribed p63 DNA with equal amounts of RNA polymerase (based on the enzymes' abilities to transcribe polydeoxyadenylate-thymidylate) (17) from the σ^{29} peak fraction of *spoIIC* and from the P^{31} peak fraction of *spoIIE*. The only discernible promoter specificity we observed in the EP^{31} preparation was σ^{37}-like and could be attributed to σ^{37} "trailing" into the P^{31}-containing fractions (data not shown).

To eliminate the influence of σ^{37}-directed transcription on our assay, we attempted to transcribe a segment of cloned *B. subtilis* DNA (pGLR4) which carries a promoter that is recognized in vitro by Eσ^{29} but not by Eσ^{37} with the enzyme preparations from the mutant cells. Eσ^{29} generates a unique "run-off" RNA of 370 bases when the template is *Eco*RI-cleaved DNA (G. L. Ray, unpublished data). Figure 6A shows that, although the pooled Eσ^{29} fractions from the *spoIIC* mutant readily generated an RNA of the correct size when analyzed on polyacrylamide gels, no detectable band of RNA was present when the transcribing enzyme was pooled EP^{31} (Fig. 6B and C). We conclude that EP^{31} is significantly less active on Eσ^{29}-recognized templates in vitro.

Although P^{31} failed to substitute for σ^{29} in vitro, we wished to investigate the possibility that P^{31} may still be able to substitute for σ^{29} in vivo. As a preliminary approach to this question, we prepared RNA from various stage II mutants that synthesized P^{31} and σ^{29} (*spoIIC*), synthesized only P^{31} (*spoIIE*), or failed to synthesize either protein (*spoIIA*). The RNA was spotted in decreasing concentrations on nitrocellulose membranes, the membrane-immobilized RNA was probed with radioactively labeled DNA that contained sequences transcribed by Eσ^{29} in vitro, and the degree of hybridization, and hence RNA abundance, was judged by autoradiography. In Fig. 7A the probe was p213-1, a previously described recombinant plasmid which carries two known *B. subtilis* genes: *veg* (a gene transcribed in vitro by Eσ^{55}) and *0.3* (a gene transcribed in vitro by Eσ^{29}) (15). We were assuming that all three stage II mutants would have equivalent amounts of the *veg* transcript, and any difference in the degree of DNA hybridization would be due to *0.3* gene RNA. The probes in Fig. 7B–F are small *B. subtilis* chromosome fragments (approximate size, 1.5 kilobases) that were identified and cloned on the basis of their containing active in vitro σ^{29} promoters (G. L. Ray, unpublished data). Although there are quantitative differences among the hybridization profiles, there is a definite trend: RNA from the *spoIIA* strain (P^{31-} σ^{29-}) (lane 1) did not contain appreciable amounts of RNA that could hybridize to the probes, whereas RNA from the *spoIIC* strain (P^{31+} σ^{29+}) (lane 2) hybridized readily to all the probes. RNA from the *spoIIE* strain (P^{31+} σ^{29-}) failed to hybridize to an appreciable extent with five of the six probes (Fig. 7A–E); however, the sixth probe, pGLR5 (Fig. 7F), hybridized to a similar extent with both the *spoIIC* and *spoIIE* RNAs. This particular DNA segment, unlike the cloned DNA fragments used in Fig. 7B–E, has regions that are transcribed at various times during growth and sporulation and hence is likely to display a hybridization profile that may not be exclusively dependent on σ^{29} activity.

The results of this hybridization experiment must be viewed with caution. The nature of the defect in each of the stage II mutants is unknown, and each probably has effects on gene regulation beyond P^{31} and σ^{29} synthesis. Nevertheless, the detection of transcripts, likely to be the in vivo counterparts of those synthesized in vitro by $E\sigma^{29}$, in cells that carry σ^{29} but not in cells that have only P^{31} is consistent with the in vitro evidence that P^{31} does not confer σ^{29}-like transcriptional specificity on RNA polymerase.

DISCUSSION

The regulation of sporulation genes by the sequential modification of RNA polymerase is an attractive, albeit hypothetical, mechanism for activating and silencing blocks of these genes simultaneously. It is, however, a mechanism of control that must itself be rigorously governed. The synthesis of a sigma factor at an inappropriate time could have lethal consequences for a cell.

The σ^{29} protein is an RNA polymerase specificity determinant whose synthesis appears to fulfill this prediction of being highly regulated: σ^{29} is abundant for a strikingly narrow period during sporulation (t_2 to t_4). While analyzing the factors that influence σ^{29} regulation, we discovered another protein, P^{31}, whose synthesis is also rigorously controlled and which has a number of properties that indicate an underlying relationship to σ^{29}. P^{31} and σ^{29} are physically similar: they share a unique antigenic site and seem to have a common peptide substructure. In addition, they are both RNA polymerase binding proteins that have the unique property of conferring an identical salt-dependent elution profile on their respective holoenzymes during DNA-cellulose chromatography. Synthesis of both P^{31} and σ^{29} is dependent on sporulation-specific regulatory factors, with P^{31} being synthesized earlier in development than σ^{29}. Moreover, we have not been able to detect σ^{29} in any

B. subtilis strain (of 15 Spo$^+$ and Spo$^-$ strains examined) or in other *Bacillus* species (*B. amyloliquefaciens*, *B. cereus*, *B. licheniformis*, and *B. natto*) unless P^{31} was also present. Although P^{31} was present in all bacteria that could synthesize σ^{29}, the reverse was not true; that is, at least one strain of *B. subtilis* (a *spoIIE* mutant) synthesized P^{31} but not σ^{29}. In spite of the fact that P^{31} shares many physical properties with σ^{29}, we have no evidence that P^{31} is a sigma factor. P^{31} appears to associate only weakly with RNA polymerase (most of the P^{31}, initially present in cell extracts, is lost during RNA polymerase purification), and there is as yet no indication that this association results in an active holoenzyme. (In our hands, EP^{31} is transcriptionally inert on a number of *B. subtilis* genes in vitro.) P^{31} and σ^{29} could, however, be distinct sigma factors with similar physical properties, and our inability to detect P^{31}-specific transcription in vitro may be the result of a lack of P^{31}-recognized promoters on any of our DNA templates.

Alternative models can be constructed with the data presently available. For example, it is intriguing to speculate that the physical similarities between P^{31} and σ^{29} reflect their being the product of a single gene, and the weak RNA polymerase binding of P^{31}, as well as its lack of "sigmalike" activity, is a consequence of its being an inactive form of the gene product. Activation could involve its conversion into σ^{29}. Clearly, synthesizing a protein that can alter RNA polymerase specificity as an inactive precursor is a way of ensuring that the expression of its coding sequence at an inappropriate time in development does not result in the disruption of cellular metabolism. The synthesis of an active gene product would be under the control of several layers of regulation, all of which must be activated before this potential major alteration in the pattern of gene expression could take place.

It is obvious that at our present level of knowledge we cannot distinguish among a number of highly speculative relationships between P^{31} and σ^{29}. Considerably more information, including the cloning and analysis of the coding sequences for P^{31} and σ^{29}, will be required before we can definitively ascribe roles to P^{31} and σ^{29}. Accumulating this information is one of the tasks that our laboratory is vigorously pursuing.

SUMMARY

As the bacterium *B. subtilis* undergoes endospore formation, it synthesizes not only a novel 29,000-dalton RNA polymerase specificity determinant, σ^{29}, but also a 31,000-dalton protein, P^{31}, that has a number of characteristics in common with σ^{29}. Both proteins have the same unique antigenic site and similar peptide substructure. In this paper we report that each of the proteins also appears capable of binding to RNA polymerase and that, after binding, the respective holoenzymes have identical salt-dependent elution profiles from DNA-cellulose. P^{31} and σ^{29} differ in that P^{31}, unlike σ^{29}, appears to associate poorly with RNA polymerase and is readily lost during RNA polymerase purification. Moreover, EP^{31} fails to transcribe known σ^{29}-recognized promoters in vitro. The properties of P^{31} and σ^{29} suggest that they could be the products of the same structural gene, with P^{31} being a relatively inactive version of that product.

We thank C. L. Truitt and D. Kolodrubetz for critical readings of the manuscript.

This work was supported by National Science Foundation grant PCM-8021300 and Robert A. Welch Foundation grant AQ-932.

LITERATURE CITED

1. **Batteiger, B., W. J. Newhall, and R. B. Jones.** 1982. The use of Tween 20 as a blocking agent in the immunological detection of proteins transferred to nitrocellulose membranes. J. Immunol. Methods **55:**297–307.
2. **Fukuda, R., and R. Doi.** 1977. Two polypeptides associated with the ribonucleic acid polymerase core of *Bacillus subtilis* during sporulation. J. Bacteriol. **129:**422–432.
3. **Gilman, M. Z., and M. J. Chamberlin.** 1983. Developmental and genetic regulation of *Bacillus subtilis* genes transcribed by σ^{28}-RNA polymerase. Cell **35:**285–293.
4. **Haldenwang, W. G., N. Lang, and R. Losick.** 1981. A sporulation-induced sigma-like regulatory protein from *Bacillus subtilis*. Cell **23:**615–624.
5. **Haldenwang, W. G., and R. Losick.** 1979. A modified RNA polymerase transcribes a cloned gene under sporulation control in *Bacillus subtilis*. Nature (London) **282:**256–260.
6. **Haldenwang, W. G., and R. Losick.** 1980. Novel RNA polymerase σ factor from *Bacillus subtilis*. Proc. Natl. Acad. Sci. U.S.A. **77:**7000–7004.
7. **Haldenwang, W. G., and C. L. Truitt.** 1982. Peptide maps of regulatory subunits of *Bacillus subtilis* RNA polymerase. J. Bacteriol. **151:**1624–1626.
8. **Hawkes, R., E. Niday, and J. Gordon.** 1982. A dot-immunobinding assay for monoclonal and other antibodies. Anal. Biochem. **119:**142–147.
9. **Johnson, D. A., J. W. Gautsch, J. R. Sportsman, and J. H. Eider.** 1984. Improved technique utilizing nonfat dry milk for analysis of proteins and nucleic acids transferred to nitrocellulose. Gene Anal. Tech. **1:**3–8.
10. **Johnson, W. C., C. P. Moran, Jr., and R. Losick.** 1983. Two RNA polymerase sigma factors from *Bacillus subtilis* discriminate between overlapping promoters for a developmentally regulated gene. Nature (London) **302:**800–804.
11. **Laemmli, U. K., and M. Favre.** 1970. Maturation of the head of bacteriophage T4. J. Mol. Biol. **80:**575–599.
12. **Linn, T., A. L. Greenleaf, and R. Losick.** 1975. RNA polymerase from sporulating *Bacillus subtilis*. Purification and properties of a modified form of enzyme containing two sporulation polypeptides. J. Biol. Chem. **250:**9256–9261.
13. **Losick, R.** 1982. Sporulation genes and their regulation, p. 179–201. *In* D. Dubnau (ed.), The molecular biology of the bacilli. Academic Press, Inc., New York.
14. **Oi, V. T., and L. A. Herzenberg.** 1980. Immunoglobulin-producing hybrid cell lines, p. 351–372. *In* B. B. Mishell and S. M. Shiigi (ed.), Selected methods in cellular immunology. W. H. Freeman Co., San Francisco.
15. **Ollington, J. F., W. G. Haldenwang, T. V. Huynh, and R. Losick.** 1981. Developmentally regulated transcription in a cloned segment of the *Bacillus subtilis* chromosome. J. Bacteriol. **147:**432–442.
16. **Schaeffer, P., J. Millet, and J. Aubert.** 1965. Catabolite repression of bacterial sporulation. Proc. Natl. Acad. Sci. U.S.A. **54:**704–711.
17. **Shorenstein, R. G., and R. Losick.** 1973. Purification and properties of the sigma subunit of ribonucleic acid polymerase from vegetative *Bacillus subtilis*. J. Biol. Chem. **248:**6163–6169.
18. **Shulman, M., C. D. Wilde, and G. Kohler.** 1978. A better cell line for making hybridoma secreting antibodies. Nature (London) **276:**269–270.
19. **Thomas, P. S.** 1980. Hybridization of denatured RNA and small DNA fragments transferred to nitrocellulose. Proc. Natl. Acad. Sci. U.S.A. **77:**5201–5205.
20. **Towbin, H., T. Staehlin, and J. Gordon.** 1979. Electrophoretic transfer of proteins from polyacrylamide gels to nitrocellulose sheets: procedure and some applications. Proc. Natl. Acad. Sci. U.S.A. **76:**4350–4354.
21. **Voller, A. D., D. Bidwell, and A. Bartlett.** 1976. Microplate enzyme immunoassays for the immunodiagnosis of virus infections, p. 506–512. *In* N. R. Rose and H. Friedman (ed.), Manual of clinical immunology. American Society for Microbiology, Washington, D.C.
22. **Wiggs, J. L., M. A. Gilman, and M. J. Chamberlin.** 1981. Heterogeneity of RNA polymerase in *Bacillus subtilis*: evidence for an additional σ factor in vegetative cells. Proc. Natl. Acad. Sci. U.S.A. **78:**2762–2766.
23. **Wong, S. L., C. W. Price, D. S. Goldfarb, and R. Doi.** 1984. The subtilis in E gene of *Bacillus subtilis* is transcribed from a σ^{37} promoter in vivo. Proc. Natl. Acad. Sci. U.S.A. **81:**1184–1188.

Effects of Base Substitutions in a Complex Promoter from *Bacillus subtilis*

REGINE E. HAY AND CHARLES P. MORAN, JR.

Department of Microbiology and Immunology, Emory University School of Medicine, Atlanta, Georgia 30322

As the bacterium *Bacillus subtilis* differentiates from the vegetative form into a dormant endospore, complex morphological and physiological changes occur which require the sequential expression of many genes. During this process, new RNA polymerase sigma subunits appear which displace one another in a sequential cascade, conferring on the RNA polymerase a changing specificity for the recognition of different classes of promoters (4). This mechanism of altering the transcriptional specificity of RNA polymerase may play the cardinal role in the temporal regulation of gene expression during endospore formation.

Several genes have been identified recently that are transcribed by more than one form of RNA polymerase which initiate transcription from overlapping or tandem promoters (3, 10). These multipromoter systems may ensure the transcription of these genes during more than one developmental period. The *ctc* gene from *B. subtilis* is transcribed from a complex promoter that is recognized in vitro by three different forms of RNA polymerase. The sigma-37 RNA polymerase ($E\sigma^{37}$) and the sigma-32 RNA polymerase ($E\sigma^{32}$) are found in growing cells and may be responsible for the activation of *ctc* transcription at the end of the exponential growth phase (3, 7). The sigma-29 RNA polymerase ($E\sigma^{29}$) appears in sporulating cells about 2 h after the end of the exponential-growth phase (2) and may continue transcription of *ctc* after the disappearance of $E\sigma^{37}$ and $E\sigma^{32}$ from sporulating cells.

Although the *ctc* promoter is utilized by three forms of RNA polymerase which initiate transcription at or near the same nucleotide (N. Lang, Ph.D. thesis, Harvard University, Cambridge, Mass., 1982), some promoters are utilized by only one form of RNA polymerase. Of the three polymerases, only $E\sigma^{37}$ recognizes the P1 promoter of *spoVG* and only $E\sigma^{32}$ utilizes the P2 promoter of *spoVG* (3). $E\sigma^{29}$, but not $E\sigma^{37}$ or $E\sigma^{32}$, recognizes the promoter for the T gene of *B. subtilis* (Lang, Ph.D. thesis). Each RNA polymerase, therefore, must recognize its cognate promoters by interacting with a different specific set of nucleotide contacts. How are the sets of nucleotides that signal recognition by each of the three forms of RNA polymerase

arranged in the *ctc* promoter? To address this question and to test the role of promoter structure in the temporal control of gene expression, we have constructed a series of single and multiple base substitutions within the *ctc* promoter. We have studied the effects of the mutations on promoter utilization in vitro by $E\sigma^{37}$ (8) and $E\sigma^{29}$ (K. Tatti and C. Moran, submitted for publication) and report here the effects of these mutations on promoter utilization by $E\sigma^{32}$. These results enable us to predict that some mutant promoters may direct a temporally altered pattern of transcription in vivo.

EXPERIMENTAL PROCEDURES

RNA polymerase ($E\sigma^{37}$ and $E\sigma^{32}$) from *B. subtilis* JH646 *spo0A* was purified by gradient elution from DNA-cellulose as described by Johnson et al. (3). $E\sigma^{32}$ was further purified by elution from heparin-Sepharose (Pharmacia Fine Chemicals, Inc.) with a gradient of KCl (0.2 to 0.8 M) in buffer 2 (10 mM Tris, pH 8.0, 1 mM EDTA, 10 mM $MgCl_2$, 0.3 mM dithiothreitol, 0.3 mg of phenylmethylsulfonyl fluoride per ml, 50 mM KCl, and 7.5% glycerol). Run-off transcription of plasmid pCB1291 with *Eco*RI was used to show that the $E\sigma^{32}$ enzyme fraction contained no detectable $E\sigma^{37}$ activity (3) which elutes in later fractions (see Fig. 2, lane r).

The protocol for run-off transcription reactions has been described previously (8). The dinucleotide-primed reactions were done in a similar manner except that the RNA polymerase was preincubated at 37°C for 10 min with the DNA template in the presence of 150 μM dinucleotide (Sigma) or 60 μM each of ATP and GTP or in the absence of nucleotides. RNA synthesis was initiated by the addition of 2 μM each of ATP, GTP, and UTP and 0.6 μM [α-^{32}P]CTP (10 μCi). After 1 min, heparin was added to prevent reinitiation. The mixture was incubated for an additional 10 min at 37°C, at which time 60 μM each of ATP, GTP, CTP, and UTP was added. After an additional 5 min, the reaction was stopped by the addition of an equal volume of stop solution (10 M urea, 0.02% bromophenol blue, and 0.02% xylene cyanole in 0.2 M EDTA–10 mM Tris borate, pH 8.3). A sample was then subjected to electrophoresis in a polyacrylamide slab gel containing 7 M urea.

RESULTS

We previously reported the use of site-directed bisulfite mutagenesis to construct a series of single and multiple base substitutions in the *ctc* promoter (8). For this purpose, we cloned a 130-base pair (bp) DNA fragment containing the *ctc* promoter into *Escherichia coli* phage M13 mp9. Base substitutions in the *ctc* promoter were identified by directly determining the nucleotide sequence of the *ctc* promoter region in each phage after bisulfite mutagenesis. The 143-bp *Eco*RI/*Hin*dIII DNA fragment containing each mutated promoter was recloned between the *Eco*RI and *Hin*dIII restriction endonuclease cleavage sites of plasmid pUC8 (9). The collection of promoter mutants that we generated included promoters with single base substitutions (transitions) at every position within the promoter region that contained a guanine residue in the nontranscribed strand of the wild-type promoter from position +3 (3 bp downstream from the start point of transcription) to position −39 (39 bp upstream from the start point of transcription) (see Fig. 1). The collection also included promoters with different combinations of multiple transitions at the positions occupied by cytosine in the nontranscribed strand of the wild-type *ctc* promoter from positions +8 to −40.

To test the function of the mutant promoters, we assayed their ability to direct transcription by $E\sigma^{32}$ in an in vitro run-off assay. Template DNA was prepared by cutting the plasmid DNA at the unique *Hin*dIII restriction endonuclease cleavage site 95 bp downstream from the start point of transcription. RNA polymerase ($E\sigma^{32}$) was incubated at 37°C with an excess of the cut template DNA before the addition of the nucleoside triphosphates. One minute after the addition of the nucleoside triphosphates, heparin was added to prevent reinitiation. The discrete run-off transcripts were visualized by autoradiography after electrophoresis in a polyacrylamide-urea gel.

Figure 2 shows the transcripts generated by $E\sigma^{32}$ from a series of mutant promoters. Single base substitutions at positions −36, −16, −15, and −14 (lanes j, l, m, and n, respectively) greatly reduce the amount of transcript generated by $E\sigma^{32}$ in this run-off assay. The absolute level of transcription directed by each mutant promoter varies slightly from one experiment to the next, but the promoters with single base substitutions at positions −36, −16, −15, and −14 always direct the least amount of transcription by $E\sigma^{32}$. The base substitutions at the other positions normally occupied by guanine or cytosine in the nontranscribed strand of the wild-type *ctc* promoter had less effect on the amount

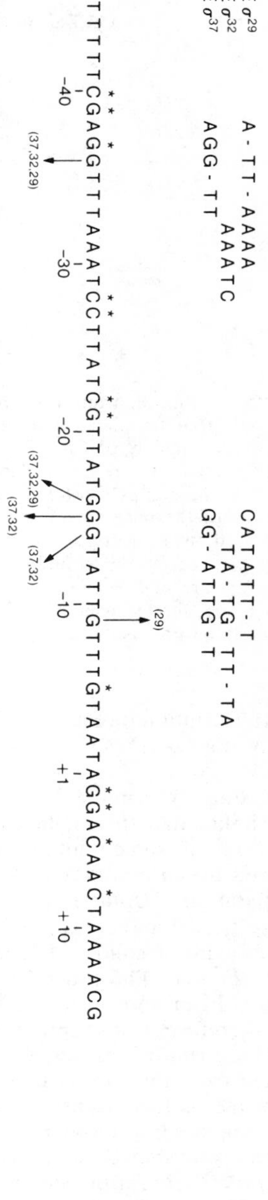

FIG. 1. Nucleotide sequence of the *ctc* promoter region in plasmid pUC8-31. Shown is the nucleotide sequence of the nontranscribed strand. Transcription proceeds from left to right, with the start point of transcription indicated as nucleotide +1. The stars represent the positions at which base substitutions (transitions) have little effect on the utilization of this promoter in the run-off assay by $E\sigma^{37}$, $E\sigma^{32}$, or $E\sigma^{29}$. The arrows indicate those positions at which single base transitions most dramatically affect the utilization of the promoter by $E\sigma^{37}$, $E\sigma^{32}$, or $E\sigma^{29}$ (arrows that point down indicate a decrease in utilization; the arrow that points up indicates an increase in utilization). The nucleotide sequences shown above the *ctc* promoter sequence are the sequences at the −10 and −35 regions which are conserved in promoters recognized by $E\sigma^{37}$, $E\sigma^{32}$, or $E\sigma^{29}$, as indicated.

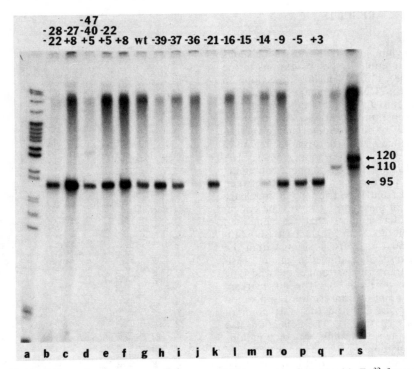

FIG. 2. Run-off transcripts from mutant promoters. RNA was transcribed with $E\sigma^{32}$ from the following plasmid DNA templates that had been cleaved at the HindIII site 95 bp downstream from the ctc promoter: pUC8-1167, lane b; pUC8-117, lane c; pUC8-115, lane d; pUC8-114, lane e; pUC8-112, lane f; pUC8-31, lane g; pUC8-39, lane h; pUC8-37, lane i; pUC8-36, lane j; pUC8-21, lane k; pUC8-16, lane l; pUC8-15, lane m; pUC8-14, lane n; pUC8-9, lane o; pUC8-5, lane p; pUC8-3, lane q. The number(s) above each lane refers to the position of the base substitution in the ctc promoter region of the template. Also shown are the transcripts generated by transcription of plasmid pCB1291 that had been cleaved at the EcoRI site, 120 and 110 bp downstream from the P1 and P2 promoters, respectively, of spoVG: transcribed by $E\sigma^{32}$ (lane r) and transcribed by a mixture of $E\sigma^{37}$ and $E\sigma^{32}$ (lane s). Single-stranded DNA molecular-weight standards, pBR322 cut with HpaII, are shown in lane a. The arrows show the positions of the 95-, 110-, and 120-base transcripts. The transcripts were visualized by autoradiography after electrophoresis into a 7 M urea–9% (wt/vol) polyacrylamide gel.

of transcription directed by the promoter in this assay (Fig. 2; lanes b, c, d, e, f, h, i, k, o, p, and q).

Mixing experiments were used to test the possibility that the inefficiency of transcription by $E\sigma^{32}$ of some mutant templates could be caused by an inhibitor of $E\sigma^{32}$ in the template preparations. HindIII-cut templates similar to those used for the experiments in Fig. 2 were mixed with plasmid pCB1291 that had been cut with EcoRI. Plasmid pCB1291 contains the spoVG P2 promoter, 110 bp from the EcoRI site (3); therefore, transcription by $E\sigma^{32}$ generates a 110-base run-off transcript. This transcript generated from the spoVG template was separated from the 95-base transcript generated from the ctc template in each reaction by electrophoresis in a polyacrylamide-urea gel. A similar amount of spoVG transcript was generated by $E\sigma^{32}$ in the presence of mutant or wild-type ctc promot-

ers (data not shown). The inefficiency of transcription by $E\sigma^{32}$ of the mutant templates was, therefore, the result of the single base substitutions at position −36, −16, −15, or −14 and was not caused by inhibitors in the template preparations. Furthermore, $E\sigma^{29}$ efficiently transcribes the mutant templates with the base substitutions at position −15 or −14 (Tatti and Moran, submitted for publication), providing additional evidence that the inefficiency of transcription of these mutant templates by $E\sigma^{32}$ is not caused by the quality of the template but by the single base substitutions in the promoter.

As discussed below, some of the base substitutions differentially affect the utilization of these promoters by $E\sigma^{37}$ and $E\sigma^{29}$. In the run-off assay reported here, the ability of $E\sigma^{32}$ to utilize this series of mutated promoters is similar to that of $E\sigma^{37}$ (8). If the two RNA polymerases position themselves at the same start point of tran-

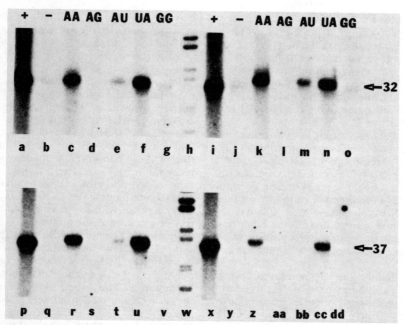

FIG. 3. Dinucleotide-primed transcription. RNA was transcribed by Eσ[32] (lanes a–o) or Eσ[37] (lanes p–dd) from wild-type (pUC8-31) (lanes a–g and p–v) and mutant (pUC8-9) (lanes i–o and x–dd) templates, which had been cleaved at the *Hin*dIII site. Transcription was primed by the dinucleotides indicated above the lane. RNA polymerase was preincubated with template in the presence of a high concentration of ATP and GTP (+) and in the absence of nucleotides (−) (see Experimental Procedures). The transcripts were visualized by autoradiography after electrophoresis into a 7 M urea–9% (wt/vol) polyacrylamide gel. Molecular-weight standards of pBR322 cut with *Hpa*II are shown in lanes h and w. The arrows show the positions of the transcripts generated by Eσ[32] and Eσ[37].

scription but interact with different sets of nucleotides, then a slight alteration of the location of the start point may differentially affect the interaction of the two polymerases with the promoter. We altered the location of the start point of transcription by using dinucleotide-primed reactions (5).

The initiation of transcription in vitro requires a high concentration of the initiating ribonucleoside triphosphate. Alternatively, a dinucleotide can prime initiation near the start point of the transcript, and elongation can proceed in the presence of a lower concentration of ribonucleoside triphosphates (5). When either Eσ[37] or Eσ[32] was used, transcription from the wild-type *ctc* promoter and a mutant promoter (transition at position −9) was primed most efficiently by the dinucleotides UpA and ApA which initiate at positions −1 and −3, respectively (Fig. 3, lanes c, f, k, n, r, u, z, and cc). The small difference in size of the transcripts, initiated by the different dinucleotides, is apparent in Fig. 3, but the sizes of these different transcripts also have been measured more precisely by electrophoresis in a high-resolution nucleotide sequencing gel (data not shown). The low level of transcription initiated by ApU (from position −2) (Fig. 3, lanes e, m, bb, and t) identified a poorly used, minor start point. The utilization of the minor start point (ApU) relative to the major start points was enhanced by the base substitution at position −9 when Eσ[32] was used to transcribe the template (Fig. 3, lane m). In comparison with its major start points, Eσ[37] used the minor start point (ApU) less efficiently than did Eσ[32] (Fig. 3, lanes t and bb).

DISCUSSION

We have used a prebinding run-off transcription assay to test the effects of base substitutions in the *ctc* promoter on utilization of this promoter by Eσ[32]. These results and the effects of the base substitutions on the utilization of the *ctc* promoter by Eσ[37] (8) and Eσ[29] (Tatti and Moran, submitted for publication) in this assay are summarized in Fig. 1. Of the base substitutions tested, the transition at position −15 causes the most severe reduction in promoter utilization by Eσ[37] and Eσ[32] but does not affect utilization of the promoter by Eσ[29]. The base substitution at position −14 also reduces the efficiency of promoter utilization by Eσ[37] and

$E\sigma^{32}$ but not by $E\sigma^{29}$. We conclude from these results that recognition of the *ctc* promoter by $E\sigma^{29}$ involves interaction of the polymerase with a different set of nucleotides from those involved in utilization of the promoter by $E\sigma^{37}$ and $E\sigma^{32}$. This conclusion is supported by the effects of the base substitution at position -9, which increases the utilization of the promoter by $E\sigma^{29}$ but not by $E\sigma^{37}$ or $E\sigma^{32}$.

The single base substitutions at positions -36 and -16 reduce the efficiency of promoter utilization by all three forms of RNA polymerase; therefore, all three forms of RNA polymerase probably interact with bases at or near positions -36 and -16. Promoter utilization by all three forms of RNA polymerase was affected less severely by base substitutions at the other positions occupied by guanine or cytosine in the nontranscribed strand of the wild-type *ctc* promoter.

Can we identify the sequences that signal promoter recognition by each of the three forms of RNA polymerase? Promoter recognition by $E\sigma^{37}$ appears to involve the interaction of the polymerase with nucleotides near the -10 region and the -35 region of the promoter (see Fig. 1). These nucleotides (shown in Fig. 1) are conserved in promoters recognized by $E\sigma^{37}$ (Lang, Ph.D. thesis), and they are positioned within the region, determined by deletion analysis, that contains the sequences which are sufficient for recognition of a promoter by $E\sigma^{37}$ (1). Furthermore, methylation protection experiments have demonstrated that bound $E\sigma^{37}$ closely contacts the nucleotides near the -10 region and the -35 region of the *ctc* promoter (6), and base substitutions at positions -36, -16, -15, or -14 severely affect the utilization of the *ctc* promoter by $E\sigma^{37}$ (8). Since it was the guanines at positions -36, -16, -15, and -14 that were dramatically protected in the methylation protection experiments (6) and it was these positions at which the base substitutions had the greatest effects, the utilization of the *ctc* promoter by $E\sigma^{37}$ probably involves direct interaction of the polymerase with these bases.

Recognition of the *ctc* promoter by $E\sigma^{29}$ appears to involve interaction of the polymerase with a different set of sequences at the -10 region and the -35 region. Figure 1 shows the nucleotide sequences which are highly conserved in four promoters that are used by $E\sigma^{29}$ (Lang, Ph.D. thesis). The bases at positions -12, -10, and -35 are invariant among the four promoters. The base substitutions that affect the utilization of the *ctc* promoter by $E\sigma^{29}$ (at positions -36, -16, and -9) do not replace conserved nucleotides, but these base substitutions may act by affecting $E\sigma^{29}$ contacts with neighboring conserved nucleotides. For example, the

base substitution at position -9 may allow a more favorable RNA polymerase contact with the invariant base at position -10, and the substitution at -36 may affect contact at position -35.

Two promoters are known to be used by $E\sigma^{32}$, *ctc*, and the P2 promoter of *spoVG* (3). Sequences near the -10 region and the -35 region are conserved between these two promoters. The effects of the base substitutions on promoter utilization by $E\sigma^{32}$ do not support or refute the role of these conserved sequences in promoter recognition by $E\sigma^{32}$. The transition at position -28, which altered a conserved base, did not affect a rate-limiting step in this assay; however, a more sensitive assay may reveal an effect of this or other base substitutions. The inefficient utilization of promoters with transitions at -36, -16, -15, or -14 may indicate that recognition of a promoter by $E\sigma^{32}$ involves interaction of the polymerase with sequences near these positions. In the run-off assay, the ability of $E\sigma^{32}$ to utilize the series of mutated promoters is similar to that of $E\sigma^{37}$. In the dinucleotide priming experiment, however, $E\sigma^{32}$ used the ApU start point more readily than did $E\sigma^{37}$, providing evidence that the matrices of contacts between the *ctc* promoter and bound $E\sigma^{37}$ or $E\sigma^{32}$ are not identical.

$E\sigma^{37}$, $E\sigma^{29}$, and $E\sigma^{32}$ appear to recognize their cognate promoters by interacting with different specific sets of nucleotide sequences. Each of these sets of recognition sequences is composed of two domains, one near the -10 region of the promoter and the other near the -35 region of the promoter. It appears that the *ctc* promoter contains three different overlapping sets of sequences at the -10 and -35 regions that signal recognition by these three forms of RNA polymerase.

The effects of the base substitutions on promoter utilization in vitro and the model for the role of the sequentially appearing sigma subunits during sporulation (4) enable us to make predictions about the utilization of the mutant promoters in vivo. For example, we expect that transcription from the mutant promoters with base substitutions at position -16 or -36 would occur at only a low level since these promoters are used inefficiently in vitro by all three RNA polymerases that are known to use the *ctc* promoter. The promoter with the base substitution at position -15 probably will not direct transcription at the end of exponential growth because the activation of transcription of *ctc* is thought to require $E\sigma^{37}$ or $E\sigma^{32}$, or both. Transcription from this promoter, however, may be activated by the appearance of $E\sigma^{29}$ about 2 h after the end of exponential growth. Transcriptional fusions constructed in vitro and transformed into *B. subtilis* have been used to show that the base substitution at position −15 does prevent the activation of transcription from this promoter at the end of exponential growth (C. Ray and C. Moran, unpublished data). Analysis of these and additional mutant *ctc* promoters should allow us to define more clearly the role of each form of RNA polymerase in the regulation of transcription of the *ctc* gene.

SUMMARY

The *ctc* promoter from *B. subtilis* is utilized by three different forms of RNA polymerase, which initiate transcription at or near the same nucleotide. To identify the nucleotides that signal recognition by each form of RNA polymerase, we have constructed a series of single and multiple base substitutions within the *ctc* promoter. We have assayed the effects of the mutations on promoter utilization in vitro by $E\sigma^{37}$ (8), $E\sigma^{29}$ (Tatti and Moran, submitted for publication), and now $E\sigma^{32}$. Several mutations differentially affect promoter utilization by $E\sigma^{37}$ and $E\sigma^{29}$. Although the mutations have similar effects on promoter utilization by $E\sigma^{37}$ and $E\sigma^{32}$ in the run-off assay, dinucleotide utilization indicates that $E\sigma^{37}$ and $E\sigma^{32}$ do not interact with the *ctc* promoter in an identical manner. These data support a model in which the *ctc* promoter contains three different but overlapping sets of nucleotides that signal recognition by the three different forms of RNA polymerase. The effects of the mutations on promoter utilization in vitro and the model for the role of the sequentially appearing sigma subunits during sporulation predict that several of the mutant promoters should direct temporally altered patterns of transcription in vivo.

We thank K. Tatti for helpful suggestions and J. Scott for critically reading this manuscript.

This work was supported by Public Health Service grant AI20319 from the National Institute of Allergy and Infectious Diseases to C.M.

LITERATURE CITED

1. **Banner, C. D. B., C. P. Moran, Jr., and R. Losick.** 1983. Deletion analysis of a complex promoter for a developmentally regulated gene from *Bacillus subtilis*. J. Mol. Biol. **168:**351–365.

2. **Haldenwang, W. G., N. Lang, and R. Losick.** 1981. A sporulation induced sigma-like regulatory protein from *B. subtilis*. Cell **23:**615–624.

3. **Johnson, C. W., C. P. Moran, Jr., and R. Losick.** 1983. Two RNA polymerase sigma factors from *Bacillus subtilis* discriminate between overlapping promoters for developmentally regulated gene. Nature (London) **302:**800–804.

4. **Losick, R.** 1981. Sigma factors, stage 0 genes and sporulation, p. 48–56. *In* H. S. Levinson, A. L. Sonenshein, and D. J. Tipper (ed.), Sporulation and germination. American Society for Microbiology, Washington, D.C.

5. **Minkley, E. G., and D. Pribnow.** 1973. Transcription of the early region of bacteriophage T7: selective initiation with dinucleotides. J. Mol. Biol. **77:**255–277.

6. **Moran, C. P., Jr., W. C. Johnson, and R. Losick.** 1982. Close contacts between σ^{37}-RNA polymerase and a *Bacillus subtilis* chromosomal promoter. J. Mol. Biol. **162:**709–713.

7. **Ollington, J. F., W. G. Haldenwang, T. V. Huynh, and R. Losick.** 1981. Developmentally regulated transcription in a cloned segment of the *Bacillus subtilis* chromosome. J. Bacteriol. **147:**432–442.

8. **Tatti, K. M., and C. P. Moran, Jr.** 1984. Promoter recognition by sigma-37 RNA polymerase from *Bacillus subtilis*. J. Mol. Biol. **175:**285–297.

9. **Vieira, J., and J. Messing.** 1982. The pUC plasmids, an M13 mp7-derived system for insertion mutagenesis and sequencing with synthetic universal primers. Gene **19:**259–268.

10. **Wong, H. C., H. E. Schnepf, and H. R. Whiteley.** 1983. Transcriptional and translational start sites for the *Bacillus thuringiensis* crystal protein gene. J. Biol. Chem. **158:**1960–1967.

The *Bacillus subtilis spo0* Regulon: Intergenic Suppression Analysis of Initiation Control

ABRAHAM LEUNG, DEAN NG, CHARLES YANG, SHERYL RUBINSTEIN, CATHERINE DEEDE, WENDY PUTNAM, AND TERRANCE LEIGHTON

Department of Microbiology and Immunology, University of California, Berkeley, California 94720

Intergenic suppression has emerged as an extremely powerful new approach to the analysis of developmental gene function in *Bacillus subtilis* (4–7, 11, 12). Studies of second-site functional suppressors which are epistatic to Spo^{ts} RNA polymerase and ribosomal protein mutations, as well as of other suppressors epistatic to *spo0* lesions, have yielded new insights into the nature and organization of the regulatory networks which control the initiation and propagation of developmental events (1, 2, 4–7, 11, 12). Apparently, disparate mutational alterations in RNA polymerase, the ribosome, and sporulation initiation (*spo0*) genes disrupt the normal pattern of gene expression such that a common aspect of proper developmental membrane structure and function is compromised (1, 2, 4–7, 11, 12). The intergenic suppressors of these Spo defects cause alterations in membrane composition and function which are presumed to be related to their ability to correct or bypass the physiological perturbations elicited by mutations in the cognate Spo genes (4, 5, 7). The view suggested by these findings is that the sporulation process is much more dependent on the interaction of diverse cellular functions than was previously appreciated. These new results imply that developmental regulation is characterized by highly interactive as well as processive events.

To investigate further the regulatory functions required for the initiation of sporulation, we have extended our previous analysis of intergenic suppressor mutations to include two new loci, the *crsA* gene (8, 10) and an additional mutation (*rvtA2*) at or near the *rvtA* gene (7). The *crsA* mutations were initially identified by Sun and Takahashi (8, 10) as a catabolite-resistant sporulation (Crs) phenotype and were subsequently shown by C. W. Price and R. H. Doi (submitted for publication) and F. Kawamura and R. H. Doi (submitted for publication) to be mutations in the major sigma factor gene of *B. subtilis* RNA polymerase (*rpoD*). The *rvtA* mutations were initially identified by Sharrock et al. (7) as intergenic suppressors of *spo0F*, *spo0B*, and *spo0E* genes. We show here that the *rpoD* mutations *crsA47* and *crsA4* are powerful intergenic suppressors of *spo0F*, *spo0B*, *spo0E*,

and *spo0K* mutations, and that a second mutation at or near the *rvtA* gene, *rvtA2*, causes a pattern of *spo0* mutant suppression quantitatively distinct from that elicited by the *rvtA11* mutation. We further show that both of these *rvtA* mutations are likely to lie within the region of the *B. subtilis* chromosome defined by the *spo0A* gene. The validity of this latter inference is established by the results of F. A. Ferrari, K. Trach, D. Le Coq, and J. A. Hoch (this volume), who have shown by nucleotide sequence studies of the *spo0A* gene that a mutation, *sof*, which was isolated by Kawamura and Saito (2), is similar if not identical to *rvtA11* and results in a base substitution alteration at amino acid codon 12 of the *spo0A* protein.

EXPERIMENTAL PROCEDURES

Procedures for the growth and sporulation of *B. subtilis* in liquid culture, strain maintenance, quantitation of spore yields, viable cell counts, and transformation, transduction, and recombination index measurements have been described (4). The sporulation sensitivity to alcohol (Ssa) and Crs phenotypes were recognized as described (7). Isogenic sets of *spo0* mutations carrying the *lys-1* or *aroD* markers were constructed by transduction, with strains RS1100 or 1A8 used as donors (7). Strain 1S8 (*spo0A9 trpC2*) was obtained from the Bacillus Genetic Stock Center. Strains Glu-47 (*crsA47 strA*) and Fru-4 (*crsA4*) were obtained from I. Takahashi. Additional *rvt* mutations were isolated and characterized as described (7).

RESULTS

Intergenic suppression of Spo0 phenotypes by the *rpoD* mutation *crsA47*. A series of isogenic *spo0* strains (7) containing the *lys-1* mutation were derived by transductional replacement of the *trpC2* gene, with the use of the *lys-1* Trp$^+$ donor strain RS1100. The *crsA47* mutation was introduced from strain Glu-47 into each of the *spo0* mutant backgrounds by cotransduction, with initial selection for Lys$^+$. The Lys$^+$ transductants which contained the *crsA* gene were recognized by their "small-colony" phenotype on minimal glucose agar plates. The *crsA* transductants represented 15 to 20% of the Lys$^+$

TABLE 1. Sporulation of *spo0* and suppressed
(*crsA*) *spo0* strains at 37°C[a]

Strain	% Wild-type sporulation
1A96	(100)
spo0F221 pheA	<0.0001
spo0F221 crsA47 pheA	100
spo0E11 pheA	3.0
spo0E crsA47 pheA	100
spo0K141 pheA	0.001
spo0K141 crsA47 pheA	76
spo0B136 pheA	<0.0001
spo0B136 crsA47 pheA	7.6
spo0H116 pheA	<0.0001
sp0H116 crsA47 pheA	<0.0001
spo0J87 pheA	0.5
spo0J87 crsA47 pheA	0.5

[a] Values, expressed as percentage of wild-type
sporulation, were calculated on the basis of a 1A96
sporulation frequency of 10^9 spores per ml in 2× SG
medium at 37°C (7).

recombinants and 70 to 75% of the AroD+
recombinants. The effects of the *crsA47* gene on
the sporulation properties of various *spo0* mutants are shown in Table 1. It is clear that the
crsA47 mutation is a very powerful intergenic
suppressor of the *spo0F*, *spo0E*, *spo0K*, and
spo0B mutations. It is unlikely that any other
mutation which might be contained in the Glu-47
strain contributes to these effects, since all of 54
CrsA Lys+ recombinants of a *spo0F221* strain
were Spo+, and all of 108 CrsA AroD+ recombinants were also Spo+. Hence, the sporulation-
suppressing phenotype and the CrsA phenotype
cannot be separated by recombination.

**Isolation and characterization of a second mu-
tation at or near the *rvtA* locus.** A large collection
of Spo+ revertants from strain RS1112 (*spo0F221
lys-1*) was isolated and screened for additional
rvt mutations in the *lys-1* region of the chromosome, as described previously (7). An additional

mutation, *rvtA2*, was identified. Three-factor
recombination analysis (transduction; see reference 7) placed this mutation at or near the site
on the *B. subtilis* chromosome occupied by the
rvtA gene (data not shown). We have carried out
a series of recombination index determinations
to ascertain whether the *rvtA2* and *rvtA11* mutations might be contained within the same gene.
The results of one such experiment are shown in
Table 2. Measurement of the recombination
index between the *rvtA2* and *rvtA11* mutations
(using the Ssa phenotype), relative to the recombination index between unlinked markers (i.e.,
AroD− and AroD+), shows that these two mutations are very closely linked. In other experiments, not shown here, we measured the recombination index of these two mutations by determining the number of Spo− recombinants obtained in a transformation involving *spo0F rvtA2*
and *spo0F rvtA11* strains. If the donor was
rvtA11 and the recipient was *rvtA2*, the measured recombination index was 0.071. In this
experiment we found no recombination of these
mutations (recombination index ≤ 0.007) if the
donor was *rvtA2* and the recipient was *rvtA11*.
Hence, these mutations are apparently located
within the same cistron.

We had previously discussed the possibility
that the *rvtA11* mutation could be contained
within the *spo0A* gene (7). We examined this
question further by carrying out transformation
of a *spo0A9* strain (1S8) with *rvtA11* DNA. Spo+
recombinants were selected, and the cotransformation frequency of the Ssa phenotype, determined by the *rvtA11* mutation, was measured.
We found that the Spo+ and Ssa phenotypes
were cotransformed at a frequency of 96% (52 of
54 Spo+ recombinants were Ssa). This experiment may underestimate the true linkage of
these alleles since the *spo0A9* strain has a significant frequency of reversion to Spo+ and is
only moderately competent for DNA-mediated
transformation. Hence, some of the Spo+ clones
which were observed may not have arisen by a
recombination event involving *spo0A9* and
rvtA11.

The *rvtA2* mutation has been introduced into a
series of isogenic *spo0* strains (7), and its effects
on the sporulation properties of these derivatives have been quantified (Table 3). The pattern

TABLE 2. Recombination index (RI) values for the determination of linkage between *rvt* mutations (4)[a]

Donor	Recipient	Recombinants/total transformants tested		% Recombinants		RI	
		AroD	Rvt	AroD	Rvt	AroD	Rvt
rvtA2	*rvtA11 trpC2 aroD*	19/750	2/798	2.5	0.25	1	0.10
rvtA11	*rvtA2 trpC2 aroD*	14/750	0/750	1.9	0	1	<0.072

[a] Primary selection was for Trp.

of suppression is similar qualitatively and quantitatively to that produced by the *rvtA11* mutation, except that there is little improvement of sporulation in the case of strain *spo0B136*.

Ssa phenotypes produced by *rvtA2*, *rvtA11*, *crsA47*, and *crsA4* mutations. Since all of the mutations studied at the *rvtA* and *crsA* loci had substantial activity as intergenic suppressors of *spo0* mutations, we were interested in determining whether these mutations, like the *rvtA11* mutation, were capable of preventing the blockage of the initiation of sporulation by aliphatic alcohols. The data in Table 4 demonstrate that all of these mutations produce an Ssa phenotype which allows these strains to sporulate efficiently in media containing 0.7 M ethanol.

DISCUSSION

The singular finding reported here is that mutations in the major sigma factor gene (*rpoD*) of *B. subtilis* RNA polymerase are powerful intergenic suppressors of *spo0B*, *spo0E*, *spo0F*, and *spo0K* mutations. It is likely that these effects are due to allele-nonspecific functional suppression since various mutations in several sporulation genes are affected, including the *spo0K141* strain which contains two unlinked *spo* mutations (R. A. Sharrock and T. Leighton, unpublished data). This interpretation is strengthened by the observation that the *rvtA11* suppressor, which is known to be allele nonspe-

TABLE 3. Sporulation of *spo0* and suppressed (*rvtA2*) *spo0* strains at 37°C[a]

Strain	% Wild-type sporulation
1A96	(100)
spo0F221 pheA	<0.0001
spo0F221 rvtA2 pheA	100
spo0E11 pheA	<0.0001
spo0E11 rvtA2 pheA	15
spo0B136 pheA	<0.0001
spo0B136 rvtA2 pheA	0.0004
spo0H116 pheA	<0.0001
spo0H116 rvtA2 pheA	<0.0001
spo0J87 pheA	2.5
spo0J87 rvtA2 pheA	0.17
spo0K141 pheA	0.0003
spo0K141 rvtA2 pheA	0.840

[a] Values, expressed as percentage of wild-type sporulation, were calculated on the basis of 1A96 sporulation frequency 10^9 spores per ml in 2× SG medium at 37°C (see 7).

TABLE 4. Effect of ethanol on the sporulation of 168, *rvtA*, and *crsA* strains[a]

Strain	% Sporulation with ethanol
168	0.005
168 *rvtA11*	22
168 *rvtA2*	11
crsA47	35
crsA4	23

[a] Values, expressed as percentage of sporulation, were calculated on the basis of 168 (wild-type) sporulation frequencies of 10^9 spores per ml in 2× SG medium at 37°C, with sporulation in the absence of ethanol 100% for all strains. Ethanol (0.7 M) was added at midlog growth phase.

cific (2, 7), also suppresses a subset of the *spo* gene mutations included in the *crsA* epistatic domain. The phenotypes produced by the *crsA* and *rvtA* mutations have an intriguing commonality. Mutations in these two genes produce sporulation suppression (Sus) phenotypes, Ssa phenotypes, Che phenotypes (chemosensory defects; T. Jones and T. Leighton, unpublished data), and Crs phenotypes (8, 10; R. H. Doi et al., this volume). Both of these genes are known to have direct interactions or effects on the *B. subtilis* transcriptional apparatus: the *rpoD* gene by its intrinsic involvement in determining the site selectivity of RNA polymerase, and the *rvtA* gene by virtue of the ability of these mutations to activate transcription at σ^{28} promoters in *spo0B*, *spo0E*, and *spo0F* strains, which are normally defective in the production of σ^{28} gene transcripts (1). The recombination index analysis of the *rvtA2* and *rvtA11* mutations suggests that these lesions are likely to be located within the same gene (7). The discovery that the *rvtA* mutations are in fact contained within the coding sequence of the *spo0A* gene, which is a positive pleiotropic regulator of sporulation initiation (Ferrari et al., this volume), suggests an interesting duality of function for this gene product. Whereas various mutations in the *spo0A* gene cause a blockage at the earliest definable stage of development, other mutations (*rvtA*) allow sporulation to occur in various genetic backgrounds (*spo0* strains) which would otherwise abort development at stage 0. One possible interpretation of these observations is that the effects of the *spo0* mutations, and their intergenic suppressors, are exerted through interactions involving the *B. subtilis* transcriptional apparatus. It is now recognized that in vivo transcription is carried out by a multicomponent complex involving RNA polymerase holoenzyme, *nusA*, *nusE*, *rho*, and undoubtedly other unidentified factors (M. C. Schmidt and M. J.

Chamberlin, submitted for publication). If the *spo0* mutations considered here interfere with either qualitative or quantitative functions of the transcriptional complex, it is not surprising that compensatory suppressor mutations would occur in the *rpoD* or *spo0A* gene. It is interesting to note in this context that the *rvtA2* mutation produces a qualitatively similar, but quantitatively distinct, pattern of suppression when compared with the *rvtA11* mutation. Consistent with this view is a recent finding of Takahashi and co-workers (9) that mutations in the RNA polymerase β subunit (*rif-11*) can suppress the Crs phenotype of *crsC*, *crsD*, *crsE*, and *crsF* mutations.

Since the *spo0A*, *spo0B*, and *spo0F* gene products are known to be present in vegetative cells (7), we presume that lesions in these genes are tolerable to such cells until the point at which the initiation of development commences. Although we do not understand the precise nature of the cellular disturbances induced by *spo0* gene mutations, we presume that the effects of these mutations are exerted at the level of membrane processes which recognize and transduce the chemical information required to initiate development (7). The obvious and similar membrane phenotypes produced by *rpoD/crsA* (8, 9) and *spo0A/rvtA* (7; Ferrari et al., this volume) mutations are consistent with the notion that suppression of *spo0* gene defects is accomplished by alterations in membrane structure and function which correct or bypass the dysfunctions induced by initiation gene mutations. Consistent with our previous interpretation of the *rvtA11* suppression of *spo0B*, *spo0E*, and *spo0F* gene defects (7), it is clear from the results presented here that cells containing a *crsA47* (*rpoD*) mutation are able to sporulate in the absence of a functional *spo0B*, *spo0E*, *spo0F*, or *spo0K* gene product.

The results described here suggest that the continued study of the *spo0* regulon, and its intergenic suppressors, may provide a unique opportunity for the investigation of transcriptional and membrane-related processes which had previously been inaccessible to detailed experimental analysis.

SUMMARY

Mutations (*crsA47* and *crsA4*) in the major sigma factor gene (*rpoD*) of *B. subtilis* RNA polymerase have been found to be powerful intergenic suppressors of *spo0B*, *spo0E*, *spo0F*, and *spo0K* mutations. The *crsA47* suppressor restores sporulation of *spo0E*, *spo0F*, and *spo0K* mutants to levels near those of wild-type bacteria and substantially improves the sporulation of a *spo0B* strain. A new intergenic suppressor mutation (*rvtA2*) at the *rvtA* locus has been isolated. This mutation causes a pattern of *spo0B*, *spo0E*, and *spo0F* suppression quantitatively distinct from that previously observed with the *rvtA11* allele. Recombination index analysis suggests that the *rvtA2* and *rvtA11* mutations are contained within the same genetic region of the *B. subtilis* chromosome defined by the *spo0A* gene. The *crsA* and *rvtA* mutations are shown to prevent the induction by aliphatic alcohols of Spo0 phenocopies in wild-type *B. subtilis* cells.

LITERATURE CITED

1. **Gilman, M. Z., and M. J. Chamberlin.** 1983. Developmental and genetic regulation of *Bacillus subtilis* genes transcribed by σ^{28} RNA polymerase. Cell **35**:285–293.
2. **Kawamura, F., and H. Saito.** 1983. Isolation and mapping of a new suppressor mutation of an early sporulation gene mutation in *Bacillus subtilis*. Mol. Gen. Genet. **192**:330–334.
3. **Piggot, P. J., and J. G. Coote.** 1976. Genetic aspects of bacterial endospore formation. Microbiol. Rev. **44**:908–962.
4. **Sharrock, R. A., and T. Leighton.** 1981. Intergenic suppressors of temperature-sensitive sporulation in *Bacillus subtilis* are allele non-specific. Mol. Gen. Genet. **183**:532–537.
5. **Sharrock, R. A., and T. Leighton.** 1982. Suppression of defective-sporulation phenotypes by the *Bacillus subtilis* mutation *rev4*. Mol. Gen. Genet. **186**:432–433.
6. **Sharrock, R. A., T. Leighton, and H. G. Wittmann.** 1981. Macrolide and aminoglycoside antibiotic resistance mutations in the *Bacillus subtilis* ribosome resulting in temperature-sensitive sporulation. Mol. Gen. Genet. **183**:538–543.
7. **Sharrock, R. A., S. Rubinstein, M. Chan, and T. Leighton.** 1984. Intergenic suppression of *spo0* phenotypes by the *Bacillus subtilis* mutation *rvtA*. Mol. Gen. Genet. **194**:260–264.
8. **Sun, D., and I. Takahashi.** 1982. Genetic mapping of catabolite-resistant mutants of *Bacillus subtilis*. Can. J. Microbiol. **28**:1242–1251.
9. **Sun, D., and I. Takahashi.** 1984. A catabolite-resistant mutation is localized in the *rpo* operon of *Bacillus subtilis*. Can. J. Microbiol. **30**:423–429.
10. **Takahashi, I.** 1979. Catabolite repression-resistant mutations of *Bacillus subtilis*. Can. J. Microbiol. **25**:1283–1287.
11. **Wayne, R. R., and T. Leighton.** 1981. Physiological suppression of *Bacillus subtilis* conditional sporulation phenotypes: RNA polymerase and ribosomal mutations. Mol. Gen. Genet. **183**:550–552.
12. **Wayne, R. R., C. W. Price, and T. Leighton.** 1981. Physiological suppression of the temperature-sensitive sporulation defect in a *Bacillus subtilis* RNA polymerase mutant. Mol. Gen. Genet. **183**:544–549.

Effect of Sporulation Mutations on Subtilisin Expression, Assayed Using a Subtilisin–β-Galactosidase Gene Fusion

EUGENIO FERRARI,[1] SANDRA M. H. HOWARD,[2] AND JAMES A. HOCH[2]

Department of Molecular Biology, Genentech, Inc., South San Francisco, California 94080,[1] and Division of Cellular Biology, Research Institute of Scripps Clinic, La Jolla, California 92037[2]

The production of alkaline protease is tightly associated with sporulation in *Bacillus subtilis*. Many investigators have tried to understand the relationship between sporulation and production of alkaline protease, but these studies have always been difficult to interpret as a result of the presence of at least two other secreted proteases (for a review, see references 10, 12, and 14). This multiplicity of proteases made it difficult to obtain and analyze specific protease mutants which could have clarified the confusion over the active or passive role of the proteases during sporulation. The effect on protease production of other genes involved in either sporulation, such as *cat* (11), *hpr* (9), and *sco* (2), or secretion, such as *sacU* (17), further clouded the issue.

With the cloning of the *B. subtilis* alkaline (16, 18) and neutral protease (19) genes, it became easier to study how proteases are regulated. The isolation of mutants carrying a deletion in either or both genes (16, 19) made it clear that the expression of at least these two major secreted proteases was only coincidental with, and not required for, the onset of sporulation. By using the cloned genes, it is possible to investigate which of the sporulation mutations affect also protease expression and whether the same type of control is exerted by the same mutation on all protease genes.

The expression of the subtilisin gene was studied in a variety of sporulation or sporulation-associated mutants. Because of the low level of protease produced in some of the stage 0 sporulation mutants, it was decided to study the regulation of the subtilisin gene by fusing it to the *lacZ* gene of *Escherichia coli* (7, 20) and assaying the β-galactosidase activity present in the cell extracts of the *B. subtilis* strains carrying such a construction. In the construction reported here the portion of the subtilisin gene which directs secretion is missing. In this way, the results obtained indicate the effect that the tested mutants exert on the alkaline protease gene at the transcriptional/translational level, avoiding any effect they might have on the secretion process.

EXPERIMENTAL PROCEDURES

Bacterial strains and plasmids. The bacterial strains used in this study are listed in Table 1. *E. coli* strains were cultured in LB broth and transformed with the procedure of Dagert and Ehrlich (6). *B. subtilis* strains were cultured on tryptose blood agar base (TBAB; Difco Laboratories), on Schaeffer's sporulation medium (15), or on minimal glucose medium and were transformed by the procedure of Anagnostopoulos and Spizizen (1). The plasmids pJH101 (8) and pS168.1 (16) have already been described. Plasmid DNA extractions were done as described by Birnboim and Doly (3).

β-Galactosidase assays. *B. subtilis* cells, carrying pSG35, were grown in Schaeffer's sporulation medium in the absence of chloramphenicol. Samples (5 ml) were taken as described in the text, pelleted by centrifugation, and washed with ice-cold Tris hydrochloride (pH 7.6). Pellets were immediately frozen in dry ice-ethanol and stored at −70°C. The frozen pellets were resuspended in 5 ml of Z buffer (13) just before the assay and were incubated at 37°C for 5 min with 100 µg of lysozyme per ml. After treatment at room temperature with 0.1% Triton X-100, samples were assayed for β-galactosidase activity by the method of Miller (13).

RESULTS AND DISCUSSION

Construction of the fusion plasmid pSG35. The plasmid pS168.1 (16), which carries the entire subtilisin gene at one end of a 6.5-kilobase (kb) *Eco*RI fragment of *B. subtilis* chromosomal DNA, after complete digestion with *Sau*3A releases a DNA fragment of approximately 1.6 kb which contains 850 base pairs (bp) of the vector plasmid, the *Eco*RI junction, and 750 bp of insert. The 750-bp fragment carries the regulatory region of the alkaline protease gene and eight codons of its translated sequence (Fig. 1). The single-strand ends of this DNA were filled in with the Klenow fragment of DNA polymerase I and deoxynucleotide triphosphates, and the 1.6-kb DNA fragment was purified by electroelution from a preparative acrylamide gel. This

TABLE 1. Bacterial strains used in this study

Strain	Genotype	Source
B. subtilis		
JH642	*trpC2 pheA1*	J. A. Hoch
BG125	*hisA1 thr-5 trpC2*	QB917, J. A. Hoch
BG2100	*trpC2 pheA1*(::pSG35)	This work
BG2136	*hisA1 thr-5 trpC2*(::pSG35)	This work
BG2137	*hisA1 trpC2 spoOAΔ204*(::pSG35)	This work
BG2138	*hisA1 trpC2 spoOAΔ204*(::pSG35)	This work
BG2139	*hisA1 trpC2 spoOB136*(::pSG35)	This work
BG2140	*hisA1 trpC2 spoOE11*(::pSG35)	This work
BG2141	*hisA1 trpC2 spoOF221*(::pSG35)	This work
BG2142	*hisA1 trpC2 spoOH81*(::pSG35)	This work
BG2143	*hisA1 trpC2 spoOJ87*(::pSG35)	This work
BG2149	*hpr-1*(::pSG35)	This work
BG2150	*hpr-2*(::pSG35)	This work
BG2151	*hpr-97*(::pSG35)	This work
BG2152	*sacA331 sacU^{h}32*(::pSG35)	This work
BG2153	*sacA321 sacS^{c}2 hisA1 trpC2*(::pSG35)	This work
E. coli		
K5070	F⁻ Δ*pro-lac supE thi-1*	H. I. Miller
MM 294	F⁻ *supE44 endA1 thi-1 hsdR4*	

fragment was inserted into an *Sma*I site of pJF751, a vector plasmid derivative of JH101 (8) which carries the *lacZ* gene and a short fragment of the *lacY* gene from plasmid pMC1403 (5) in place of the tetracycline resistance gene (Fig. 1; F. A. Ferrari, K. Trach, and J. A. Hoch, manuscript in preparation). The ligation mixture was used to transform *E. coli* K5070 and was plated on LB plates containing X-gal (5-bromo-4-chloro-3-indolyl-1-β-D-galactoside). From some of the transformants which gave rise to blue colonies, plasmid DNA was extracted, and the proper insertion of the fragment was checked by *Eco*RI and *Eco*RI-*Bam*HI digestions. Any plasmid carrying the insert in the right orientation should release an 850-bp *Eco*RI fragment and two fragments of 850 and 750 bp after digestion with *Eco*RI and *Bam*HI.

One of these plasmids, pSG3, was digested with *Eco*RI, religated to itself, and transformed into *E. coli* 294 to eliminate the 850-bp *Eco*RI fragment of the pS168.1 vector. In one of the derivatives, pSG35, the exact placement in frame of the fusion was also verified by sequencing the junction region.

Plasmid pSG35 does not contain a replicative function recognized by *B. subtilis* and upon transformation integrates in the chromosome at the subtilisin region. β-Galactosidase synthesis is under the control of any known or unknown control elements which lay upstream from the gene or, in this case, upstream from the eighth codon of the structural gene.

Expression of the fused *lacZ* in *hpr* and *sacU^h* mutants. A DNA preparation of pSG35 was used to transform the wild-type strain JH642, selecting for Cm^r. A transformant was chosen and streaked for a single colony, and chromosomal DNA extracted from it was used to transform to Cm^r strains carrying *hpr1*, *hpr2*, *hpr97*, *sacS^{c}2*, and *sacU^{h}32* mutations. The transformants were purified to single colonies. A fresh streak on TBAB plates of the parental strain and of each mutant carrying pSG35 was used to inoculate Schaeffer's sporulation medium. The cells were grown at 37°C and growth was monitored spectrophotometrically. Samples were taken at 1-h intervals, and β-galactosidase activity was assayed as described in Experimental Procedures.

Figure 2 shows the results obtained with the strains containing *hpr* mutations. The initial rate of synthesis of β-galactosidase activity in the mutants is at least 10-fold higher than in the parental strain. Considering that in this system any increase due to secretion is avoided as a result of the lack of a signal sequence on the fusion protein, one possible conclusion is that the *hpr* mutations have a direct effect on the expression of the subtilisin gene at either the transcriptional or translational level, or both. This tentative conclusion needs to be verified by assay of mRNA levels. The *hpr* mutations do not render the cells constitutive for subtilisin production, however. The enzyme is repressed during vegetative growth and appears after the end of vegetative growth. Thus, factors other than the *hpr* gene product must also be involved in the control of subtilisin transcription.

A *sacU^h* mutant, which has been reported to increase the yield of several enzymes, including subtilisin secreted by *B. subtilis*, does not have any effect on the regulation of the subtilisin–β-galactosidase fusion gene at the transcriptional or translational level (Fig. 3). This result strongly suggests that the increased yield of secreted proteases in such mutants is due only to a better secretion efficiency.

Construction of *spo0* mutants carrying pSG35. To test the effect of *spo0* mutations on subtilisin synthesis, a special set of strains carrying pSG35 was constructed. It has been shown that, when some sporulation genes under *spo0* control are fused to *lacZ* and this fusion is present in the chromosome in as little as three to four copies (20; P. J. Piggot, J. W. Chapman, and C. A. M.

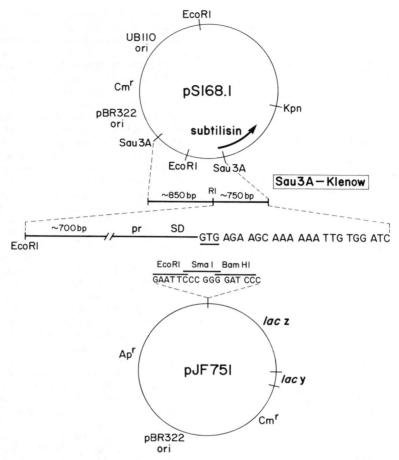

FIG. 1. Construction of the plasmid carrying the subtilisin-*lacZ* fusion. The strategy used for the construction of the fusion plasmid is outlined in the text.

Curtis, this volume), the *spo0* control that it is subject to when present in a single copy is escaped. As a consequence of the Cm^r selection while introducing pSG35 into *hpr* and *sac* mutants, we had to monitor the copy number of the integrated plasmid through Southern blot analysis of the DNA extracted from chosen transformants (data not shown). For this reason we followed a different strategy with the construction of stage 0 sporulation mutants carrying the fusion construction. The plasmid pSG35 was used to transform BG125 and one of the transformants, BG2136, after single-colony isolation on TBAB with chloramphenicol, was grown for a few generations in the absence of chloramphenicol and the copy number of the plasmid was checked by Southern blotting. All subsequent growth and manipulation were carried on without selective pressure. We reasoned that, as a result of the low frequency of excision of a nonreplicative integrated plasmid (16, 19; A.

Galizzi, personal communication), the probability of having a culture in which a significant portion of the population had lost the plasmid was very low. Furthermore, the absence of chloramphenicol selection should maintain the copy number at or near one.

spo0 mutations were transferred by congression in BG2136, and after single-colony isolation the newly constructed strains were tested for the presence of pSG35 on plates of TBAB plus chloramphenicol. All subsequent growth and manipulation were carried out in the absence of chloramphenicol selective pressure. At the end of every experiment a sample of the culture was plated on TBAB plates, and after overnight incubation at 37°C approximately 100 colonies were tested for Cm^r. We never detected loss of the plasmid.

Effect of some stage 0 sporulation mutations on the synthesis of subtilisin-*lacZ* fusion protein. From a fresh streak on TBAB plates the *spo0*

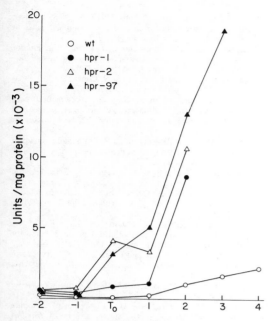

FIG. 2. Effect of the *hpr* mutations on the expression of the subtilisin-*lacZ* fusion. The ordinate is the specific activity of β-galactosidase expressed in Miller units, plotted over the time expressed in hours on the abscissa. T_0 is the transition point from vegetative growth to stationary phase.

mutants carrying pSG35 were inoculated into Schaeffer's sporulation medium and grown at 37°C. Samples of 5 ml were taken at 60- to 90-min intervals and treated as described in Experimental Procedures. The amount of cellular protein in each sample was determined, and the β-galactosidase specific activity was plotted against time of culture. The results (Fig. 4) show that the *spo0J* strain, which has a level of β-galactosidase activity similar to that of the Spo⁺ control, has little or no effect on the expression of subtilisin. This is in agreement with previous observations that the *spo0J* mutation does not affect the level of proteases produced at the onset of sporulation. The *spo0A* mutation exerts the most stringent control on the expression of the subtilisin-*lacZ* fusion. This is not surprising since the *spo0A* mutants have, among all the stage 0 mutants, the most drastic effect on protease production (4). Also, in view of recent observations which seem to indicate that the *spo0A* gene product has a major role in the sporulation process, one would expect that almost all the genes whose expression is affected by sporulation are somehow controlled, directly or indirectly, by the *spo0A* locus.

Surprisingly, all the other stage 0 mutants tested effectively repress the expression of the fused gene to the same extent, which is interme-

diate between the Spo⁺/*spo0J* level and the *spo0A* level. On plates of LB broth containing skim milk the control of protease production exerted by *spo0B* and *spo0F* mutants seems to be more effective (similar to *spo0A*) than by *spo0E* or *spo0H* mutants. However, this type of plate assay is more sensitive to the neutral protease than to the alkaline protease (16, 19). This result might indicate that the two major proteases are controlled in a slightly different way, by different stage 0 sporulation mutants either at the transcriptional/translational level or at the secretional level. However, the results indicate that mutations in *spo0* genes affect the transcription or translation of subtilisin directly. There may be a further effect on subtilisin secretion in some *spo0* mutants that manifests itself as a lower extracellular concentration.

SUMMARY

A fusion at the 5' end of the *B. subtilis* subtilisin gene and β-galactosidase was accomplished in which the first eight amino acids of the "prepro" subtilisin polypeptide were fused to the amino-terminal end of the *lacZ* gene in the vector pJF751, which cannot replicate in *B. subtilis*. The derivative plasmid, pSG35, was inserted by transformation in a variety of mutants which affect either sporulation or the level of enzymes secreted by *B. subtilis*. It was found that *hpr* mutations dramatically increased the rate of fusion protein synthesis without altering the timing of synthesis in the developmental cycle or rendering synthesis constitutive. The *sacU^h* mutation had no effect on fusion protein synthesis, suggesting that its role is related to enzyme secretion. All of the *spo0* mutations, except the *spo0J* mutation, were found to lower

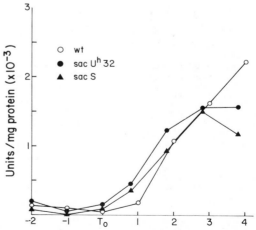

FIG. 3. Effect of the *sacU^h* and *sacS* mutations on the expression of the subtilisin-*lacZ* fusion. The parameters used are the same as in Fig. 2.

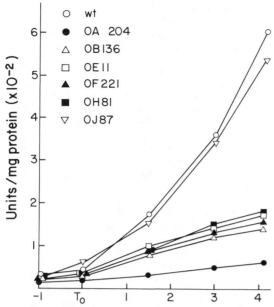

FIG. 4. Effect of some *spo0* mutations on the expression of the subtilisin-*lacZ* fusion. The parameters used are the same as in Fig. 2. The data relative to the *spo0A* reflect the average of experiments done with two independently isolated transformants.

the rate of subtilisin synthesis, with the lowest rate observed in a *spo0A* mutant.

E.F. would like to thank D. J. Henner for helpful discussions.

This work was supported, in part, by Public Health Service grant GM19416 from the National Institute of General Medical Sciences. This is publication number 3748-BCR from the Research Institute of Scripps Clinic.

LITERATURE CITED

1. **Anagnostopoulos, C., and J. Spizizen.** 1961. Requirements for transformation in *Bacillus subtilis*. J. Bacteriol. **81:**741–746.
2. **Balassa, G., B. Dod, V. Jeannoda, P. Milhaud, J. Zucca, J. C. F. Sousa, and M. T. Silva.** 1978. Pleiotropic control mutations affecting the sporulation of *Bacillus subtilis*. Ann. Microbiol. (Paris) **129B:**537–549.
3. **Birnboim, H. C., and J. Doly.** 1979. A rapid alkaline extraction procedure for screening recombinant plasmid DNA. Nucleic Acids Res. **7:**1513–1522.
4. **Brehm, P. S., S. P. Staal, and J. A. Hoch.** 1973. Phenotypes of pleiotropic-negative sporulation mutants of *Bacillus subtilis*. J. Bacteriol. **115:**1063–1070.
5. **Casabadan, M. J., J.Chou, and S. N. Cohen.** 1980. In vitro gene fusions that join enzymatically active β-galactosidase segment and amino-terminal fragments of exogenous proteins: *Escherichia coli* plasmid vectors for the detection and cloning of translational initiation signals. J. Bacteriol. **143:**971–980.
6. **Dagert, M., and S. D. Ehrlich.** 1979. Prolonged incubation in calcium chloride improves the competence of *Escherichia coli* cells. Gene **6:**23–28.
7. **Donnelly, C. E., and A. L. Sonenshein.** 1982. Genetic fusion of *E. coli* lac gene to a *B. subtilis* promoter, p. 63–72. *In* A. T. Ganesan, S. Chang, and J. A. Hoch (ed.), Molecular cloning and gene regulation in Bacilli. Academic Press, Inc., New York.
8. **Ferrari, F. A., A. Nguyen, D. Lang, and J. A. Hoch.** 1983. Construction and properties of an integrable plasmid for *Bacillus subtilis*. J. Bacteriol. **154:**1513–1515.
9. **Higerd, T. B., J. A. Hoch, and J. Spizizen.** 1972. Hyperprotease-producing mutants of *Bacillus subtilis*. J. Bacteriol. **112:**1026–1028.
10. **Hoch, J. A.** 1976. Genetics of bacterial sporulation. Adv. Genet. **18:**69–98.
11. **Ito, J., and J. Spizizen.** 1973. Genetic studies of catabolite repression insensitive sporulation mutants of *Bacillus subtilis*. Colloq. Int. C.N.R.S. **227:**81–82.
12. **Maurizi, M. R., and R. L. Switzer.** 1980. Proteolysis in bacterial sporulation. Curr. Top. Cell. Regul. **16:**163–224.
13. **Miller, J. H.** 1972. Experiments in molecular genetics. Cold Spring Harbor Laboratory, Cold Spring Harbor, N.Y.
14. **Piggot, P. J., and J. G. Coote.** 1976. Genetic aspects of bacterial endospore formation. Bacteriol. Rev. **40:**908–962.
15. **Schaeffer, P., J. Millet, and J. Aubert.** 1965. Catabolite repression of bacterial sporulation. Proc. Natl. Acad. Sci. U.S.A. **54:**704–711.
16. **Stahl, M. L., and E. Ferrari.** 1984. Replacement of the *Bacillus subtilis* subtilisin structural gene with an in vitro-derived deletion mutation. J. Bacteriol. **158:**411–418.
17. **Steinmetz, M., F. Kunst, and R. Dedonder.** 1976. Mapping of mutation affecting synthesis of excocellular enzymes in *Bacillus subtilis*. Mol. Gen. Genet. **148:**281–285.
18. **Wong, S. L., C. W. Price, D. S. Goldfarb, and R. H. Doi.** 1984. The subtilisin E gene of *Bacillus subtilis* is transcribed from a promoter *in vivo*. Proc. Natl. Acad. Sci. U.S.A. **81:**1184–1188.
19. **Yang, M. Y., E. Ferrari, and D. J. Henner.** 1984. Cloning of the neutral protease gene of *Bacillus subtilis* and the use of the cloned gene to create an in vitro-derived deletion mutation. J. Bacteriol. **160:**15–21.
20. **Zuber, P., and R. Losick.** 1983. Use of a *lacZ* fusion to study the role of the *spo0* genes of *Bacillus subtilis* in developmental regulation. Cell **35:**275–283.

Metabolic Control of Gene Expression and Sporulation

Recent Progress in Metabolic Regulation of Sporulation

ABRAHAM L. SONENSHEIN

Department of Molecular Biology and Microbiology, Tufts University Schools of Medicine, Veterinary Medicine, and Dental Medicine and Sackler School of Graduate Biomedical Sciences, Boston, Massachusetts 02111

It is useful for experimental design and discussion to consider the process of endospore formation as occurring in two discrete phases: (i) an initiation phase, during which growing cells recognize and respond to the fact that the environment no longer supports rapid growth, and (ii) a differentiation phase, induced by the initiation phase, during which genes whose products are required for sporulation are expressed at different times in a defined and genetically determined order. This simplification allows one to frame questions about how each phase is determined and regulated even though it obscures a number of issues, such as the role of growth genes in sporulation, the significance of genes that are expressed during sporulation but whose products are not required for sporulation, and the possible interdependence of biochemical events that lead to morphological changes and those that lead to changes in gene expression.

Since 1969 the differentiation phase of sporulation has been attributed at least in part to the action of polypeptides (sigma factors) that bind to and modify the promoter specificity of RNA polymerase (19). It is only in recent years, however, that definitive biochemical evidence has been obtained that proves that forms of RNA polymerase found in sporulating cells transcribe in vitro certain genes that are expressed during sporulation and do not transcribe other genes that are expressed only during growth. This has been reviewed recently (18) and is the subject of several articles in the present volume. I only point out that most of the genes used for these transcription studies do not have known products and that their products are not necessarily required for sporulation. The ultimate proof of the theory that sporulation-specific genes are transcribed at different times because their promoter sites can only be recognized by sigma factors active at those times will come from studies with mutants with lesions in the genes for these sigma factors. If a mutation in the structural gene for a particular sigma factor knocks out transcription of a particular class of genes in vivo and prevents sporulation, one could conclude that that sigma factor was necessary for transcription of at least one gene whose product was essential for sporulation. Rapid progress in this area is likely since the cloned *spoIIG* gene, which has extensive homology with the gene for σ^{55}, may code for σ^{29} (J. Szulmajster and P. Stragier, personal communication; Y. Kobayashi, personal communication). Moreover, several sporulation-specific templates are now becoming available, including the *spoIIA* and *spoVA* loci (D. Savva and J. Mandelstam, this volume) and the *spoIID* and *spoIIIC* genes (M. Rosenkrantz, S. Rong, and A. L. Sonenshein, manuscript in preparation).

The initiation phase has also been reviewed in recent years (10, 11, 25). The purpose of the present analysis is not to cover old ground again but rather to bring up to date the previous reviews and to try to anticipate the directions of advances in the near future. Most of the work reviewed has been done with the best-studied sporeformer, *Bacillus subtilis*.

It has long been known that rapidly growing cells can be induced to sporulate en masse by allowing them to exhaust a nutrient broth medium or by resuspending them in a medium containing a poorly metabolizable carbon, nitrogen, or phosphorus source. This is convenient for experimentation but probably a poor reflection of the situation of *B. subtilis* in its natural habitat, the soil. There the population probably grows slowly. Under these conditions, a significant fraction of the cells initiate sporulation while the population grows (23). The probability that a given cell will initiate sporulation in any given cell generation is inversely related to the growth rate of the population (2), which is in turn a reflection of available nutrients. This suggests that one or more metabolites have threshold values that determine whether a cell can initiate sporulation. Evidence has also accumulated that cells are competent to initiate only at a specific stage of the DNA replication cycle (4). This leads to the notion that during every round of DNA replication the cell chooses between growth and sporulation depending on whether a key metabolite(s) is above or below a certain threshold value. If one admits this general scheme, it becomes clear that the central questions of initiation are: what is the key metabolite(s), what protein(s) in the cell responds to its concentration(s), and by what mechanism does the receptor protein regulate the expression of the first genes expressed once the signal to initiate has been received? This initial activation phenomenon presumably leads to a cascade of gene expression, at least part of which is mediated by changes in the sigma factor component of RNA polymerase.

Given that there is little or no information available at present as to the mechanism regulating metabolism of carbon, nitrogen, or phosphorus compounds in *Bacillus* spp. or any other gram-positive organism, there are several possible experimental approaches to solving the initiation problem. One might focus on the key metabolites, hoping to identify such compounds and subsequently to find regulatory proteins to which they bind. A second approach is to define the genes that are turned on first during sporulation induction and use those genes to assay for regulatory proteins and factors. An alternative to either of the first two approaches is to study general mechanisms of carbon, nitrogen, and phosphorus regulation in growing cells and then ask whether regulatory mechanisms that become known are relevant to sporulation. The advantage of this approach is that growth genes subject to such regulation are known; its major disadvantage is that the mechanisms adduced may have no relation whatever to those governing sporulation. It is likely that a full understanding of how sporulation initiates will depend on application of all three approaches listed above, as well as others.

KEY METABOLITES

GTP, a key metabolite. The search for key metabolites whose intracellular concentrations might signal the onset of sporulation has a long and checkered history (10, 25). Several groups sought by mutational analysis to identify a particular carbon-containing metabolite or a particular nitrogen-containing metabolite which could serve as a monitor of availability of adequate nutritional sources. No clear-cut answer emerged from these studies inasmuch as genetic blocks at many steps in metabolism interfered with sporulation. At the same time it was recognized that the carbon, nitrogen, and phosphorus signals need not be independent; a single molecule (such as a nucleotide) containing all three elements could serve as a simultaneous monitor for all three conditions. Such a unified theory of initiation control emerges from the clearest work to date on the subject. Freese and colleagues have shown that initiation of sporulation is always accompanied by a substantial, if transient, decrease in the intracellular pools of GDP and

GTP (16). Moreover, conditions that reduce these pools induce sporulation even when cells are in a medium containing excess sources of carbon, nitrogen, and phosphorus. GDP and GTP pools have been altered experimentally by reducing the guanine source available to a guanine auxotroph (12), by adding to growing cells an inhibitor of GTP synthesis (31), or by induction of the stringent response (21, 22). (The stringent response is thought to be responsible for initiation of sporulation during exhaustion of nutrient broth medium and after resuspension of growing cells in the medium of Sterlini and Mandelstam [16].)

The beauty of this analysis is the unitary explanation it offers for how sporulation initiates. The conclusion that GDP and GTP pools drop whenever sporulation initiates is inescapable, and the conclusion that sporulation is induced by forced fluctuations in these pools is entirely convincing. Nonetheless, some questions remain at both the experimental and theoretical levels.

Partial purine deprivation results at best in 20% of the cells converting to spores (12), activation of the stringent response leads to 1 to 30% sporulation (22), and exposure of cells to decoyinine, an inhibitor of GMP synthetase, results in 20 to 50% sporulation (10). These numbers reflect increases of several orders of magnitude over what would have occurred in the absence of intervention. On the other hand, during exhaustion of nutrient broth medium and after resuspension of growing cells in Sterlini-Mandelstam medium, it is typically observed that 80 to 90% of the cells become spores. This suggests that reduction of the GTP pool may not be the only signal to initiate sporulation. The size of the decrease in the GTP pool is also uncertain. In *relA*[+] strains undergoing the stringent response, it has been described as 3- to 5-fold in various defined media, whereas for *relA* mutants it was 2- to 2.5-fold and less rapid (22). (In nutrient broth sporulation medium, a rapid 7-fold drop was seen for both *relA*[+] and *relA* strains. Both sporulated well, although the *relA* strain did so more slowly [16].) The change in GTP pool is thus not greatly different between *relA*[+] and *relA* strains during the stringent response. These results make it difficult to know whether the cell senses an absolute concentration of GTP as a threshold value for sporulation or whether it senses a relative drop of some critical magnitude. The latter is more difficult to understand mechanistically. The most careful measurements of the GTP pool to date indicate that in growing cells it is 800 μM, and after addition of decoyinine it drops to 200 to 300 μM (H. Wabiko, K. Ochi, D. Nguyen, and E. Freese, manuscript in preparation).

How does the GTP effect work? Although the details remain to be clarified, one can accept the general appropriateness of the concept that a decrease in the GTP pool is perceived by the cell as at least one signal to initiate sporulation. Two simple models of gene regulation can then be postulated. In the first, a GTP-binding protein is a repressor of one or more key genes which must function for sporulation to initiate. In the second model, a protein able to bind GTP is an activator of these key genes; the effect of GTP is to prevent this activation. Known GTP-binding proteins include guanylate cyclase, protein synthesis factors IF2 and EFTu, adenylosuccinate synthetase, and DNA-dependent RNA polymerase. Moreover, GTP is a precursor via guanosine of 5-methyl tetrahydrofolic acid and is a cofactor for the 2-ketoglutarate dehydrogenase complex. For each of these it is possible to construct a scheme that would explain how a decrease in GTP would alter cellular metabolism and gene expression. For example, a decrease in GTP might reduce synthesis of cyclic GMP (1) or interrupt protein synthesis, causing synthesis of an alarmone, or decrease the overall rate of RNA synthesis as well as the relative rate of transcription from particular promoters (30). Early blocked sporulation mutants have long been thought to define genes whose products play a role in sensing environmental change. One or more of the *spo0* gene products may be a GTP-binding protein. The fact that addition of decoyinine does not allow *spo0* mutants to sporulate (13) indicates that products of these genes function in steps that follow the decrease in GTP.

It is difficult to speculate about the possible role of GTP or some derivative of it without any knowledge of what genes are likely to be affected by changes in its pool. Identifying such genes is not straightforward, however. Possible candidates include genes thought to function early during sporulation. Whereas the *spo0* gene products may be the agents of the GTP effect, the *spo0* genes are not likely to be its targets inasmuch as at least some of these genes are expressed during vegetative growth. Other early blocked sporulation mutants are also not candidates, since their mutations can be bypassed by artificial reduction in the GTP pool (13). Mutants with defects in the tricarboxylic acid cycle are a special case. This is because such mutants fail to sporulate in the first place because of inability to supply sufficient energy for sporulation. Addition of decoyinine to a complete medium allows sporulation to ensue, using energy derived from glycolysis (13), thus obviating the need for an active tricarboxylic acid cycle. The genes for aconitase and citrate synthase do appear to be targets of the GTP effect, apparently because

activation of 2-ketoglutarate dehydrogenase by reduction in GTP causes a decrease in the pool of 2-ketoglutarate (see below) (29). Genes that are targets of the GTP effect and essential for sporulation under all conditions should be expressed only after the onset of sporulation, and mutations in them should not be bypassed by reduction in the GTP pool.

One can hope to find a new class of genes that is expressed very soon after reduction of GTP, but not before, by hybridizing pulse-labeled RNA to a restriction digest of chromosomal DNA in the presence of various competitor RNAs (e.g., RNA from cells growing in excess nutrients). Only genes that are turned on rapidly under all conditions of sporulation induction should be considered. Once such genes have been purified and analyzed, their promoter regions could serve as sites at which binding of putative regulatory proteins could be tested. They could also be used to purify such regulatory proteins. One such gene may be *spoVG* (P. Zuber, this volume).

MODEL SYSTEMS OF METABOLIC REGULATION OF GENE EXPRESSION

Carbon source regulation. Given our lack of knowledge of how GTP might act or what genes it might act upon, it seems logical to try to understand examples of metabolic regulation other than the initiation of sporulation. The hope is that the mechanism by which various carbon sources, for example, affect the expression of certain "catabolite-sensitive" genes might shed light on how the earliest expressed sporulation genes repond to the same environmental conditions. In gram-negative bacteria, glucose catabolite repression results from a deficiency in cyclic AMP. The association of cyclic AMP with catabolite activator protein allows that protein to bind to promoter regions of catabolite-sensitive genes and stimulate transcription. This exact mechanism cannot pertain to *B. subtilis*, however, since this organism lacks detectable cyclic AMP under aerobic conditions (24). Thus, the mechanism of catabolite repression in *Bacillus* is still a mystery. Nonetheless, significant progress has been made in analyzing several catabolite-sensitive systems.

Production of α-amylase in complex medium occurs at the end of the logarithmic growth phase in *B. subtilis*, but its appearance is prevented if glucose is present in the medium. A mutant has been isolated that expresses the α-amylase gene even in the presence of glucose, but it does so only at the end of log phase. (W. Nicholson and G. Chambliss, manuscript in preparation). This suggests that there are two mechanisms governing α-amylase expression, one of which is mediated by glucose and the other of which is a reflection of some other component of the environment. The mutation in question has been mapped to the region just upstream from the α-amylase structural gene and shows its effect only in *cis* (W. Nicholson and G. Chambliss, manuscript in preparation). It thus appears to be in a site at which glucose regulation is imposed. The phenotype of this mutant is consistent with the glucose effect being a negative regulatory phenomenon (W. Nicholson and G. Chambliss, personal communication). It should be possible to use this mutant to find compensatory mutations in a gene coding for a putative protein that interacts with this site. Such a protein would presumably be the mediator of glucose repression and would be likely to act at other promoter sites as well.

Aconitase appears early during sporulation, whether induced by exhaustion of a nutrient broth sporulation medium or by addition of decoyinine to growing cells (26). When sporulation is induced by carbon deprivation, aconitase activity is in fact essential for sporulation, to meet the cell's energy demand. Aconitase activity is also found in cells growing in a minimal medium. In this case, its appearance is prevented by the simultaneous presence in the medium of a rapidly metabolizable carbon source, such as glucose or maltose, and a source of 2-ketoglutarate, such as glutamate or glutamine (14). This is a reflection of the dual role of the tricarboxylic acid cycle; it is involved in both production of energy and synthesis of carbon skeletons for biosynthesis. Only when alternative sources of energy and 2-ketoglutarate are present simultaneously is aconitase activity greatly reduced. The specific activity of aconitase has been shown to vary with the intracellular pools of pyruvate and 2-ketoglutarate (7), suggesting that these compounds or derivatives of them might be the effectors of this regulation. This relationship holds in wild-type cells growing in different media (7), in mutant cells that accumulate or fail to synthesize one or both compounds (7), and in cells induced to sporulate by addition of decoyinine (29).

Study of the cloned aconitase gene has revealed that its transcription in vivo in cells growing in minimal medium correlates well with enzyme activity. When glucose and glutamine are present in the medium, very little of the aconitase transcript can be found, but when either of the "repressing" compounds is removed, transcription is greatly increased (Fig. 1). The variation in aconitase mRNA in the various media tested was sufficient to explain the differences in aconitase specific activity seen under the same conditions (M. S. Rosenkrantz, Ph.D. thesis, Tufts University, Boston, Mass., 1984). In nutrient broth sporulation medium,

aconitase mRNA appears at the end of log-phase growth, although enzymatic activity appears only several hours later. This difference has proved to be due to synthesis of inactive aconitase protein in late log phase; it becomes activated during sporulation (D. W. Dingman and A. L. Sonenshein, manuscript in preparation). Thus, expression of the aconitase gene, like expression of α-amylase, is a presporulation or early sporulation event. By constructing a fusion of the aconitase promoter and the *Escherichia coli lacZ* gene, it has been possible to design selections for mutants that have become insensitive to glucose-glutamine repression. These will be used to search for compensatory mutations in a hypothetical regulatory gene whose product acts at the aconitase promoter region.

Among other catabolite-repressed genes currently under intensive study are two linked genes involved in gluconate metabolism (Y. Fujita et al., this volume). It should be noted, however, that not all glucose-repressed genes are appropriate models for early sporulation genes. The fact that addition of decoyinine to cells in glucose-containing medium causes induction of some glucose-repressed genes (e.g., aconitase and citrate synthase) (26), but not of others (e.g., inositol dehydrogenase, sorbitol dehydrogenase, acetoin dehydrogenase) (17), means that the link between the general mechanism of glucose repression and repression of sporulation is not absolute. It may turn out that only those genes which are repressed by the combination of glucose and glutamate (citrate synthase, aconitase, histidase [7], etc.) are regulated by the same mechanism that regulates sporulation.

Nitrogen source regulation. In gram-negative bacteria, genes whose products are involved in production of ammonia and glutamine are regulated in response to the availability of ammonia by the *ntrA* and *ntrC* gene products (20). The genes subject to this regulation include those that code for glutamine synthetase (*glnA*), urease, nitrogen fixation enzymes, and a variety of amino acid-degrading enzymes (20). In *B. subtilis* the only enzymes whose specific activities are known to vary with the nitrogen source are glutamate synthase and glutamine synthetase. For *gltA* and *gltB*, the genes that appear to code for the two subunits of glutamate synthase, and *glnA*, the steady-state level of mRNA in cells grown in different nitrogen sources is proportional to enzymatic activity (8; D. Bohannon, M. S. Rosenkrantz, and A. L. Sonenshein, manuscript in preparation). Thus, regulation by the nitrogen source can be fully accounted for by the amount of transcript present. For the *glnA* gene, we have tentatively identified a regulatory

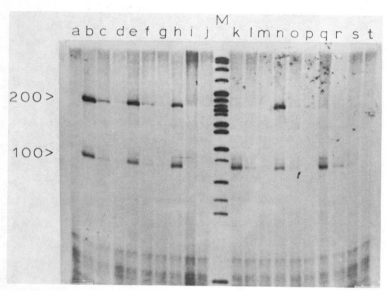

FIG. 1. Levels of *citB* transcript under various growth conditions. RNA was isolated from mid-log-phase cells of *B. subtilis* strain SMY grown in minimal medium containing glucose or citrate, or both, as carbon source and ammonium chloride or glutamine as nitrogen source. RNA was hybridized to pMR41 (Rosenkrantz, Ph.D thesis, Tufts University, 1984), a plasmid that contains the *citB* promoter, which had been cut with *Hin*dIII and 5′-end labeled with polynucleotide kinase. The same hybridization reaction contained a control DNA, pPH9 (S. F. J. LeGrice, C.-C. Shih, and A. L. Sonenshein, manuscript in preparation), which contains the promoter for the *veg* gene. After hybridization, the nucleic acids were treated with nuclease S1, denatured, and subjected to electrophoresis in a gel of polyacrylamide containing 8 M urea. The 200-base protected fragment corresponds to the *citB* transcript; the 100-base protected fragment is the *veg* transcript. The type and amount of RNA contained in each lane are as follows: lanes a and t, 9 μg of *Saccharomyces cerevisiae* RNA; lanes b, c, and d, 9, 3, and 1 μg of citrate-ammonia RNA; lanes e, f, and g, 9, 3, and 1 μg of citrate-glutamine RNA; lanes h, i, and j, 9, 3, and 1 μg of glucose-ammonia RNA; lanes k, l, and m, 9, 3, and 1 μg of glucose-glutamine RNA; lanes n, o, and p, 9, 3, and 1 μg of citrate-glucose-ammonia RNA; lanes q, r, and s, 9, 3, and 1 μg of citrate-glucose-glutamine RNA. Lanes marked M contained pBR322 DNA cut with *Hpa*II and end labeled.

protein. It is glutamine synthetase, the product of *glnA*, which appears to be a *trans*-acting repressor. When we cloned the *B. subtilis glnA* gene in *E. coli*, we observed not only that glutamine synthetase was active in the foreign host but that it was regulated by the nitrogen source as well (8; H. J. Schreier, S. H. Fisher, and A. L. Sonenshein, manuscript in preparation). That is, in media in which the nitrogen source was in excess the specific activity of glutamine synthetase was reduced 10-fold. Mutations in *E. coli* genes known to be involved in nitrogen regulation did not prevent this regulation (Schreier et al., in preparation). By fusing the promoter region of the *B. subtilis glnA* gene to *lacZ*, we were able to show that expression of β-galactosidase became dependent upon the nitrogen source as long as the *E. coli* cell contained intact copies of *B. subtilis glnA* coding region (Fig. 2 and Table 1). Deletions of the cloned *glnA* region that ended within the coding region greatly reduced or abolished the negative regulatory activity of the wild-type *glnA* region; mutations outside the coding region had no effect on regulation (Fig. 2 and Table 1). These results have led us to conclude that the protein coded for by the *glnA* gene is a negative regulator of *glnA* transcription. This is in keeping with the suggestion made originally by Dean et al. (3), who observed that a particular mutation in the *glnA* region led to overproduction of glutamine synthetase in *B. subtilis*. Whether glutamine synthetase acts as a regulator for any other genes remains to be determined, although the properties of some *glnA* mutant strains suggest that it does (9).

Phosphorus source regulation. The study of regulation by the phosphorus source in *Bacillus* has been limited to the alkaline phosphatase system. This enzyme exists in at least two states (intracellular and extracellular), but only one species of protein has been detected by size, activity, and antigenicity. Alkaline phosphatase can be induced in vegetative cells by starvation for phosphate and is normally produced during sporulation as well, no matter how sporulation is initiated. Again, the alkaline phosphatase proteins produced cannot be distinguished. Cloning

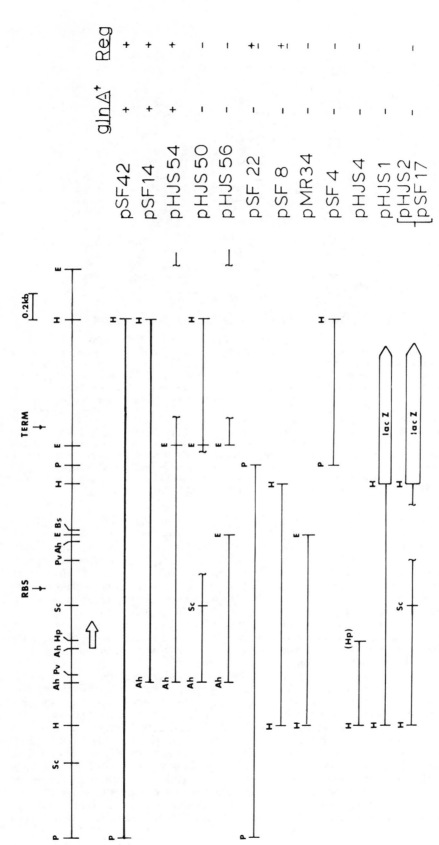

FIG. 2. Physical maps of plasmids containing DNA from the *glnA* region. The topmost line shows a restriction map of the region surrounding the *glnA* gene of *B. subtilis* (8). The notations RBS and TERM refer to the apparent start and stop points for synthesis of glutamine synthetase. The open arrow indicates the start point and direction of transcription (8). The lower lines show the *B. subtilis* DNA contained in various subclones of the *glnA* region. On the right-hand side of the figure are indicated the ability of each of these plasmids to complement a mutant of *E. coli* deleted for its own *glnA* gene and the ability of each plasmid to act in *trans* as a negative regulator of the *glnA* promoter.

190

of a region of *B. licheniformis* DNA that is able to express alkaline phosphatase activity in *E. coli* (15) has revealed the existence of two tandem transcription units, both of which code for seemingly indistinguishable alkaline phosphatase activities. Interestingly, one of the genes can be transcribed in vitro by the $E\sigma^{55}$ form of RNA polymerase, whereas the other gene is transcribed by the $E\sigma^{37}$ form (F. M. Hulett, this volume). Whether these genes correspond to the vegetative and sporulation forms of alkaline phosphatase remains to be determined, as does the nature of the regulation by phosphate.

In summary, the analysis of genes regulated by the carbon, nitrogen, or phosphorus source has progressed to the point where studies of regulation at the molecular level are possible. It remains to be seen whether the regulator of carbon source metabolism (possibly a protein that binds to pyruvate or 2-ketoglutarate or both) or of glutamine synthetase (possibly glutamine synthetase bound to a nitrogen metabolite) or of alkaline phosphatase is in any way related to GTP accumulation or regulation of sporulation.

MUTANTS THAT SPORULATE IN EXCESS NUTRIENTS

If one assumes that excess nutrients prevent sporulation by activating a repressor whose primary function is to block expression of a key initiator gene for sporulation, it should be possible to isolate mutants that sporulate despite the presence of excess nutrients. Such mutants have been found. In some cases, at least, their defect is a failure to metabolize a particular nutrient. For instance, a glutamine synthetase mutant of *B. megaterium* sporulated in excess ammonia, but not in the presence of glutamine, presumably because it was unable to synthesize glutamine. A second mutation allowed this strain to sporulate even in excess glutamine. The second lesion proved to be in the purine biosynthetic pathway (5), leading to the suggestion that the ultimate defect that turned on sporulation was failure to make enough GTP (10).

Mutants with deficiencies in pyruvate carboxylase or fumarase are unable to make enough oxalacetate to maintain a sufficient pool of aspartate and therefore show a partial stringent response (6). This is enough to cause a higher-than-normal frequency of spore formation in complete medium.

A third class of mutants has been studied by Takahashi. These mutants (said to have *crs* mutations) sporulate in the presence of 1% glucose. That is, 75% of a population of a *crs* mutant form spores in glucose minimal medium, whereas for wild-type cells only 1% do so (28). These mutants are also relatively insensitive to

TABLE 1. Regulation of expression from the *B. subtilis glnA* promoter in *E. coli*[a]

Strain	Glutamine synthetase sp act		β-Galactosidase sp act	
	Limiting N	Excess N	Limiting N	Excess N
YMC11	<1	<1	<1	<1
YMC11(pSF14)	100[b]	6	<1	<1
YMC11(pSF17)	<1	<1	100[b]	79
YMC11(pSF17)(pSF14)	100	7	94	14
YMC11(pSF17)(pHJS50)	<1	<1	209	100
YMC11(pSF17)(pSF22)	<1	<1	107	35
YMC11(pSF17)(pSF8)	<1	<1	100	49
YMC11(pSF17)(pMR34)	<1	<1	186	108

[a] *E. coli* strain YMC11 ($\Delta glnALG$ Δlac) was transformed with various plasmids, and the activities of glutamine synthetase and β-galactosidase were measured. Cells were grown in minimal glucose medium containing either excess nitrogen (0.2% ammonium chloride and 0.2% glutamine) or limiting nitrogen (0.2% glutamine).

[b] Specific activities of glutamine synthetase and β-galactosidase have been normalized to the values obtained for strains YMC11(pSF14) and YMC11(pSF17), respectively.

glucose repression of acetoin dehydrogenase (28). At least six *crs* loci are known, two of which correspond to RNA polymerase genes. One group, *crsE*, lies within the *rpoB-rpoC* region (27); the other, *crsA*, maps at *rpoD* (26), within the coding sequence for the major vegetative sigma factor (R. H. Doi et al., this volume). It is not known whether these mutations have a direct effect on expression of catabolite-sensitive genes or whether they cause the cell to metabolize glucose poorly, perhaps by reducing the expression of genes necessary for glucose uptake and utilization. The fact that the *crsA* mutants grow slowly with glucose as sole carbon source favors the latter possibility.

SPORULATION AS AN ADAPTIVE RESPONSE

Many species of bacteria (probably all) respond to environmental crises. These may be crises of nutritional imbalance, such as amino acid deprivation, which leads to the stringent response, or of survivability, such as radiation damage, which induces the SOS response. In all of these cases, the cell is geared to activate multiple genes and to deactivate those genes when the crisis has passed. It is unclear to what extent the mechanisms of crisis management are common within the same bacterium or among different species. The fact that the heat shock response in *E. coli* is mediated by a novel sigma factor (A. D. Grossman, J. W. Erickson, and C. A. Gross, Cell, in press) suggests that differential gene expression during sporulation (and phage infection) in *B. subtilis* (18) is mediated by a mechanism that has some generality. In fact, recent results of Chamberlin et al. (this volume) imply that the σ^{28} factor of *B. subtilis* is functionally related to the *E. coli* heat shock sigma factor.

Like other bacteria, *B. subtilis* responds to carbon, nitrogen, or phosphorus limitation, amino acid deprivation, damaging radiation, and heat shock. Its ability to adapt to nutritional alterations, however, is much more limited than is that of *E. coli*. Whereas the latter organism can utilize a very large number of carbohydrates and amino acids as carbon sources and a wide variety of nitrogen-containing compounds as sources of ammonia, *B. subtilis* lacks the genes to do many of these tasks. Instead, *B. subtilis* is geared to sporulate when rapidly metabolizable carbon and nitrogen sources run out. This lack of adaptability is primitive in that sporulation is antithetical to growth, but the process itself is highly sophisticated in the context of the bacterial world. That *B. subtilis* devotes so much of its genetic capacity to coding for sporulation-related functions testifies to the importance that evolution of this organism has attached to protective adaptation at the expense of proliferation. It also suggests that the mechanism that turns on sporulation in response to nutritional limitation will prove to differ significantly from the mechanisms that allow cells to switch from metabolism of favored carbon and nitrogen sources to metabolism of less-favored compounds.

I thank G. Chambliss, E. Freese, W. Haldenwang, F. M. Hulett, P. Stragier, and J. Szulmajster for providing information about their results prior to publication, E. Freese for helpful discussions, and H. J. Schreier for critically reviewing the manuscript.

Research in my laboratory reported here was supported by Public Health Service grant R01-GM19168 from the National Institute of General Medical Sciences.

LITERATURE CITED

1. Cook, W. R., V. F. Kalb, Jr., A. A. Peace, and R. W. Bernlohr. 1980. Is cyclic guanosine 3′,5′-monophosphate a cell cycle regulator? J. Bacteriol. 141:1450–1453.
2. Dawes, I. W., and J. Mandelstam. 1970. Sporulation of *Bacillus subtilis* in continuous culture. J. Bacteriol. 103:529–535.
3. Dean, D. R., J. A. Hoch, and A. I. Aronson. 1977. Alteration of the *Bacillus subtilis* glutamine synthetase results in overproduction of the enzyme. J. Bacteriol. 131:981–987.
4. Dunn, G., P. Jeffs, N. H. Mann, D. M. Torgersen, and M. Young. 1978. The relationship between DNA replication and the induction of sporulation in *Bacillus subtilis*. J. Gen. Microbiol. 108:189–195.
5. Elmerich, C., and J.-P. Aubert. 1975. Involvement of glutamine synthetase and the purine nucleotide pathway in repression of bacterial sporulation, p. 385–390. *In* P. Gerhardt, R. N. Costilow, and H. L. Sadoff (ed.), Spores

VI. American Society for Microbiology, Washington, D.C.
6. Endo, T., H. O. Ishihara, and E. Freese. 1983. Properties of a *Bacillus subtilis* mutant able to sporulate continually during growth in synthetic medium. J. Gen. Microbiol. 129:17–30.
7. Fisher, S. H., and B. Magasanik. 1984. 2-Ketoglutarate and the regulation of aconitase and histidase formation in *Bacillus subtilis*. 158:379–382.
8. Fisher, S. H., M. S. Rosenkrantz, and A. L. Sonenshein. 1984. Glutamine synthetase gene of *Bacillus subtilis*. Gene 32:427–438.
9. Fisher, S. H., and A. L. Sonenshein. 1984. *Bacillus subtilis* glutamine synthetase mutants pleiotropically altered in glucose catabolite repression. J. Bacteriol. 157:612–621.
10. Freese, E. 1981. Initiation of bacterial sporulation, p. 1–12. *In* H. Levinson, A. L. Sonenshein, and D. J. Tipper (ed.), Sporulation and germination. American Society for Microbiology, Washington, D.C.
11. Freese, E., and J. Heinze. 1984. Metabolic and genetic control of bacterial sporulation, p. 101–172. *In* A. Hurst, G. Gould, and J. Dring (ed.), The bacterial spore, vol. 2. Academic Press, London.
12. Freese, E., J. E. Heinze, and E. M. Galliers. 1979. Partial purine deprivation causes sporulation of *Bacillus subtilis* in the presence of excess ammonia, glucose and phosphate. J. Gen. Microbiol. 115:193–205.
13. Freese, E. B., N. Vasantha, and E. Freese. 1979. Induction of sporulation in developmental mutants of *Bacillus subtilis*. Mol. Gen. Genet. 170:67–74.
14. Hanson, R. S., and D. P. Cox. 1967. Effect of different nutritional conditions on the synthesis of tricarboxylic acid cycle enzymes. J. Bacteriol. 93:1777–1787.
15. Hulett, F. M. 1984. Cloning and characterization of the *Bacillus licheniformis* gene coding for alkaline phosphatase. J. Bacteriol. 158:978–982.
16. Lopez, J. M., A. Dromerick, and E. Freese. 1981. Response of guanosine 5′-triphosphate concentration to nutritional changes and its significance for *Bacillus subtilis* sporulation. J. Bacteriol. 146:605–613.
17. Lopez, J. M., B. Uratani-Wong, and E. Freese. 1980. Catabolite repression of enzyme synthesis does not prevent sporulation. J. Bacteriol. 141:1447–1449.
18. Losick, R., and J. Pero. 1981. Cascades of sigma factors. Cell 25:582–584.
19. Losick, R., and A. L. Sonenshein. 1969. Change in the template specificity of RNA polymerase during sporulation of *Bacillus subtilis*. Nature (London) 224:35–37.
20. Magasanik, B. 1982. Genetic control of nitrogen assimilation in bacteria. Annu. Rev. Genet. 16:135–168.
21. Ochi, K., J. C. Kandala, and E. Freese. 1981. Initiation of *Bacillus subtilis* sporulation by the stringent response to partial amino acid deprivation. J. Biol. Chem. 256:6866–6875.
22. Ochi, K., J. Kandala, and E. Freese. 1982. Evidence that *Bacillus subtilis* sporulation induced by the stringent response is caused by a decrease in GTP or GDP. J. Bacteriol. 151:1062–1065.
23. Schaeffer, P., J. Millet., and J.-P. Aubert. 1965. Catabolic repression of bacterial sporulation. Proc. Natl. Acad. Sci. U.S.A. 54:704–711.
24. Setlow, P. 1973. Inability to detect cyclic AMP in vegetative or sporulating cells or dormant spores on *Bacillus megaterium*. Biochem. Biophys. Res. Commun. 52:365–372.
25. Sonenshein, A. L., and K. Campbell. 1978. Control of gene expression during sporulation. p. 179–192. *In* G. Chambliss and J. C. Vary (ed.), Spores VII. American Society for Microbiology, Washington, D.C.
26. Sun, D., and I. Takahashi. 1982. Genetic mapping of catabolite-resistant mutants of *Bacillus subtilis*. Can. J. Microbiol. 28:1242–1251.
27. Sun, D., and I. Takahashi. 1984. A catabolite-resistance mutation is localized in the *rpo* operon of *Bacillus subtilis*.

Can. J. Microbiol. **30**:423–429.

28. **Takahashi, I.** 1979. Catabolite repression-resistant mutants of *Bacillus subtilis*. Can. J. Microbiol. **25**:1283–1287.

29. **Uratani-Wong, B., J. M. Lopez, and E. Freese.** 1981. Induction of citric acid cycle enzymes during initiation of sporulation by guanine nucleotide deprivation. J. Bacteriol. **146**:337–344.

30. **Vasantha, N., E. M. Galliers, and J. N. Hansen.** 1984. Effect of purine and pyrimidine limitations on RNA synthesis in *Bacillus subtilis*. J. Bacteriol. **158**:884–889.

31. **Zain-ul-Abedin, J. M. Lopez, and E. Freese.** 1983. Induction of bacterial differentiation by adenine- and adenosine-analogs and inhibitors of nucleic acid synthesis. Nucleosides Nucleotides **2**:257–274.

Metabolic Initiation of Spore Development

E. FREESE, E. B. FREESE, E. R. ALLEN, Z. OLEMPSKA-BEER, C. ORREGO,† A. VARMA, AND H. WABIKO‡

Laboratory of Molecular Biology, National Institute of Neurological and Communicative Disorders and Stroke, Bethesda, Maryland 20205

Many microorganisms and some cell types in higher organisms start to differentiate when an essential nutrient becomes scarce. Eventually, the differentiated cells of different organisms have quite different morphology. Nevertheless, we considered it likely that most organisms use the same intracellular molecule as a signal for the recognition of starvation because they would not have to reinvent the recognition mechanism during their individual evolution.

For simplicity's sake, we shall use the concept "sporulation of yeast" to mean, for all experiments reported here, the sequential process of meiosis and sporulation, resulting in the formation of tetrads of spores. In all yeast experiments, we have used the homothallic *Saccharomyces cerevisiae* strain Y55, which we obtained from H. O. Halvorson, or its derivatives, which we either obtained from J. E. Haber or constructed ourselves.

We had learned from work with *Bacillus subtilis* that the metabolic machinery had to work much more precisely for differentiation than for growth (3). There are so many possibilities to interfere with development that it was very difficult to learn much about the metabolic processes specifically needed for sporulation by using compounds or mutants that prevent sporulation. Therefore, we looked for different conditions under which sporulation could be initiated.

It was known that *S. cerevisiae* sporulated after adaptation to gluconeogenic growth and transfer to potassium acetate, which was called the sporulation medium (16). Addition of ammonia prevented this sporulation, which presumably results from nitrogen deprivation (2, 6). Freese et al. (6) showed that sporulation could be initiated in a growth medium simply by decreasing the concentration of the carbon, nitrogen, phosphorus, or sulfur source. However, sporulation occurred only when the compound whose concentration was reduced was inefficiently taken up; i.e., the K_m for its transport was high (which usually implies the absence of

active transport). Only in that case did the intracellular concentration of the compound increase in parallel with the extracellular concentration.

As a demonstration, this is shown here for carbon starvation in the presence of all other nutrients (Fig. 1 and 2). When limiting amounts of acetate or ethanol were used as the sole carbon source, growth continued until the compound had been completely used up and then stopped abruptly because the yeast cells could actively take up these compounds (Fig. 1). Consequently, no sporulation ensued, presumably because all metabolism stopped when the compound had been exhausted. In contrast, when cells were grown on carbon compounds which were inefficiently taken up, such as pyruvate or dihydroxyacetone, the rate of growth depended on the concentration of the compound. At an intermediate concentration of the compound the cells sporulated well (Fig. 2).

We had made the same observation, that partial but not complete starvation of certain nutrients enabled sporulation, also in *B. subtilis*. In retrospect, this is not surprising, because cells have to produce new RNA, protein, and other molecules during differentiation (4). There are some exceptions to this rule: certain microorganisms can differentiate in distilled water, in a phosphate solution, or in the presence of only a carbon source. However, we are convinced that all these exceptions are due to the fact that the nutrients not available in the medium have previously accumulated in the cells as such or in the form of storage compounds (allantoin, poly-β-hydroxybutyrate, polyphosphate) or can be made available by internal turnover of macromolecules or by de novo synthesis if auxotrophs are used. For this purpose, proteases, RNases, phosphatases, and enzymes degrading carbon-containing polymers are of obvious importance during differentiation. (Whether a particular degradative enzyme or other cell component is necessary for differentiation depends on the sporulation medium and conditions and on the ability of the cell to take up nutrients supplied in the medium.)

Attempting to determine whether sporulation of *S. cerevisiae* was controlled similarly to that of *B. subtilis*, our laboratory had shown that

---

† Present address: Biology Department, Tufts University, Medford, MA 02155.

‡Present address: Department of Biochemistry, University of Wyoming, Laramie, WY 82071.

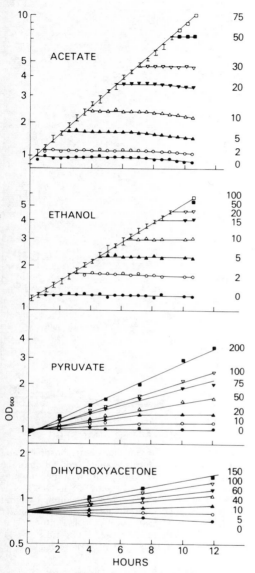

FIG. 1. Growth of *S. cerevisiae* on different carbon sources. Strain Y55 was grown from a low inoculum in synthetic medium (MN) supplemented with 100 mM potassium acetate or ethanol, 300 mM potassium pyruvate, or 150 mM dihydroxyacetone. At an optical density at 600 nm (OD$_{600}$) of about 1, the cells were washed twice on a membrane filter with MES buffer (pH 5.5) and then transferred to synthetic medium containing different concentrations of the same carbon source; the millimolar concentrations of the carbon sources are indicated at the right of the curves. [MES buffer = 2-(*N*-morpholino)ethanesulfonic acid adjusted to pH 5.5 with KOH. MN = 200 mM MES buffer plus 6.7 g of yeast nitrogen base without amino acids (Difco) per liter.] Data are from Freese et al. (6).

partial purine deprivation by amethopterin initiates yeast sporulation (5). However, because *S. cerevisiae* normally cannot convert extracellular thymine or thymidine to dTMP, the possibility that amethopterin caused a deprivation of dTMP could not be excluded. Subsequently, Freese et al. (7) used purine-requiring auxotrophs to prove that partial purine deprivation initiates yeast sporulation. Because adenine and hypoxanthine are actively transported, the authors supplied purines at the required slow rate by the addition of the purine intermediate 5-amino-4-

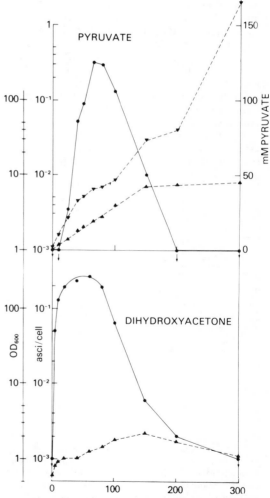

FIG. 2. Sporulation of *S. cerevisiae* on different carbon sources. Y55 cells were grown and transferred as described for Fig. 1. At 20 h after cell transfer, the OD$_{600}$ (▲), the frequency of asci per cell (●), and the remaining carbon source (▼) were measured. Data are from Freese et al. (6).

imidazolecarboxamide. Since partial purine deprivation generally decreases the synthesis of guanine nucleotides more than that of adenine nucleotides, it appeared likely that the observed sporulation resulted from the deprivation of guanine nucleotides. This was demonstrated by the use of a guanine auxotroph which could grow at different rates when different guanine concentrations were used. The mutant was leaky because it grew somewhat even without guanine in the medium. Figure 3 shows that guanine deprivation of this mutant enabled sporulation. The more guanine present, the better the cells grew and the less sporulation was obtained. Sporulation could also be induced by high concentrations of Virazole (ribavirin), which specifically inhibits IMP dehydrogenase. In contrast, partial starvation of uracil mutants for uracil or uridine did not induce yeast sporulation. Measurements of nucleotide pools verified that after guanine starvation GTP dropped rapidly, whereas ATP and CTP did not change much and UTP increased (Fig. 4). In contrast, after uracil starvation, the UTP pool decreased whereas the GTP pool increased.

Apart from its effect on polymer synthesis, GTP deficiency also caused a deficiency in methionine and its metabolic product S-ad-

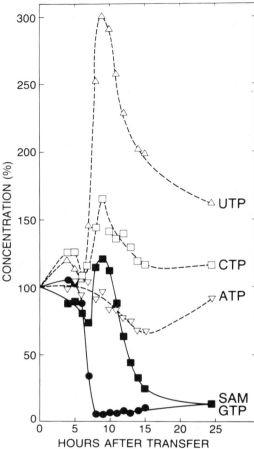

FIG. 4. Changes in nucleoside triphosphate and SAM concentrations during guanine deprivation of the *gua-1* mutant. Cells were grown in MNA medium plus 150 μM guanine (see Fig. 3 legend) to an OD_{600} of 1, washed, and suspended in MNA medium (without guanine). The concentrations just before cell transfer were (picomoles/OD_{600}): ATP, 738.4 $\pm$ 10.4; GTP, 343.3 $\pm$ 7.1; CTP, 122.3 $\pm$ 2.3; UTP, 121.9 $\pm$ 1.4; SAM, 218.3 $\pm$ 7.8. Data are from Freese et al. (7).

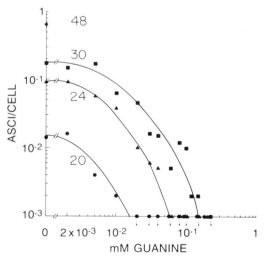

FIG. 3. Sporulation of the *gua-1* auxotroph after guanine deprivation. The cells were grown in synthetic medium (MN) plus 100 mM potassium acetate, pH 5.5 (MNA), supplemented with 100 μM guanine. When the OD_{600} was 1, the cells were washed and suspended in MNA, and 10-ml samples were transferred to flasks containing different amounts of guanine. Sporulation was measured after 20 (●), 24 (▲), 30 (■), and 48 (♦) h (only the sample without guanine was evaluated after 48 h); 100% of the asci were tetrads. Data are from Freese et al. (7).

enosylmethionine (SAM), for GTP is a precursor of folic acid which in turn is used to methylate the methionine precursor, homocysteine. Conceivably, sporulation could therefore result from deprivation of methionine or SAM rather than of GTP. This appeared even more likely when it was shown that partial deprivation of sulfate (7) or of methionine in *met* auxotrophs (A. Varma and E. B. Freese, unpublished data) also caused some sporulation. This sporulation induction was observed only for methionine and not for amino acid starvation of other auxotrophs. Under all these conditions of sulfur deprivation, the intracellular concentration of SAM decreased as expected, but the concentration of other nucleo-

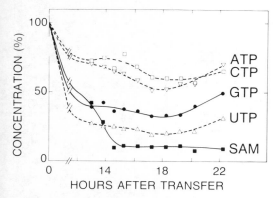

FIG. 5. Effect of sulfur deprivation on the intracellular concentration of nucleoside triphosphates and SAM. Y55 cells were grown in MNA medium (see Fig. 3 legend) to an OD_{600} of 0.5, diluted 10-fold in MNA medium, and grown to an OD_{600} of 0.2. They were collected on a membrane filter, washed twice with MES buffer, and suspended in the sulfate-free MNA medium. Just before cell transfer to sulfate-free medium, two samples were taken, and at different times thereafter, one sample was taken. The averages for the zero time values were (picomoles/OD_{600}): ATP, 826.8 $\pm$ 16.6; GTP, 189.9 $\pm$ 5.4; CTP, 157.3 $\pm$ 1.2; UTP, 214.2 $\pm$ 8.3; SAM, 402.9 $\pm$ 13.2. Data are from Freese et al. (7).

tides, in particular that of GTP, also decreased (Fig. 5) (7). It was not easy to resolve whether sporulation was initiated by the decrease of methionine or SAM or by that of GTP; guanine addition could not prevent the GTP decrease because SAM deficiency prevented guanine uptake. Conversely, addition of methionine to a culture of the *gua* auxotroph produced high intracellular concentrations of methionine and SAM, which inhibited sporulation. To solve this puzzle, we isolated a double mutant that was auxotrophic for guanine and lacked the activity of one of the two SAM synthetases. When this mutant was starved for guanine and at the same time supplied with 0.1 mM methionine, the GTP concentration decreased as before, but both the methionyl-tRNA and the SAM concentrations remained essentially constant or increased slightly (Fig. 6). Nevertheless, these cells still sporulated well (insert in Fig. 6). Thus, we conclude that the specific decrease of GTP (or GDP) alone suffices to induce yeast sporulation. The relatively weak sporulation effect caused by methionine deficiency can be explained by the resulting decrease of GTP.

In addition to the causal relationship between the decrease of guanine nucleotides and sporulation, we have found that the concentration of GTP decreased also under all other sporulation conditions which we have examined, including the sporulation caused by carbon and nitrogen deprivation. The concentration of the other nucleoside triphosphates increased in some cases and decreased in others.

The data can be summarized by the scheme in Fig. 7. A partial deprivation of any of the boxed-in compounds causes a decrease in GTP

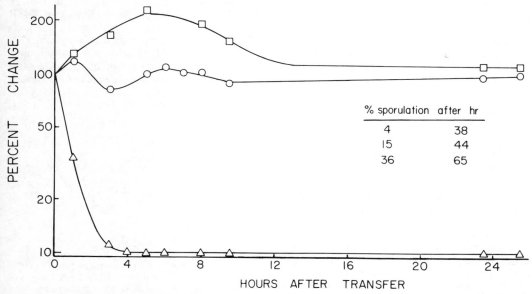

FIG. 6. Changes of GTP ($\triangle$), SAM ($\square$), and methionyl-tRNA ($\bigcirc$) after guanine removal from an *eth-10 gua* auxotroph. Cells were grown to an OD_{600} of 1 in MNA medium (see Fig. 3 legend) containing 200 μM guanine and 100 μM methionine, washed, and transferred to MNA + 100 μM methionine.

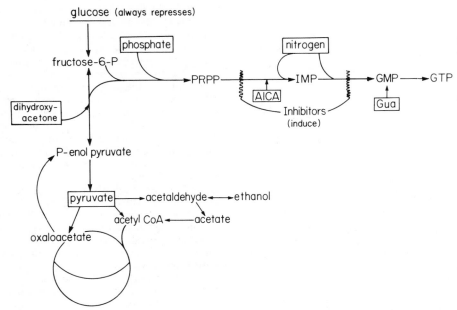

FIG. 7. Scheme of the biochemical control of the initiation of meiosis and sporulation in *S. cerevisiae*. Partial deprivation of boxed-in compounds or partial inhibition of the purine path causes a decrease in GTP and thereby initiates meiosis and ensuing sporulation.

(and in the growth rate) and thereby initiates sporulation. No matter what medium is used, the presence of glucose always prevents (re-presses) sporulation (7). If it were possible to generate a partial intracellular deprivation of acetate or ethanol, this should also cause sporu-lation. The effect of sulfur and methionine dep-rivation is not included in the graph because it is not known which biochemical reaction is inter-fered with when these compounds are deficient. (Matsumoto et al. have found that temperature-sensitive cyclic AMP-deficient mutants [*cyr1-2*] can sporulate to some extent after several days of growth in a medium that initially contained glucose, whereas the parent strain does not [11]. We propose that this sporulation is due to partial carbon deprivation which occurs when the cells attempt to use the ethanol and acetate produced from glucose as the carbon source. The cyclic AMP-deficient mutants grow extremely slowly on these two carbon compounds, presumably because the gluconeogenic path requires cyclic AMP [i.e., protein phosphorylation] for maximal activity.)

The initiation of *B. subtilis* sporulation has been discussed in detail previously (3). All evi-dence indicates that again GTP is the compound controlling massive sporulation. The evidence can be summarized by the scheme in Fig. 8. When the concentration of any boxed-in com-pound is limiting so that the growth rate is reduced (if necessary in mutants), the concen-tration of GTP decreases and massive sporula-tion takes place. An important difference be-tween *S. cerevisiae* and *B. subtilis* is that the latter can sporulate even in the presence of glucose or any other carbon source if the con-centration of GTP is somehow decreased. It is easy to obtain this GTP decrease in *B. subtilis* by the addition of specific inhibitors of GMP synthesis such as mycophenolic acid (5), which inhibits IMP dehydrogenase, or decoyinine (12) or psicofuranine (21), which specifically inhibit GMP synthetase. These compounds have little or no effect in *S. cerevisiae* because they cannot enter yeast cells. Sporulation can also be initi-ated by partial deprivation of one or more amino

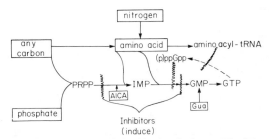

FIG. 8. Scheme of the biochemical control of the initiation of sporulation in *B. subtilis*. Partial depriva-tion of boxed-in compounds or partial inhibition of the purine path by external inhibitors or via internally produced (p)ppGpp decreases GTP and thereby initi-ates asymmetric septation and sporulation.

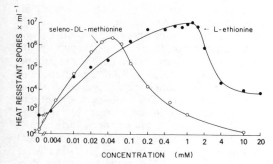

FIG. 9. Increase of sporulation caused by amino acid analogs. The relaxed strain 61885 (*ethA1 ilvBΔ1 kauA relA1*) was inoculated at a low OD_{600} (0.001) into S7 medium (14) containing 100 mM D-glucose and 1 mM each of isoleucine and valine, and the culture was shaken at 37°C. When the OD_{600} reached 0.5, 10-ml samples were placed into 125-ml flasks containing various amounts of L-ethionine or seleno-DL-methionine. After 13 h of further incubation, the titers of heat-resistant (75°C for 20 min) spores were determined. Data are from Ochi and Freese (14).

acids; this results in the production of uncharged tRNA, which in turn causes the formation of pppGpp and ppGpp [(p)ppGpp]. This "stringent response" is also the reason cells sporulate after cell transfer from casein hydrolysate to glutamate medium, i.e., the well known Sterlini-Mandelstam (17) conditions (10). (p)ppGpp inhibits IMP dehydrogenase and thereby decreases GMP synthesis. A mycophenolate-resistant mutant still produces (p)ppGpp but does not produce a sufficient GTP decrease and does not sporulate during amino acid deprivation (15). However, the addition of decoyinine, which inhibits the synthesis of guanine nucleotides independently of the stringent response, again initiates sporulation. Relaxed (*relA*) mutants, which are unable to produce (p)ppGpp, do not sporulate under conditions of amino acid deprivation, but they sporulate well after the addition of decoyinine or mycophenolic acid.

(Certain antibiotics which interfere with protein or RNA synthesis can abolish sporulation caused by the stringent response at concentrations at which they inhibit growth by much less than 50% [14]. Under these conditions, the formation of (p)ppGpp is greatly reduced, presumably because amino acyl-tRNA gets uncharged at a reduced rate. Again, sporulation can be restored by the addition of decoyinine.)

In our laboratory some mutants have been isolated which sporulate spontaneously throughout growth at a higher frequency than the standard strain in a medium containing all required nutrients. The mutants, which we have called *spd*, in analogy to an *spd* mutation of Dawes in

S. cerevisiae (19) can be subdivided into two types. One *spd* type causes an increased sporulation frequency only in stringent but not in relaxed strains; Endo et al. (1) have shown for one such mutation that it causes a slight deficiency of aspartic acid, which suffices to produce a slight stringent response although it does not affect growth. The other type of *spd* mutation functions also in relaxed (*relA*) strains; its biochemical property has not been identified. The continual sporulation of such an *spd* mutant could be avoided by the addition of methionine or SAM, which suggests that it may depend on undermethylation (H. Wabiko, K. Ochi, D. M. Nguyen, and E. Freese, manuscript in preparation).

To investigate this latter effect further, sporulation was measured in the presence of methionine analogs; good sporulation was observed with intermediate concentrations of ethionine and seleno-methionine (Fig. 9). However, we later observed that the MIC of DL-ethionine for this strain (2 mM) was higher than that for the standard strain (60001; 0.1 mM), whose sporulation could not be induced by any ethionine concentration. We call the responsible mutation *ethA1*. It has pleiotropic properties which will be described elsewhere. In contrast to starvation of guanine nucleotides, which causes massive sporulation 7 h later, ethionine addition enables continual sporulation during growth; i.e., it increases the probability with which cells enter sporulation during each cell cycle. This is shown in Fig. 10 for a culture whose optical density was maintained between 0.25 and 0.5 by dilution with decoyinine- or ethionine-contain-

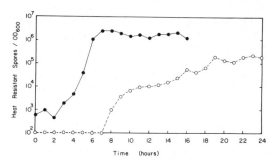

FIG. 10. Effect on sporulation of decoyinine or ethionine addition to cells maintained at constant turbidity. Cells were grown in synthetic medium (S7) (14) plus 2% glucose (1 mM each of isoleucine and valine were added to strain 61885; L-tryptophan [100 μg/ml] was added to strain 62381). When the OD_{600} was 0.5, decoyinine (400 μg/ml) was added to the culture of strain 61885 (*ilvBΔ1 kauA relA1 ethA1*) (●) or ethionine (4 mM) was added to the culture of strain 62381 (*ethA1 trpC2*) (○). The cultures were maintained at an OD_{600} between 0.25 and 0.5 by periodic dilution with the same supplemented media.

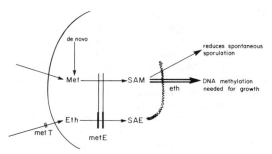

FIG. 11. Biochemical consequences of the *eth* and *met* mutations. Met, L-Methionine; Eth, L-ethionine; *metT* causes absence of ethionine transport; *metE* causes deficiency of SAM synthetase. The double line indicates an increase in DNA methylase activity due to the *eth* mutation. The wavy line indicates inhibition of the pathways.

ing medium, and the spore titer was measured at different times after the initial exposure to decoyinine or ethionine.

We wanted to know whether ethionine induces sporulation by itself, after its incorporation into proteins, or after its conversion to *S*-adenosylethionine (SAE). For this purpose mutants highly resistant to ethionine (10 mM) were isolated, and their biochemical change was determined (13; Wabiko et al., in preparation). The results are summarized in Fig. 11. Some of these mutants (genotype *metT*) are deficient in ethionine transport but still transport methionine actively. Other mutants (genotype *metE*) produce only about 2% of the normal SAM synthetase activity. This activity suffices to maintain an intracellular SAM concentration of about 40% of the normal amount (in *metE1* mutants; Table 1) and to enable normal growth and sporulation. In the presence of ethionine (2 mM), the *metE1* and *eth* mutants produce more SAM but less SAE than the standard strain (60001; Table 1); this shows that the growth inhibition of the *eth* as well as the *eth*⁺ strains is not caused

by ethionine itself but by SAE, probably in competition with SAM.

Using restriction of phage φ105 by the *Bsu*R restriction system (8), we have shown that the *eth* mutation enables, in the absence of ethionine, higher methylation of GGCC sequences in DNA than that found in the parental strain (Table 2). The higher DNA modification activity was observed even in the absence of prophage SPβ. (This temperate phage codes for a *Bsu*R-specific methyltransferase which is expressed upon induction of the prophage [18].) Presumably, SAE inhibits DNA methylation and thus growth; as a consequence of an *eth* mutation, the cells become partially resistant to ethionine by increasing the concentration of the specific methylase revealed in our studies (indicated in Fig. 11 by the double line from SAM to methylated DNA). In the *eth* mutant, SAE can then cause sporulation without excessively interfering with normal cellular metabolism. This ability of SAE to increase the spontaneous sporulation frequency is probably due to the inhibition of some other methylation reaction (Fig. 11); in this case we do not know whether the methylation affects DNA or some other molecule. However, we do know that the induction of sporulation is due to the formation of SAE because an *ethA1 metE* double mutant, in which the SAE/SAM ratio is fivefold lower (Table 1), no longer shows increased sporulation at any ethionine concentration (Fig. 12).

We asked ourselves how it is possible to reconcile the induction of massive sporulation by GTP deprivation with the increased frequency of continual sporulation caused by SAE or found in certain *spd* mutants. For this purpose, we propose a model which summarizes the data and can be used to initiate further experiments (Fig. 13). Remember that we have no direct evidence for this model! The model proposes that a protein attaches to a controlling region of DNA which is normally methylated. The configuration of the protein is directly or

TABLE 1. Intracellular concentrations (in picomoles/OD₆₀₀) of SAM and SAE in *B. subtilis*[a]

Strain no.	Relevant genotype		Concn of SAM without ethionine	Concn with 2 mM DL-ethionine	
	eth	*metE1*		SAM	SAE
60001	+	+	44.9	2.3	176.0
62381	−	+	56.0	37.6	51.5
62378	+	−	18.7	16.1	10.0
62496	−	−	27.3	27.0	6.7

[a] Cells were grown at 37°C in synthetic medium (S7 + 2% glucose + 100 μg of tryptophan per ml) to an OD₆₀₀ of 0.5. To half of the cultures DL-ethionine was added (2 mM final concentration), and the cultures were further incubated for 2 h. Samples (25-ml) were collected on membrane filters (10-cm diameter, 0.45-μm pore size, Schleicher & Schuell Co.), and the cells were then extracted with 1.5 ml of 0.5 M ice-cold formic acid. SAM and SAE were measured by high-pressure liquid chromatography.

TABLE 2. Protection against R restriction of phage φ105c30 propagated in an *eth* mutant of *B. subtilis*[a]

Propagation in strain:	Genotype	PFU/ml on:		Restriction ratio	Protection ratio
		r_R^-	r_R^+		
60001	*trpC2*	2.7×10^9	9.6×10^4	3.6×10^{-5}	
62381	*trpC2 ethA1*	5.7×10^9	8.6×10^6	1.5×10^{-3}	42

[a] Bacterial strains were grown in M medium (20) with 0.1% glucose plus calcium, magnesium, and manganese. The culture (1×10^8 to 2×10^8 cells per ml) was infected with phage (multiplicity of infection, ca. 0.1). Lysis was complete in 3 to 4 h. Phage lysates were titered on M plates with a pair of isogenic indicator strains, one member (r_R^+) containing the *BsuR* restriction/modification system (9).

indirectly controlled by the concentration of GTP so that when the GTP concentration decreases the protein tends to dissociate from the DNA, enabling massive sporulation. The continual sporulation caused by ethionine addition would then result either from DNA ethylation instead of methylation or, more likely, from a lack of methylation due to interference of SAE with a DNA methylase. These altered methyl groups would not prevent binding of the protein to DNA but only decrease its affinity. At a critical time in each cell cycle, the cell would then have an increased probability with which it would enter asymmetric septation and sporula-

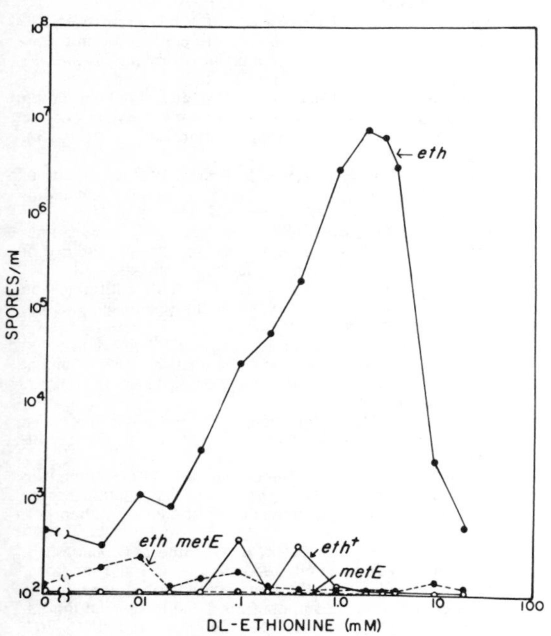

FIG. 12. Effect of ethionine on sporulation. Cells were grown in synthetic medium (S7) (14) containing 2% glucose and 100 µg of L-tryptophan per ml. When the OD$_{600}$ was 0.5, 5-ml portions were added to 125-ml flasks containing various concentrations of DL-ethionine. Heat-resistant spore titers were measured after 13 h of further incubation. Strains used were: 60001 *eth+ trpC2* (O—O), 62381 *ethA1 trpC2* (●—●), 62378 *metE1 trpC2* (O---O), 62496 *ethA1 metE1 trpC2* (●---●).

tion rather than symmetric septation and cell division. For *S. cerevisiae* there is so far no evidence that methylation plays any role in the control of sporulation, but the effect of GTP deficiency could be the same as that in *B. subtilis*.

The protein controlling sporulation could be a repressor, and the GTP-controlled portion could be a co-repressor. It is not likely that GTP itself would be the co-repressor because a decrease of the intracellular concentration of GTP from 600 to 200 µM induces sporulation in *B. subtilis*, and the concentrations are of similar magnitude in *S. cerevisiae*. These concentrations are much too high for a strongly binding co-repressor. Rather, they are in the range where GTP can affect the rate of RNA synthesis or act as an inhibitor, e.g., of an ATP-dependent reaction. Conceivably, the protein could be RNA polymerase itself. This enzyme could normally have a high affinity to a DNA location enabling cell division but upon the decrease of GTP could change its preferential attachment to another promoter location so that only one of the two promoters can be expressed at a time. The next challenge is to identify the particular protein whose attachment to DNA responds to the GTP concentration and to identify the DNA site where its attachment controls sporulation.

SUMMARY

The sporulation of *B. subtilis* and the meiosis and ensuing sporulation of *S. cerevisiae* can be initiated by partial deprivation of various compounds. All these conditions lead to a decrease of GTP (and GDP), whereas other nucleotides as well as SAM increase under some and decrease under other starvation conditions. The specific deprivation of guanine nucleotides either in *gua* auxotrophs or caused by specific inhibitors also induces sporulation. In *B. subtilis*, the stringent response (production of ppGpp) causes sporulation also because it produces a decrease of guanine nucleotides. In this organism, the methionine analog ethionine continually induces sporulation of a small fraction of cells. This effect is observed only in *eth* mutants that are partially resistant to ethionine and show increased methylation of GGCC sites in DNA; it

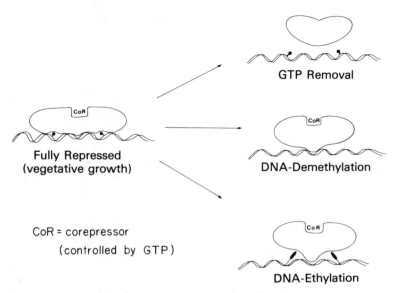

FIG. 13. Model of the control of differentiation by GTP and by methylation of DNA.

results from the production of SAE, which presumably inhibits methylation. A model is presented to summarize the data.

With this paper we would like to honor the late Elisabeth Bautz Freese, who spent her last years in intensive work attempting to determine the similarities between the initiation of sporulation of *B. subtilis* and the initiation of meiosis and the ensuing sporulation of *S. cerevisiae*. It is one consolation for us that she succeeded.

LITERATURE CITED

1. **Endo, T., H. Ishihara, and E. Freese.** 1983. Properties of a *Bacillus subtilis* mutant able to sporulate continually during growth in synthetic medium. J. Gen. Microbiol. **129:**17–30.
2. **Fowell, R. R.** 1969. Sporulation and hybridization of yeasts, p. 303–386. *In* A. H. Rose and J. S. Harrison (ed.), The yeasts, vol. 1. Academic Press, Inc., New York.
3. **Freese, E.** 1981. Initiation of bacterial sporulation, p. 1–12. *In* H. S. Levinson, A. L. Sonenshein, and D. J. Tipper (ed.), Sporulation and germination. American Society for Microbiology, Washington, D.C.
4. **Freese, E., and J. Heinze.** 1984. Metabolic and genetic control of bacterial sporulation, p. 101–173. *In* A. Hurst, G. Gould, and J. Dring (ed.), The bacterial spore, vol. 2. Academic Press, London.
5. **Freese, E., J. M. Lopez, and E. B. Freese.** 1979. Initiation of bacterial and yeast sporulation by partial deprivation of guanine nucleotides, p. 127–143. *In* D. Richter and G. Loch (ed.), Regulation of macromolecular synthesis by low molecular weight mediators. Academic Press, Inc., New York.
6. **Freese, E. B., M. I. Chu, and E. Freese.** 1982. Initiation of yeast sporulation by partial carbon, nitrogen, or phosphate deprivation. J. Bacteriol. **149:**840–851.
7. **Freese, E. B., Z. Olempska-Beer, A. Hartig, and E. Freese.** 1984. Initiation of meiosis and sporulation of *Saccharomyces cerevisiae* by sulfur or guanine deprivation. Dev. Biol. **102:**438–451.
8. **Guenthert, U., K. Storm, and R. Bald.** 1978. Restriction and modification in *Bacillus subtilis*. Localization of the methylated nucleotide in the *Bsu*RI recognition sequence. Eur. J. Biochem. **90:**581–583.
9. **Ikawa, S., T. Shibata, and T. Ando.** 1980. Genetic studies on endodeoxyribonucleases in *Bacillus subtilis*: modification and restriction systems in transformants of *Bacillus subtilis* 168. Mol. Gen. Genet. **177:**359–368.
10. **Lopez, J. M., A. Dromerick, and E. Freese.** 1981. Response of guanosine-5'-triphosphate concentration to nutritional changes and its significance for *Bacillus subtilis* sporulation. J. Bacteriol. **146:**605–613.
11. **Matsumoto, K., I. Uno, and T. Ishikawa.** 1983. Initiation of meiosis in yeast mutants defective in adenylate cyclase and cyclic AMP-dependent protein kinase. Cell **32:**417–423.
12. **Mitani, T., J. E. Heinze, and E. Freese.** 1977. Induction of sporulation in *Bacillus subtilis* by decoyinine or hadacidin. Biochem. Biophys. Res. Commun. **77:**1118–1125.
13. **Ochi, K., and E. Freese.** 1982. A decrease in *S*-adenosylmethionine synthetase activity increases the probability of spontaneous sporulation. J. Bacteriol. **152:**400–410.
14. **Ochi, K., and E. Freese.** 1983. Effect of antibiotics on sporulation caused by the stringent response. J. Gen. Microbiol. **129:**3709–3720.
15. **Ochi, K., J. Kandala, and E. Freese.** 1982. Evidence that *Bacillus subtilis* sporulation induced by the stringent response is caused by the decrease in GTP or GDP. J. Bacteriol. **151:**1062–1065.
16. **Roth, R., and H. O. Halvorson.** 1969. Sporulation of yeast harvested during logarithmic growth. J. Bacteriol. **98:**831–832.
17. **Sterlini, J. M., and J. Mandelstam.** 1969. Commitment to sporulation in *Bacillus subtilis* and its relationship to development of actinomycin resistance. Biochem. J. **113:**29–37.
18. **Trautner, T. A., B. Pawlek, U. Guenthert, U. Canosi, J. Jentsch, and M. Freund.** 1980. Restriction and modification in *Bacillus subtilis*: identification of a gene in the temperate phage SPβ coding for a *Bsu*R specific modification methyltransferase. Mol. Gen. Genet. **180:**361–367.
19. **Vezinhet, F., J. H. Kinnaird, and J. W. Dawes.** 1979. The physiology of mutants derepressed for sporulation in *Saccharomyces cerevisiae*. J. Gen. Microbiol. **115:**391–402.
20. **Yehle, C. O., and R. H. Doi.** 1967. Differential expression of bacteriophage genomes in vegetative and sporulating cells of *Bacillus subtilis*. J. Virol. **1:**935–947.
21. **Zain-ul-Abedin, J. Lopez, and E. Freese.** 1983. Induction of sporulation in *Bacillus subtilis* by adenosine analogs and inhibitors of nucleic acid synthesis. Nucleosides Nucleotides **2:**257–274.

Organization and Cloning of a Gluconate (*gnt*) Operon of *Bacillus subtilis*

YASUTARO FUJITA, JUN-ICHI NIHASHI, AND TAMIE FUJITA

Department of Biochemistry, Hamamatsu University School of Medicine, 3600 Handa-cho, Hamamatsu, Japan 431-31

In the genus *Bacillus*, catabolite repression is commonly observed not only in enzyme synthesis but also in the onset of sporulation (3). Several research groups (9, 11, 13) have implied the existence of a common mechanism regulating gene expression in those systems of *Bacillus subtilis* under catabolite repression. However, catabolite repression in this genus cannot be explained by a mechanism involving a cyclic AMP receptor protein-cyclic AMP complex that has been well documented in enteric bacteria (12), because *Bacillus* species contain neither detectable cyclic AMP nor adenyl cyclase (7, 14). Consequently, catabolite repression in the genus *Bacillus* is one of the unsolved systems for regulation of gene expression. Our efforts to elucidate the mechanism of catabolite repression have focused on the investigation of the gluconate utilization system of *B. subtilis*.

After entering the *B. subtilis* cell, gluconate is phosphorylated to gluconate-6-phosphate by gluconate kinase. Gluconate-6-phosphate is metabolized through the pentose cycle because *B. subtilis* lacks the Entner-Doudoroff pathway. Therefore, only two enzymes may be needed for utilization of gluconate as a carbon source; one is gluconate permease, which transports gluconate into the cell, and the other is gluconate kinase. Both enzymes are induced in response to gluconate, and their inductions are repressed in the presence of rapidly metabolizable carbohydrates such as glucose (2, 11). We previously reported that carbohydrates must be metabolized to sugar phosphates in the upper part of the Embden-Meyerhof pathway to exert their repressive effects on gluconate kinase induction (11) and that the genes for gluconate utilization are cloned in a temperate phage, ρ11 (6). In this paper, we report the genetic mapping and characterization of the *gnt* mutations that render the cell unable to grow on gluconate, and we also report that an *Eco*RI fragment cloned in ρ11 was recloned in another temperate phage, φ105, in which an intact gluconate operon was encoded. Moreover, the genetic map of the *gnt* mutations was correlated with their physical locations in restriction enzyme fragments generated from the fragment cloned in φ105.

EXPERIMENTAL PROCEDURES

Strain 60015 (*trpC2 metC7*) is our standard strain. Among Gnt⁻ mutants, isolation of strains 61656 (Δ*igf hisA1 leuA8 metB5 trpC2*), YF127 (*gnt-4 trpC2 metC7*), and YF029 (*gnt-9 purA16 leuA8 metB5 hisA3*) has been reported (4, 5). Strain YF160 (*gnt-10 trpC2 metC7*) was isolated from strain 60015 that had been treated with ethyl methanesulfonate, by the method previously described (5). Strains YF161 (*gnt-23 trpC2 metC7*) and YF162 (*gnt-26 trpC2 metC7*) were isolated as follows. After strain 60015 was treated with ethyl methanesulfonate, cells were grown in a liquid minimal medium (S6) containing glucose as sole carbon source and were plated on a solid minimal medium (N) containing gluconate (0.1%) and glucose (0.01%) as carbon source. Among colonies which appeared, 1,500 tiny colonies were isolated and tested for growth on glucose and gluconate. Only two colonies showed the Gnt⁻ phenotype, and these were called YF161 and YF162.

PBS1-mediated transduction and DNA-mediated transformation were performed as described previously (5). Preparation of phage φ105 and its DNA and φ105-mediated specialized transduction were carried out as described by Iijima et al. (8).

RESULTS

Mapping and characterization of *gnt* mutations. Mapping of the *gnt-4* mutation by PBS1-mediated transduction has been reported (5). The map order from *sacA* to *purA* is as follows: *sacA*, *thiC*, *hsrE*, *iol-6*, *gnt-4*, *fdp-74*, *hsrB*, *ts199*, and *purA* (Fig. 1). The Δ*igf* deletion covers loci from *iol-6* to *hsrB*. The *gnt-9* mutation was also located inside the Δ*igf* deletion because the DNA of strain 61656 (Δ*igf*) could not transform strain YF029 (*gnt-9*) to the wild type. PBS1 transduction crosses revealed a map order of *iol-6*, *gnt-9*, and *fdp-74*. When strain YF029 (*gnt-9 purA16*) was tranduced with PBS1 propagated in strain YF127 (*gnt-4*) and Pur⁺ recombinants were selected, two Gnt⁺ colonies were found among 161 Pur⁺ recombinants. These results suggested that *gnt-9* is very close to *gnt-4* and located between *gnt-4* and *fdp-74*. All of the *gnt* mutations of our collection, including *gnt-10*,

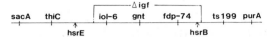

FIG. 1. Chromosomal localization of *gnt* mutations of *B. subtilis* by PBS1-mediated transduction. The details of this study were reported by Fujita and Fujita (5).

gnt-23, and *gnt-26* that had been newly isolated, were finally mapped by two-factor transformation crosses between Gnt⁻ strains (Fig. 2). The crosses revealed a map order of *gnt-10*, *gnt-4*, *gnt-26*, *gnt-23*, and *gnt-9*.

Since the enzymes involved in gluconate metabolism are the gluconate kinase and permease, the inducibility of these two enzymes in the Gnt⁻ mutants was investigated (Table 1). As expected, strain 61656 (Δ*igf*) could induce neither the kinase nor the permease because the Δ*igf* mutation is a large deletion covering the *gnt* locus. Strain YF029 (*gnt-9*) also could not induce the two enzymes. As the *gnt-9* mutation was thought to be a point mutation, this strain seemed to be impaired in a function necessary for induction of both enzymes. Strain YF160 (*gnt-10*) had abnormal induction only of the kinase, whereas strains YF161 (*gnt-23*) and YF162 (*gnt-26*) were impaired only in the induction of the permease. The *gnt-10* mutation seemed to be located within the structural gene for the kinase and the *gnt-23* and *gnt-26* mutations were likely to be within that for the permease. Strain YF127 (*gnt-4*) could not induce the kinase at all, but unexpectedly this strain constitutively synthesized the permease.

Cloning of the *gnt*⁺ fragment in a temperate phage, φ105. An *Eco*RI fragment from *B. subtilis* Marburg 168 (*trpC2*) DNA containing the genes for gluconate utilization has been cloned in a temperate phage, ρ11, by prophage transformation with strain 61656 as recipient (6). However, it was difficult to isolate the *gnt*⁺ fragment by agarose-gel electrophoresis because the *Eco*RI digestion of ρ11 DNA produces more than 25 fragments (10). Hence, the *gnt*⁺ fragment was recloned in another temperate phage, φ105, which is relatively small and is convenient for isolation of this fragment (8). The DNA (5 μg) of ρ11*hisA*⁺*gnt*⁺E (a constructed ρ11 possessing the *gnt*⁺ fragment) and the DNA (10 μg) of φ105 were each completely digested with *Eco*RI. Each digest was mixed and ligated with T4 DNA ligase. The ligated DNA was used to transform strain 61656 (Δ*igf* and lysogenic with φ105) to Gnt⁺. All of the 250 Gnt⁺ transformants obtained were considered to result from the integration of the *gnt*⁺ fragment in the φ105 prophage genome, because of the absence of homology between this fragment and the chro-

mosome of the Δ*igf* strain. Five transformants were treated with mitomycin C and at the same time infected with the wild-type φ105. All lysates obtained exhibited the ability to transduce strain 61656 to Gnt⁺. One of the Gnt⁺ transductants was used for further experiments after single-colony isolation. This transductant could induce a *gnt*⁺ specialized transducing φ105 (φ105*gnt*⁺) without helper, but the φ105*gnt*⁺ induced was unable to transduce strain 61656 to Gnt+ without helper. Phage φ105*gnt*⁺ could transduce not only the Δ*igf* mutant but also all the *gnt* mutants (*gnt-4*, *gnt-9*, *gnt-10*, *gnt-23*, and *gnt-26*) to Gnt⁺.

The inducibility of the gluconate kinase and permease in strain 61656 (Δ*igf*) lysogenic with φ105*gnt*⁺ [61656 (φ105*gnt*⁺)] was compared with that in our standard strain 60015 (Table 2). The two enzymes were induced in strain 61656 (φ105*gnt*⁺) by addition of gluconate to the medium, and their induction was repressed by addition of glucose. The results clearly indicated that the structural genes of the gluconate kinase and permease and regulatory regions for their expression were cloned in φ105.

The DNA of φ105*gnt*⁺ was analyzed by agarose-gel electrophoresis. Agarose-gel electrophoretic patterns of *Eco*RI digests of φ105 and φ105*gnt*⁺ DNAs are shown in Fig. 3. The digest of φ105 DNA produced nine fragments (from A to I), whereas that of φ105*gnt*⁺ DNA contained a new fragment of 4.5 megadaltons (MDa) (= the *gnt*⁺ fragment, indicated as *gnt*⁺) instead of three fragments (E, G, and I). The *Eco*RI cleavage map of φ105*gnt*⁺ DNA is shown with that of φ105 (Fig. 3).

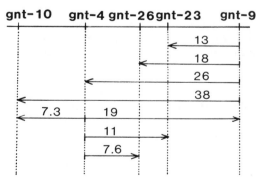

FIG. 2. Map of the *gnt* locus deduced from transformation crosses. In crosses, YF141 (*gnt-4 trpC2 metB5 hisA1*) and YF029 (*gnt-9 purA16 leuA8 metB5 trpC2*) were used as recipients. The recipient is indicated as a straight line in each cross. Distances are expressed as recombination indexes (1). (The ratio of *gnt*⁺ to *his*⁺ [YF141 as recipient] or *leu*⁺ [YF029 as recipient] transformants is a measure of the relative frequency of transformation to gluconate utilization.)

TABLE 1. Induction of gluconate permease and gluconate kinase in Gnt⁻ mutants[a]

Strain	Gluconate permease (cpm/min per OD)		Gluconate kinase (nmol/min per mg)	
	Not induced	Induced	Not induced	Induced
60015 (gnt+)	1,136	12,392	0.4	25.4
61656 (Δigf)	146	148	<0.1	<0.1
YF029 (gnt-9)	88	150	<0.1	3.3
YF127 (gnt-4)	34,454	26,136	<0.1	<0.1
YF160 (gnt-10)	2,562	12,816	<0.1	3.9
YF161 (gnt-23)	62	476	0.1	19.0
YF162 (gnt-26)	372	5,252	0.1	35.1

[a] The gluconate kinase was induced and assayed as described previously (11). The gluconate permease was induced under the same conditions used for gluconate kinase induction. The permease was assayed by the method of Dowds et al. (2) with our slight modifications. OD, Optical density.

Construction of restriction enzyme map of the gnt⁺ fragment and physical localization of the gnt mutations. The restriction enzyme map of the gnt⁺ fragment was constructed by use of BglII, HindIII, MluI, PstI, PvuII, SmaI, and StuI (Fig. 4). Since the gnt⁺ fragment had the ability to transform all the Gnt⁻ mutants except the Δigf deletion mutant to the wild type, we attempted to localize physically each gnt mutation onto a particular restriction enzyme fragment by analyzing which fragment exhibits the activity to transform the respective gnt mutant (Table 3). The BglII-A, MluI-A, PstI-A, SmaI-A, StuI-A, and HindIII-A fragments were able to transform the gnt-4 mutant to the wild type, and the BglII-A, MluI-A, PstI-A, SmaI-B, StuI-B, and HindIII-A fragments could transform the gnt-9 mutant. The gnt-10 mutant was transformed with the BglII-B, MluI-B, PstI-B, SmaI-A, StuI-A, and HindIII-A fragments, whereas the gnt-23 and gnt-26 mutants were transformed with the BglII-A, MluI-A, PstI-A, SmaI-A, StuI-B, and HindIII-A fragments. From these results, each gnt mutation was localized onto the respective restriction enzyme fragment (Fig. 5).

Cloning of the HindIII-A fragment with pC194 and φ105 and its expression. To investigate the regulation of the expression of the cloned genes when they are multiple copies, we attempted to clone the gnt⁺ fragment in a plasmid, pUB110, but failed, presumably because the fragment was too large to be cloned. We then decided to clone the HindIII-A fragment (2.4 MDa) in another plasmid, pC194, because this fragment was smaller yet contained all the gnt mutations. The HindIII-A fragment (2 μg), electrophoretically isolated from a HindIII digest of the gnt⁺ fragment, was ligated with pC194 (2 μg) that had been digested with HindIII. The ligated DNA was transferred to a competent culture of strain YF158 (gnt-4 recE4 trpC2) to obtain Gnt⁺ transformants. Of 187 Gnt⁺ colonies obtained, 10 colonies were analyzed for their plasmid content. Two kinds of plasmid (pCG1 and pCG8) were found in which the HindIII-A fragment was cloned in opposite orientations. Strain 61656 (Δigf) transformed with each plasmid was used for further experiments. The inducibility of the gluconate permease and kinase in strain 61656 (Δigf) harboring pCG1 or pCG8 [61656 (pCG1 or pCG8)] was investigated. As shown in Table 4, the two enzymes were constitutively synthesized in either 61656 (pCG1) or 61656 (pCG8). Specific activities of the enzymes synthesized in 61656 (pCG8) were approximately two times higher than those in 61656 (pCG1).

The constitutive synthesis of gluconate permease and kinase in this experiment might be due to the high copy number of the HindIII-A fragment cloned in pC194. To test this possibility, the HindIII-A fragment was cloned in the genome of φ105 that can be integrated as a single copy in the host chromosome, and the expression of the genes in this fragment was investi-

TABLE 2. Enzyme inductions and their repression by glucose in the Δigf strain lysogenic with φ105gnt⁺[a]

Strain	Gluconate permease (cpm/min per OD)			Gluconate kinase (nmol/min per mg)		
	Not induced	Induced	Induced + glucose	Not induced	Induced	Induced + glucose
60015 (gnt⁺)	712	10,004	628	0.1	22.8	0.6
6156 (Δigf) (φ105gnt⁺)	608	8,656	556	0.9	28.5	0.1

[a] Glucose was added at a concentration of 10 mM to repress enzyme induction. OD, Optical density.

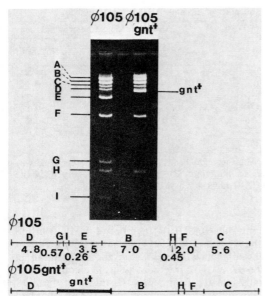

FIG. 3. Agarose-gel electrophoresis of *Eco*RI digest of φ105 and φ105*gnt*⁺ DNAs. The DNA (2 μg) was digested with *Eco*RI and electrophoresed. *Eco*RI cleavage maps of φ105 and φ105*gnt*⁺ are shown. The map of φ105DNA was recently corrected (Y. Kobayashi, personal communication). The *gnt*⁺ fragment (4.5 MDa) cloned is indicated as a boldface bar. The insertion of the *gnt*⁺ fragment between the fragments B and D was ascertained as follows. Phage φ105 DNA did not contain any *Bgl*II cleavage site, whereas φ105*gnt*⁺ DNA had one *Bgl*II cleavage site in the *gnt*⁺ fragment cloned. After digestion of φ105*gnt*⁺ DNA with *Bgl*II, the smaller fragment was isolated by agarose-gel electrophoresis. The *Eco*RI digestion of this fragment produced the fragment D and the *Bgl*II-B fragment indicated in Fig. 4.

gated. The *Hin*dIII-A fragment was cloned by prophage transformation with 61656 (Δ*igf*) as recipient by almost the same procedures described above. From two Gnt⁺ transducing phages (φ105*gnt*⁺H2 and φ105*gnt*⁺H4) obtained, the DNAs were prepared and restriction enzyme digests were analyzed by agarose-gel electrophoresis. The electrophoretic patterns of the *Eco*RI digest and the *Eco*RI and *Hin*dIII double digest of φ105*gnt*⁺H2 were different from those of φ105*gnt*⁺H4, but double digests of both DNAs gave a band of 2.4 MDa that was considered to be the *Hin*dIII-A fragment (data not shown). The results indicated that both φ105*gnt*⁺H2 and H4 possess the *Hin*dIII-A fragment, but that it is cloned in different *Hin*dIII sites of φ105 in the two phages. Table 4 shows that strain 61656 lysogenic with either φ105*gnt*⁺H2 or H4 synthesized the gluconate kinase and permease constitutively. The results imply that the constitutive synthesis of the two

enzymes was not due to the multiple copies of the fragment present in the strains carrying pCG1 or pCG8.

DISCUSSION

All of the mutations affecting the gluconate permease and kinase and the regulation of their biosynthesis were found clustered between *iol-6* and *fdp-74* (Fig. 1). On the basis of two-factor crosses between Gnt⁻ mutants (Fig. 2), we distinguished on the map the *gnt-10* mutation and the *gnt-23* and *gnt-26* mutations in the structural genes of the kinase and the permease, respectively, and the *gnt-9* mutation in the regulatory region. The existence of the *gnt-9* mutation affecting either the permease or the kinase and the fact that the Δ*igf* strain lysogenic with φ105*gnt*⁺H2 or H4 constitutively synthesized the two enzymes strongly suggested that the two structural genes belong to a single transcription unit, a gluconate (*gnt*) operon. The presence of the *gnt-4* mutation affecting the kinase and causing constitutive synthesis of the permease is not clearly understood. However, at present we prefer the possible explanation that the gluconate kinase protein itself is involved in the regulation of the gluconate operon so that an impaired gluconate kinase produced in the *gnt-4* mutant might cause the constitutive synthesis of the permease.

We have tentatively reported that the structural gene of the gluconate kinase and regulatory regions for its expression are cloned in a phage,

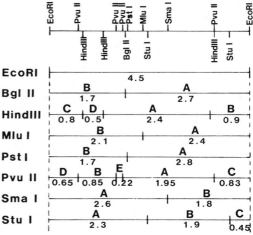

FIG. 4. Restriction enzyme map of the *gnt*⁺ fragment. Fragments generated by *Bgl*II, *Hin*dIII, *Mlu*I, *Pst*I, *Pvu*II, *Sma*I, and *Stu*I digestion are alphabetically named; their sizes are expressed as MDa. These fragments were used for transformation experiments to localize *gnt* mutations.

TABLE 3. Gnt$^+$ transforming activities of restriction enzyme fragments of the gnt$^+$ fragment[a]

Donor fragment	No. of transformants and revertants per 0.1 ml of competent culture with recipient genotype:				
	gnt-4 (YF127)	gnt-9 (YF029)	gnt-10 (YF160)	gnt-23 (YF161)	gnt-26 (YF162)
BglII-A	600	1,300	25	810	1,100
BglII-B	1	95	380	186	76
MluI-A	65	550	37	450	660
MluI-B	8	128	900	210	110
PstI-A	442	1,170	44	280	300
PstI-B	2	83	440	175	57
SmaI-A	1,000	83	4,000	230	180
SmaI-B	7	216	18	190	91
StuI-A	192	95	2,800	220	56
StuI-B	1	212	38	250	174
HindIII-A	630	340	1,500	770	910
No addition	0	80	8	193	95

[a] The gnt$^+$ fragment (1 μg) was digested with BglII, MluI, PstI, SmaI, StuI, and HindIII. The digests were electrophoresed in low-melting-point agarose gel, and the bands strained with ethidium bromide were excised. After melting the gels, a portion (10 μl) of the gel containing each fragment was used for transformation. Independent transformation experiments were performed for each recipient.

ρ11 (6). An EcoRI gnt$^+$ fragment generated from the DNA of a specialized gnt$^+$ transducing ρ11 that was constructed was recloned in another temperate phage, φ105. Since the regulation of gluconate permease and kinase induction in the Δigf strain lysogenic with φ105gnt$^+$ was indistinguishable from that in the wild-type strain (Table 2), and φ105gnt$^+$ was able to complement the Δigf deletion as well as all the gnt mutations to Gnt$^+$, this strongly indicates that an intact gluconate operon was cloned in φ105. To our knowledge, this work is the first to clone with a B. subtilis host-vector system an intact operon whose expression is easily induced at any stage of cell growth and is under catabolite repression.

The gnt$^+$ fragment cannot complement the iol-6 and fdp-74 mutations (6) but does complement all the gnt mutations. From these results and those obtained from the genetic crosses, the following order from iol-6 to fdp-74 was deduced: iol-6, gnt-10, gnt-4, gnt-26, gnt-23, gnt-9,

and fdp-74. All of the gnt mutations were successfully localized onto restriction enzyme fragments generated from the gnt$^+$ fragment (Fig. 5). The physical location of the mutations coincided reasonably with that deduced from their genetic mapping (Fig. 2).

The HindIII-A fragment was able to transform all the gnt mutants to the wild type. However, the Δigf strain lysogenic with φ105 bearing this fragment constitutively synthesized the gluconate permease and kinase. This suggests that the HindIII-A fragment probably contains the structural genes of the two enzymes but does not contain a part of the regulatory region of the gluconate operon. The entire nucleotide sequence of the HindIII-A fragment (2.4 MDa) was determined (Y. Fujita, T. Fujita, and Y. Aratani, unpublished data), and two open reading frames of over 400 codons were identified, which are probably the coding frames for the gluconate permease and kinase. The promoter region of this operon is currently being investigated.

FIG. 5. Localization of gnt mutations within the HindIII-A fragment. This is deduced from the results of the transformation experiments shown in Table 3.

TABLE 4. Induction of gluconate permease and gluconate kinase in Δigf strains bearing plasmid or φ105 containing the HindIII-A fragment

Plasmid or phage	Gluconate permease (cpm/min per OD)		Gluconate kinase (nmol/min per mg)	
	Not induced	Induced	Not induced	Induced
pCG1	5,364	4,024	28.6	15.9
pCG8	16,084	9,150	73.6	49.5
φ105gnt$^+$H2	2,764	2,847	16.2	11.0
φ105gnt$^+$H4	3,710	3,050	25.0	20.2

SUMMARY

We investigated the gluconate utilization system that is under catabolite repression. The enzymes involved in gluconate utilization seem to be gluconate permease and gluconate kinase, both of which are induced by gluconate. Several mutants unable to grow on gluconate were isolated. These mutations (*gnt*) were clustered between *iol-6* and *fdp-74* on the *B. subtilis* chromosome (a map order of *gnt-10*, *gnt-4*, *gnt-26*, *gnt-23*, and *gnt-9*). The *gnt-10* mutation seemed to be located within the structural gene of the kinase, and the *gnt-23* and *gnt-26* mutations seemed to be within that of the permease. The *gnt-9* mutant was impaired in a regulatory function for the expression of the two genes. An *Eco*RI fragment containing an intact gluconate (*gnt*) operon consisting of these two structural genes was cloned in φ105 by prophage transformation. The cloned fragment (4.5 MDa) was physically mapped. To correlate the genetic map with the physical map, the *gnt* mutants were transformed with each restriction enzyme fragment to examine which fragment contains the transforming activity. The physical location of the mutations coincided with that deduced from their genetic mapping. The *Hin*dIII-A fragment (2.4 MDa) where all the *gnt* mutations were localized was subcloned in pC194. The cloned fragment did not contain a part of the regulatory region of the gluconate operon but contained the structural genes for the gluconate permease and kinase.

We thank H. Saito and F. Kawamura for valuable suggestions. M. Fujita encouraged us throughout the course of this work.

This work was supported in part by a Grant-in Aid for Scientific Research from the Ministry of Education, Science and Culture of Japan.

LITERATURE CITED

1. **Chasin, L. A., and B. Magasanik.** 1968. Induction and repression of the histidine-degrading enzymes of *Bacillus subtilis*. J. Biol. Chem. **243:**5165–5178.
2. **Dowds, B., L. Baxter, and M. Mckillen.** 1978. Catabolite repression in *Bacillus subtilis*. Biochim. Biophys. Acta **541:**18–34.
3. **Freese, E., and Y. Fujita.** 1976. Control of enzyme synthesis during growth and sporulation, p. 164–184. *In* D. Schlessinger (ed.), Microbiology—1976. American Society for Microbiology, Washington, D.C.
4. **Fujita, Y., and E. Freese.** 1981. Isolation and properties of a *Bacillus subtilis* mutant unable to produce fructose-bisphosphatase. J. Bacteriol. **145:**760–767.
5. **Fujita, Y., and T. Fujita.** 1983. Genetic analysis of a pleiotropic deletion mutation (Δ*igf*) in *Bacillus subtilis*. J. Bacteriol. **154:**864–869.
6. **Fujita, Y., T. Fujita, F. Kawamura, and H. Saito.** 1983. Efficient cloning of genes for utilization of D-gluconate of *Bacillus subtilis* in phage ρ11. Agric. Biol. Chem. **47:**1679–1682.
7. **Ide, M.** 1971. Adenyl cyclase of bacteria. Arch. Biochem. Biophys. **144:**262–268.
8. **Iijima, T., F. Kawamura, H. Saito, and Y. Ikeda.** 1980. A specialized transducing phage constructed from *Bacillus subtilis* phage φ105. Gene **9:**115–126.
9. **Lopez, J. M., and B. Thomas.** 1977. Role of sugar uptake and metabolic intermediates on catabolite repression in *Bacillus subtilis*. J. Bacteriol. **129:**217–224.
10. **Mizukami, T., F. Kawamura, H. Takahashi, and H. Saito.** 1980. A physical map of the genome of the *Bacillus subtilis* temperate phage ρ11. Gene **11:**157–162.
11. **Nihashi, J., and Y. Fujita.** 1984. Catabolite repression of inositol dehydrogenase and gluconate kinase syntheses in *Bacillus subtilis*. Biochim. Biophys. Acta **798:**88–95.
12. **Pastan, I., and S. Adhya.** 1976. Cyclic adenosine 5'-monophosphate in *Escherichia coli*. Bacteriol. Rev. **40:**527–551.
13. **Sasajima, K., and T. Kumada.** 1981. Change in the regulation of enzyme synthesis under catabolite repression in *Bacillus subtilis* pleiotropic mutant lacking transketolase. Agric. Biol. Chem. **45:**2005–2012.
14. **Setlow, P.** 1973. Inability to detect cyclic AMP in vegetative or sporulating cells or dormant spores of *Bacillus megaterium*. Biochem. Biophys. Res. Commun. **52:**365–372.

Endotoxin Formation in Sporulating Organisms

Protein Toxins of *Bacillis* spp.

H. E. SCHNEPF AND H. R. WHITELEY

Department of Microbiology and Immunology, University of Washington, Seattle, Washington 98195

Gram-positive endospore-forming bacteria produce a large number of proteinaceous toxins which are important both to industry and to public health. The majority of the medically important toxins are elaborated by members of the genus *Clostridium*: *C. botulinum*, *C. tetani*, and *C. perfringens*. The toxins synthesized by these bacteria have been described in several recent reviews (17, 38, 53) and will not be considered here. The genus *Bacillus* contains the remaining toxin-producing endospore-forming bacteria. The *B. anthracis* toxin and, to a lesser extent, *B. cereus* toxins are of medical importance in human and animal infections, and the crystalline toxins of *B. thuringiensis* are useful for the control of economically important agriculture pests and may become significant in the control of insect vectors of human diseases. *B. sphaericus* is also under consideration as a biological control agent, and it has now also been shown to produce a parasporal crystalline inclusion. *B. popilliae*, *B. lentimorbus*, and *B. larvae* are obligate pathogens of certain insects and, although they have not been shown to produce toxins, are nevertheless important economically. The latter organisms are also of interest because of the involvement of spores in their respective diseases and because they require unusual conditions for sporulation.

The growth conditions for the efficient production of many of these toxins have been well defined, but little is known about the regulatory mechanisms controlling toxin synthesis with the exception of the *B. thuringiensis* Lepidopteran crystalline toxin. In this paper, we review the toxins produced by the above-named members of the genus *Bacillus* and the current information on the regulation of toxin synthesis by *B. thuringiensis*.

TOXINS OF THE *B. CEREUS* GROUP OF BACILLI

B. thuringiensis can be distinguished from *B. cereus* and *B. anthracis* primarily by the synthesis of a sporulation-specific insecticidal protein, the delta-endotoxin, which crystallizes as an inclusion in the cell (6). The crystal proteins, which are the main insecticidal component of *B. thuringiensis* cultures, have different insecticidal

spectra depending on the bacterial strain (16). Although *B. thuringiensis* preparations are considered to be safe for use as insecticides and not to be pathogenic for mammals (9), two recent infections involving this organism have been reported (47, 59). *B. anthracis* is a mammalian pathogen, and *B. cereus* has been shown to be responsible for outbreaks of food poisoning (43) and occasionally has been isolated from human wound and ophthalmic infections and from cases of bovine mastitis (56). Aside from the morphological similarities, the close relationship of these three species is demonstrated by a similarity in the distribution of some of the toxins (reviewed below), the transfer of plasmids between *B. cereus* and *B. thuringiensis* via an apparently conjugationlike mechanism (20; also B. C. Carlton and J. M. Gonzalez, Jr., this volume), and by the transduction of plasmids among all three species (46). This is particularly significant in view of the fact that at least one *B. anthracis* virulence factor (58) and the *B. thuringiensis* crystal protein gene (61) are plasmid encoded.

B. thuringiensis Lepidopteran Toxin

Protoxin and toxin polypeptides. During sporulation, strains of *B. thuringiensis* which are toxic to Lepidopteran insects synthesize proteins of ca. M_r 130,000 to 160,000 (11) which form parasporal crystalline inclusions. These inclusions contain the main insecticidal activity of these bacteria to Lepidoptera and account for 20 to 30% of the dry weight of sporulated cultures (35). Crystals from these strains are dissolved in the reduced, alkaline conditions of the midgut of susceptible insects, or by similar in vitro conditions, to release the crystal protein protoxin of M_r 130,000 to 160,000. These protoxin molecules can undergo cleavage, either in vitro or in the insect midgut, to produce polypeptides of M_r 80,000 to 30,000 which are toxic to Lepidopteran larvae and cytolytic to cultured Lepidopteran cell lines (reviewed in 23). Two well-characterized crystal proteins are polypeptides of M_r 134,000 to 136,000 from subspecies *kurstaki* (7) and *thuringiensis* (24), which are converted by proteolysis to toxic polypeptides of M_r 70,000 to 55,000. Nagamatsu et al. (42) have determined the NH_2 termini of the M_r 145,000 protoxin and M_r 59,000 toxin of subsp. *dendrolimus* crystals;

the toxin NH_2 terminus was residue 29 of the protoxin by comparison with the deduced sequence of Wong et al. (62).

Investigations of the cytopathology of crystal protein intoxication of Lepidopteran larvae have shown that the insect gut cells swell, vacuolate, and burst. Abnormalities in treated cells are observed within minutes of toxin addition. The lack of a latent period, combined with the above observations, suggests a cytolytic mode of action. A similar course of events has been observed when cultured Lepidopteran cell lines are treated with the toxin (37). It has been shown that N-acetyl galactosamine and lectins which will bind this sugar can protect the insect cells from cytolysis (29; D. J. Ellar et al., this volume).

More than 20 different varieties of *B. thuringiensis* have been isolated and classified on the basis of flagellar antigens. Different varieties and even strains of a given variety have different spectra of insecticidal activities (16). It is not known whether the differences in the insecticidal activities are due to differences in crystal composition (i.e., different toxins) or to differences in the solubilization and proteolysis of the protoxin in the midgut of different insects. Alternatively, some specific molecular target (e.g., a binding or effector site) may vary in structure among the different insect species, resulting in altered susceptibility of the insect to the toxin. The cloning of a crystal protein gene with toxicity to *Noctuidae* larvae (A. Klier et al., this volume) and the observation that crystalline toxins from different *B. thuringiensis* strains show different degrees of toxicity to cultured cells from different Lepidopteran insects (D. J. Ellar, personal communication) suggest that the mechanism of the insecticidal spectra may soon be elucidated.

Relationship of the crystal protein to spore coat proteins. A number of reports have indicated that crystals and spore coats or exosporium contain common antigens or similarly sized polypeptides, or both (3, 49, 51). The presence of the crystal protein on the spore surface has been correlated with altered germination properties of several crystalliferous strains of *B. thuringiensis* (4). In addition, it has been shown that spore extracts of *B. cereus* and *B. thuringiensis* are toxic to Lepidopteran larvae (51). However, the demonstration that *B. cereus* exosporium antigens (J. P. DesRosier, Ph.D. thesis, University of Washington, Seattle, 1980) and *B. thuringiensis* spore coat proteins (2) are present on the surface of *B. thuringiensis* crystals raises an alternative possibility for the origin of the common components in crystals and spores and suggests that the nature of these common components deserves reexamination.

Regulation of protoxin synthesis. Several investigations of the timing of protoxin synthesis (6, 50, 62) have shown that crystals, the crystal protein, and crystal protein mRNA are produced beginning at about stage II of sporulation and that synthesis continues until late in sporulation. Crystals are detected in most sporulation mutants blocked at stage II or later (50), but most mutants blocked at stage 0 of sporulation fail to make crystals (40); acrystalliferous mutants, on the other hand, sporulate normally. This indicates that, although crystal production is dispensable for sporulation, at least some of the factors regulating sporulation are required for crystal synthesis.

The cloning of the crystal protein gene from a plasmid in *B. thuringiensis* into *Escherichia coli* has allowed a determination of the DNA sequence of the gene, including the promoter region (48, 62). The translational start site was established by the determination of eight of the amino acids at the NH_2 terminus of the protein. Nine bases of the 11-base-pair sequence located three bases upstream from the initiator ATG codon (GATGAGGTAA) are complementary to the 3' end of the *Bacillus* rRNA (complementary bases are underlined), implying an efficient translation (39) of the crystal protein gene mRNA.

Whereas the crystal protein gene is expressed at all stages of growth in the recombinant *E. coli* strain (62), it is transcribed only during sporulation in *B. subtilis* (27; W. R. Widner and H. R. Whiteley, unpublished data) and in *B. thuringiensis* (27, 62). S1 nuclease mapping has shown that transcription in *B. thuringiensis* is initiated from two overlapping promoters; one (Bt I) is utilized beginning at an early stage of sporulation, whereas transcription from the other (Bt II) begins at about the midpoint of the sporulation process. The Bt I promoter shows homology to the *B. subtilis* *spoVG* and *spoVC* gene promoters (these genes are also activated early during sporulation) in the −10 region but not in the −35 region. The Bt I and *spoVG* promoters also contain highly adenine-plus-thymine-rich regions of 20 to 30 base pairs located about 50 bases upstream from the transcriptional start sites. The adenine-plus-thymine-rich region is believed to play a role in the regulation of the *spoVG* gene and may, therefore, also play a role in the regulation of the crystal protein gene. However, the lack of homology in the −35 regions of the Bt I and *spoVG* promoters and the differences in the pattern of expression of these promoters in *B. subtilis* (5, 66; Widner and Whiteley, unpublished data) suggests that they are regulated somewhat differently and that they may be recognized by RNA polymerases containing different sigma factors.

A comparison of the crystal protein Bt II

promoter region with the promoter of the *B. subtilis* 0.3-kilobase sporulation-specific gene indicates significant homologies at ca. −6 and −26 base pairs from the respective initiation sites (52; H. R. Whiteley, H. E. Schnepf, J. W. Kronstad, and H. C. Wong, *in* A. T. Ganesan and J. A. Hoch, ed., *Genetics and Biotechnology of Bacilli*, in press). Additionally, transcription of the two promoters begins at about the same stage of sporulation, suggesting that they might be recognized by sigma factors having a similar specificity.

A different transcriptional start site has been identified by Klier et al. (28) for a crystal protein gene cloned from subsp. *thuringiensis*. Interestingly, a sequence very similar to the latter start site was found within the coding sequence of the crystal protein gene from subsp. *kurstaki* (H. E. Schnepf, H. C. Wong, and H. R. Whiteley, manuscript in preparation).

It has been proposed that the stability of the crystal protein mRNA contributes to the massive synthesis of this protein (45). It has also been proposed that mRNA stability may be determined, at least in part, by the structure of the transcriptional terminator (22). The transcriptional terminator of the crystal protein gene consists of a potentially very stable stem-and-loop structure (Schnepf et al., in preparation). Deletion of the terminator region sharply decreases the amount of crystal protein expressed in *E. coli*, and fusion of the terminator to the 3′ end of such deletions increases the expression of the crystal protein antigen (H. E. Schnepf and H. R. Whiteley, manuscript in preparation; H. R. Whiteley, J. W. Kronstad, and H. E. Schnepf, this volume). Similar results were obtained by H. C. Wong and S. Chang (this volume) by fusing the crystal protein gene terminator to foreign genes in both *E. coli* and *B. subtilis*. Although this suggests a role for the crystal protein terminator in gene expression, estimates of mRNA stability will be required to determine unambiguously whether the increased expression is related to mRNA half-life.

It should be noted that the expression of some crystal protein genes may be regulated by additional, more complex regulatory mechanisms, such as those described by Minnich and Aronson (41).

Location of the crystal protein gene. A gene encoding a crystal protein which is toxic to Lepidopterans has been cloned from a large plasmid in subsp. *kurstaki* and has been shown to be present on plasmids in a number of other varieties (61). Results obtained from curing and plasmid transfer experiments have generally agreed with the results of the cloning and hybridization experiments (Carlton and Gonzalez, this volume). Interestingly, some strains were found

to have two or three related crystal protein genes present on different plasmids (32; Whiteley et al., this volume), and at least in one variety, the gene may be present on the chromosome. Other crystal protein genes have been cloned from subsp. *kurstaki* (21) and from subsp. *thuringiensis* (27), and copies of these genes have been found on plasmids and the bacterial chromosome; however, the chromosomal genes are apparently not expressed.

The presence of the crystal protein gene on plasmids of a number of different sizes has stimulated interest in the mechanism by which this gene moves from replicon to replicon. Results from two independent investigations (Whiteley et al., this volume; Klier et al., this volume) have indicated that transposonlike elements are in close association with the crystal protein gene in many cases. Although transposition of the crystal protein genes has not been demonstrated directly, the involvement of the transposonlike sequences in direct transposition or as effectors of more complex recombinational events causing plasmid rearrangement appears likely.

Crystalline Toxins Active on Dipterans

Two toxins active against mosquitoes have been isolated. One toxin, found in subsp. *kyushuensis* (44) and subsp. *israelensis* (19), is active against mosquitoes and certain other Diptera (e.g., blackflies). Crystals from subsp. *israelensis* contain polypeptides of several sizes (57). Although the smallest of these (M_r 25,000) has been shown to be the principal toxic component (64; Ellar et al., this volume), the toxicity of the isolated peptide is significantly lower than that of isolated crystals. The dissolved crystals are cytolytic to a number of insect and mammalian cell lines. This toxin is thought to bind to membranes containing several types of phospholipids having unsaturated acyl residues and to act as a protein surfactant (54). The gene encoding this toxin resides on a 75-megadalton plasmid in *B. thuringiensis* and has been cloned (Ellar et al., this volume).

Another toxin is found only in some isolates of several varieties of *B. thuringiensis*. It is active against both mosquitoes and Lepidopterans, is of M_r 65,000 (65), and is formed into cuboidal bodies within (or adjacent to) crystals of the M_r 134,000 Lepidopteran toxin described above (25). The mode of action for the M_r 65,000 toxin has not yet been reported.

Other Insecticidal Toxins Produced by *B. thuringiensis* and *B. cereus*

At least two insecticidal compounds in addition to the crystalline toxins may be reponsible for the opportunistic infection of insects by these bacteria. Additionally, they may contribute to the insecticidal activity of *B. thuringiensis* preparations toward Lepidopterans which are relatively resistant to the crystalline Lepidopteran toxin (16).

Alpha- and beta-exotoxins. The alpha-exotoxin of *B. thuringiensis* is a heat-labile insecticidal activity which was thought to be phospholipase C. However, some isolates which do not produce phospholipase C have been shown to produce alpha-exotoxin, leaving the identity of this activity in doubt (16). Turnbull (55) has speculated that some of the problems encountered in characterizing the alpha-exotoxin of *B. thuringiensis* may be due to difficulties in separating the phospholipase, hemolytic, and diarrheagenic/dermonecrotic activities of *B. thuringiensis*. Whether the alpha-exotoxin is the same as one of the above-mentioned proteins remains to be determined.

The beta-exotoxin is a heat-stable adenine nucleotide derivative which is toxic to a large number of cell types. Since it is detoxified in the digestive tract, it is not toxic to mammals by ingestion; however, it is lethal to many types of insects, particularly flies. This compound is produced by some isolates of *B. cereus* and by some *B. thuringiensis* strains which are not used as commercial pesticides (33). It should be noted, however, that some methods of preparing insecticidal powders of *B. thuringiensis* may also partially precipitate beta-exotoxin (16).

Role of the spore. Spores and spore extracts of *B. cereus* and *B. thuringiensis* are toxic to Lepidopteran larvae. As discussed above, the Lepidopteran protoxin is present on the surface of spores of at least some strains of *B. thuringiensis*; however, the nature of the toxic material present on the surface of *B. cereus* and *B. thuringiensis* spores deserves reexamination, in part because additional toxic components may be present. Burges et al. (10) have shown that spores of *B. thuringiensis* subsp. *galleriae* are either toxic or contain an adjuvant of crystal toxicity, and it is known that the presence of spores substantially enhances the insecticidal activity of crystal-containing *B. thuringiensis* preparations against some Lepidopterans which are relatively insensitive to the Lepidopteran crystalline toxin (37). Neither the toxic material from the spores nor the mechanism by which spores enhance the toxicity of crystal preparations has been described in detail.

Phospholipase C

All three species of *B. cereus*-like organisms produce phospholipase (55), but the enzymatic activity has been extensively analyzed only in *B. cereus*. The latter synthesizes a phosphatidylcholine hydrolase of M_r 23,000, a phos-phatidylinositol hydrolase of M_r 29,000, and a sphingomyelinase of M_r 24,000. Unlike the *C. perfringens* alpha-toxin phospholipase C, which is highly hemolytic, only the sphingomyelinase of *B. cereus* has been reported to be hemolytic, and then only in proportion to the amount of sphingomyelin in the erythrocyte membrane. A serological neutralization study has indicated that *B. cereus* may contain more of the phospholipase C activities than *B. anthracis* and, possibly, *B. thuringiensis*. Detailed studies comparing the hemolysins, the diarrheagenic/dermonecrotic toxin, and phospholipases indicate that the phospholipases of *B. cereus* are not particularly toxic in animals and are of questionable significance in pathogenesis.

Hemolysins

Hemolysins are produced by *B. cereus* and *B. thuringiensis* but not by *B. anthracis*. The former bacteria each synthesize two hemolysins: a sulfhydryl-activated cytolysin of M_r 45,000 to 59,000 (cereolysin, thuringiolysin) and a second non-thiol-activated hemolysin (hemolysin II) of M_r 29,000 to 34,000 (55). Cereolysin and thuringiolysin are similar to other thiol-activated cytolysins (1) with respect to inhibition by cholesterol and oxygen. Cereolysin is lethal to mice, although it is inhibited by human serum and is therefore of questionable significance in human infections. The cereolysin gene has been cloned and is presumably chromosomally encoded (30).

Diarrheagenic/Dermonecrotizing Toxin

B. cereus synthesizes a diarrheagenic/dermonecrotizing toxin of ca. M_r 50,000, but the epidemiological data correlating this activity to outbreaks of diarrhea are controversial (55). However, increased production of this toxin has been correlated with the severity of nondiarrheal infections from which *B. cereus* has been isolated (56). The toxin is produced in late logarithmic growth, but only when cultured in appropriate media (e.g., brain heart infusion broth containing glucose). Activities attributed to this toxin include fluid accumulation in rabbit ileal loops, vascular permeability upon intradermal injection in rabbits, and lethality upon intravenous injection in mice. The toxin has been detected in culture supernatants of many *B. thuringiensis* strains, but not in *B. anthracis*. In vascular permeability tests, serum raised to the *B. cereus* toxin neutralized the *B. thuringiensis* activity. The mode of action of this toxin is not well characterized; however, it may have adenyl cyclase stimulating activity; it also appears to be cytolytic.

Emetic Toxin

Several cases of *B. cereus*-induced food poisoning have been well documented (43), and a low-molecular-weight (M_r <5,000) fraction has been shown to cause an emetic response when fed to monkeys. Production of this emetic toxin was not detected in assays of several isolates of *B. cereus* which were not associated with food poisoning (55). It seems unlikely, therefore, that *B. thuringiensis* or *B. anthracis* would produce this toxin, but this has not been tested directly.

Anthrax Toxin

B. anthracis synthesizes an exotoxin which is a primary determinant of virulence when produced in conjunction with a polyglutamic acid capsule (63). This toxin consists of three components (34), each of ca. M_r 80,000: (i) protective antigen (PA), which is required for the action of each of the other two components, presumably by interacting with cellular binding sites; (ii) edema factor (EF), which is a calmodulin-activated adenylate cyclase; and (iii) lethal factor (LF), whose mode of action has not yet been established. None of these factors is active alone; however, PA and EF together produce edema when injected intradermally in guinea pigs, and PA and LF together are lethal to mice when injected intravenously. Anthrax toxin is produced during late exponential growth, and specific culture conditions are required for its production. While the location of the EF and LF genes is not known, the gene for PA is plasmid encoded and has been cloned (58). Cloning of the genes coding for EF and LF can be anticipated. When specific probes for all three factors become available, it should be possible to determine whether DNA sequences homologous to any of these factors are present in *B. cereus* and *B. thuringiensis*.

OTHER BACILLI THAT ARE INSECTICIDAL OR CRYSTALLIFEROUS, OR BOTH

B. sphaericus

Isolates of *B. sphaericus* can be divided into two classes on the basis of DNA homology and high or low toxicity to mosquitoes (12). When grown in the appropriate media, the high-toxicity strains produce parasporal crystals which are enclosed within the exosporium and are toxic to mosquitoes (26). Proteins of ca. M_r 35,000 and 55,000 are the predominant components of toxic spore extracts (13). The cloning of a mosquitocidal determinant from a highly toxic strain of *B. sphaericus* on a plasmid in *E. coli* has been reported (18, 36). The insert of *B. sphaericus* DNA containing the determinant hybridized to different restriction fragments in highly toxic

and poorly toxic strains of *B. sphaericus*. The *B. sphaericus* DNA insert was also shown to encode polypeptides of ca. M_r 21,000, 19,000, 15,000, and 14,000; DNA encoding all four polypeptides is apparently needed for toxicity (J. Louis and J. Szulmajster, this volume). Additional experimentation will be required to determine the relationship of the in vitro-synthesized peptides to the peptides found in spore extracts and to define the insecticidal components of this organism.

B. popilliae, *B. lentimorbus*, and *B. larvae*

B. popilliae and *B. lentimorbus* cause a progressive invasive infection (milky disease) in scarabaeid beetles, and fully infectious cultures of these bacteria are obtained only after propagation in insects (8, 31). *B. larvae*, the causative agent of American foulbrood of honeybees, is also difficult to grow and sporulate in the laboratory (8). The normal course of infection for all three bacteria involves ingestion of the spores by the insect, followed by germination of the spores and penetration of the bacteria through the insect gut to the hemolymph. The bacteria proliferate and sporulate in the insect larvae, which then die. The mechanism by which these bacteria cause the death of the insects, including the possible role of toxins, has not been determined. Interestingly, *B. larvae* requires poor aeration for efficient sporulation, unlike most other *Bacillus* spp. (15). Spores of this organism are also capable of selective germination in bee larvae rather than in adult bees which they cannot infect. *B. popilliae* produces a parasporal crystalline inclusion which is enclosed within its exosporium. Although this inclusion is reportedly toxic to Japanese beetle larvae (60), its production is not sufficient for infection (31).

Other Crystalliferous Bacilli

Aside from *B. sphaericus*, *B. popilliae*, and toxic isolates of *B. thuringiensis*, there are a number of *Bacillus* spp. which produce parasporal crystals which either are nontoxic or are toxic to an as yet undetermined organism. Surprisingly, this group includes a number of varieties of *B. thuringiensis* (14). The inclusion can be deposited either within the exosporium (*B. thuringiensis* subsp. *finitimus*, *B. cereus* subspp. *fowler* and *lewin*) or outside of the exosporium (non-*finitimus* subspecies of *B. thuringiensis* and *B. medusa*; 2). Unlike the other crystalliferous species of *Bacillus*, *B. medusa* begins synthesis of its inclusion during late logarithmic growth.

CONCLUDING REMARKS

Bacillus spp. produce a number of toxins which are of medical and economic importance.

These toxins can be classified, for the most part, into one of two groups. First are the insecticidal crystalline toxins of *B. thuringiensis* and *B. sphaericus*. Second are the soluble secreted toxins which are likely to play a role as virulence factors in invasive infections and food poisoning of mammals and in opportunistic infections of insects. These include anthrax toxin and the diarrheagenic/dermonecrotic, emetic, and, possibly, the hemolytic factors of *B. cereus* and *B. thuringiensis*.

The cloning of crystal protein genes from *B. thuringiensis* has allowed detailed investigations of the structure and regulation of this sporulation-specific, insecticidal protein. Additional studies are needed to determine how many different genes are present both on plasmids and on the chromosome of different subspecies (i.e., genes which do not hybridize with the currently available probes) and why some chromosomal copies of the crystal protein gene are silent. A more complete understanding of the regulation of this gene will require the isolation of RNA polymerases containing the appropriate sigma factors to transcribe this gene, analyses of the promoter structures of several additional genes, and an assessment of the possible existence of other regulatory mechanisms such as repressors, positive transcriptional regulatory proteins, or the more complex regulation observed by Minnich and Aronson (41). Such analyses should reveal whether all crystal protein genes are regulated by the same or different mechanisms. Other questions of interest concern the relationship between crystal protein synthesis and spore coat protein production and the structure of the apparently nontoxic crystal proteins synthesized by some subspecies of *B. thuringiensis*.

The biochemical nature, mode of action, and spectrum of toxicity of the Lepidopteran toxin are incompletely understood. However, the molecular structure of the Lepidopteran toxin is being defined at the DNA and protein sequence level, and binding sites on insect cells are being identified. Crystal proteins with different insecticidal spectra are being cloned, and comparative data from sequencing and toxicological studies should provide explanations for differences in insecticidal spectra.

Of the mosquitocidal *Bacillus* spp., substantial progress has been made on *B. thuringiensis* subsp. *israelensis* in terms of identifying the toxin, cloning its gene, and elucidating its mode of action, and similar progress can be expected in the analysis of the toxic activity(s) of *B. sphaericus*. Studies of these toxins are particularly important in view of their potential use to control vectors of disease.

In view of the increasing use of *B. thuringiensis* as an insecticide and the increased awareness of the involvement of *B. cereus* in infections, additional studies are needed to determine what factors contribute to the virulence of (or susceptibility to) *B. cereus* and to what extent these factors apply to *B. thuringiensis*. In particular, the diarrheagenic/dermonecrotic factor deserves further study since it is already suspected of being a major determinant of *B. cereus* infections and is also produced by *B. thuringiensis*. Although there have been only two reports recently of infections attributed to *B. thuringiensis*, it may be desirable to use recombinant DNA techniques either to move the crystalline toxin gene to strains lacking the diarrheagenic/dermonecrotic toxin or to eliminate production of the latter toxin in *B. thuringiensis*, for example, by insertional mutagenesis.

A number of characteristics distinguish *B. anthracis* from *B. cereus* and *B. thuringiensis* despite the many features they share and their ability to exchange genetic information. Notably, virulent strains of *B. anthracis* synthesize a distinctive toxin and a polyglutamic acid capsule while lacking hemolysins and penicillinase. Although it seems unlikely that *B. thuringiensis* isolates could be converted genetically to cause anthrax, it would be worthwhile to determine whether any of the factors which contribute to the virulence of *B. anthracis* could be carried by isolates of *B. thuringiensis*.

LITERATURE CITED

1. **Alouf, J. E.** 1977. Cell membrane and cytolytic bacterial toxins, p. 220–270. *In* P. Cuatrecasas (ed.), The specificity and action of animal, bacterial and plant toxins. Chapman and Hall, London.

2. **Aronson, A. I., and P. Fitz-James.** 1976. Structure and morphogenesis of the bacterial spore coat. Bacteriol. Rev. **40:**360–402.

3. **Aronson, A. I., and N. K. Pandey.** 1978. Comparative structural and functional aspects of spore coats, p. 54–61. *In* G. Chambliss and J. C. Vary (ed.), Spores VII. American Society for Microbiology, Washington, D.C.

4. **Aronson, A. I., D. J. Tyrell, P. C. Fitz-James, and L. A. Bulla, Jr.** 1982. Relationship of the synthesis of spore coat protein and parasporal crystal protein in *Bacillus thuringiensis*. J. Bacteriol. **151:**399–410.

5. **Banner, C. D. B., C. P. Moran, Jr., and R. Losick.** 1983. Deletion analysis of a complex promoter for a developmentally regulated gene from *Bacillus subtilis*. J. Mol. Biol. **168:**351–365.

6. **Bechtel, D. B., and L. A. Bulla, Jr.** 1976. Electron microscope study of sporulation and parasporal crystal formation in *Bacillus thuringiensis*. J. Bacteriol. **127:**1472–1481.

7. **Bulla, L. A., Jr., K. J. Kramer, D. J. Cox, B. L. Jones, L. I. Davison, and G. L. Lookhart.** 1981. Purification and characterization of the entomocidal protoxin of *Bacillus thuringiensis*. J. Biol. Chem. **256:**3000–3004.

8. **Bulla, L. A., Jr., R. A. Rhodes, and G. St. Julian.** 1975. Bacteria as insect pathogens. Annu. Rev. Microbiol. **29:**163–190.

9. **Burges, H. D.** 1981. Safety, safety testing and quality control of microbial pesticides, p. 737–767. *In* H. D. Burges (ed.), Microbial control of pests and plant diseases

1970–1980. Academic Press, London.

10. **Burges, H. D., E. M. Thomson, and R. A. Latchford.** 1976. Importance of spores and delta-endotoxin protein crystals of *Bacillus thuringiensis* in *Galleria mellonella*. J. Invertebr. Pathol. **27:**87–94.

11. **Calabrese, D. M., K. W. Nickerson, and L. C. Lane.** 1980. A comparison of protein crystal subunit sizes in *Bacillus thuringiensis*. Can. J. Microbiol. **26:**1006–1010.

12. **Davidson, E. W.** 1981. Bacterial diseases of insects caused by toxin-producing *Bacilli* other than *Bacillus thuringiensis*, p. 269–291. *In* E. W. Davidson (ed.), Pathogenesis of invertebrate microbial diseases. Allanheld, Osmun and Co., Totowa, N.J.

13. **Davidson, E. W.** 1983. Alkaline extraction of toxin from spores of the mosquito pathogen, *Bacillus sphaericus* strain 1593. Can. J. Microbiol. **29:**271–275.

14. **DeLucca, A. J., II, M. S. Palmgren, and H. deBarjac.** 1984. A new serovar of *Bacillus thuringiensis* fom grain dust: *Bacillus thuringiensis* serovar *colmeri* (serovar 21). J. Invertebr. Pathol. **43:**437–438.

15. **Dingman, D. W., and D. P. Stahly.** 1983. Medium promoting sporulation of *Bacillus larvae* and metabolism of medium components. Appl. Environ. Microbiol. **46:**860–869.

16. **Dulmage, H. T., and Cooperators.** 1981. Insecticidal activity of isolates of *Bacillus thuringiensis* and their potential for pest control, p. 193–222. *In* H. D. Burges (ed.), Microbial control of pests and plant diseases 1970–1980. Academic Press, London.

17. **Eidels, L., R. L. Proia, and D. A. Hart.** 1983. Membrane receptors for bacterial toxins. Microbiol. Rev. **47:**596–620.

18. **Ganesan, S., H. Kamdar, K. Jayaraman, and J. Szulmajster.** 1983. Cloning and expression in *Escherichia coli* of a DNA fragment from *Bacillus sphaericus* coding for biocidal activity against mosquito larvae. Mol. Gen. Genet. **189:**181–183.

19. **Goldberg, L. J., and J. Margalit.** 1977. A bacterial spore demonstrating rapid larvicidal activity against *Anopheles sergentii*, *Uranotaenia unquiculata*, *Culex univitattus*, *Aedes aegypti* and *Culex pipiens*. Mosq. News **37:**355–358.

20. **Gonzalez, J. M., Jr., B. J. Brown, and B. C. Carlton.** 1982. Transfer of *Bacillus thuringiensis* plasmids coding for delta-endotoxin among strains of *B. thuringiensis* and *B. cereus*. Proc. Natl. Acad. Sci. U.S.A. **79:**6951–6955.

21. **Held, G. A., L. A. Bulla, Jr., E. Farrari, J. Hoch, A. I. Aronson, and S. A. Minnich.** 1982. Cloning and localization of the lepidopteran protoxin gene of *Bacillus thuringiensis* subsp. *kurstaki*. Proc. Natl. Acad. Sci. U.S.A. **79:**6065–6069.

22. **Holmes, W. M., T. Platt, and M. Rosenberg.** 1983. Termination of transcription in *E. coli*. Cell **32:**1029–1032.

23. **Huber, H. E., and P. Luthy.** 1981. *Bacillus thuringiensis* delta-endotoxin composition and activation, p. 209–234. *In* E. W. Davidson (ed.), Pathogenesis of invertebrate microbial diseases. Allanheld, Osmun and Co., Totowa, N.J.

24. **Huber, H. E., P. Luthy, H.-R. Rudolf, J.-L. Cordier.** 1981. The subunits of the parasporal crystal of *Bacillus thuringiensis*: size, linkage, and toxicity. Arch. Microbiol. **129:**14–18.

25. **Iizuka, T., and T. Yamamoto.** 1983. Possible location of the mosquitocidal protein in the crystal preparation of *Bacillus thuringiensis* subsp. *kurstaki*. FEMS Microbiol. Lett. **19:**187–192.

26. **Kalfon, A., J.-F. Charles, C. Bourgouin, and H. deBarjac.** 1984. Sporulation of *Bacillus sphaericus* 2297: an electron microscope study of crystal-like inclusion biogenesis and toxicity to mosquito larvae. J. Gen. Microbiol. **130:**893–900.

27. **Klier, A., F. Fargette, J. Ribier, and G. Rapoport.** 1982. Cloning and expression of the crystal protein genes from *Bacillus thuringiensis* strain *berliner* 1715. EMBO J. **1:**791–799.

28. **Klier, A., C. Parsot, and G. Rapoport.** 1983. *In vitro*

29. **Knowles, B. H., W. E. Thomas, and D. J. Ellar.** 1984. Lectin-like binding of *Bacillus thuringiensis* var. *kurstaki* Lepidopteran-specific toxin as an initial step in insecticidal action. FEBS Lett. **168:**197–202.

30. **Kreft, J., H. Berger, M. Hartlein, B. Muller, G. Weidinger, and W. Goebel.** 1983. Cloning and expression in *Escherichia coli* and *Bacillus subtilis* of the hemolysin (cereolysin) determinant from *Bacillus cereus*. J. Bacteriol. **155:**681–689.

31. **Krieg, A.** 1981. The genus *Bacillus*: insect pathogens, p. 1743–1755. *In* M. P. Starr, H. Stolp, H. G. Truper, A. Ballows, and H. Schlegl (ed.), The prokaryotes, vol. 2. Springer-Verlag, New York.

32. **Kronstad, J. W., H. E. Schnepf, and H. R. Whiteley.** 1983. Diversity of locations for the *Bacillus thuringiensis* crystal protein gene. J. Bacteriol. **154:**419–428.

33. **Lecadet, M.-M., and H. deBarjac.** 1981. *Bacillus thuringiensis* beta-exotoxin, p. 293–321. *In* E. W. Davidson (ed.), Pathogenesis of invertebrate microbial diseases. Allanheld, Osmun and Co., Totowa, N.J.

34. **Leppla, S. H.** 1982. Anthrax toxin edema factor: a bacterial adenylate cyclase that increases cyclic AMP concentrations in eukaryotic cells. Proc. Natl. Acad. Sci. U.S.A. **79:**3162–3166.

35. **Lilley, M., R. N. Ruffel, and H. J. Somerville.** 1980. Purification of the insecticidal toxin in crystals of *Bacillus thuringiensis*. J. Gen. Microbiol. **118:**1–11.

36. **Louis, J., K. Jayaraman, and J. Szulmajster.** 1984. Biocide gene(s) and biocidal activity in different strains of *Bacillus sphaericus*. Expression of the gene(s) in *E. coli* maxicells. Mol. Gen. Genet. **195:**23–28.

37. **Luthy, P., and H. R. Ebersold.** 1981. *Bacillus thuringiensis* delta-endotoxin: histopathology and molecular mode of action, p. 235–267. *In* E. W. Davidson (ed.), Pathogenesis of invertebrate microbial diseases. Allanheld, Osmun and Co. Totowa, N.J.

38. **McDonel, J. L.** 1980. *Clostridium perfringens* toxins (type A, B, C, D, E). Pharmacol. Ther. **10:**617–655.

39. **McLaughlin, J. R., C. L. Murray, and J. C. Rabinowitz.** 1981. Unique features in the ribosome binding site sequence of the Gram-positive *Staphylococcus aureus* beta-lactamase gene. J. Biol. Chem. **256:**11283–11291.

40. **Meenakshi, K., and K. Jayaraman.** 1979. On the formation of crystal proteins during sporulation in *Bacillus thuringiensis* var. *thuringiensis*. Arch. Microbiol. **120:**9–14.

41. **Minnich, S. A., and A. I. Aronson.** 1984. Regulation of protoxin synthesis in *Bacillus thuringiensis*. J. Bacteriol. **158:**447–454.

42. **Nagamatsu, H., Y. Itai, C. Hatanaka, G. Fumatsu, and K. Hayashi.** 1984. A toxic fragment from the entomocidal crystal protein of *Bacillus thuringiensis*. Agric. Biol. Chem. **48:**611–619.

43. **Norris, J. R., R. C. W. Berkeley, N. A. Logan, and A. G. O'Donnel.** 1981. The genera *Bacillus* and *Sporolactobacillus*, p. 1711–1742. *In* M. P. Starr, H. Stolp, H. G. Truper, A. Balows, and H. Schlegl (ed.), The prokaryotes, vol. 2. Springer-Verlag, New York.

44. **Ohba, M., and K. Aizawa.** 1978. Serological identification of *Bacillus thuringiensis* possessing 11a:11c flagellar antigenic structure: *Bacillus thuringiensis* subsp. *kyushuensis*. J. Invertebr. Pathol. **33:**387–388.

45. **Petit-Glatron, M.-F., and G. Rapoport.** 1975. *In vitro* and *in vivo* evidence for the existence of stable messenger ribonucleic acids in sporulating cells of *Bacillus thuringiensis*, p. 255–264. *In* P. Gerhardt, R. N. Costilow, and H. M. Sadoff (ed.), Spores VI. American Society for Microbiology, Washington, D.C.

46. **Ruhfel, R. E., N. J. Robillard, and C. B. Thorne.** 1984. Interspecies transduction of plasmids among *Bacillus anthracis*, *B. cereus*, and *B. thuringiensis*. J. Bacteriol. **157:**708–711.

47. **Samples, J. R., and H. Buettner.** 1983. Corneal ulcer

caused by a biological insecticide (*Bacillus thuringiensis*). Am. J. Ophthalmol. **95**:258–260.

48. **Schnepf, H. E., and H. R. Whiteley.** 1981. Cloning and expression of the *Bacillus thuringiensis* crystal protein gene in *Escherichia coli*. Proc. Natl. Acad. Sci. U.S.A. **78**:2893–2897.

49. **Short, J. A., P. D. Walker, R. O. Thomson, and H. J. Somerville.** 1974. The fine structure of *B. finitimus* and *B. thuringiensis* spores with special reference to the location of the crystal antigen. J. Gen. Microbiol. **84**:261–276.

50. **Somerville, H. J.** 1971. Formation of the parasporal inclusion of *Bacillus thuringiensis*. Eur. J. Biochem. **18**:226–237.

51. **Somerville, H. J., and H. V. Pockett.** 1975. An insect toxin from spores of *Bacillus thuringiensis* and *Bacillus cereus*. J. Gen. Microbiol. **87**:359–369.

52. **Stephens, M. A., N. Lang, K. Sandman, and R. Losick.** 1984. A promoter whose utilization is temporally regulated during sporulation in *Bacillus subtilis*. J. Mol. Biol. **176**:333–348.

53. **Sugiyama, H.** 1980. *Clostridium botulinum* neurotoxin. Microbiol. Rev. **44**:419–448.

54. **Thomas, W. E., and D. J. Ellar.** 1983. Mechanism of action of *Bacillus thuringiensis* var. *israelensis* insecticidal delta-endotoxin. FEBS Lett. **154**:362–368.

55. **Turnbull, P. C. B.** 1981. *Bacillus cereus* toxins. Pharmacol. Ther. **13**:453–505.

56. **Turnbull, P. C. B., and J. M. Kramer.** 1983. Non-gastrointestinal *Bacillus cereus* infections: an analysis of exotoxin production by strains isolated over a two-year period. J. Clin. Pathol. **36**:1091–1096.

57. **Tyrell, D. J., L. A. Bulla, R. E. Andrews, Jr., K. J. Kramer, L. I. Davidson, and P. Nordin.** 1981. Comparative biochemistry of entomocidal parasporal crystals of selected *Bacillus thuringiensis* strains. J. Bacteriol.

145:1052–1062.

58. **Vodkin, M. H., and S. H. Leppla.** 1983. Cloning of the protective antigen gene of *Bacillus anthracis*. Cell **34**:693–697.

59. **Warren, R. E., D. Rubenstein, D. J. Ellar, J. M. Kramer, and R. J. Gilbert.** 1984. *Bacillus thuringiensis* var. *israelensis*: protoxin activation and safety. Lancet **i**:678–679.

60. **Weiner, B. A.** 1978. Isolation and partial characterization of the parasporal body of *Bacillus popilliae*. Can. J. Microbiol. **24**:1557–1561.

61. **Whiteley, H. R., J. W. Kronstad, H. E. Schnepf, and J. P. DesRosier.** 1982. Cloning the crystal protein gene of *Bacillus thuringiensis* in *Escherichia coli*, p. 131–144. *In* A. T. Ganesan, S. Chang, and J. A. Hoch (ed.), Molecular cloning and gene regulation in *Bacilli*. Academic Press, Inc., New York.

62. **Wong, H. C., H. E. Schnepf, and H. R. Whiteley.** 1983. Transcriptional and translational start sites for the *Bacillus thuringiensis* crystal protein gene. J. Biol. Chem. **258**:1960–1967.

63. **Wright, G. G.** 1975. Anthrax toxin, p. 292–295. *In* D. Schlessinger (ed.), Microbiology—1975. American Society for Microbiology, Washington, D.C.

64. **Yamamoto, T., T. Iizuka, and J. N. Aronson.** 1983. Mosquitocidal protein of *Bacillus thuringiensis* subsp. *israelensis*: identification and partial isolation of the protein. Curr. Microbiol. **9**:279–284.

65. **Yamamoto, T., and R. E. McLaughlin.** 1981. Isolation of a protein from the parasporal crystal of *Bacillus thuringiensis* var. *kurstaki* toxic to the mosquito larva, *Aedes taeniorhynchus*. Biochem. Biophys. Res. Commun. **103**:414–421.

66. **Zuber, P., and R. Losick.** 1983. Use of a *lacZ* fusion to study the role of *spo*0 genes in developmental regulation in *Bacillus subtilis*. Cell **35**:275–283.

Cloning and Expression in *Escherichia coli* of the Crystal Protein Gene from *Bacillus thuringiensis* Strain *aizawa* 7-29 and Comparison of the Structural Organization of Genes from Different Serotypes

ANDRÉ KLIER,[1] DIDIER LERECLUS,[1] JACQUES RIBIER,[2] CATHERINE BOURGOUIN,[3] GHISLAINE MENOU,[1] MARGUERITE-M. LECADET,[1] AND GEORGES RAPOPORT[1]

Laboratoire de Biochimie Microbienne, Institut Pasteur, 75724 Paris Cedex 15,[1] Laboratoire de Biologie Cellulaire Végétale, Université Paris VII, 75251 Paris Cedex 05,[2] and Laboratoire de Lutte Biologique, Institut Pasteur, 75724 Paris Cedex 15,[3] France

Strains of the gram-positive bacterium *Bacillus thuringiensis* produce several insecticidal toxins, one of which, the delta endotoxin, is of commercial importance because of its high toxicity to many Lepidopteran larvae, except the Noctuidae (for reviews, see references 2 and 5). The delta endotoxin is a protein that appears only during the sporulation phase as a crystalline inclusion (1, 16).

It has been shown that various strains of *B. thuringiensis* harbor plasmids, some of which seem to be related to the presence of the parasporal crystal (8, 12). Recently, Gonzalez et al. (7) discovered that some *B. thuringiensis* plasmids could be transferred between two strains during growth in mixed cultures. They have demonstrated that some of these transmissible plasmids carry the structural gene of the delta endotoxin crystal. A 3-megadalton (MDa) DNA sequence (a transposonlike sequence termed Th sequence), isolated originally from a 54-MDa plasmid of strain *kurstaki* KTo after mating involving the plasmid pAMβ1 from *Streptococcus faecalis*, has been found to be related to several large host plasmids of different strains of *B. thuringiensis* (13, 13a).

Another approach to locate the crystal protein gene is through cloning, and several papers have reported cloning of the crystal protein gene from strains *kurstaki* and *berliner* (9, 10, 17). We have shown the presence of two copies of the gene in strain *berliner* 1715: one is located on a large host plasmid (42 MDa), and the other seems to be located in the chromosomal DNA. The situation is similar in the *subtoxicus* strain. In other strains the crystal protein gene appears to be located either on a large host plasmid (*kurstaki*, *sotto*, *tolworthi*) or on the chromosome (*dendrolimus*) (10).

One possible mechanism for extending the host range specificity of *B. thuringiensis* against Lepidoptera, including the Noctuidae, is to screen all the isolates available for their biological activity and to clone the gene of interest.

Such an approach has been attempted and is reported in this work. In this paper we also present evidence that the different crystal protein genes fall into two classes which differ primarily in the restriction map corresponding to the C-terminal part of the protein and in the 3' flanking region. The close association of the 3-MDa Th sequence with several crystal protein genes is also demonstrated, and a model for the structural organization of these elements is proposed.

EXPERIMENTAL PROCEDURES

Bacterial strains. Wild-type strains of *B. thuringiensis* of various serotypes originated from the OILB Centre (Institut Pasteur). Strain *aizawa* 7-29 was identified among 900 isolates of 19 different H serotypes tested (A. R. Kalfon, Thèse, Université Paris VII, Paris, France, 1984). Growth of *Escherichia coli* and *B. thuringiensis* was performed as reported previously (10).

Plasmid construction and analysis. All enzymes were used as recommended by the manufacturer. Genetic engineering was carried out by standard methods as described in the laboratory manual of Maniatis et al. (14).

Plasmids pBT15-88 and pBT42-1 were constructed as reported previously (10). The plasmid pBT45-1 was obtained by cloning *Bam*HI fragments from the 45-MDa host plasmid of strain *aizawa* 7-29 into the *Bam*HI site of the vector pUC8 (18). Identification of the recombinant plasmid carrying the crystal gene was made by hybridization, using as a probe the 2-kilobase (kb) *Pvu*II fragment of plasmid pBT15-88 which corresponds to an internal part of the crystal gene from strain *berliner* 1715. The other plasmids, pBT30-1, pBT47-1, and pBT56-1, were constructed and identified in the same manner from host plasmids of strains *sotto*, *kurstaki* HD1, and *subtoxicus*, respectively, with pHV33 as a cloning vector (15). The plasmid pMT9 was recovered after mating of strain *kurstaki* KTo

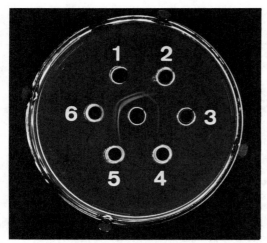

FIG. 1. Detection of the crystal antigen in *E. coli* JM83 (pBT45-1) clone a1 by double immunodiffusion. The center well contains antibodies against the *B. thuringiensis berliner* 1715 crystal protein. Well 1, Dissolved crystals from *berliner* 1715; well 2, extract of JM83 (pBT45-1) clone a1 after induction by 10^{-2} M IPTG; well 3, dissolved crystals from *aizawa* 7-29; well 4, extract of clone a1 without IPTG; well 5, extract of *E. coli* JM83 alone; well 6, extract of *E. coli* HB101 (pBT42-1).

was cloned into the unique *Bam*HI site of the expression vector pUC8, giving rise to plasmid pBT45-1. After transformation of *E. coli* JM83, one recombinant clone, referred to as clone a1, was studied in more detail.

To prove that the 13-kb *Bam*HI DNA fragment contains the crystal protein gene, extracts from clone a1 were tested by biochemical and biological means. Figure 1 shows the results of a typical double immunodiffusion experiment. An immunoprecipitate band was observed with rabbit antibodies directed against the crystal protein of the *berliner* 1715 strain (Fig. 1, well 2). The presence of this precipitation line was dependent upon induction by isopropyl-β-D-thiogalactopyranoside (IPTG) of the *E. coli* cells (Fig. 1; compare well 2 and well 4). The same line of precipitation was detected with dissolved crystals of the original strain *aizawa* 7-29 (Fig. 1, well 3). Clearly, the *E. coli* cells harboring pBT45-1 synthesized a polypeptide related to the crystal protein of strain *aizawa* 7-29. It is noteworthy that the antigenic determinant found in *aizawa* 7-29 is somewhat different from the two major determinants present in the *berliner* 1715 strain or in the extract of *E. coli* containing the cloned crystal protein gene pBT42-1 (Fig. 1, wells 1 and 6).

containing the plasmid pAMβ1 from *S. faecalis* with a crystalless strain *kurstaki* HD1. It is a deleted derivative of the plasmid pAMβ1T (13a).

Protein analysis. Sodium dodecyl sulfate-polyacrylamide gel electrophoresis and immunodetection were carried out as described previously (10).

Electron microscopy. The bacteria were fixed as described by Charles and de Barjac (3). Ultrathin sections were stained with lead citrate. The method used in the formation of homoduplex or heteroduplex DNA molecules was that described by Davis et al. (6). The micrographs correspond to 32,000× magnification of the DNA molecules.

RESULTS AND DISCUSSION

Cloning and expression in *E. coli* of the crystal protein gene from strain *aizawa* 7-29 active against *Spodoptera littoralis*. The purified host plasmids from an isolate of serotype 7 which was active against *S. littoralis* were hybridized with the 2-kb *Pvu*II fragment of pBT15-88 harboring the chromosomal crystal gene from strain *berliner* 1715. Among the five major plasmids detected in such a strain (*aizawa* 7-29), only one reacted strongly with the probe. This 45-MDa plasmid contained a *Bam*HI fragment of 13 kb and a *Pvu*II fragment of 8 kb which hybridized after Southern blotting with the probe described above (data not shown). The 13-kb *Bam*HI fragment

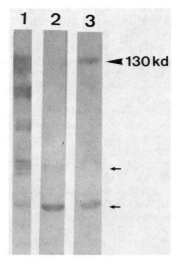

FIG. 2. Size of the polypeptide antigen synthesized in JM83 (pBT45-1) clone a1. Autoradiogram of a solid-phase radioimmunoassay of polypeptides reacting with antibodies against the *berliner* 1715 crystal protein after sodium dodecyl sulfate-polyacrylamide gel electrophoresis. Lane 1, Dissolved crystals from strain *aizawa* 7-29; lane 2, extract of clone a1 without IPTG; lane 3, extract of clone a1 induced by 10^{-2} M IPTG. The main arrow indicates the migration of the intact crystal subunit (130 kilodaltons). Small arrows show the presence of nonspecific polypeptides.

Extracts prepared from clone a1 were also electrophoresed on a sodium dodecyl sulfate-polyacrylamide gel, transferred on nitrocellulose filters, and treated with specific antibodies and finally with ^{125}I-labeled protein A from *Staphylococcus aureus* (Fig. 2). Several polypeptides gave a reaction with the antibodies, and one had a molecular weight of 130,000 as did the original crystal protein subunit (Fig. 2, lane 3). This polypeptide was missing in the extract of clone a1 not induced by IPTG (Fig. 2, lane 2). This result indicates that the crystal protein gene is orientated in the same direction as the *lacZ* promoter of the pUC8 vector.

In *E. coli* containing pBT45-1 and producing crystal protein antigen, phase-contrast microscopy revealed the presence of a dark body in each cell. The size of the inclusions was estimated by electron microscopy to about 0.3 μm; however, they did not display a crystallike or regular organization (Fig. 3). They were quite similar to those already described in *E. coli* cells containing the plasmid pBT42-1 (11).

Extracts of the recombinant clone a1 were then tested for their toxicity against *Spodoptera littoralis* larvae. The results of the bioassays are reported in Table 1. First, it is clear that the recombinant clone a1 possesses activity against L1 instar larvae of *S. littoralis* and that the expression of the toxin in *E. coli* is under the control of the *lacZ* promoter, since the death rate was significantly increased after induction by IPTG. Second, the results indicate that the other strains used which are normally less active against the Noctuidae, such as *berliner*, *sotto*, *subtoxicus*, and *kurstaki*, show lower toxicity than that produced by clone a1 or the original

TABLE 1. Toxicity assays on *Spodoptera littoralis* larvae[a]

Source of extract	Vol tested (ml)	No. of dead larvae after:	
		3 days	6 days
Controls			
Water	0.2	0	0
E. coli JM83	0.2	3	3
	0.4	3	3
E. coli HB101	0.2	3	4
B. thuringiensis			
berliner 1715 (H1)	0.2	7	9
kurstaki HD1 (H3a, 3b)	0.2	7	8
sotto (H4a, 4b)	0.2	5	6
subtoxicus (H6)	0.2	4	6
aizawa 7-29 (H7)	0.2	20	20
E. coli			
HB101 (pBT 42-1)	0.2	4	7
	0.4	6	8
HB101 (pBT 47-1)	0.2	2	2
	0.4	3	5
HB101 (pBT 30-1)	0.2	2	3
	0.4	4	4
HB101 (pBT 56-1)	0.2	0	1
	0.4	4	4
JM83 (pBT 45-1)	0.2	7	9
	0.4	7	11
JM83 (pBT 45-1)	0.2	13	17*
+ 10^{-2} M IPTG	0.4	16	19*

[a] Overnight cultures of the different bacterial strains in rich medium were concentrated 10-fold and disrupted by ultrasonic desintegration. Extracts of 0.2 or 0.4 ml were sprayed on petri dishes (30 cm^2) containing 10 ml of solid nutrient medium. The assays were performed with 20 *S. littoralis* larvae (first instar) for each dilution. The dead larvae were scored after 3 or 6 days. The asterisks indicate that the survivors were affected in their growth and development.

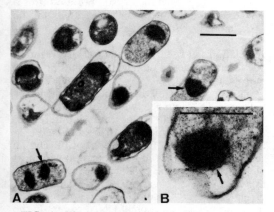

FIG. 3. Electron micrograph of *E. coli* JM83 (pBT45-1) clone a1 induced with IPTG and producing the crystal protein from strain *aizawa* 7-29. Typical inclusions are indicated by arrows. (A) ×10,000, bar = 1 μm; (B) ×40,000, bar = 0.5 μm. (By courtesy of J.-F. Charles, Institut Pasteur.)

strain *aizawa* 7-29. Extracts of recombinant clones of *E. coli* harboring the crystal protein genes of plasmid origin of the former strains of *B. thuringiensis* (pBT42-1, pBT30-1, pBT56-1, and pBT47-1) also have little effect in the assay. Furthermore, the larvae which were resistant to treatment with the extract of clone a1 were nevertheless severely affected in their growth and development, in contrast to the other resistant larvae, which reached the L3 to L4 instar.

Structural organization of the crystal protein genes and their association with the Th sequence. The restriction maps of the six cloned crystal protein genes of different *B. thuringiensis* serotypes, one of chromosomal origin (pBT15-88) and five of plasmid origin (pBT42-1, 47-1, 30-1, 56-1, and 45-1) were compared after digestion by *Pvu*II, *Eco*RI, or *Hind*III. Hybridization after Southern blotting was performed with the 2-kb *Pvu*II fragment of pBT15-88 as a probe. Host plasmid and chromosomal DNAs from strains *kurstaki* KTo (serotype 3a, 3b), *dendrolimus*

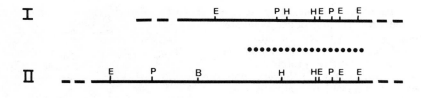

FIG. 4. Simplified restriction maps of the crystal protein genes. The dotted line indicates the position of the coding sequence of the crystal gene as described by Wong et al. (19). B, *Bam*HI; E, *Eco*RI, H, *Hin*dIII; P, *Pvu*II.

(serotype 4a, 4b), *subtoxicus* (serotype 6), *entomocidus* (serotype 6), and *aizawa* 7-29 (serotype 7) were treated in the same manner.

Two kinds of restriction maps were obtained for the different crystal protein genes examined

TABLE 2. Organization of the crystal protein genes of *B. thuringiensis*[a]

Strain and H serotype	Origin of the DNA used	Localization of the crystal genes in the *Pvu*II fragments of:		Association of the Th sequence with the crystal genes
		2 kb (type I)	8 kb (type II)	
berliner 1715 (1)	pBT15-88 (C)	+		−
	pBT42-1 (P)		+	+
kurstaki HD1 (3a, 3b)	pBT47-1 (P)	+		ND
kurstaki KTo (3a, 3b)	C	ND	ND	+
	P		+	+
sotto (4a, 4b)	pBT30-1 (P)	+		+
dendrolimus (4a, 4b)	C	ND	ND	ND
	P	−	−	−
subtoxicus (6)	C	+		ND
	pBT56-1 (P)		+	−
entomocidus (6)	C	+		−
aizawa 7-29 (7)	C	+		−
	pBT45-1 (P)		+	+

[a] The crystal genes of six different origins were cloned as described in the text. The presence of the crystal genes was examined in the plasmid and chromosomal DNAs after treatment with *Pvu*II, *Eco*RI, or *Hin*dIII, Southern blotting, and hybridization with the 2-kb *Pvu*II fragment of plasmid pBT 15-88 used as a probe. To detect the Th sequence, the plasmid DNA was treated with both *Kpn*I and *Pvu*II and the chromosomal DNA was restricted with *Bam*HI. Hybridization was performed after Southern blotting, using as a probe the 2-MDa *Kpn*I-*Pvu*II fragment which corresponds to the internal part of the Th sequence. Plus and minus signs indicate a positive or negative signal with the corresponding probes; P, DNA from host plasmid origin; C, DNA from chromosomal origin. ND, Not done.

(Fig. 4 and Table 2). They differ essentially in the position of one *Pvu*II site and one *Eco*RI site. In the type I structure a 2-kb *Pvu*II fragment is present, whereas in structure II it is replaced by a fragment of 8 kb. The difference in the position of the *Eco*RI sites has already been used to distinguish between the two genes of plasmid and chromosomal origins (10). Another minor difference resides in the distance between the two *Hin*dIII sites present: in structure I the *Hin*dIII fragment is 1.2 kb long, whereas in structure II it has a size of 1.3 kb. Structure I is very similar to that described by Wong et al. (19) for the organization of the crystal protein gene of strain *kurstaki* HD1, and the position and orientation of the crystal protein genes were deduced from the reported physical map and DNA sequence corresponding to the N-terminal portion of this protein (19).

It is tempting to postulate that structures I and II are very similar in their DNA sequences coding for the N-terminal part of the crystal protein and differ primarily in the region coding for the C-terminal portion of the protein and its 3′ flanking region. This agrees with the results of Chestukhina et al. (4), who suggested that the N-terminal region of the crystal protein forms a stable compact domain corresponding to the toxic moiety of the molecule. It is also clear from the results in Table 2 that there is no correlation between the structural organization of the crystal genes and their plasmid or chromosomal origins.

The association of the crystal protein genes with the Th sequence which hybridized with several host plasmids from strains of different *B. thuringiensis* serotypes was also examined (Table 2) and was found in 5 cases out of 10. The Th sequence was observed in association with host plasmids carrying the crystal protein genes mentioned in Table 2, except in the 56-MDa plasmid of the *subtoxicus* strain. It was not detected in the restricted chromosomal DNA, except in the *kurstaki* KTo strain.

Structural organization of the Th sequence and of the crystal protein gene in the 42-MDa plasmid of strain *berliner* 1715. On the basis of the previously reported map of pBT42-1 (10), addi-

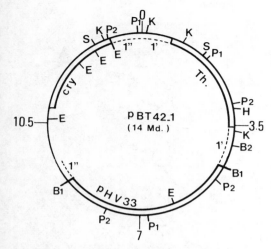

FIG. 5. Localization of the crystal gene (*cry*) and of the Th sequence on plasmid pBT42-1. B1, *Bam*HI; B2, *Bgl*II; E, *Eco*RI; H, *Hind*III; K, *Kpn*I; P1, *Pst*I; P2, *Pvu*II. Sequences marked 1' and 1'' are discussed in the text. The numbers 0, 3.5, 7, and 10.5 indicate the distance in MDa on the restriction map.

tional restriction endonuclease and hybridization analyses indicated that the Th sequence has been cloned in pBT42-1 together with the crystal gene. The physical map presented in Fig. 5 shows that the Th sequence is separated from the crystal gene by an approximately 1.3-MDa DNA sequence. DNA-DNA hybridization revealed homologies between this DNA segment and the external junction fragments of the crystal gene and of the Th sequence (fragments 1' and 1'').

Figure 6 shows the localization of the Th sequence on pBT42-1 and pMT9 by homoduplex analysis. The plasmid pMT9 was derived from the *S. faecalis* plasmid pAMβ1 after insertion of the Th sequence and deletion in vivo (see Experimental Procedures). Electron microscopic examination of the self-annealed pBT42-1 revealed the formation of three single-stranded loops joined by two double-stranded stems (Fig. 6A). The cloverleaf structure observed can be explained by the association of the two 1' segments and of the two 1'' segments. This hypothesis implies that the 1' and 1'' segments are present in inverted orientation on both sides of the crystal gene and of the Th sequence. The reannealed single-strand structure obtained with an *Eco*RI restriction fragment of pMT9 shows a loop, a double-stranded stem, and two linear junction fragments (Fig. 6B). Figure 6C represents the heteroduplex formed between pBT42-1 and the linearized pMT9. Consequently, it appears that the double-stranded loop (a) corresponds to the Th sequence (Fig. 6C). Similarly, heteroduplex analysis carried out with pBT42-1

and pHV33 (not shown) indicated that the loop marked b in Fig. 6C corresponds to the pHV33 vector. We deduce, therefore, that the loop marked c contains the crystal gene.

The localization of the Th sequence and of the crystal gene was next examined in the 42-MDa host plasmid of strain *berliner* 1715 (pBT42) from which the hybrid plasmid pBT42-1 was derived. The self-annealed single strand of pBT42 (Fig. 7A) generates three single-stranded loops separated by two double-stranded stems. We assume that the smaller distal loop (3 MDa) corresponds to the Th sequence and that its junction stem (1.3 MDa) marked IR1 represents the sum of the 1' and 1'' DNA segments (0.9 + 0.4 MDa). In this respect, we hypothesize that in pBT42 the entire IR1 sequence is repeated in inverted orientation on both sides of the crystal gene and of the Th sequence.

The interpretation of the results led us to propose the scheme in Fig. 7A and the model in

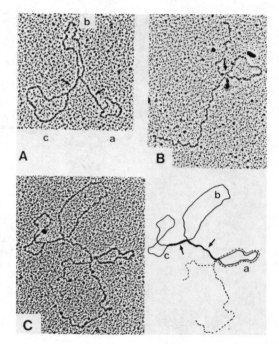

FIG. 6. Localization of the Th sequence on pBT42-1 and pMT9 by homoduplex and heteroduplex analysis. (A) Self-annealed single strand of pBT42-1; (B) self-annealed single strand of *Eco*RI-treated pMT9; (C) heteroduplex between pBT42-1 and *Eco*RI-treated pMT9. The letters a, b, and c are defined in the text. On the tracing, arrows indicate double-stranded stems. The dotted line represents the *Eco*RI-treated pMT9. The single-stranded linear segments correspond to the *Eco*RI junction fragments of pMT9. The double-stranded loop marked a results from the association of the two DNA segments corresponding to the Th sequences of pBT42-1 and pMT9.

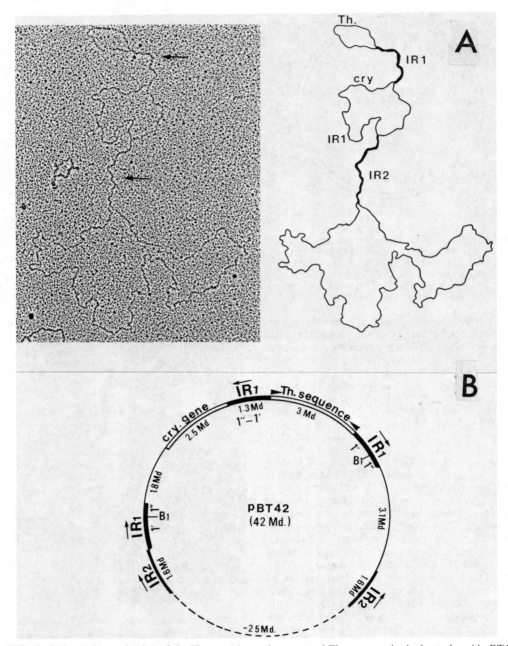

FIG. 7. Structural organization of the IR, crystal protein gene, and Th sequence in the host plasmid pBT42. (A) Electron micrograph of the self-annealed single strand of pBT42, with the proposed corresponding sketch. (B) Model proposed. The size of each sequence is indicated in MDa (Md). The dashed line represents the unexamined part of pBT42 (25 MDa). The small arrowheads localized at each extremity of the Th sequence represent the inverted repeats forming the short double-stranded stem (0.2 MDa) during homoduplex formation. B1 indicates the *Bam*HI sites dividing IR1 into two fragments, 1' (0.9 MDa) and 1'' (0.4 MDa), which delimit the 9.5-MDa DNA fragment cloned in pBT42-1.

Fig. 7B for the structural organization of the IR sequence, the crystal gene, and the Th sequence in the native 42-MDa plasmid. This model implies the presence of three copies of IR1, which would explain the cloverleaf structure of the self-annealed pBT42-1 (Fig. 6) and the 1.3-MDa double-stranded stem obtained with pBT42 (Fig. 7A). As can be deduced from the reannealed single-strand structure of pBT42-1, the crystal gene whose size corresponds to 2.5 MDa (10) is located inside the 4.3-MDa sequence, itself flanked by two IR1s and which formed loop c in Fig. 6. The pairing of these two IR1s is not observed with pBT42 (Fig. 7A), since no loop of 4.3 MDa could be seen. To explain this fact we suggest that the association of the two IR1s flanking the Th sequence occurs preferentially as a result of the smaller size of its internal sequence (3 MDa). This implies that, during the reannealing of single strands of pBT42, the crystal gene and the adjacent IR1 (in orientation 1′ to 1″) would give a single-stranded fragment of 5.6 MDa (4.3 MDa + 1.3 MDa). The expected size is in agreement with the longer internal single-stranded sequence (5.8 MDa) found in the structure of pBT42 shown in Fig. 7A. This observation implies the presence of a second pair of inverted repeated sequences, one of which is contiguous to the IR1 at the extremity of the 5.6-MDa sequence. These inverted repeats that form the internal stem of 1.6 MDa (Fig. 7A) are designated IR2. The model presented in Fig. 7B strongly suggests that the crystal protein gene and the Th sequence are part of a transposonlike structure flanked by two elements of IR sequences.

SUMMARY

The crystal gene from an isolate of H7 serotype (strain *aizawa* 7-29) active against *S. littoralis* larvae was cloned in pUC8 and expressed in *E. coli*, where it retains its biological specificity.

The comparison of the structural organization of the crystal protein genes of plasmid and chromosomal origins from different serotypes, six of which have been cloned in *E. coli*, revealed the existence of two classes of genes which are very similar in the restriction map corresponding to the N-terminal part of the protein and which differ essentially in the 3′ region.

In several cases a transposonlike sequence was found associated with the crystal gene in the same host plasmid, and a model for their structural organization is proposed.

This work was supported by the Centre National de la Recherche Scientifique (ATP Microbiologie 1983) and the Fondation pour la Recherche Médicale.

ADDENDUM IN PROOF

Whiteley et al. report in this volume the complex arrangement of inverted repeat sequences around the crystal protein gene.

LITERATURE CITED

1. **Bulla, L. A., Jr., K. J. Kramer, and L. I. Davidson.** 1977. Characterization of the entomocidal parasporal crystal of *Bacillus thuringiensis*. J. Bacteriol. **130**:375–383.
2. **Burges, H. D. (ed.).** 1981. Microbial control of pests and plant diseases 1970–1980. Academic Press, London.
3. **Charles, J.-F., and H. de Barjac.** 1982. Sporulation et cristallogénèse de *Bacillus thuringiensis* var. *israelensis* en microscopie électronique. Ann. Microbiol. (Paris) **133A**:425–442.
4. **Chestukhina, G. G., L. I. Kostina, A. L. Mikhailova, S. A. Tyurin, F. S. Klepikova, and V. M. Stepanov.** 1982. The main features of *Bacillus thuringiensis* δ-endotoxin molecular structure. Arch. Microbiol. **132**:159–162.
5. **Davidson, E. W. (ed.).** 1981. Pathogenesis of invertebrate microbial diseases. Allanheld, Osmun & Co., Totowa, N.J.
6. **Davis, R. W., M. Simon, and N. Davidson.** 1971. Electron microscope heteroduplex methods for mapping regions of base sequence homology in nucleic acids. Methods Enzymol. **21**:413–428.
7. **Gonzalez, J. M., Jr., B. J. Brown, and B. C. Carlton.** 1982. Transfer of *Bacillus thuringiensis* plasmids coding for δ-endotoxin among strains of *B. thuringiensis* and *B. cereus*. Proc. Natl. Acad. Sci. U.S.A. **79**:6951–6955.
8. **Gonzalez, J. M., Jr., H. T. Dulmage, and B. C. Carlton.** 1981. Correlation between specific plasmids and δ-endotoxin production in *Bacillus thuringiensis*. Plasmid **5**:351–365.
9. **Held, G. A., L. A. Bulla, Jr., E. Ferrari, J. Hoch, A. I. Aronson, and S. A. Minnich.** 1982. Cloning and localization of the lepidopteran protoxin gene of *Bacillus thuringiensis* subsp. *kurstaki*. Proc. Natl. Acad. Sci. U.S.A. **79**:6065–6069.
10. **Klier, A., F. Fargette, J. Ribier, and G. Rapoport.** 1982. Cloning and expression of the crystal protein genes from *Bacillus thuringiensis* strain *berliner* 1715. EMBO J. **1**:791–799.
11. **Klier, A., and G. Rapoport.** 1984. Cloning and heterospecific expression of the crystal protein genes from *B. thuringiensis*, p. 65–72. *In* J. E. Alouf, F. J. Fehrenbach, J. H. Freer, and J. Jeljaszewicz (ed.), Bacterial toxins. Academic Press, London.
12. **Lereclus, D., M.-M. Lecadet, J. Ribier, and R. Dedonder.** 1982. Molecular relationships among plasmids of *Bacillus thuringiensis*: conserved sequences through 11 crystalliferous strains. Mol. Gen. Genet. **186**:391–398.
13. **Lereclus, D., G. Menou, and M.-M. Lecadet.** 1983. Isolation of a DNA sequence related to several plasmids from *Bacillus thuringiensis* after a mating involving the *Streptococcus faecalis* plasmid pAMβ1. Mol. Gen. Genet. **191**:307–313.
13a. **Lereclus, D., J. Ribier, A. Klier, G. Menou, and M.-M. Lecadet.** 1984. A transposon-like structure related to the δ-endotoxin gene of *Bacillus thuringiensis*. EMBO J. **3**:2561–2567.
14. **Maniatis, T., E. F. Fritsch, and J. Sambrook.** 1981. Molecular cloning. A laboratory manual. Cold Spring Harbor Laboratory, Cold Spring Harbor, N.Y.
15. **Primrose, S. B., and S. D. Ehrlich.** 1981. Isolation of plasmid deletion mutants and study of their instability. Plasmid **6**:193–201.
16. **Ribier, J., and M.-M. Lecadet.** 1973. Etude ultrastructurale et cinétique de la sporulation de *Bacillus thuringiensis* var. *berliner* 1715. Remarques sur la formation de l'inclusion parasporale. Ann. Microbiol. (Paris) **124A**:311–344.

17. **Schnepf, H. E., and H. R. Whiteley.** 1981. Cloning and expression of the *Bacillus thuringiensis* crystal protein gene in *Escherichia coli*. Proc. Natl. Acad. Sci. U.S.A. **78:**2893–2897.

18. **Vieira, J., and J. Messing.** 1982. The pUC plasmids, an M13mp7-derived system for insertion mutagenesis and sequencing with synthetic universal primers. Gene **19:**259–268.

19. **Wong, H. C., H. E. Schnepf, and H. R. Whiteley.** 1983. Transcriptional and translational start sites for the *Bacillus thuringiensis* crystal protein gene. J. Biol. Chem. **258:**1960–1967.

Structure and Expression of Cloned *Bacillus thuringiensis* Toxin Genes

H. R. WHITELEY, J. W. KRONSTAD, AND H. E. SCHNEPF

Department of Microbiology and Immunology, University of Washington, Seattle, Washington 98195

Sporulating cultures of *Bacillus thuringiensis* synthesize one or more large intracellular entomocidal crystals. Many subspecies have been isolated, and most are toxic to Lepidopteran larvae. Synthesis of the crystals begins at about stage II of sporulation and extends through stage V (2); in sporulated cultures, the crystals can account for 20 to 30% of the dry weight (14). The crystals are composed of one to three peptides of M_r 130,000 to 160,000, although smaller peptides have also been reported for some subspecies (3, 5). The crystal protein, which may also be part of the spore coat of some subspecies (1, 11), is a protoxin which can be converted by proteolysis both in vitro and in vivo to a toxic peptide of ca. M_r 70,000. The sporulation-dependent synthesis, the production of the protoxin in large amounts, and the ease of isolating crystals make this protein an attractive marker for the study of gene expression during sporulation.

As a first step in studying the mechanisms regulating the synthesis of the crystal protein, we cloned a gene from a large plasmid in *B. thuringiensis* subsp. *kurstaki* HD1-Dipel into *Escherichia coli* (17). The gene was located on the insert in pBR322 by Tn5 mutagenesis and was mapped by use of restriction endonucleases (19). Intragenic DNA fragments were subcloned to provide probes for identifying the 5' and 3' ends of the gene by S1 nuclease mapping, and the DNA sequence was determined for the entire gene and ca. 450 base pairs upstream from the initiation sites (19; H. E. Schnepf, H. C. Wong, and H. R. Whiteley, manuscript in preparation). Two developmentally regulated adjacent transcriptional start sites were identified by S1 nuclease mapping; the presumed promoter regions for these sites showed no base sequence homology to the −10 and −35 regions of promoters recognized by the major vegetative RNA polymerase of *B. subtilis*, but a significant degree of homology was found with a sporulation-specific gene (14a, 15). Gene-specific probes were also used in a hybridization survey to determine the distribution of homologous genes in 22 strains of *B. thuringiensis* (10). In the present communication, two additional facets of the structure of the gene are discussed: (i) evidence for the presence of inverted repeated elements flanking the gene in three strains and (ii) experiments to delineate the DNA sequences coding for the toxic M_r 70,000 peptide within the larger protoxin gene.

EXPERIMENTAL PROCEDURES

Methods used for growing cultures, preparing plasmids, cloning, restriction mapping, hybridization, immunoblotting toxicity assays, and the electrophoretic separation of plasmids, peptides, and restriction fragments have been described (10, 17, 19). Inverted repeated sequences were detected by the method of Ohtsubo and Ohtsubo (16). A complete description of the plasmids constructed for the deletion of internal repeat (IR) elements is given elsewhere (10a); plasmids used in the localization of the toxic peptide within the protoxin molecule will also be described elsewhere (H. E. Schnepf and H. R. Whiteley, manuscript in preparation).

RESULTS AND DISCUSSION

Multiple crystal protein genes. Intragenic fragments of the cloned protoxin gene from *B. thuringiensis* subsp. *kurstaki* HD1-Dipel were used as hybridization probes to determine the distribution of homologous sequences in 22 strains of *B. thuringiensis* (10). In 17 of 18 strains which reacted with the probe, hybridization was with plasmids, and in seven of these, more than one plasmid was hybridized. A more detailed investigation was made of three strains of subspecies *kurstaki* and three strains of subspecies *thuringiensis* in which one to three plasmids were hybridized. Held et al. (8) and Klier et al. (9) have reported that in one *kurstaki* strain and one *thuringiensis* strain, respectively, protoxin genes are present on the chromosome as well as on plasmids. Comparisons of restriction endonuclease digests of total DNA and purified plasmid DNA of the three *kurstaki* and three *thuringiensis* strains we selected gave no evidence for the presence of a chromosomal gene homologous to our probe. Unexpectedly, these experiments revealed that almost all of the hybridized *Hin*dIII fragments fell into discrete size classes and that in some strains more than one fragment was hybridized. The numbers and sizes of the hybridized *Hin*dIII fragments were as follows: three fragments (6.6, 5.3, and 4.5

kilobases [kb]) in *kurstaki* strain HD1; two fragments (6.6 and 4.5 kb) in *kurstaki* HD1-Dipel, one fragment (6.6 kb) in *kurstaki* HD73, three fragments in *thuringiensis berliner* 1715 (5.3, 4.5, and 2.3 kb), and two fragments (5.3 and 4.5 kb) in *thuringiensis* HD290 and HD120.

To compare the genes in three of the size classes, we selected the gene cloned from *kurstaki* HD1-Dipel as a representative of the "4.5 gene class" and cloned three additional genes (H. R. Whiteley, H. E. Schnepf, J. W. Kronstad, and H. C. Wong, in press). The "6.6 gene" was obtained from the ca. 75-kb plasmid of *kurstaki* HD73, a "5.3 gene" was cloned from the ca. 66-kb plasmid of *kurstaki* HD1, and another "5.3 gene" was cloned from the ca. 86-kb plasmid doublet from *thuringiensis* HD2. Mapping the three size classes of HindIII fragments by digestion with restriction endonucleases showed that the genes were very similar but not identical.

Conflicting reports concerning the size and number of proteins in crystals of different subspecies have resulted, in large part, from the fact that preparations of crystals are usually contaminated with proteases. Some investigators (18) have reported that crystals consist of a single polypeptide, and others (3, 5) have detected one to three polypeptides. Immunoblotting analyses of the crystal protein peptides synthesized by *E. coli* strains harboring the cloned genes showed that the 4.5 and 6.6 genes encoded peptides with very similar electrophoretic mobilities corresponding to ca. M_r 135,000, whereas the peptide encoded by the 5.3 gene migrated more rapidly (ca. M_r 130,000). In agreement with these findings, electrophoresis of solubilized preparations of crystals showed a single band of M_r 135,000 in *kurstaki* HD73 (6.6 gene), a band of M_r 130,000 in *thuringiensis* HD2 (5.3 gene), a single broad band in HD1-Dipel (6.6 and 4.5 genes), and a broad band of ca. M_r 135,000 plus one of M_r 130,000 in *kurstaki* HD1 (6.6, 5.3, and 4.5 genes). Thus, the composition of the crystals agrees with the composition predicted from the distribution of genes defined by the three size classes of HindIII fragments, indicating that some strains of *B. thuringiensis* contain and express more than one crystal protein gene.

IR sequences in three strains of *B. thuringiensis*. The finding that these related genes are located on different plasmids in the strains mentioned above, the finding that the same or related crystal protein genes in other subspecies are present on other large plasmids and possibly on the chromosome in at least one subspecies, and the reports from other investigators (8, 9) that protoxin genes are located on the chromosome indicate that there must be a substantial amount of plasmid rearrangement or transposi-

tion, or both, in the various subspecies. Several lines of evidence support the existence of such mechanisms in *B. thuringiensis*. First, curing experiments (7; B. C. Carlton and J. M. Gonzalez, in press) suggest that *B. thuringiensis* plasmids may undergo numerous deletions and insertions and that plasmid DNA may be inserted into the chromosome. Second, hybridization experiments indicate a widespread distribution of common sequences in different plasmids (12). Lastly, Lereclus et al. (13) found that, when *Streptococcus faecalis* plasmid pAMβ1 was transferred into *B. thuringiensis* and then reisolated, the plasmid contained a segment of *B. thuringiensis* plasmid DNA; this DNA hybridized to large plasmids from several other subspecies.

Transposable elements frequently have direct or inverted repeated sequences at their termini (4). To test for the presence of such repeated sequences, we have made a detailed study of the ca. 75-kb plasmid of *kurstaki* HD73, which carries the 6.6 gene described above (10a). This plasmid was purified, linearized, denatured, allowed to renature briefly, and digested with S1 nuclease to determine whether there were any inverted repeat "snap-back" elements (16). Two fragments of 2,150 and 1,750 base pairs were found and mapped by digestion with restriction enzymes. Smaller fragments from within each of the internal repeat (IR) elements were subcloned and used as hybridization probes. Three copies of IR2150 and two copies of IR1750 were found around the protoxin gene. In addition, two other partially homologous versions of IR1750 were identified, and these were also located near the gene. The complex arrangement of these elements and their orientations are shown in Fig. 1A.

Comparisons of the patterns of hybridization of subcloned probes derived from IR2150 and IR1750 to BglII, PstI, and HindIII digests of purified preparations of the ca. 75-kb plasmid of *kurstaki* HD73 with similar digests of total DNA from this strain clearly showed the presence of multiple copies of both IR2150 and IR1750 on the chromosome (10a). Hybridization of the subcloned probes to 15 strains of *B. thuringiensis* demonstrated the presence of the IR sequences on all plasmids carrying the crystal protein gene except in subspecies *alesti*.

Figures 1B and C compare the arrangement of IR2150 and IR1750 sequences around the two 5.3 genes cloned from *kurstaki* HD1 and *thuringiensis* HD2 with that found in the 6.6 gene (Fig. 1A). Alignment by restriction sites shows a remarkable similarity in the arrangement of IR elements and suggests a deletion of ca. 7 kb near the 3' end of two 5.3 genes relative to the 6.6 gene. Hybridization experiments provided con-

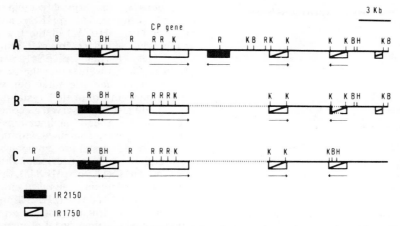

FIG. 1. Restriction maps of DNA fragments containing the crystal protein gene cloned from *B. thuringiensis* subspp. *kurstaki* HD73 (A), *kurstaki* HD1 (B), and *thuringiensis* HD2 (C) showing the locations of inverted repeated sequences. A third copy of IR2150 is located outside the mapped region shown in A and B. The following abbreviations are used: B, *Bgl*II; R, *Eco*RI; H, *Bam*HI; K, *Kpn*I. The dashed lines indicate deleted regions.

firmation that the dotted segment shown in Fig. 1B and C is missing in the 5.3 genes. The deletion may extend into the 3′ end of the crystal protein gene, possibly accounting for the smaller size of the peptide produced by the strains carrying this gene. Comparisons of restriction patterns obtained by digestion of purified preparations of the ca. 66-, 75-, and 86 (doublet)-kb plasmids (from *kurstaki* HD1, *kurstaki* HD73, and *thuringiensis* HD2, respectively) indicate that the ca. 66- and 75-kb plasmids are, in fact, very similar except for the deleted region and that the 86-kb plasmid doublet is quite different from either of the other two (data not shown). Thus, the similar organization of inverted repeat DNA around the three genes is particularly striking and implies that this crystal protein gene and its flanking sequences have moved to a different replicon either as an intact transposon or through a more complex set of recombinational events involving the IR2150 and IR1750 elements.

Portion of the protoxin which is responsible for toxicity. Although it has been proposed that proteolysis generates a toxic peptide of ca. M_r 70,000 from the N-terminal half of the protoxin molecule, different N-terminal amino acids and multiple termini have been reported for the toxic peptide (5, 14), and the portion of the gene coding for the 3′ end of the toxic peptide has not been identified. We have investigated the location of the toxic peptide within the protoxin molecule cloned from *kurstaki* HD1-Dipel by constructing a series of plasmids having increasing amounts of crystal protein gene DNA extending from the 5′ end; these plasmids had different

5′ and 3′ end modifications. Extracts of *E. coli* strains bearing the plasmids were analyzed for toxicity to larvae of the tobacco hornworm, and immunoblotting was used to identify antigenic peptides.

The 3′ end of the toxic peptide was located by first constructing a vector, pHES16, which contains the first 564 codons of the gene; extracts of *E. coli* strains bearing this plasmid were not lethal to larvae (Fig. 2). A second vector, pHES19, was constructed in which pHES16 was extended by the addition of an adjacent stretch of DNA (the *Hin*dIII-E fragment of the crystal protein gene; 19); extracts of this strain were toxic to the larvae. Additional vectors, containing different lengths of the *Hin*dIII-E fragment, were prepared and tested for toxicity as shown in Fig. 2.

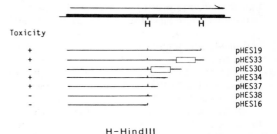

FIG. 2. Expanded restriction endonuclease map of the crystal protein gene (thickened line) showing the direction of transcription (arrow). Below are shown the portions of the crystal protein gene remaining in the indicated plasmids and the toxicity of extracts of *E. coli* strains containing these plasmids. Boxed segments of the lines indicate Tn5 sequences.

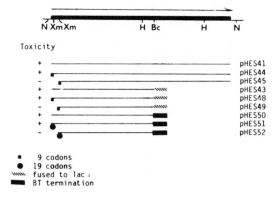

FIG. 3. Expanded restriction endonuclease map of the crystal protein gene (thickened line) showing the direction of transcription (arrow). The lines below the map show the amount of crystal protein gene remaining in the indicated plasmids and the toxicity of extracts of *E. coli* strains containing these plasmids. Boxes and circles at the left end indicate translational fusions to *lacZ* at codon 9 or 19 of the crystal protein gene; rectangles at the right end indicate fusions to the *lac* α peptide and the crystal protein terminator sequence.

Immunoblot analyses (data not shown) demonstrated that the *E. coli* strains which synthesized a toxic peptide contained a crystal protein antigen of ca. M_r 70,000 and an additional longer peptide whose size was roughly proportional to the length of the inserted crystal protein DNA. Nontoxic strains shown in Fig. 2 contained a peptide of ca. M_r 58,000; this peptide was detected poorly, possibly indicating increased susceptibility to proteolysis or loss of a major antigenic determinant.

These results are in agreement with previous work showing that the N terminus was present on a toxic peptide (5). They also demonstrate that toxicity requires the presence of sequences between codon 603 (the length of the crystal protein gene DNA in plasmid pHES38) and codon 645 (the length of the crystal protein gene DNA in plasmid pHES37). The latter region is also the site of increased susceptibility to proteolysis, leading to the production of an N-terminal M_r 70,000 peptide.

Changes at the 5′ and 3′ ends of DNA encoding the toxic peptide. To delineate the toxic peptide more accurately, we tested two types of 5′ end modifications (diagrammed in Fig. 3) at two locations: (i) the first 10 codons of the crystal protein gene were replaced with 9 codons (pHES44 and pHES48) or 19 codons (pHES51) of the β-galactosidase and polylinker system of pUC8 and pUC13, respectively, or (ii) the first

50 codons of the crystal protein gene were replaced with 9 codons (pHES45 and pHES49) or 19 codons (pHES52) of the β-galactosidase and polylinker system of pUC8 and pUC13, respectively. Plasmids pHES41, pHES43, and pHES50 (Fig. 3) were under the transcriptional control of the *lac* promoter but were under translational control of the crystal protein ribosome binding site and served as controls for the effects of the 5′-end modifications. These 5′ modifications were combined with the following 3′ ends: (i) the crystal protein coding sequence, stop codon, and transcriptional terminator (pHES41, pHES44, and pHES45), (ii) codon 645 of the crystal protein gene fused either in phase (pHES43) or out of phase (pHES48 and pHES49) with the reading frame of the *lac* α peptide, and (iii) codon 645 of the crystal protein gene fused in phase to the final 74 codons and the transcriptional terminator of the crystal protein gene (pHES50, pHES51, and pHES52). Toxicity assays (Fig. 3) showed the plasmids having the translational start site of the crystal protein gene or the β-galactosidase fusion to codon 10 of the crystal protein gene were lethal regardless of the modification the the 3′ end of the coding sequence; in contrast, all fusions to codon 50 of the crystal protein gene were nontoxic.

Immunoblot assays (data not shown) demonstrated that extracts of *E. coli* having plasmids with the crystal protein terminator and all those with the codon 10 fusion contained an antigenic peptide of ca. M_r 70,000. Small amounts of this peptide could also be detected when the plasmid had a codon 50 fusion and either the crystal protein terminator or the *lac* α peptide at the 3′ end. Interestingly, all three extracts having plasmids with the final 74 codons and the crystal protein terminator fused to codon 645 contained a prominent M_r 50,000 peptide, suggesting that sequences at the 3′ end affect proteolysis, presumably by influencing the conformation of the polypeptide.

Another unexpected result of these experiments was the finding that larger amounts of antigenic peptides were present in extracts of *E. coli* strains having the fusion to the final 74 codons and the crystal protein terminator than in extracts of strains having the *lac* α peptide fusion. Preliminary experiments to measure the rate of protein degradation indicated that there were no marked differences in strains having different 3′ end structures. Although we have not ruled out possible effects of translation, the increased expression of crystal protein antigens in strains fused to the terminator may reflect differences in the stability of mRNAs. These observations, as well as those of H. C. Wong and S. Chang (this volume) support the proposal by Court et al. (6) that terminators may regulate gene expression by influencing the stability of mRNAs.

SUMMARY

Investigations of three strains from each of two subspecies of *B. thuringiensis* disclosed that different strains contain one, two, or three closely related genes located on different plasmids and that one of the three genes codes for a smaller peptide than the other two. The peptide composition of crystals isolated from the different strains agrees with that expected from the genetic composition, indicating expression of more than one crystal protein gene in some strains. Detailed mapping and hybridization analyses showed that three cloned genes were flanked by similar complex arrangements of two different inverted repeat sequences, possibly indicating that these sequences are involved in the transposition of the crystal protein gene, either as an intact transposon or through multiple recombination events involving the inverted repeat sequences.

Construction of a series of vectors containing various lengths of crystal protein gene DNA, toxicity assays, and immunoblotting were used to identify sequences which encode the toxic portion of the protoxin. These experiments showed that the synthesis of toxic peptide by recombinant *E. coli* strains requires DNA sequences located between codons 10 and 50 at the 5' end and between codons 603 and 645 at the 3' end. Evidence was obtained that modifications to the 3' end apparently influence the extent of gene expression and the conformation of the peptides.

This research was supported by Public Health Service grants GM-20784 and GM-26100 from the National Institute of General Medical Sciences, by grant PCM-8315859 from the National Science Foundation, and by a grant from the Cetus Corp. J.W.K. was supported by the Molecular and Cellular Biology Training Program, Public Health Service National Research Award GM007270-08 from the National Institutes of Health. H.R.W. is recipient of Public Health Service Research Career Award K6-GM-442 from the National Institute of General Medical Sciences.

LITERATURE CITED

1. **Aronson, A. I., D. J. Tyrell, P. C. Fitz-James, and L. A. Bulla, Jr.** 1981. Relationship of the synthesis of spore coat protein and parasporal crystal protein in *Bacillus thuringiensis*. J. Bacteriol. **151**:399–410.
2. **Bechtel, D. B., and L. A. Bulla, Jr.** 1976. Electron microscope study of sporulation and parasporal crystal formation in *Bacillus thuringiensis*. J. Bacteriol. **127**:1472–1481.
3. **Calabrese, D. M., K. W. Nickerson, and L. C. Lane.** 1980. A comparison of protein crystal subunit sizes in *Bacillus thuringiensis*. Can. J. Microbiol. **26**:1006–1010.
4. **Calos, M. P., and J. H. Miller.** 1980. Transposable elements. Cell **20**:579–595.
5. **Chestukhina, G. G., L. I. Kostina, A. L. Mikhailova, S. A. Tyurin, F. S. Klepikova, and V. M. Stepanov.** 1982. The main feature of *Bacillus thuringiensis*-endotoxin molecular structure. Arch. Microbiol. **132**:159–162.
6. **Court, D., T. F. Huang, and A. B. Oppenheim.** 1983. Deletion analysis of the retroregulatory site for the lambda *int* gene. J. Mol. Biol. **166**:233–240.
7. **Gonzalez, J. M., H. T. Dulmage, and B. C. Carlton.** 1981. Correlation between specific plasmids and delta-endotoxin production in *Bacillus thuringiensis*. Plasmid **5**:351–365.
8. **Held, G. A., L. A. Bulla, Jr., E. Ferrari, J. Hoch, A. I. Aronson, and S. A. Minnich.** 1982. Cloning and localization of the lepidopteran protoxin gene of *Bacillus thuringiensis* subsp. *kurstaki*. Proc. Natl. Acad. Sci. U.S.A. **79**:6065–6069.
9. **Klier, A., F. Fargette, J. Ribier, and G. Rapoport.** 1982. Cloning and expression of the crystal protein genes from *Bacillus thuringiensis* strain *berliner* 1715. EMBO J. **1**:791–799.
10. **Kronstad, J. W., H. E. Schnepf, and H. R. Whiteley.** 1983. Diversity of locations for the *Bacillus thuringiensis* crystal protein gene. J. Bacteriol. **154**:419–428.
10a. **Kronstad, J. W., and H. R. Whiteley.** 1984. Inverted repeat sequences flank a *Bacillus thuringiensis* crystal protein gene. J. Bacteriol. **160**:95–102.
11. **Lecadet, M. M., G. Chevrier, and R. Dedonder.** 1972. Analysis of a protein fraction in the spore coats of *Bacillus thuringiensis*. Eur. J. Biochem. **25**:349–358.
12. **Lereclus, D., M. M. Lecadet, J. Ribier, and R. Dedonder.** 1982. Molecular relationships among plasmids of *Bacillus thuringiensis*: conserved sequences through 11 crystalliferous strains. Mol. Gen. Genet. **186**:391–398.
13. **Lereclus, D., G. Menow, and M. M. Lecadet.** 1983. Isolation of a DNA sequence related to several plasmids from *Bacillus thuringiensis* after mating involving *Streptococcus faecalis* plasmid pAMβ1. Mol. Gen. Genet. **191**:307–313.
14. **Lilley, M., R. N. Ruffel, and H. J. Somerville.** 1980. Purification of the insecticidal toxin in crystals of *Bacillus thuringiensis*. J. Gen. Microbiol. **118**:1–11.
14a. **Losick, R., and P. Youngman.** 1984. Endospore formation in *Bacillus*, p. 63–88. *In* R. Losick and L. Shapiro (ed.), Microbial development. Cold Spring Harbor Laboratory, Cold Spring Harbor, N.Y.
15. **Moran, C. P., N. Lang, and R. Losick.** 1981. Nucleotide sequence of a *Bacillus subtilis* promoter recognized by *Bacillus subtilis* RNA polymerase containing sigma-37. Nucleic Acids Res. **9**:5979–5990.
16. **Ohtsubo, H., and E. Ohtsubo.** 1976. Isolation of inverted repeat sequences including IS1, IS2, and IS3 in *Escherichia coli* plasmids. Proc. Natl. Acad. Sci. U.S.A. **73**:2316–2320.
17. **Schnepf, H. E., and H. R. Whiteley.** 1981. Cloning and expression of the *Bacillus thuringiensis* crystal protein gene in *Escherichia coli*. Proc. Natl. Acad. Sci. U.S.A. **78**:2893–2897.
18. **Tyrell, D. J., L. A. Bulla, Jr., R. E. Andrews, K. J. Kramer, L. I. Davidson, and P. Nordin.** 1981. Comparative biochemistry of entomocidal parasporal crystals of selected *Bacillus thuringiensis* strains. J. Bacteriol. **145**:1052–1062.
19. **Wong, H. C., H. E. Schnepf, and H. R. Whiteley.** 1983. Transcriptional and translational start sites for the *Bacillus thuringiensis* crystal protein gene. J. Biol. Chem. **258**:1960–1967.

Biochemistry, Genetics, and Mode of Action of *Bacillus thuringiensis* δ-Endotoxins

DAVID J. ELLAR, WENDY E. THOMAS, BARBARA H. KNOWLES, SALLY WARD, JOHN TODD, FRANCIS DROBNIEWSKI, JANE LEWIS, TREVOR SAWYER, DAVID LAST, AND COLIN NICHOLS

Department of Biochemistry, University of Cambridge, Cambridge CB2 1QW, United Kingdom

Strains within the *Bacillus thuringiensis* species produce protein δ-endotoxins that are toxic to a wide variety of lepidopteran and some dipteran larvae (35). These δ-endotoxins are synthesized during sporulation as parasporal protein inclusions (1, 43). Most of the 20 serotypes of *B. thuringiensis* are toxic to lepidopterans and synthesize a bipyramidal crystalline endotoxin that is refractile when viewed by phase-contrast microscopy. The *B. thuringiensis* subsp. *israelensis* δ-endotoxin is much more irregular in shape (25) and is extremely toxic to the larvae of mosquitoes and blackflies (5, 15). Partly because of their potential as biological insecticides, the study of these δ-endotoxins is of considerable scientific and commercial interest.

Recent studies (13, 16, 17, 20, 26, 27, 34, 49) have demonstrated the presence of a complex array of plasmids in most of the insecticidal *B. thuringiensis* strains. Mutant strains cured of one or more plasmids are frequently no longer able to synthesize the δ-endotoxin. Direct evidence for the existence of plasmid-borne toxin genes in *B. thuringiensis* strains HD-1, HD-73, HD-2, and HD-8 was obtained by analysis of a large number of acrystalliferous mutants (17, 20).

Reports of the isolation and expression of the δ-endotoxin gene of *B. thuringiensis* subsp. *kurstaki* (34, 42) and subsp. *berliner* (28) indicate that the δ-endotoxin gene may be variously located on plasmid DNA, chromosomal DNA, or both (18, 23, 28, 30, 42). For *B. thuringiensis* subsp. *israelensis*, two independent studies of strains cured of one or more plasmids concluded that δ-endotoxin synthesis is critically dependent on the presence of a 72- to 75-megadalton (MDa) plasmid (19, 49). Using the newly discovered capacity of *B. thuringiensis* for plasmid transfer by a conjugationlike mechanism, Gonzalez and Carlton (18) showed that one or more plasmids in several *B. thuringiensis* strains code for the δ-endotoxin structural gene. However, to date, transfer of plasmids from *B. thuringiensis* subsp. *israelensis* to other *B. thuringiensis* serotypes has not been achieved (18), and it therefore remains possible that the 72- to 75-MDa plasmid encodes a regulator of δ-endotoxin production, rather than the structural gene. Using the purified plasmid, we have attempted to resolve this question.

Despite the considerable scientific and industrial interest in these endotoxins, their mechanism of action at the molecular level has proved elusive. Native crystal δ-endotoxin is a water-insoluble protoxin which is solubilized in the alkaline conditions of the larval gut and activated by gut proteases. In vitro activation of the δ-endotoxin has been achieved by incubation of the crystals at high pH under reducing conditions and with insect gut proteases (14, 33). Histological studies in vivo have shown the primary target to be the larval midgut epithelial cells which swell and lyse, causing severe disruption of the gut wall (8, 10, 22, 32, 41). In vitro the activated toxin causes cytolysis of certain lepidopteran cell lines and larval midgut cells (6, 9, 38, 39, 45). Cytological effects have been reported as early as 1 min after exposure to the toxin (6, 12, 41) and are characterized by a rapid general breakdown of permeability barriers to small ions, dyes, and internal markers (8, 39).

Native crystals from different serotypes vary considerably in their polypeptide composition (Fig. 1; 7), and this is undoubtedly responsible at least in part for their differing insect specificity or potency, or both. Of the several polypeptides present in some crystals, it is possible that only one is the active δ-endotoxin; alternatively, each polypeptide may possess a distinct host range. The latter possibility was confirmed when two polypeptide species from the HD-1 strain of *B. thuringiensis* subsp. *kurstaki* were shown to be differentially active against mosquito and lepidopteran larvae (51). By analogy with other bacterial toxins, it is reasonable to argue that insect-specific factors, such as the individual gut environment and specific toxin receptors on the gut epithelial surfaces, play a role in defining the host range. As this report shows, the availability of susceptible insect cell lines has played a large part in allowing us to probe these aspects of δ-endotoxin structure and function.

EXPERIMENTAL PROCEDURES

Except where indicated, the sources of bacterial strains and insect cell lines, the growth and sporulation conditions, the purification and acti-

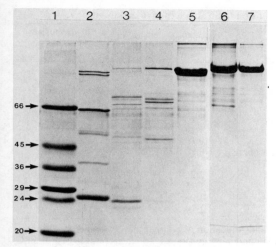

FIG. 1. SDS–10% polyacrylamide gel electrophoresis, Coomassie blue stained. Track 1, Molecular mass standards (kDa); tracks 2–7, native crystal δ-endotoxin from different strains of *B. thuringiensis* (track 2, subsp. *israelensis*; track 3, subsp. *kyushuensis*; track 4, subsp. *darmstadtiensis*; track 5, subsp. *colmeri*; track 6, subsp. *kurstaki*; track 7, subsp. *dakota*).

vation of the crystal δ-endotoxins, and the generation of cured strains have been described previously (29, 44–46, 49).

Preparation of larval gut extracts, activated toxin preparations, liposomes, cell lipids, and bacterial protoplasts was carried out by published methods (29, 45, 46) except where indicated.

In vivo and in vitro assays for toxicity and hemolytic assays were as described previously (45).

Except where indicated in the text, published methods (29, 45) were used for electrophoresis and protein and carbohydrate assays.

Isolation of plasmid DNA. pUC12 plasmid (37) was prepared from *Escherichia coli* JM101 (37) by a lysozyme-detergent lysis method (36) with the following modifications (J. Karn, personal communication): (i) Triton X-100 was used to a final concentration of 1% instead of sodium dodecyl sulfate (SDS), and (ii) NaCl was omitted after Triton addition. Total plasmid DNA from *B. thuringiensis* subsp. *israelensis* was prepared by the method of Casse et al. (3). The 72- to 75-MDa plasmid was subsequently purified from total plasmid DNA by preparative vertical electrophoresis on 0.5% low-gelling-temperature agarose slabs (Seaplaque, FMC Colloids Ltd.) 3 mm thick. Gel bands were visualized and excised as previously described (2, 50). Plasmid DNA from excised bands were purified by phenol-chloroform extraction and ethanol precipitation. Total covalently closed circular DNA for

use in the in vitro transcription-translation system was purified on cesium chloride gradients as previously described (36).

Cloning of DNA. Vector (pUC12) and *B. thuringiensis* subsp. *israelensis* DNA (72- to 75-MDa plasmid) was digested to completion with *Hin*dIII (New England Biolabs). Amounts of 25 ng of restricted vector and 100 ng of restricted 72- to 75-MDa plasmid were ligated with T4 DNA ligase (New England Biolabs) at 15°C for 16 h. Portions of the ligation mix were used to transform *E. coli* JM101 (21), and transformants were selected on L agar (36) containing 170 μg of ampicillin per ml.

Preparation of antibodies. The protein δ-endotoxin from *B. thuringiensis* subsp. *israelensis* (26,000 M_r) was isolated from purified crystals by preparative gel electrophoresis (to be described elsewhere). Purified δ-endotoxin was mixed with Freund complete adjuvant, and antibodies were raised by subcutaneous injection of this material into New Zealand White rabbits.

Analysis of recombinants. Recombinant clones were analyzed by use of an in vitro transcription-translation system (24). DNA from individual recombinants or recombinant groups was extracted by a small-scale lysozyme-Triton plasmid preparation method (36; J. Karn, personal communication), and up to 5 μg of DNA was added to the *E. coli* system. Products from the in vitro system were analyzed by use of 13% acrylamide gels (31) and fluorography (4).

In vitro toxicity assays for cloned toxin. Recombinants were grown for 16 h in L broth (36) containing 100 μg of ampicillin per ml and were harvested by centrifugation. Pellets from 1-liter cultures were resuspended in 12 ml of 50 mM Na_2CO_3-HCl (pH 10.5) and disrupted by sonication. The resulting lysate was incubated at 37°C for 1 h, and then saturated ammonium sulfate was added to a final concentration of 30%. The precipitate was pelleted by centrifugation and suspended with 5 ml of 50 mM Na_2CO_3-HCl (pH 10.5). A 50- to 100-μl amount of this suspension was added to a 4-cm petri dish containing *Aedes albopictus* cells as previously described (45, 46).

In vivo toxicity assays of cloned toxin. Recombinants were assayed for in vivo toxicity by a modification of the method of Tyrell et al. (48). Recombinants were grown for 16 h at 37°C in L broth (36) containing 100 μg of ampicillin per ml and were harvested by centrifugation. Pellets from 100 ml of culture were resuspended in 6 ml of distilled water; 1.5 ml of this suspension was added to the cup containing 25 *A. aegypti* larvae.

RESULTS AND DISCUSSION

Mechanism of action of *B. thuringiensis* subsp. *israelensis* δ-endotoxin. The discovery of the

TABLE 1. Effect of various lipid preparations on toxin action in vivo and in vitro

Liposome composition[a]	Toxin/lipid ratio (wt/wt)	Cytopathic effect[b]	In vivo toxicity[c] (no. dead/no. used)
A. albopictus cell lipid	1:10	None	0/9
B. megaterium KM cell lipid	1:50	Lysis	—
Phosphatidylcholine	1:10	None	0/2
Phosphatidylcholine:cholesterol:stearylamine	1:10	None	0/11
Phosphatidylcholine:cholesterol:dicetylphosphate	1:10	None	0/11
Sphingomyelin:cholesterol:stearylamine	1:10	None	0/11
Sphingomyelin:cholesterol:dicetylphosphate	0:10	None	0/8
Soybean phosphatidylethanolamine:cholesterol:dicetylphosphate	0:10	None	0/10
Phosphatidylserine:cholesterol:dicetylphosphate	1:10	Lysis	—
Phosphatidylserine:cholesterol:dicetylphosphate	1:25	None	—
Cardiolipin:cholesterol:dicetylphosphate	1:25	Lysis	20/20
Phosphatidylinositol:cholesterol:dicetylphosphate	1:25	Lysis	4/4
Cerebroside:cholesterol:dicetylphosphate	1:50	Lysis	—

[a] Molar ratios were 2:1.5:0.5. Preparation, dispersion, and sonication of all lipid and liposome preparations were carried out at 30°C except in the case of cerebroside-containing liposomes, which were prepared at 60°C and subsequently assayed at 20°C.

[b] Cytopathology observed after exposure of A. albopictus cells to a solution of 5 μg of δ-endotoxin per ml after it had been incubated with the appropriate lipid.

[c] Subcutaneous inoculation of suckling mice; an equivalent of 25 μg of δ-endotoxin per g of mouse was used.

cellular target for B. thuringiensis subsp. israelensis δ-endotoxin had its origins in the observation (45) that a soluble preparation of subsp. israelensis toxin caused rapid lysis of insect and mammalian cells in vitro, but did not affect bacterial protoplasts. This toxin preparation was also hemolytic to a range of erythrocytes. These observations raised the possibility that the subsp. israelensis toxin causes lysis directly through interaction with a plasma membrane component.

Subsequent fractionation of mosquito cells to identify possible toxin receptors revealed that preincubation of the toxin with an excess of phospholipids purified from cultured A. albopictus cells rendered it inactive, but phospholipids from B. megaterium were ineffective even at high lipid-toxin ratios (Table 1). The surprisingly broad eucaryotic specificity of the solubilized subsp. israelensis endotoxin suggested that certain ubiquitous eucaryotic plasma membrane phospholipids or sterols might be the target. In contrast to insect and mammalian cells, the Bacillus membranes lack phosphatidyl choline, sphingomyelin, cholesterol, or significant amounts of unsaturated fatty acids, suggesting that one or more of these components is recognized by the toxin. This hypothesis was tested in a series of experiments in which sonicated lipid dispersions (liposomes) were examined for their ability to neutralize toxicity in subsequent in vitro and in vivo assays. The results (Table 1; 46) showed that the toxin binds avidly to liposomes containing phosphatidyl choline, sphingomyelin, or phosphatidyl ethanolamine, binds less strongly to phosphatidyl serine, and

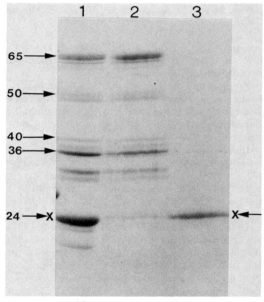

FIG. 2. SDS–13% polyacrylamide gel electrophoresis, Coomassie blue stained. Track 1, Soluble subsp. israelensis δ-endotoxin obtained by incubation of native purified crystal δ-endotoxin in 50 mM Na₂CO₃-HCl (pH 10.5); track 2, polypeptides remaining in the supernatant; track 3, polypeptides sedimenting with liposomes after centrifugation of a 1:10 (wt/wt) mixture of the soluble subsp. israelensis δ-endotoxin in track 1 with phosphatidyl choline (egg) liposomes that had been incubated at 37°C for 2 h. Endogenous proteolysis during the alkali solubilization of the native crystal in this experiment has resulted in degradation of the native israelensis 26,000 M_r toxin polypeptide to 24,000 M_r. Arrows denote molecular mass (kDa).

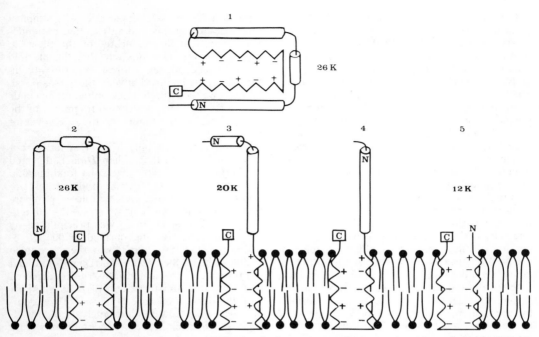

FIG. 3. Diagram summarizing the conformational changes occurring upon insertion of the 26,000 M_r alkali-soluble subsp. *israelensis* δ-endotoxin into lipid vesicles and the subsequent proteolytic processing after exposure of the vesicles to external trypsin.

shows no affinity for phosphatidyl inositol, cardiolipin, cerebroside glycolipid, or cholesterol. Confirmation of the strength of toxin binding came from the finding that, when phosphatidyl choline liposomes were centrifuged after incubation with the toxin, the supernatant was devoid of toxin activity (46). Gel electrophoresis of the liposome pellet and supernatant revealed that only one of the subsp. *israelensis* crystal polypeptides (26,000 M_r) had become inserted into the lipid bilayer (Fig. 2). These findings provided strong evidence for the identification of the 26,000 M_r component as the toxic polypeptide.

To determine the extent of insertion of the 26,000 M_r polypeptide into the membrane, we allowed a solubilized toxin preparation to bind to artificial membrane vesicles, removed unbound toxin, and then exposed the vesicles to trypsin to digest away any exposed portion of the 26,000 M_r polypeptide. When the vesicles were recovered and examined by electrophoresis and peptide mapping, a 12,000 M_r fragment of the δ-endotoxin was found to be protected from trypsin digestion by insertion into the lipid. Figure 3 summarizes these experiments. Upon contact with the membrane surface, the previously shielded hydrophobic terminal region partitions into the lipid bilayer, leaving the remaining half of the toxin exposed and unfolded where it is the substrate for a series of proteolytic processing steps that (depending on the protease

specificity) can ultimately trim away all the exposed moiety.

Against this background we attempted in other experiments to identify the structural features in the phospholipids that are responsible for endotoxin binding. The results (46) revealed two crucial determinants: (i) the presence of unsaturated fatty acid substituents and (ii) the correct

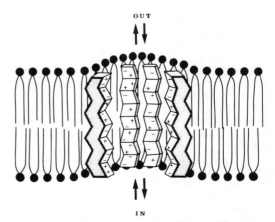

FIG. 4. Model illustrating a possible arrangement of transmembrane amphipathic helical polypeptide segments from one or more molecules of subsp. *israelensis* δ-endotoxin. (See text for details.)

phospholipid headgroup, exemplified by the di-polar ionic phosphatidyl choline or by phospha-tidyl ethanolamine. Anionic phospholipids (even when highly unsaturated) were unable to bind the toxin. The broad eucaryotic specificity of the activated toxin is immediately explained by these results, since unsaturated phosphatidyl choline, sphingomyelin, and phosphatidyl ethanolamine are the major phospholipid species in most higher cells. We suggested (46) that in susceptible in-sects insertion of the δ-endotoxin into the plasma membranes of gut epithelial cells causes a de-tergentlike rearrangement of membrane lipids, leading to disruption of membrane integrity and eventually to cytolysis.

Further experiments are in progress to map the topography of the membrane lesion pro-duced by toxin binding and to probe the stoichi-ometry of the toxin-phospholipid complexes. Figure 4 illustrates one speculative model com-bining the possibilities of toxin oligomerization with the protein structural feature known as the amphipathic helix (11). The latter can be formed by appropriate folding of a polypeptide with hydrophobic residues at every third or fourth position in such a way as to position all the hydrophobic residues on one helix face, while the other face is hydrophilic. This generates the amphipathic helices illustrated in Fig. 4, which have an affinity for both polar and nonpolar environments. Self-association of intramolecu-lar or intermolecular helical segments can yield the transmembrane structure shown in the model, in which there is internal compensation of the charged helix face and maximal lipid solubility by virtue of the externally directed hydrophobic face.

Since the experiments summarized in Fig. 3 indicate that the 12,000 M_r toxin domain binds avidly to phospholipid, it is conceivable that the four amphipathic helical segments depicted in the model could be derived from one 12,000 M_r segment. Alternatively, the hydrophilic mem-brane channel may be created by the self-asso-ciation of amphipathic helices from several 12,000 M_r fragments. Creation of a number of these hydrophilic channels by the toxin could prove cytolytic by subverting the selective per-meability of the cell membrane. The model is a considerable oversimplification and completely neglects other potential toxic mechanisms, such as the possibility that a toxin in the form of an amphipathic helix could still seriously perturb critical membrane functions through an associa-tion with only the outer half of the lipid bilayer.

Molecular genetics of *B. thuringiensis* **subsp.** *israelensis* **δ-endotoxin.** In previous reports we (49) and others (19) demonstrated that δ-endotoxin synthesis in *B. thuringiensis* subsp. *israelensis* is critically dependent upon the pres-ence of a 72- to 75-MDa plasmid. The simplest interpretation of this finding was that the endo-toxin structural gene is located on this plasmid, although it was also possible that the plasmid merely encoded a regulator of δ-endotoxin syn-thesis. To distinguish between these two possi-bilities, we constructed recombinant plasmids by inserting *Hin*dIII restriction fragments of the subsp. *israelensis* plasmid into the *E. coli* vector pUC12.

Preliminary screening with restriction en-zymes had shown that, when *Hin*dIII-digested *B. thuringiensis* subsp. *israelensis* total plasmid DNA was added to an *E. coli* in vitro transcrip-tion-translation system, a single novel polypep-tide identical in molecular weight to the authen-tic subsp. *israelensis* toxin was precipitated by antiserum raised to the purified 26,000 M_r δ-endotoxin (Fig. 5). *Hin*dIII was therefore used to digest the purified 72- to 75-MDa plasmid, and the products were ligated into *Hin*dIII-digested pUC12. After transformation of *E. coli* JM101 with the ligation mixture, a library of 450 colo-nies was selected. Random analysis of this li-brary showed that 77% were recombinants (data not shown).

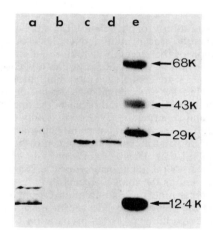

FIG. 5. Fluorographs of SDS–13% polyacrylamide gels of the ^{35}S-labeled polypeptides synthesized in the *E. coli* transcription-translation system primed with either *Hin*dIII-digested total *B. thuringiensis* subsp. *israelensis* plasmid DNA or plasmid pIP174 DNA and supplemented with L-[^{35}S]methionine: lane a, total products from pIP174 DNA; lane b, material precipi-tated from a by addition of preimmune serum; lane c, material precipitated from a by addition of antibody raised against the 26,000 M_r authentic *israelensis* δ-endotoxin; lane d, material precipitated by antibody against the 26,000 M_r authentic *israelensis* δ-endotoxin from the *E. coli* transcription-translation system primed with *Hin*dIII-digested total *B. thuringiensis* subsp. *israelensis* plasmid DNA; lane e, molecular weight standards.

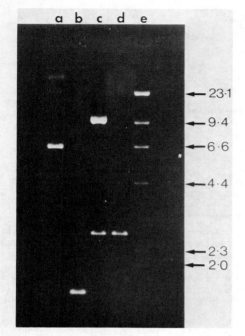

FIG. 6. Agarose gel electrophoresis of covalently closed circular forms of pIP174, pUC12, and their *Hin*dIII digestion products. Lane a, pIP174 (covalently closed circular form); lane b, pUC12 (covalently closed circular form); lane c, *Hin*dIII-digested pIP174; lane d, *Hin*dIII-digested pUC12; lane e, *Hin*dIII-digested λ DNA with fragment sizes on the right margin (in kb).

Plasmid DNA was then extracted from the clone library and screened for toxin production by immunoprecipitation with the use of the in vitro transcription-translation system (49a). By using this approach, two toxin-coding recombinants, pIP173 and pIP174, were identified and used for subsequent analysis. Figure 6 shows the results of horizontal agarose gel electrophoresis of the covalently closed circular forms of

pIP174 (lane a) and pUC12 (lane b) extracted from individual recombinants, together with their *Hin*dIII digestion products. Digestion of pIP174 yielded a 9.7-kilobase (kb) insert (lane c) in addition to the 2.7-kb fragment derived from pUC12 (lane d). Identical results were obtained with *Hin*dIII-digested pIP173 (data not shown).

Although the immunoprecipitation of the in vitro-synthesized cloned product showed clearly that a 26,000 M_r polypeptide antigenically related to the subsp. *israelensis* δ-endotoxin was encoded by the 9.7-kb insert, additional experiments were needed to confirm that the polypeptide was biologically active. Lysates were therefore prepared from 1-liter cultures of colony 174 and control *E. coli* JM101 colonies containing pUC12 lacking any insert and were assayed for toxicity in vitro. Protein extracted from colony 174 caused cytolysis of *A. albopictus* cells indistinguishable from that previously described for authentic δ-endotoxin (45, 46, 49a). *Aedes* cells exposed to an equivalent protein extract from *E. coli* containing the vector pUC12 alone were unaffected, even after prolonged exposure (24 h). The authenticity of the pIP174-encoded polypeptide was further confirmed by demonstrating that the toxicity of the 174 lysate could be neutralized either by antiserum directed against the 26,000 M_r native δ-endotoxin or by preincubation of the lysate with sonicated preparations of those phospholipids previously shown to be the cell membrane receptors for authentic δ-endotoxin (46). Final confirmation of the biological authenticity of the cloned product was obtained from in vivo bioassays. Twenty-five second-instar *A. aegypti* larvae were killed in 4 h when fed an amount of *E. coli* containing pIP174 equivalent to 25 ml of original culture. In control experiments larvae fed equivalent amounts of *E. coli* JM101 containing pUC12 with no insert were unaffected.

Addition of a β-galactosidase inducer, isopropyl-β-D-thiogalactopyranoside, to cultures of *E. coli* JM101 containing pIP174 did not

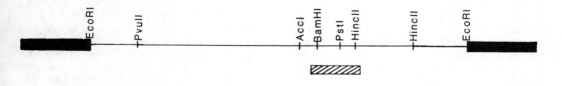

FIG. 7. Restriction map of the 5.4-kb insert of pIPEco5. Thick lines represent pUC12 DNA. The boxed area represents the location of the δ-endotoxin gene, determined by analysis of subclones of pIPEco5 in the *E. coli* in vitro system.

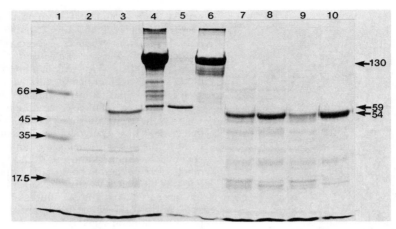

FIG. 8. SDS–10% polyacrylamide gel electrophoresis, Coomassie blue stained. Track 1, Molecular weight standards (kDa); track 2, 10 μl of *P. brassicae* gut extract; track 3, 50 μg of soluble subsp. *kurstaki* crystal protein incubated with 10 μl of *P. brassicae* gut extract for 15 min, 37°C; track 4, 50 μg of native subsp. *kurstaki* crystal δ-endotoxin; track 5, insoluble subsp. *kurstaki* crystal protein; track 6, soluble subsp. *kurstaki* crystal protein obtained by incubation of 50 μg of native crystal δ-endotoxin in 50 mM Na$_2$CO$_3$-HCl (pH 9.5) and 10 mM dithiothreitol; track 7, 50 μg of soluble subsp. *kurstaki* crystal protein after 20 days, 20°C; track 8, 50 μg of soluble crystal protein after 33 days, 20°C; track 9, 50 μg of soluble crystal protein after 45 days, 20°C; track 10, 50 μg of soluble crystal protein after 62 days, 20°C.

result in an increase in toxin production measured in vitro. This suggests that expression of the δ-endotoxin gene in *E. coli* is under the control of *B. thuringiensis* promoter sequences rather than the vector β-galactosidase promoter. More recent work has led to the isolation of a 5.4-kb *Eco*RI fragment derived from pIP173. A recombinant plasmid (pIPEco5) consisting of this 5.4-kb *Eco*RI insert in pUC12 directs the synthesis of the 26,000 M_r δ-endotoxin in the *E. coli* lysate. A preliminary restriction map of pIPEco5 is shown in Fig. 7. In subcloning experiments to date we have located the promoter region and almost all the structural gene on the smaller *Pvu*II-*Hinc*II fragment (3.1 kb) shown in Fig. 7. The availability of the δ-endotoxin primary sequence will be invaluable in assessing the various models for toxin-phospholipid interaction.

Mechanism of action of *B. thuringiensis* subsp. *kurstaki* lepidopteran toxin. Crystals from the HD-1 strain of *B. thuringiensis* subsp. *kurstaki* contain at least two distinct toxic moieties, a lepodipteran-specific toxin (P1) and a "mosquito factor" (P2) toxic to both lepidoptera and diptera (51) that can be separated under alkaline reducing conditions (45). Figure 8 (tracks 5 and 6) shows the separation of the insoluble 63,000 M_r mosquito toxin (P2) from the alkali-soluble lepidopteran-specific P1 (45, 51). Activation of the 126,000 M_r P1 endotoxin by either larval gut proteases (track 3) or endogenous proteases (tracks 7 to 10) yielded a major polypeptide of 54,000 M_r. When activated by either procedure, the soluble toxin retained its lepidopteran spec-

ificity (29). Thus, the activated toxin was lethal to *Pieris brassicae* larvae in vivo, but no toxic effects were observed on injection of suckling mice. Similarly activated toxin at 50 μg/ml caused 50% lysis of CF1 cells in 60 min in vitro, as assessed by vital staining with trypan blue, but human erythrocytes were unaffected by 100 μg of toxin per ml after 3 h. Controls containing 50 mM Na$_2$CO$_3$-HCl (pH 9.5) and 10 mM dithiothreitol, with *P. brassicae* gut extract and fetal calf serum where appropriate, showed no toxic effects in vivo or in vitro (29).

As discussed above, the ubiquitous presence in eucaryotic plasma membranes of the phospholipids that serve as plasma membrane receptors for the potent mosquitocidal *B. thuringiensis* subsp. *israelensis* δ-endotoxin explains the finding that subsp. *israelensis* δ-endotoxin activated in vitro is cytolytic to a wide variety of cell types, including mammalian cells (45). In contrast, the observation that, even after activation, the *B. thuringiensis* subsp. *kurstaki* P1 δ-endotoxin affects only lepidopteran larvae and cell lines suggests that the membrane receptor for this toxin may be cell specific either in whole or in part. The identity of this proposed specific plasma membrane receptor was investigated initially by assessing the ability of various molecules to neutralize the cytolytic effect of activated P1 toxin against *Choristoneura fumiferana* CF1 cells in vitro (29).

Preincubation with various lipid preparations was used to investigate the possibility that the lepidopteran-specific subsp. *kurstaki* P1 toxin caused cytolysis by interaction with plasma

TABLE 2. Effect of preincubation with monosaccharides on toxicity of soluble δ-endotoxin in vitro

Monosaccharide[a]	Concn (mM)	Cytopathic effect[b]
N-Acetyl-D-glucosamine	400	Lysis
N-Acetyl-D-galactosamine	125	Complete protection
	25	Partial protection
N-Acetylneuraminic acid	125	Complete protection
	25	Partial protection
D-Galactose	250	Lysis
L-Fucose	250	Lysis
Muramic acid	75	Lysis
Galactosamine	250	Lysis
D-Glucose	250	Lysis
D-Mannose	250	Lysis

[a] Monosaccharides and activated soluble δ-endotoxin were incubated together for 60 min at 20°C before addition to cells.

[b] Cytopathology observed in CF1 cells at a δ-endotoxin concentration of 50 μg/ml.

membrane phospholipids, as observed for the *B. thuringiensis* subsp. *israelensis* mosquitocidal δ-endotoxin (46). The inability of any of the lipid preparations used (29) to neutralize toxicity against CF1 cells suggested that the specificity of toxin-membrane interactions was determined by a different membrane component. In a series of experiments designed to explore the possibility of a carbohydrate-containing P1 receptor, we tested the ability of various lectins and monosaccharides to neutralize the toxin. The results (Table 2) indicated that the toxin was completely inactivated by prior incubation with N-acetylgalactosamine and N-acetylneuraminic acid, but not by a range of other monosaccharides tested.

Of the lectins tested for toxin-neutralizing ability, only wheat germ agglutinin and soybean agglutinin partially protected CF1 cells from the toxin (Table 3). This effect was noted both with prior incubation of cells with lectin and of activated toxin with lectin. The ability of wheat

germ agglutinin to inhibit toxicity was abolished when the lectin was incubated with 250 mM N-acetylglucosamine prior to incubation with cells or toxin.

Many other toxins and biological ligands bind to specific glycoprotein or glycolipid receptors in the plasma membrane. The above results with the P1 endotoxin led us to suggest (29) that binding of the lepidopteran-specific toxin to a specific glycoconjugate on the plasma membrane of susceptible cells is an essential feature of its cytolytic action. Since N-acetylgalactosamine completely neutralized the toxin, it seems likely to be part of the receptor. N-Acetylneuraminic acid is not likely to be involved since this monosaccharide is absent from all insects tested so far (29). Wheat germ agglutinin and soybean agglutinin both bind terminal N-acetylgalactosamine; therefore, their observed ability to protect CF1 cells from the toxin may be explained by competition between lectin and toxin for the same binding site on the cell surface. Specific binding of N-acetylgalactosamine to the carbohydrate recognition site of the endotoxin in the preincubation experiments blocks subsequent attachment to the cell surface receptor. The inability of N-acetylglucosamine to protect the cells (Table 2) is interesting in two ways. First, it rules out a nonspecific effect of amino sugars, and second, it indicates the extremely rigorous stereospecificity of the toxin, since, as Fig. 9 shows, the only difference between N-acetylgalactosamine and N-acetylglucosamine is the disposition of the hydroxyl group on carbon 4. N-Acetylneuraminic acid has no hydroxyl group in the analogous position.

Those toxins which must cross the plasma membrane, such as diphtheria toxin, generally show a time lag before the first symptoms of toxicity are observed. In contrast, the first effects of *B. thuringiensis* subsp. *kurstaki* δ-endotoxin occur within 1 min (6, 12, 41). The precise mechanism of action of the lepidopteran-specific subsp. *kurstaki* P1 toxin has yet to be elucidated, but it appears that after binding to a plasma membrane receptor the membrane is

TABLE 3. Effect of preincubation with lectins on toxicity of soluble δ-endotoxin in vitro

Lectin[a]	Sugar binding specificity	Cytopathic effect[b]
Concanavalin A	α-D-Mannose > α-D-glucose	Lysis
Peanut agglutinin	β-D-Galactose	Lysis
Soybean agglutinin	N-Acetylgalactosamine	Protection
Wheat germ agglutinin	N-Acetylglucosamine > N-acetylneuraminic acid > N-acetylgalactosamine	Protection
Ulex europaeus agglutinin	α-L-Fucose	Lysis

[a] Lectins and activated soluble δ-endotoxin were incubated together for 60 min at 20°C before addition to cells.

[b] Cytopathology observed in CF1 cells at a δ-endotoxin concentration of 50 μg/ml.

rapidly made leaky to small ions and larger molecules.

Studies with other *B. thuringiensis* strains. Progress of the type described above has provided a platform for us to begin a larger comparative study of a number of *B. thuringiensis* strains from different serotypes with reportedly differing insect specificity. This approach makes use of antibodies to purified toxin polypeptides and cloned toxin genes as probes in conjunction with the in vitro cell assay. As an example of the results, we have found (F. Drobniewski, T. Sawyer, and D. J. Ellar, unpublished data) that, although the δ-endotoxin from a derivative of a known serotype 10 strain (40) with potent mosquitocidal activity closely resembles serotype 14 subsp. *israelensis* in its mechanism of action, it does not cross-react with antisera to the purified subsp. *israelensis* δ-endotoxin. Similar studies have shown that, among other mosquitocidal toxins that we have purified, some (from *kurstaki* P2 and *colmeri*) are immunologically unrelated to that of *israelensis*, whereas others (from *kyushuensis*) are weakly cross-reactive. These and other data on the mechanism of action of these different endotoxins indicate

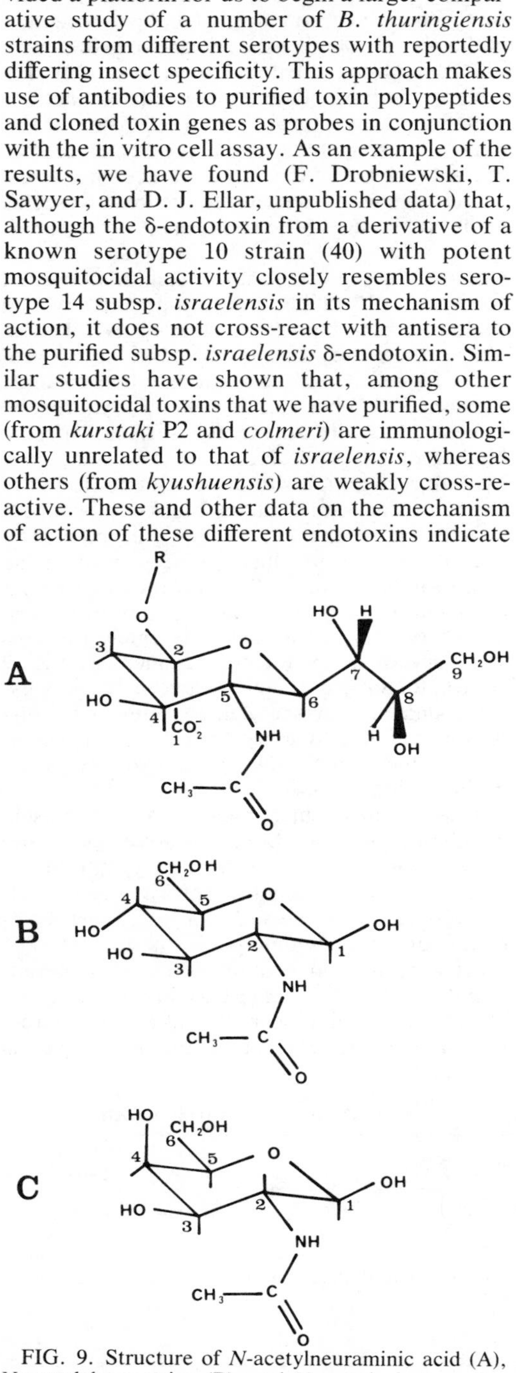

that several mosquitocidal strategies are exploited by *B. thuringiensis*. The availability of primary sequence data for many of these toxins, especially those which are immunologically similar but show different 50% lethal concentrations, should provide valuable data on the toxin active site residues.

Other workers (7) have reported that a number of *B. thuringiensis* strains produce apparently normal δ-endotoxin crystals that have not proved to be toxic to any insect species so far tested in vivo. In some cases (*dakota*) we have been able to activate the protoxin in vitro and demonstrate activity in the in vitro cell assay (D. Last, B. Knowles, C. Nichols, and D. J. Ellar, unpublished data). It thus appears that, in some cases at least, absence of in vivo toxicity in the test insects may reflect a failure in protoxin processing in the gut.

For some of the lepidopteran toxins, such as *thuringiensis* HD-2 and *kurstaki* P1 toxins, analysis of glycoproteins and glycolipids in plasma membranes from susceptible insect cells is likely to uncover the membrane receptor(s). Toxins like those of *kurstaki* P2 and *colmeri* that are active against mosquitoes and lepidopterans both in vivo and in vitro are an especially interesting group by comparison with the *israelensis* toxin, which, despite its ubiquitous phospholipid target, is inactive in the lepidopteran gut although rapidly lethal when injected into the hemocoel. The latter observation highlights the possibility that individual insect gut environments may in part be an explanation for the differing insecticidal potency of *B. thuringiensis* strains. The observation (W. E. Thomas, B. Knowles, J. Lewis, and D. J. Ellar, unpublished data) that freshly isolated lepidopteran gut cells display a degree of resistance to the *israelensis* toxin, whereas cultured lepidopteran cell lines are rapidly lysed (45), suggests that resistance in this case may be determined by some feature of the gut cell plasma membrane. In view of the need for the *israelensis* endotoxin to make close contact with the lipid bilayer for toxin insertion, it is conceivable that steric hindrance or charge interactions caused by epithelial cell surface components may play a role in resistance. These results also illustrate the need for caution in interpreting the results from in vitro assays performed with stable insect cell lines which may lack features normally present in the differentiated native gut cells.

SUMMARY

Specific phospholipids have been identified as the membrane receptors for the *B. thuringiensis* subsp. *israelensis* insecticidal δ-endotoxin. Experiments with artificial membranes show that the toxin binds to phosphatidyl choline, sphin-

FIG. 9. Structure of *N*-acetylneuraminic acid (A), *N*-acetylglucosamine (B), and *N*-acetylgalactosamine (C).

gomyelin, and phosphatidyl ethanolamine, provided these lipids contain unsaturated fatty acyl substituents. A possible insecticidal mechanism is proposed in which toxin insertion into the epithelial cell membrane creates a nonspecific hydrophilic leak channel, leading to a breakdown of membrane integrity and rapid cytolysis. *Hind*III fragments of the 72- to 75-MDa plasmid containing the *israelensis* δ-endotoxin gene have been cloned in pUC12. Two recombinants (pIP174 and pIP173) producing the 26,000 M_r δ-endotoxin have been identified by screening a clone library in an *E. coli* in vitro transcription-translation system. The 26,000 M_r polypeptide synthesized in vivo from pIP174 transformed into *E. coli* JM101 was lethal to mosquito larvae and cytotoxic to mosquito cells in vitro. An investigation of the in vitro-activated lepidopteran-specific P1 protoxin from *B. thuringiensis* subsp. *kurstaki* showed that its toxicity towards *C. fumiferana* CF1 cells was specifically inhibited by preincubation with *N*-acetylgalactosamine and by the lectins soybean agglutinin and wheat germ agglutinin, which bind *N*-acetylgalactosamine. These results suggest that the lepidopteran membrane receptor for this P1 toxin is a glycoconjugate containing terminal *N*-acetylgalactosamine.

This work was supported by grants from the Agriculture and Food Research Council, the Science and Engineering Research Council, the Medical Research Council, The Nuffield Foundation, and the Oppenheimer Fund.

We thank J. Gray and A. Phillips for supplies of the *E. coli* transcription-translation system and them and C. Howe for their helpful suggestions. Supplies of *A. aegypti* eggs were kindly provided by D. Funnell, and *C. fumiferana* cells were the gift of S. Sohi.

LITERATURE CITED

1. **Bulla, L. A., Jr., D. B. Bechtel, K. J. Kramer, Y. I. Shethna, A. I. Aronson, and P. C. Fitz-James.** 1980. Ultrastructure, physiology and biochemistry of *Bacillus thuringiensis*. Crit. Rev. Microbiol. **8:**147–204.

2. **Burns, D. M., and I. R. Beacham.** 1983. A method for the ligation of DNA following isolation from low melting temperature agarose. Anal. Biochem. **135:**48–51.

3. **Casse, F., C. Boucher, J. S. Juliott, M. Michel, and J. Dénairié.** 1979. Identification and characterization of large plasmids in *Rhizobium meliloti* using agarose gel electrophoresis. J. Gen. Microbiol. **113:**229–242.

4. **Chamberlain, J. P.** 1979. Fluorographic detection of radioactivity in polyacrylamide gels with water-soluble fluor, sodium salicylate. Anal. Biochem. **98:**132–135.

5. **De Barjac, H.** 1978. Une nouvelle variété de *Bacillus thuringiensis* trés toxique pour les moustiques: *Bacillus thuringiensis* var. *israelensis* sérotype 14. C. R. Acad. Sci. Ser. D **286:**797–800.

6. **De Lello, E. E., W. K. Hanton, S. T. Bishoff, and D. W. Misch.** 1984. Histopathological effects of *Bacillus thuringiensis* on the midgut of tobacco hornworm larvae (*Manduca sexta*): low doses compared with fasting. J. Invertebr. Pathol. **43:**169–181.

7. **De Lucca, A. J., M. S. Palmgren, and H. De Barjac.** 1984. A new serovar of *Bacillus thuringiensis* from grain dust: *Bacillus thuringiensis* serovar comeri (Serovar 21). J.

Invertebr. Pathol. **43:**437–438.

8. **Ebersold, H. R., P. Lüthy, P. Geiser, and L. Ettlinger.** 1978. The action of the δ-endotoxin of *Bacillus thuringiensis*: an electron microscopic study. Experientia **34:**1672.

9. **Ebersold, H. R., P. Lüthy, and H. E. Huber.** 1980. Membrane damaging effect of the δ-endotoxin of *Bacillus thuringiensis*. Experientia **36:**495.

10. **Endo, Y., and J. Nishiitsutsuji-Uwo.** 1980. Mode of action of *Bacillus thuringiensis* δ-endotoxin: histopathological changes in the silkworm midgut. J. Invertebr. Pathol. **36:**90–103.

11. **Epand, R. M.** 1983. The amphipathic helix: its possible role in the interaction of glucagon and other peptide hormones with membrane receptor sites. Trends Biochem. Sci. **8:**205–207.

12. **Fast, P. G., and T. P. Donaghue.** 1971. The δ-endotoxin of *Bacillus thuringiensis*. II. On the mode of action. J. Invertebr. Pathol. **18:**135–138.

13. **Faust, R. M., K. Abe, G. A. Held, T. Iizuka, L. A. Bulla, and C. L. Meyers.** 1983. Evidence for plasmid-associated crystal toxin production of *Bacillus thuringiensis* subsp. *israeliensis*. Plasmid **9:**98–103.

14. **Faust, R. M., G. M. Hallam, and R. S. Travers.** 1974. Degradation of the parasporal crystal produced by *Bacillus thuringiensis* var. *kurstaki*. J. Invertebr. Pathol. **24:**365–373.

15. **Goldberg, L. J., and J. Margalitt.** 1977. A bacterial spore demonstrating rapid larvicidal activity against *Anopheles sergentii, Uranotaenia unguiculata, Culex univittatus, Aedes aegypti* and *Culex pipiens*. Mosq. News **37:**355–358.

16. **Gonzalez, J. M., Jr., B. J. Brown, and B. C. Carlton.** 1982. Transfer of *Bacillus thuringiensis* plasmids coding for δ-endotoxin among strains of *B. thuringiensis* and *B. cereus*. Proc. Natl. Acad. Sci. U.S.A. **79:**6951–6955.

17. **Gonzalez, J. M., Jr., and B. C. Carlton.** 1980. Patterns of plasmid in crystalliferous and acrystalliferous strains of *Bacillus thuringiensis*. Plasmid **3:**92–98.

18. **Gonzalez, J. M., Jr., and B. C. Carlton.** 1982. Plasmid transfer in *Bacillus thuringiensis*, p. 85–95. *In* U. N. Streips, S. H. Goodgal, W. R. Guild, and G. A. Wilson (ed.), Genetic exchange, a celebration and a new generation. Marcel Dekker, New York.

19. **Gonzalez, J. M., Jr., and B. C. Carlton.** 1984. A large transmissible plasmid is required for crystal toxin production in *Bacillus thuringiensis* variety *israeliensis*. Plasmid **11:**28–38.

20. **Gonzalez, J. M., Jr., H. T. Dulmage, and B. C. Carlton.** 1981. Correlation between specific plasmids and δ-endotoxin production in *Bacillus thuringiensis*. Plasmid **5:**351–365.

21. **Hanahan, D.** 1983. Studies on transformation of *Escherichia coli* with plasmids. J. Mol. Biol. **166:**557–580.

22. **Heimpel, A. M., and T. A. Angus.** 1960. Bacterial insecticides. Bacteriol. Rev. **24:**266–288.

23. **Held, G. A., L. A. Bulla, Jr., E. Ferrari, J. Hoch, A. I. Aronson, and S. A. Minnich.** 1982. Cloning and localization of the lepidopteran protoxin gene of *Bacillus thuringiensis* subsp. *kurstaki*. Proc. Natl. Acad. Sci. U.S.A. **79:**6065–6069.

24. **Howe, C. J., C. M. Bowman, T. A. Dyer, and J. C. Gray.** 1982. Localization of wheat chloroplast genes for the beta and epsilon subunits of ATP synthase. Mol. Gen. Genet. **186:**525–530.

25. **Huber, H. E., and P. Lüthy.** 1981. *Bacillus thuringiensis* delta-endotoxin: composition and activation, p. 209–234. *In* E. W. Davidson (ed.), Pathogenesis of invertebrate microbial diseases. Allanheld Osmun & Co., Totowa, N.J.

26. **Jarrett, P.** 1983. Comparison of plasmids from twelve isolates of *Bacillus thuringiensis* H-serotype 7. FEMS Microbiol. Lett. **16:**55–60.

27. **Kamdar, H., and K. Jayaraman.** 1983. Spontaneous loss of a high molecular weight plasmid and the biocide of *Bacillus thuringiensis* var. *israeliensis*. Biochem. Bi-

ophys. Res. Commun. 110:477–482.

28. **Klier, A., F. Fargette, J. Ribier, and G. Rapoport.** 1982. Cloning and expression of the crystal protein genes from *Bacillus thuringiensis* strain *berliner* 1715. EMBO J. 1:791–799.

29. **Knowles, B. H., W. E. Thomas, and D. J. Ellar.** 1984. Lectin-like binding of *Bacillus thuringiensis* var. *kurstaki* lepidopteran-specific toxin is an initial step in insecticidal action. FEBS Lett. 168:197–202.

30. **Kronstadt, J. W., H. E. Schnepf, and H. R. Whiteley.** 1983. Diversity of locations for *Bacillus thuringiensis* crystal protein genes. J. Bacteriol. 154:419–428.

31. **Laemmli, U. K.** 1970. Cleavage of structural proteins during the assembly of the head of bacteriophage T4. Nature (London) 227:680–685.

32. **Lahkim-Tsror, L., C. Pascar-Gluzman, J. Margalitt, and Z. Barak.** 1983. Larvicidal activity of *Bacillus thuringiensis* subsp. *israeliensis* serovar H14 in *Aedes aegypti*: histopathological studies. J. Invertebr. Pathol. 41:104–116.

33. **Lecadet, M. M., and D. M. Martouret.** 1967. Hydrolyse enzymatique de l'inclusion parasporale de *Bacillus thuringiensis* par les proteases de *Bombyx mori*. C.R. Acad. Sci. Ser. D 265:1543–1546.

34. **Lereclus, D., M. M. Lecadet, J. Ribier, and R. Dedonder.** 1982. Molecular relationships among plasmids of *Bacillus thuringiensis*: conserved sequences through 11 crystalliferous strains. Mol. Gen. Genet. 186:391–398.

35. **Lüthy, P.** 1980. Insecticidal toxins of *Bacillus thuringiensis*. FEMS Microbiol. Lett. 8:1–7.

36. **Maniatis, T., E. F. Fritsch, and J. Sambrook.** 1982. Molecular cloning, a laboratory manual. Cold Spring Harbor Laboratory, Cold Spring Harbor, N.Y.

37. **Messing, J.** 1983. New M13 vectors for cloning. Methods Enzymol. 101(Part C):20–78.

38. **Murphy, D. W., S. S. Sohi, and P. G. Fast.** 1976. *Bacillus thuringiensis* enzyme digested δ-endotoxin: effect on cultured insect cells. Science 194:954–956.

39. **Nishiitsutsuji-Uwo, J., Y. Endo, and M. Himeno.** 1980. Effect of *Bacillus subtilis* δ-endotoxin on insect and mammalian cells *in vitro*. Appl. Entomol. Zool. 15:133–139.

40. **Padua, L. E., M. Ohba, and K. Aizawa.** 1980. The isolates of *Bacillus thuringiensis* serotype 10 with a highly preferential toxicity to mosquito larvae. J. Invertebr. Pathol. 36:180–186.

41. **Percy, J., and P. G. Fast.** 1983. *Bacillus thuringiensis* crystal toxin: ultrastructural studies of its effect on silk-worm midgut cells. J. Invertebr. Pathol. 41:86–98.

42. **Schnepf, H. E., and H. R. Whiteley.** 1981. Cloning and expression of the *Bacillus thuringiensis* crystal protein in *Escherichia coli*. Proc. Natl. Acad. Sci. U.S.A. 78:2893–2897.

43. **Somerville, H. J.** 1978. Insect toxin in spores and protein crystal of *Bacillus thuringiensis*. Trends Biochem. Sci. 3:108–110.

44. **Stewart, G. S. A. B., K. Johnstone, E. Hagelberg, and D. J. Ellar.** 1981. Commitment of bacterial spores to germinate. A measure of the trigger reaction. Biochem. J. 196:101–106.

45. **Thomas, W. E., and D. J. Ellar.** 1983. *Bacillus thuringiensis* var. *israeliensis* crystal δ-endotoxin: effects on insect and mammalian cells *in vitro* and *in vivo*. J. Cell Sci. 60:181–197.

46. **Thomas, W. E., and D. J. Ellar.** 1983. Mechanism of action of *Bacillus thuringiensis* var. *israeliensis* insecticidal δ-endotoxin. FEBS Lett. 154:362–368.

47. **Tyrell, D. J., L. A. Bulla, Jr., R. E. Andrews, Jr., K. J. Kramer, L. I. Davidson, and P. Nordin.** 1981. Comparative biochemistry of entomocidal parasporal crystals of selected *Bacillus thuringiensis* strains. J. Bacteriol. 145:1052–1062.

48. **Tyrell, D. J., L. J. Davidson, L. A. Bulla, Jr., and W. A. Ramoska.** 1979. Toxicity of parasporal crystals of *Bacillus thuringiensis* subsp. *israelensis* to mosquitos. Appl. Environ. Microbiol. 38:656–658.

49. **Ward, E. S., and D. J. Ellar.** 1983. Assignment of the δ-endotoxin gene of *Bacillus thuringiensis* var. *israeliensis* to a specific plasmid by curing analysis. FEBS Lett. 158:45–49.

49a.**Ward, E. S., D. J. Ellar, and J. A. Todd.** 1984. Cloning and expression in *Escherichia coli* of the insecticidal δ-endotoxin gene of *Bacillus thuringiensis* var. *israelensis*. FEBS Lett. 175:377–382.

50. **Weislander, L.** 1979. A simple method to recover intact high molecular weight RNA and DNA after electrophoretic separation in low gelling temperature agarose gels. Anal. Biochem. 98:305–309.

51. **Yamamoto, T., and R. E. McLaughlin.** 1981. Isolation of a protein from the parasporal crystal of *Bacillus thuringiensis* var. *kurstaki* toxic to the mosquito larva, *Aedes taeniorhynchus*. Biochem. Biophys. Res. Commun. 103:414–421.

Biocidal Gene(s) of *Bacillus sphaericus* Cloned and Expressed in *Escherichia coli*

JOHN LOUIS AND JEKISIEL SZULMAJSTER

Laboratorie d'Enzymologie du Centre National de la Recherche Scientifique, 91190 Gif-sur-Yvette, France

Bacterial insecticides as alternatives to chemical pesticides for controlling insect-carried diseases have great potential in pest management programs, provided their environmental impact is fully evaluated. Several sporeforming bacteria which are insect pathogens have been shown to be effective microbial insecticides and are widely used as such in a number of tropical countries.

Bacillus sphaericus is one of the sporeforming bacteria that synthesize a potent entomocidal toxin (16), and early work resulted in the selection and characterization of strains presenting high pathogenicity and in the characterization of the nutritional requirements and physiology of these strains (8, 9). The insecticidal strains of *B. sphaericus* exhibit a spectrum of activity toward three major genera of mosquito larvae, *Culex*, *Anopheles*, and *Aedes* (1, 3). This activity appears to be associated with the cell wall-membrane fractions of sporulating cells and with the mature spores (3). Attempts to purify the active principle of one of the most potent and most stable strains, 1593 (13), to study its biochemical properties led only to limited results (2, 12, 13, 18).

Further progress toward a better understanding of the biogenesis of the larvicidal toxin and its mode of action at the molecular level was made by our recent cloning of the larvicidal genetic determinant of *B. sphaericus* 1593 (4). The cloning of this genetic determinant into the shuttle vector pHV33 (14) enabled us to express the biocidal activity in *Escherichia coli*, which may also be advantageous for maximization of its production.

We report here the characterization of the biocidal gene(s) and the analysis of its encoded polypeptides expressed in *E. coli* maxicells. We show further that all four polypeptides encoded by these genes are necessary for the expression of total larvicidal activity.

EXPERIMENTAL PROCEDURES

Bacterial strains, plasmids, growth conditions for *E. coli* strains, DNA manipulations, and hybridization procedures have been described elsewhere (11). *B. sphaericus* was grown in Hungerer and Tipper medium (6).

Bioassays were performed with extracts from cells harvested after 48 h of growth and prepared as described (4). Third-instar larvae of *Culex tritaenirynchus* or *C. pipiens pipiens* were used, and 50% lethal concentration (LC$_{50}$) values were determined after 24 h.

RESULTS AND DISCUSSION

Endonuclease cleavage map of a *Sau*3A DNA fragment carrying the larvicidal determinant. The strategy employed for cloning the larvicidal determinant from *B. sphaericus* 1593M into the shuttle vector pHV33 was described previously (4). Figure 1 shows the endonuclease restriction map of the 3.8-kilobase (kb) insert carrying the biocidal activity. No sites have been found for the following endonucleases: *Apa*I, *Ara*I, *Bal*I, *Bam*HI, *Bcl*I, *Bst*EII, *Bgl*I, *Cla*I, *Hpa*I, *Kpn*I, *Nru*I, *Pst*I, *Pvu*I, *Pvu*II, *Sac*II, *Sma*I, *Sph*I, *Xho*I, and *Xma*III.

Expression and characterization of the larvicidal gene products. Using *E. coli* maxicells (11, 15) followed by electrophoretic analysis of the synthesized proteins, we previously showed that the 3.8-kb DNA insert in pGspO3 encodes four polypeptides with molecular weights of 21,000, 19,000, 15,000, and 12,000 (11). The first two proteins are synthesized in relatively larger amounts.

Are all four proteins required for the biocidal activity of *B. sphaericus*? Attempts have been made to reduce the size of the insert in pGspO3 by subcloning, without losing the biocidal activity. The 3.8-kb cloned DNA was cleaved with *Eco*RV-*Bgl*II, *Bgl*II-*Sph*I, *Eco*RV-*Eco*RI, or *Hpa*II-*Hpa*II (Fig. 2). Appropriate DNA fragments were religated with plasmid pHV33, and the screened subclones were expressed in *E. coli* maxicells as described (10) and were tested for larvicidal activity. Of the four subclones tested, the one carrying the 2.8-kb *Eco*RV-*Eco*RI DNA fragment (C) had 4 to 5% of the original larvicidal activity (Table 1). The expression of the subclones in *E. coli* maxicells (Fig. 3) showed that fragment A codes for the synthesis of the 15- and 12-kilodalton polypeptides and fragment B codes for the 21-kilodalton polypeptide. This further suggests that cleavage at the *Bgl*II site in the 3.8-kb insert interferes with the expression of the 19-kilodalton polypeptide.

Nucleotide sequence analysis of the 3.8-kb insert, now in progress in this laboratory, will

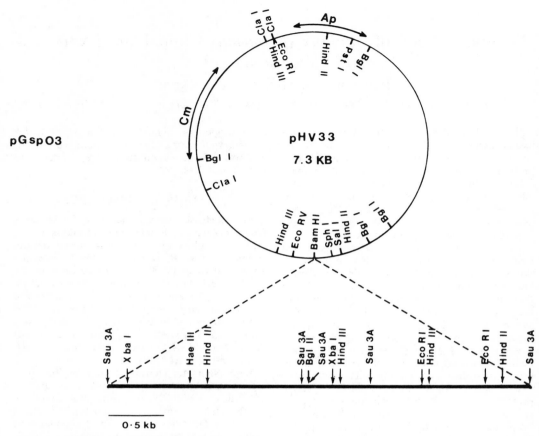

FIG, 1. Endonuclease cleavage map of the 3.8-kb *Sau*3A DNA fragment. Endonuclease restriction of chromosomal and plasmid DNA was carried as instructed by the suppliers.

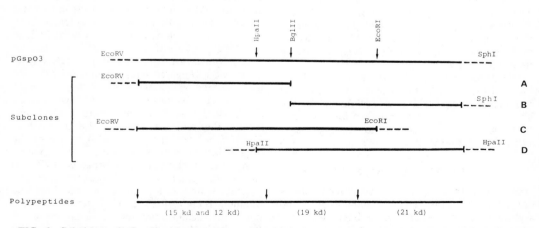

FIG. 2. Subclones of pGspO3. After treatment of the 3.8-kb insert with (A) *Eco*RV-*Bgl*II, (B) *Bgl*II-*Sph*I, (C) *Eco*RV-*Eco*RI, or (D) *Hpa*II-*Hpa*II, the restricted fragments were religated to the pHV33 plasmid and the subclones were screened on LB agar plates containing the appropriate antibiotics. The dashed lines denote pHV33 sequences.

TABLE 1. Strains, larvicidal activity, and gene copy number[a]

Strains	LC_{50}[b] (µg/ml)	Genomic sequences complementary to the pGsp03 probe
B. sphaericus		
1593M	1×10^{-3}–2×10^{-3}	2–3
2362	1×10^{-3}–2×10^{-3}	2–3
2297	1×10^{-3}–2×10^{-3}	4–5
1881	3–5	1
1691	2–4	1
1404	4–11	1
Rem 4	0	0[c]
SS-II	0	0
E. coli		
C600/pGsp03	50×10^{-3}–70×10^{-3}	
C600/SC (A)	0	
C600/SC (B)	0	
C600/SC (C)	20×10^{-2}–30×10^{-2}	
C600/SC (D)	0	

[a] The gene copy number was determined by dot hybridization. Denatured DNA of the strains listed were spotted on nitrocellulose filters in linear decreasing concentrations and were hybridized to the labeled probe as described (7, 11). The evaluation of the genomic sequences complementary to the pGsp03 probe was calculated as previously reported (11).
[b] LC_{50} corresponds to the protein concentration which kills 50% of third instar of *C. tritaenirynchus* or *C. pipiens pipiens* larvae population in 24 h.
[c] Below the level of detection.

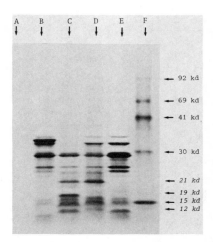

FIG. 3. Sodium dodecyl sulfate-polyacrylamide gel electrophoresis pattern of the polypeptides synthesized by *E. coli* maxicells (CSR603) carrying the recombinant plasmid pGsp03 and its derivatives. Cells were grown and labeled as described (11, 15). Samples were loaded onto a 5 to 20% linear gradient of sodium dodecyl sulfate-polyacrylamide gel. The gels were dried and autoradiographed (11). Lanes: A, control CSR603; B, CSR603(pHV33); C, CSR603(pGsp03); D, CSR603-sub *Bgl*II-*Sph*I; E, CSR603-sub *Eco*RV-*Bgl*II; F, molecular size standards in kilodaltons.

provide more information on the structure of the larvicidal genes and on the number and nature of the encoded polypeptides. However, on the basis of the present results, it is conceivable that all four polypeptides encoded by the DNA insert, possibly forming a protein complex, are necessary for the expression of full larvicidal activity.

DNA sequence homology and levels of toxicity in various strains of *B. sphaericus*. Considerable diversity is observed among *B. sphaericus* strains with respect to their levels of toxicity and target spectrum. Studies on the genetic relatedness of a great number of strains, both toxic and nontoxic, by DNA-DNA hybridization provided little information on the genetic specificity of the strains synthesizing the larvicidal toxin (10). We approached this problem by hybridizing *Eco*RI-digested DNA from six *B. sphaericus* strains with various levels of toxicity to the cloned larvicidal gene(s) isolated from *B. sphaericus* 1593M (Fig. 4). In general, the potent strains analyzed were characterized by the presence of DNA sequences not found in the nontoxic

strains, although some differences were observed between variants of these strains. Thus, the highly toxic strain 2297 seems to lack the 9- and 5.8-kb sequences, and a sequence of 7.4 kb appears instead. However, experiments designed to test the relative strength of the hybrids have shown that the 9-kb sequence forms the least stable hybrid under the stringent washing conditions used (11). This sequence thus appears to contribute very little to the level of toxicity of these strains. These results strongly suggest that the sequences of 6.4 to 6.6, 5.8 (or 7.4), and 1.6 kb, with some variations, might be essential for the expression of a high level of toxicity and that the reduced potency of the other strains analyzed might be the result of deletions or some other kind of chromosomal rearrangements. The two nontoxic strains analyzed showed either a complete lack of hybridization to the probe (SSII-I) or hybridization to only one sequence of 2.3 kb (Rem-4).

Gene copy number. Using the dot hybridization method (7) and the empirical formula described elsewhere (11), we have been able to calculate the gene copy number of the strains shown in Table 1. On the basis of the extent of label (hybridization signals) detectable in the dots of different DNA concentrations, it was estimated that the most potent strains, 1593, 2297, and 2362, carry two to five gene copies and

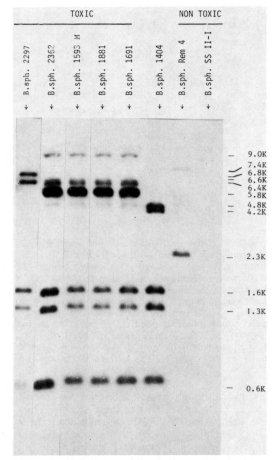

FIG. 4. Hybridization analysis of *Eco*RI-restricted DNA from different *B. sphaericus* strains with the specific ^{32}P-labeled probe. The *Eco*RI-digested DNA was electrophoresed on 0.8% agarose gel, denatured, transferred to nitrocellulose filters (17), and hybridized with the ^{32}P-labeled DNA probe (11).

the less potent strains, 1881, 1691, and 1404, carry at least one.

Stability of pGspO3. In the course of these studies we have observed that the pGspO3 recombinant plasmid is stably maintained only in lyophilized form but loses its biocidal activity if maintained by successive transfers on agar plates, although in the latter case no deletions were observed by electrophoretic analysis of the plasmid. When transferred into *B. subtilis*, the *B. sphaericus* DNA insert of pGspO3 and often also a part of the pHV33 plasmid was rapidly deleted, and this occurred independently of the Rec character of the recipient *B. subtilis* strain. However, when recloned into the *Staphylococcus aureus* plasmid pUB110 (5), the DNA insert

remained stable for at least 6 months, and no loss of biocidal activity was observed.

The knowledge of the genetic mechanisms(s) controlling the production of the larvicidal toxin of *B. sphaericus* and the possibility of the genetic engineering of this toxin hold out great promise for the development of highly potent strains of this microbial insect pathogen and the expansion of its target spectrum.

We thank Regine Barbey for excellent technical assistance. This research was supported by grants from the Directors Initiative Fund of the UNDP/World Bank/World Health Organization Special Programme for Research and Training in Tropical Diseases, from the Centre National de la Recherche Scientifique (ATP Microbiology, 1983), and from the Fondation pour la Recherche Médicale.

LITERATURE CITED

1. **Davidson, E. W.** 1981. Bacterial diseases of insects caused by toxin-producing Bacilli other than *Bacillus thuringiensis*, p. 269–291. *In* E. W. Davidson and A. Held (ed.), Pathogenesis of invertebrate microbial diseases. Allenheld, Osmun, & Co., Totowa, N.J.
2. **Davidson, E. W.** 1982. Alkaline extraction of toxin from spores of the mosquito pathogen, *Bacillus sphaericus* strain 1593. Can. J. Microbiol. **29**:271–275.
3. **Davidson, E. W., P. Myers, and A. A. Yousten.** 1978. Cellular location of insecticidal activity in *Bacillus sphaericus*. Proceedings of the International Congress on Invertebrate Pathology, Prague, Czechoslovakia, p. 47.
4. **Ganesan, S., K. Haresch, K. Jayaraman, and J. Szulmajster.** 1983. Cloning and expression in *Escherichia coli* of a DNA fragment from *Bacillus sphaericus* coding for biocidal activity against mosquito larvae. Mol. Gen. Genet. **184**:181–183.
5. **Gryczan, T. J., S. Contente, and D. Dubnau.** 1978. Characterization of *Staphylococcus aureus* plasmids introduced by transformation into *Bacillus subtilis*. J. Bacteriol. **134**:318–329.
6. **Hungerer, K. D., and D. J. Tipper.** 1969. Cell wall polymers of *Bacillus sphaericus* 9602. I. Structure of the vegetative cell wall peptidoglycans. Biochemistry **8**:3577–3586.
7. **Kafatos, F. C., C. W. Jones, and A. Efstratiadis.** 1979. Determination of nucleic acid sequence homologies and relative concentrations by a dot hybridisation procedure. Nucleic Acids Res. **7**:1541–1552.
8. **Kellen, W., T. Clark, J. Lindengren, B. Ho, M. Rogoff, and S. Singer.** 1965. *Bacillus sphaericus* Neide as a pathogen of mosquitoes. J. Invertebr. Pathol. **7**:442–448.
9. **Knight, B. C. J. G., and H. Proom.** 1950. A comparative survey of the nutrition and physiology of mesophilic species in the genus *Bacillus*. J. Gen. Microbiol. **4**:508–538.
10. **Krych, V., J. Johnson, and A. Yousten.** 1980. Deoxyribonucleic acid homologies among strains of *Bacillus sphaericus*. Int. J. Syst. Bacteriol. **30**:476–484.
11. **Louis, J., K. Jayaraman, and J. Szulmajster.** 1984. Biocide gene(s) and biocidal activity in different strains of *Bacillus sphaericus*. Expression of the gene(s) in *E. coli* maxicells. Mol. Gen. Genet. **195**:23–28.
12. **Myers, P. S., and A. A. Yousten.** 1980. Localization of a mosquito larval toxin of *Bacillus sphaericus* 1593. Appl. Environ. Microbiol. **39**:1205–1211.
13. **Myers, P., A. A. Yousten, and E. W. Davidson.** 1979. Comparative studies of the mosquito larval toxin of *Bacillus sphaericus* SSII.I and 1593. Can. J. Microbiol. **25**:1227–1231.
14. **Primrose, S. B., and S. D. Ehrlich.** 1981. Isolation of plasmid deletion mutants and study of their instability.

Plasmid **6**:193–201.

15. **Sancar, A., A. M. Hack, and W. P. Rupp.** 1979. Simple method for identification of plasmid-coded proteins. J. Bacteriol. **137**:692–693.

16. **Singer, C.** 1973. Insecticidal activity of recent bacterial isolates and their toxins against mosquito larvae. Nature (London) **244**:110–111.

17. **Southern, E. M.** 1975. Detection of specific sequences among DNA fragments separated by gel electrophoresis. J. Mol. Biol. **98**:503–517.

18. **Tinelli, R., and C. Bourgouin.** 1982. Larvicidal toxin from *Bacillus sphaericus* spores. FEBS Lett. **142**:155–158.

Plasmids and Delta-Endotoxin Production in Different Subspecies of *Bacillus thuringiensis*

BRUCE C. CARLTON† AND JOSÉ M. GONZÁLEZ, JR.†

Department of Molecular and Population Genetics, The University of Georgia, Athens, Georgia 30602

Natural isolates of *Bacillus thuringiensis* have been classified into over 20 subspecies (varieties) and subvarieties (28 to date) according to their flagellar serotypes (H-antigens) and other biochemical properties (1). Preparations of *B. thuringiensis* spores and crystals have been used agriculturally as bioinsecticides against larvae of lepidopteran pests and, more recently, against some dipteran larvae (4). The insecticidal activity resides in the proteinaceous inclusion, generally bipyramidal, that is formed during sporulation and is known as the parasporal body or crystal; the protein toxin itself is known as the delta-endotoxin (7). Plasmids are ubiquitous in *B. thuringiensis* strains, usually in complex arrays (B. C. Carlton and J. M. González, Jr., *in* D. Dubnau, ed., *The Molecular Biology of the Bacilli*, vol. 2, in press), and some of these plasmids are capable of transfer between *B. thuringiensis* strains and into *B. cereus* (10). Delta-endotoxin genes have been assigned to specific *B. thuringiensis* plasmids by studies involving isolation of acrystalliferous (Cry⁻) *B. thuringiensis* variants by plasmid curing (12, 23), cloning of toxin genes (14, 21), and transfer of toxin plasmids into Cry⁻ *B. thuringiensis* and *B. cereus* strains, which were thereby converted into crystalliferous (Cry⁺) toxin producers (9).

Intravarietal similarities among the complex *B. thuringiensis* plasmid arrays were suggested in reports on different isolates of the same subspecies (12, 14). A general survey of the plasmid content of all *B. thuringiensis* subspecies would be useful as a first step towards identifying new transmissible toxin plasmids, which might be combined in a single *B. thuringiensis* recipient to generate strains with multiple toxin plasmids (9). Furthermore, it might prove feasible to use some of the more complex plasmid patterns as an aid to classification of *B. thuringiensis* strains. Toward these ends, the plasmid arrays of *B. thuringiensis* strains representing every subspecies were visualized and compared.

MATERIALS AND METHODS

Most of the *B. thuringiensis* strains were ob-

tained from the collection of H. T. Dulmage, Cotton Insects Research, U.S.D.A.-S.E.A., Brownsville, Tex., and carry serial numbers with the prefix "HD-." The catalog of HD strains (5) contains information on strains available and their sources. A few strains were obtained from H. deBarjac (Institut Pasteur, Paris), D. H. Dean (Ohio State University, Columbus), and A. J. DeLucca II (Southern Regional Research Center, U.S.D.A., New Orleans, La.).

Plasmid patterns were visualized on 0.5% vertical agarose gels using Tris-borate running buffer and a high running voltage, by our modification (9) of the method of Eckhardt (6). Sizes of unknown plasmids were estimated by comparison with the plasmids of strain HD-2, which were measured previously by electron microscopy (12).

RESULTS AND DISCUSSION

Plasmid arrays of *B. thuringiensis* strains. Strains representing all known *B. thuringiensis* subspecies, including the recent isolates of serotypes 17 (19), 18 and 19 (20), 20 (22), and 21 (3), were examined for plasmid DNA. As shown in Fig. 1, all strains contained plasmid DNA ranging from 1 to 12 plasmids per strain. Generally, the plasmid array of each *B. thuringiensis* serotype or subspecies was unique and could be distinguished from that of other strains, although there were cases of internal diversity of plasmid content within serotypes. Some new examples of strains carrying plasmids that were present mostly in the open circular (OC) or linear form (12) were also found.

The results in Fig. 1a are typical of the variations in plasmid number and size range observed. The plasmids range from 2 to 12 per strain and from ~1.4 to ~150 megadaltons (MDa) in size. There are no obvious resemblances in plasmid arrays among the subspecies represented. When strains of the same subspecies are compared, however, such as HD-4 and *anduze* (both subsp. H-3a *alesti*), similarities in plasmid content are obvious. The five largest plasmids in HD-4 and *anduze* are of identical sizes. The smaller plasmids are less similar, but still rather alike. In the three H-3ab *kurstaki* strains, however, the small plasmids are similar and the large plasmids are different. Figure 1a also shows a

† Present address: Ecogen Inc., Lawrenceville, NJ 08648.

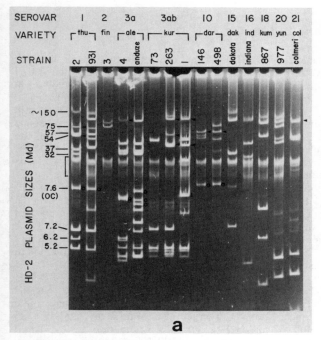
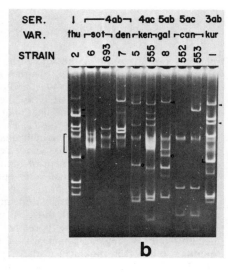
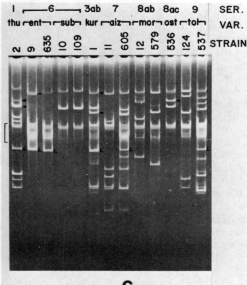
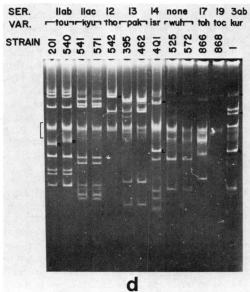

FIG. 1. Modified Eckhardt lysate-electrophoresis (9) on 0.5% agarose of cells of representative strains of *B. thuringiensis*, demonstrating the presence of complex plasmid arrays in most of them. The subspecies represented are as follows. (a) H-1, *thuringiensis* (thu); H-2, *finitimus* (fin); H-3a, *alesti* (ale); H-3ab, *kurstaki* (kur); H-10, *darmstadiensis* (dar); H-15, *dakota* (dak); H-16, *indiana* (ind); H-18, *kumamotoensis* (kum); H-20, *yunnanensis* (yun); H-21, *colmeri* (col). Serovar (serotype), subspecies, and strain designations are listed above each gel lane. The masses of the HD-2 plasmids, listed on the left margin, are intended as size standards. A broad band corresponding to linearized plasmids and chromosomal fragments is present in each lane; its general position is indicated by the bracket on the left margin. The open circles (○) indicate some of the plasmids (such as those in HD-2 and HD-4) which are present mostly or only in the OC form; the (L) indicates the linear plasmidlike DNA element (LDE), ~10 MDa in size, present in HD-1 and other strains below. The arrowheads indicate individual plasmids implicated in toxin production in the course of our studies. (b) Subspecies H-4ab *sotto* (sot) and *dendrolimus* (den); H-4ac *kenyae* (ken); H-5ab *galleriae* (gal); H-5ac *canadensis* (can); HD-2 and HD-1 are included for comparison. Symbols are as in (a). (c) Subspecies H-6 *entomocidus* (ent) and *subtoxicus* (sub); H-7 *aizawai* (aiz); H-8ab *morrisoni* (mor); H-8ac *ostriniae* (ost); H-9 *tolworthi* (tol); HD-2 and HD-1 are included for comparison. Symbols are as in (a). (d) Subspecies H-11ab *toumanoffi* (tou); H-11ac *kyushuensis* (kyu); H-12 *thompsoni* (tho); H-13 *pakistani* (pak); H-14 *israelensis* (isr); *wuhanensis* (wuh); H-17 *tohokuensis* (toh); H-19 *tochigiensis* (toc). HD-1 is included for comparison; symbols are as in (a).

plasmid present only in the OC form (HD-2, HD-931), several present mostly as the OC form (HD-4, *anduze*, HD-146, HD-498), and one, the "linear DNA element" (LDE), as the linear form (HD-1). The plasmids implicated in crystal toxin production in our work (9, 11, 12; unpublished data) are indicated by arrowheads. Isolation of single colonies from the received cultures sometimes aided the assignment of toxin genes to a specific plasmid, because in some cultures a significant part of the population consisted of Cry⁻ variants. These were isolated and their plasmids were compared with those of the Cry⁺ cells of the same or a closely related strain; usually, only a single plasmid was absent or modified (altered in mobility) in the Cry⁻ variant. Most of these plasmids encode a phase-refractile crystal under most standard media and conditions, such as NSM agar (16) at 30°C; two of them, the HD-2 54-MDa plasmid and the HD-1 44-MDa plasmid are unusual in encoding a crystal only in certain media or temperatures (González and Carlton, manuscript in preparation). Individual *B. thuringiensis* strains have been found to carry one, two, and even three toxin plasmids.

Figure 1b further demonstrates that plasmid patterns can be used to distinguish individual strains of the same flagellar serotype. The H-4ab strains HD-693 (biotype *sotto*) and HD-7 (biotype *dendrolimus*) have distinct plasmid arrays, as do HD-5 and HD-555 (both H-4ac *kenyae*, one from Kenya, the other from India). Plasmids present mostly as the OC form are present in strains HD-5 (H-4ac *kenyae*) and HD-8 (H-5ab *galleriae*). It is interesting that the received cultures of HD-6, HD-552, and HD-553 were found to be totally Cry⁻. The loss of crystal production in HD-6 might be linked with changes in the single plasmid seen in the related strain, HD-693, which is present in HD-6 is in a smaller (deleted?) form. The single plasmid of a *sotto* strain (probably identical to HD-693) was implicated in toxin production by Kronstad et al. (14). The tendency of H-5ac *canadensis* strains to become Cry⁻ upon subculturing has been previously reported (2).

Figure 1c demonstrates the presence of complex plasmid arrays in *B. thuringiensis* serotypes H-6 through H-10. HD-9 and HD-635 are derivatives of a single original isolate (H. T. Dulmage, personal communication). HD-9 makes large, blunt bipyramidal crystals with pronounced toxicity against *Aedes aegypti* larvae, and HD-635 makes tiny crystals, nontoxic to *A. aegypti* (unpublished data), and also lacks the ~105- and 29-MDa plasmids present in HD-9 (the other large plasmid and LDE appear to be present in both strains). Derivatives of HD-9 (not shown) have been isolated that lack the

29-MDa plasmid but retain their large, mosquito-toxic crystals, so that it can be deduced that the ~105-MDa plasmid of HD-9 is involved in delta-endotoxin production. HD-9 may also carry a second toxin plasmid since HD-635 still produces (tiny) crystal inclusions. The presence of LDE molecules with the same mobility as the LDE of HD-1 is evident in HD-6, HD-635, HD-605 (H-7 *aizawai*), and HD-537 (H-9 *tolworthi*).

Figure 1d shows several pairs of *B. thuringiensis* strains, most of them from a single original isolate, which have begun to diverge in plasmid content. HD-201 and HD-540 are from the same isolate, but during passage through different laboratories, their OC plasmids have diverged in size. Likewise, part of the population of HD-395 retains the same plasmid content as HD-462, whereas part has started to mutate. Strains HD-541 and HD-571, on the other hand, have not yet diverged in plasmid array, and HD-525 and HD-572 are also the same.

All strains of H-14 *israelensis* (*B. thuringiensis* subsp. *israelensis*) examined had nearly identical plasmid patterns; the differences could be explained as simple loss of one or more plasmids from the single original isolate of Goldberg and Margalit (8), which was the source of ONR-60A, 4Q1, 4Q2, HD-500, HD-567, and other *B. thuringiensis* subsp. *israelensis* strains (D. H. Dean, personal communication). Comparison of the plasmid arrays of these strains suggested that 4Q1, the strain shown in Fig. 1d, is closest to the original isolate. Presumably, the loss of a ~125-MDa plasmid (present in 4Q1) produced the *B. thuringiensis* subsp. *israelensis* strain HD-567 (not shown), in which a 75-MDa plasmid has been implicated in toxin production by means of plasmid curing and plasmid transfer (11). Loss of a 4.9-MDa plasmid (present in 4Q1) probably led to the plasmid array of HD-500, another *B. thuringiensis* subsp. *israelensis* strain in which the 75-MDa plasmid has been implicated in toxin synthesis by curing (1).

Figure 1d also shows two cultures of strain *wuhanensis*, a nonmotile *B. thuringiensis* strain from China. Their plasmid arrays are the same and are also almost identical to that of HD-8 (H-5ab *galleriae*) in Fig. 1b. This close resemblance suggests a common origin for H-5ab *galleriae* and the *wuhanensis* isolates. Finally, the plasmids of H-17 *tohokuensis* (HD-866) and H-19 *tochigiensis* (HD-868) strains are shown. HD-868 is unusual in possessing only a single plasmid, of rather large size (>100 MDa).

The number and variety of each strain in which a plasmid has been implicated in toxin production, and the size of the plasmid, are listed in Table 1. It is evident from Table 1 that all plasmids implicated in toxin production in

TABLE 1. Correlations between specific plasmids and delta-endotoxin production in various strains of *B. thuringiensis*

Prototype strain no.[a]	H-serotype (flagellar antigens) and subspecies	No. of plasmids	Size of plasmid (MDa) implicated in toxin production[b]	Type of implicating evidence (and references)[c]
HD-2	1, *thuringiensis*	10	75	A, T (1, 9, 12)
			[54]	T (González and Carlton, in preparation)
HD-225	1, *thuringiensis*	11	~110 (P1, P2)	D (González and Carlton, in preparation)
HD-73	3ab, *kurstaki*	6	50	A, T (9, 12)
HD-74	3ab, *kurstaki*	7	55	A, T (1; González and Carlton, in preparation)
HD-1	3ab, *kurstaki*	12	~115 (P1, P2)	A, D (González and Carlton, in preparation)
			[44 (P1)]	A, T (González and Carlton, in preparation)
HD-263	3ab, *kurstaki*	11	~110 (P1, P2)	A (González and Carlton, in preparation; Yamamoto et al., in preparation)
			60 (P1)	A, I (González and Carlton, in preparation)
			44 (P1)	T (9)
HD-4	3a, *alesti*	10	~105	A (12)
HD-8	5ab, *galleriae*	4	~130	D (12)
HD-9	6, *entomocidus*	4	~105	A (this study)
HD-536	8ac, *ostriniae*	4	68	A, T (1; this study)
HD-13	9, *tolworthi*	7	~110	A (González and Carlton, in preparation)
			44	A (González and Carlton, in preparation)
HD-146	10, *darmstadiensis*	5	70	A (this study)
HD-498	10, *darmstadiensis*	5	65	A (this study)
HD-542	12, *thompsoni*	4	~100	A (1)
HD-567	14, *israelensis*	9	75	A, T (11)
HD-500	14, *israelensis*	9	75	A (1)
BT Col	21, *colmeri*	6	~100	D (this study)

[a] Refer to numbers in the strain collection of H. T. Dulmage, who provided the original cultures, except for BT Col, the *B. thuringiensis* subsp. *colmeri* isolate, provided by A. J. DeLucca II.

[b] Plasmids in brackets code for crystal synthesis conditionally only (dependent on growth medium, or temperature, or both). P1 and P2 distinguish between plasmids coding for P1 toxin only and plasmids encoding both P1 and P2 toxins, in strains that produce both types of crystal (González and Carlton, in preparation; Yamamoto et al., in preparation).

[c] A, Absence of the plasmid in Cry⁻ variant(s); D, deleted form of the plasmid detected in Cry⁻ variant(s); I, form of plasmid with insertion detected in Cry⁻ variant(s); T, transfer of the plasmid into a Cry⁻ *B. thuringiensis* or *B. cereus* recipient converted it to Cry⁺.

our studies were invariably large (>40 MDa). No plasmid <40 MDa has yet been implicated. Other laboratories have also implicated large *B. thuringiensis* plasmids (14, 23), but no small plasmid has been convincingly associated with production of crystal toxin (Carlton and González, in press).

Plasmid-toxin association in HD-1. Strain HD-1 (H-3ab, *kurstaki*), the strain of commercial importance in the United States, has also proved to be interesting scientifically. Our initial study (12) suggested that a 29-MDa plasmid was involved in toxin production in HD-1; this conclusion was based on isolation of a single Cry⁻ mutant (HD1-25), in which the 29-MDa plasmid had acquired an insertion of 3 to 4 MDa of DNA. Further analysis of this strain and of other HD-1 derivatives has led us to conclude that the 29-MDa plasmid does not code for the toxin crystal and that instead a large (~115-MDa)

plasmid is involved in toxin crystal production. Some representative variants of HD-1 are discussed here, both as an example of the misleading results that may occur during genetic study of *B. thuringiensis* and of how *B. thuringiensis* plasmid arrays may change over time.

Figure 2 shows the plasmid patterns of the prototype strain, HD1-1, and of several derivatives, both Cry⁺ and Cry⁻, having altered plasmid arrays. HD1-1 lost the 44-, 51-, 9.6-, and ~130-MDa plasmids spontaneously (in four sequential steps; the intermediate strains, not shown, were Cry⁺) to generate the Cry⁺ variant HD1-14. HD1-14 then apparently lost the ~115-MDa plasmid to produce the Cry⁺ variant HD1-15. However, HD1-15 proved unusual in that it generated several variants, both Cry⁺ and Cry⁻, containing new plasmids. One of them was HD1-18, in which a very faint band of ~160-MDa (not visible in Fig. 2) appeared.

HD1-18 then mutated to give the Cry⁻ variant HD1-25, in which the ~160- and 29-MDa plasmids were replaced by plasmids of ~105 and 34 MDa. Other derivatives of HD1-15 harboring new plasmids are HD1-19, HD1-20, and HD1-21. HD1-19 is Cry⁺ and is possibly a true "revertant," since its new plasmid is ~115 MDa. HD1-20 is Cry⁻ and has a new plasmid of 88 MDa, whereas HD1-21 is Cry⁺ and has a new 80-MDa plasmid. A line of descent entirely separate from that leading to HD1-14 produced the variants HD1-7 and HD1-9. Strain HD1-1 generated HD1-7 (Cry⁺) through loss of the LDE and the 51-, 5.4-, 9.6-, and ~130-MDa plasmids, some by growth at 37 or 42°C and some by spontaneous curing. The strains representing the four intermediate stages are not shown. Growth of HD1-7 at 42°C generated the Cry⁻ mutant HD1-9, which has lost the ~115-MDa plasmid.

In our earlier study of HD-1 (12), we believed that the ~115-MDa plasmid had been lost in HD1-15; cells of the HD1-19 type were detected in the HD1-15 culture but were thought to be contaminants. We now believe that the ~115-MDa plasmid was not lost in HD1-15 since new plasmids, possibly derived from it, predictably arise in the HD1-15 cell population, a phenomenon not seen in other variants of HD1-1. Some of the new plasmids are larger than the original ~115-MDa plasmid; others are smaller and may be deleted versions. Some of the strains with new plasmids are Cry⁺ and some are Cry⁻. New plasmids smaller than ~115 MDa are not necessarily associated with the Cry⁻ phenotype since the Cry⁺ variant HD1-21 contains a new 80-MDa plasmid that is smaller than the new 88-MDa plasmid of (Cry⁻) HD1-20. Strain HD1-15 is unstable and generates spontaneous Cry⁻ mutants (such as HD1-20) at a detectable frequency of 0.1 to 1%, much higher than the rate of mutation to Cry⁻ in other HD-1 variants, where it is 0.01% or less. In contrast to the Cry⁻ mutants derived from HD1-15, the Cry⁻ mutant HD1-9 appears to have arisen by simple loss of the ~115-MDa plasmid.

Our new data implicate the ~115-MDa plasmid in crystal production in HD-1. The unusual variant HD1-15 may be analogous to the HD-567 variant HD567-26 (11), in that the probable toxin plasmid is undetectable on gels but remains present and is capable of generating smaller versions of itself, as shown in Fig. 2. We also conclude that the insertion into the 29-MDa plasmid in HD1-25 is probably fortuitous, and that the plasmid has no role in toxin production. The curing data in Fig. 2 also indicate that the ~130-, 51-, 44-, 9.6-, and 5.4-MDa plasmids and LDE can be lost without any obvious effect on crystal production. These results, and others

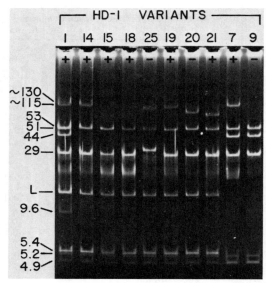

FIG. 2. Agarose gel plasmid patterns of representative mutants of strain HD-1, demonstrating the correlation of the ~115-MDa plasmid with crystalline toxin production. Molecular masses of the HD1-1 plasmids (in MDa) are listed on the left-hand margin and were obtained by using the plasmids of strain HD-2 as standards; L marks the position of the LDE. The 53- and 51-MDa plasmid doublet was earlier reported as a single 52-MDa plasmid (11). A 1.4-MDa plasmid is also present in all variants shown here, but was run off the gel during electrophoresis.

presented previously, indicate that plasmid curing, although suggesting the locations of toxin genes on plasmids, is not definite proof that a given *B. thuringiensis* plasmid carries the structural genes for delta-endotoxin (1, 11, 12), especially if there are difficulties in detecting the larger toxin plasmids (12; Carlton and González, in press). Transfer of a putative toxin plasmid from a Cry⁺ *B. thuringiensis* donor into a Cry⁻ *B. thuringiensis* or *B. cereus* recipient, thereby converting the recipient into a Cry⁺ strain which produces delta-endotoxin antigenically like that of the donor strain, provides stronger evidence that the transmitted plasmid carries the toxin structural genes (1, 9–11).

The 44-MDa plasmid, although not required for toxin production, may contribute to crystal formation. Minnich and Aronson (17) recently reported that our Cry⁻ mutant HD1-9 formed crystals when grown at 25°C, but not at 30°C. We have confirmed their observation; however, variants such as HD1-25 and HD1-20, which lack the 44-MDa plasmid and contain what is possibly a deleted form of the ~115-MDa plasmid, are Cry⁻ at both 25 and 30°C. Variant HD1-10 (not shown), derived from HD1-9 by loss of the 44-MDa plasmid, is also Cry⁻ at both temperatures. Recent results indicate that HD-1 toxin production is controlled by toxin genes on both the 44- and ~115-MDa plasmids (T. Yamamoto, J. M. González, Jr., and B. C. Carlton, manuscript in preparation).

Unusual nutritional requirements of *B. thuringiensis* strains. In the course of this study, some quirks of individual *B. thuringiensis* subspecies became evident. For example, when cells were streaked on SCG agar (12) for Eckhardt slot-lysate electrophoresis, strains of certain subspecies failed to grow. Supplementation of SCG agar with additional nutrients (tryptophan, glutamine, vitamins, etc.) led to full growth of the auxotrophic strains and satisfactory plasmid visualization. Requirements for tryptophan, nicotinic acid, and thiamine were identified in various strains. All H-5ab *galleriae* strains, such as HD-8 (Fig. 1b), required nicotinic acid. The requirement of nicotinic acid by H-5ab *galleriae* strains has been reported (13, 15). The two aflagellar (*wuhanensis*) strains, HD-525 and HD-572 (Fig. 1d), also required nicotinic acid. This finding, together with their similarity in plasmid array to HD-8, suggests that *wuhanensis* probably arose from an H-5ab *galleriae* strain through loss of flagella. The two H-10 *darmstadiensis* strains, HD-146 and HD-498 (Fig. 1a), both required tryptophan. Finally, the H-17 *tohokuensis* strain, HD-866 (Fig. 1d), required thiamine for growth.

Strains of subspecies H-4ac and H-6 formed unstable spores on NSM agar. Although sporulation appeared normal, the spores began to become phase dark and to lose their viability within days. However, it was observed that spores grown on SCG agar were much more stable. Subsequently, these strains were grown on NSM agar containing glucose (0.1 to 0.2%) and were found to produce stable spores (judged by their ability to remain phase bright after lysis of the sporangia). These preliminary results indicate that some *B. thuringiensis* strains may require glucose (or perhaps any fermentable carbohydrate) for normal sporulation.

The requirement of H-10 strains for tryptophan is not surprising, since there have been many reports of the inability of *B. thuringiensis* to grow in defined media unless one or more amino acids are added (15, 18). However, this is not true for all *B. thuringiensis* strains. Preliminary experiments (results not shown) have revealed that strain HD-2 (H-1) can grow (although slowly), sporulate, and make crystals in a simple defined medium containing glucose, sodium phosphate, ammonium chloride, and small amounts of other salts. Therefore, at least HD-2 appears to have no amino acid, vitamin, or other auxotrophic requirements. On the other hand, other strains, such as HD-1 (H-3ab), were tested

on the same minimal medium and grew very poorly, suggesting that true amino acid auxotrophies are present in *B. thuringiensis* strains other than those of serotype H-10. When adequate glucose-ammonium salts minimal media for *B. thuringiensis* strains are developed, the patterns of amino acid, vitamin, and other requirements, together with the complex *B. thuringiensis* plasmid arrays, may prove to be distinctive enough to play a role in the classification and identification of strains of this important industrial microorganism.

SUMMARY

A large number of strains of *B. thuringiensis*, including every known subspecies, were examined for the presence of extrachromosomal DNA by agarose gel electrophoresis. All Cry⁺ strains contained at least one large plasmid of 30 MDa or more; most strains harbored complex plasmid arrays of three or more size classes. Strains of different subspecies had distinct plasmid arrays. Strains of the same subspecies usually showed similarities in their plasmid arrays, but sometimes the plasmid arrays were very different. Plasmids implicated in crystalline delta-endotoxin production by plasmid curing or plasmid transfer were invariably large (>40 MDa). An unusual mutant of strain HD-1 (subsp. H-3ab *kurstaki*) is discussed as an example of the misleading variants that sometimes arise during attempts at curing. In this Cry⁺ mutant, a toxin plasmid was only apparently "cured," as deduced from the occurrence of "revertants" in which the plasmid reappeared.

Some subspecies of *B. thuringiensis* exhibited characteristic nutritional requirements; for example, strains of subsp. H-10 (*darmstadiensis*) required tryptophan for growth. Some strains appeared to require the presence of glucose in the nutrient medium for production of stable spores. The complex plasmid arrays and auxotrophic requirements of individual *B. thuringiensis* strains may be useful in *B. thuringiensis* classification and identification, as well as providing convenient markers for genetic studies.

We thank H. T. Dulmage, H. deBarjac, D. H. Dean, and A. J. DeLucca II for providing the *B. thuringiensis* strains.

This work was supported by a grant from Shell Development Co., Modesto, Calif.

LITERATURE CITED

1. **Carlton, B. C., and J. M. González, Jr.** 1984. Plasmid-associated delta-endotoxin production in *Bacillus thuringiensis*, p. 387–400. *In* A. T. Ganesan and J. A. Hoch (ed.), Genetics and biotechnology of bacilli. Academic Press, Inc., New York.
1a. **deBarjac, H.** 1981. Identification of H-serotypes of *Bacillus thuringiensis*, p. 35–43. *In* H. D. Burges (ed.), Microbial control of pests and plant diseases, 1970–1980. Academic Press, Inc., New York.

2. **deBarjac, H., and A. Bonnefoi.** 1972. Presence of antigenic subfactors in serotype V of *Bacillus thuringiensis* with the description of a new type: *B. thuringiensis* var. *canadensis*. J. Invertebr. Pathol. **20:**212–213.

3. **DeLucca, A. J., II, M. S. Palmgren, and H. deBarjac.** 1984. A new serovar of *Bacillus thuringiensis* from grain dust; *Bacillus thuringiensis* serovar *colmeri* (serovar 21). J. Invertebr. Pathol. **43:**437–438.

4. **Dulmage, H. T., and K. Aizawa.** 1982. Distribution of *Bacillus thuringiensis* in nature, p. 209–237. *In* E. Kurstak (ed.), Microbial and viral pesticides. Marcel Dekker, New York.

5. **Dulmage, H. T., C. C. Beegle, H. de Barjac, D. Reich, G. Donaldson, and J. Krywienczyk.** 1982. *Bacillus thuringiensis* cultures available from the U.S. Department of Agriculture. U.S.D.A.-A.R.S. Agricultural Reviews and Manuals, ARM-S-30/Oct. 1982. U.S. Department of Agriculture, Agriculture Research Service, New Orleans.

6. **Eckhardt, T.** 1978. A rapid method for the identification of plasmid desoxyribonucleic acid in bacteria. Plasmid **1:**584–588.

7. **Fast, P. G.** 1981. The crystal toxin of *Bacillus thuringiensis*, p. 223–248. *In* H. D. Burges (ed.), Microbial control of pests and plant diseases, 1970–1980. Academic Press, Inc., New York.

8. **Goldberg, L. J., and J. Margalit.** 1977. A bacterial spore demonstrating rapid larvicidal activity against *Anopheles sergentii*, *Uranotaenia unguiculata*, *Culex univitattus*, *Aedes aegypti* and *Culex pipiens*. Mosq. News **37:**355–358.

9. **González, J. M., Jr., B. J. Brown, and B. C. Carlton.** 1982. Transfer of *Bacillus thuringiensis* plasmids coding for δ-endotoxin among strains of *B. thuringiensis* and *B. cereus*. Proc. Natl. Acad. Sci. U.S.A. **79:**6951–6955.

10. **González, J. M., Jr., and B. C. Carlton.** 1982. Plasmid transfer in *Bacillus thuringiensis*, p. 85–95. *In* U. N. Streips, S. H. Goodgal, W. R. Guild, and G. A. Wilson (ed.), Genetic exchange: a celebration and a new generation. Marcel Dekker, New York.

11. **González, J. M., Jr., and B. C. Carlton.** 1984. A large transmissible plasmid is required for crystal toxin production in *Bacillus thuringiensis* var. *israelensis*. Plasmid **11:**28–38.

12. **González, J. M., Jr., H. T. Dulmage, and B. C. Carlton.** 1981. Correlation between specific plasmids and δ-endotoxin production in *Bacillus thuringiensis*. Plasmid **5:**351–365.

13. **Karabekov, B. P., A. G. Chakhmakhchyan, and M. G. Oganesyan.** 1982. Transmissive genetic factors in *Bacillus thuringiensis*. I. Reasons for differences in some biochemical properties in wild type strains of *Bacillus thuringiensis* var. *galleriae*. Genetika **18:**1062–1066.

14. **Kronstad, J. W., H. E. Schnepf, and H. R. Whiteley.** 1983. Diversity of locations for *Bacillus thuringiensis* crystal protein genes. J. Bacteriol. **154:**419–428.

15. **Kuznetsov, L. E., M. P. Khovrychev, and Z. V. Sakharova.** 1984. A synthetic medium for culturing *Bacillus thuringiensis*. Mikrobiologiya **52:**663–666.

16. **Li, E., and A. Yousten.** 1975. Metalloprotease from *Bacillus thuringiensis*. Appl. Microbiol. **30:**354–361.

17. **Minnich, S. A., and A. I. Aronson.** 1984. Regulation of protoxin synthesis in *Bacillus thuringiensis*. J. Bacteriol. **158:**447–454.

18. **Nickerson, K. W., and L. A. Bulla, Jr.** 1974. Physiology of sporeforming bacteria associated with insects: minimal nutritional requirements for growth, sporulation, and parasporal crystal formation of *Bacillus thuringiensis*. Appl. Microbiol. **28:**124–128.

19. **Ohba, M., K. Aizawa, and S. Shimizu.** 1981. A new subspecies of *Bacillus thuringiensis* isolated in Japan: *Bacillus thuringiensis* subsp. *tohokuensis* (serotype 17). J. Invertebr. Pathol. **38:**307–309.

20. **Ohba, M., K. Ono, K. Aizawa, and S. Iwanami.** 1981. Two new subspecies of *Bacillus thuringiensis* isolated in Japan: *Bacillus thuringiensis* subsp. *kumamotoensis* (serotype 18) and *Bacillus thuringiensis* subsp. *tochigiensis* (serotype 19). J. Invertebr. Pathol. **38:**184–190.

21. **Schnepf, H. E., and H. R. Whiteley.** 1981. Cloning and expression of the *Bacillus thuringiensis* crystal protein gene in *Escherichia coli*. Proc. Natl. Acad. Sci. U.S.A. **78:**2893–2897.

22. **Wang, W.-Y., Q.-F. Wang, X.-P. Zhang, and Y.-W. Chen.** 1979. *Bacillus thuringiensis* var. *yunnanensis*, a new variety of *Bacillus thuringiensis*. Acta Microbiol. Sin. **19:**117–121.

23. **Ward, E. S., and D. J. Ellar.** 1983. Assignment of the δ-endotoxin gene of *Bacillus thuringiensis* var. *israelensis* to a specific plasmid by curing analysis. FEBS Lett. **158:**45–49.

Differential Expression of Two Homologous Genes Coding for Spore-Specific Proteins in *Myxococcus xanthus*

MARTIN TEINTZE, TEIICHI FURUICHI, ROLLAND THOMAS, MASAYORI INOUYE, AND SUMIKO INOUYE

Department of Biochemistry, State University of New York at Stony Brook, Stony Brook, New York 11794

Myxococcus xanthus is a rod-shaped, gram-negative bacterium that undergoes a unique developmental cycle. Upon starvation on a solid surface, the cells aggregate to form fruiting bodies, and then some of them convert to round or ovoid spores while the majority of cells lyse (for review, see 14). During differentiation there are many changes in the pattern of protein synthesis, the most striking of which is the appearance of protein S (7, 8). Synthesis of protein S is induced early in development and increases until it reaches 15% of total protein synthesis at the stage of mound formation. It accumulates in the cytoplasm as a soluble protein until the onset of sporulation, after which it is found assembled on the surface of the spores (8). Protein S can be removed from the spores by extraction with 1 N NaCl or 10 mM EDTA and can be reassembled on the spores by adding a 10 mM excess of Ca^{2+} ion or removing the NaCl by dialysis. Protein S has been purified and crystallized (11), and a preliminary X-ray crystallographic study has been performed (13).

The gene has been cloned, and it was found that there are two very homologous genes (designated *ops* and *tps*) in the same orientation separated by a short spacer region (10, 12). When the amino acid sequences predicted from the DNA sequences of the two genes were compared with the protein S sequence, only the downstream (*tps*) gene was found to code for protein S (10, 22). The protein S mRNA has a very long half-life (19) and is found only in developing cells (9, 19); the major site for initiation of transcription is 51 bases upstream from the initiation codon for the *tps* gene, and the probable sequence of the development-specific promoter has been identified (9). The *ops* gene product and its mRNA have eluded detection until now.

Deletion mutants have been constructed from which the protein S (*tps*) gene or the region containing both the *ops* and *tps* genes has been removed (15). When the *tps* gene alone was deleted, no differences in fruiting body formation or myxospore yield were observed between the mutant and wild-type cells. However, when

both genes were deleted together with the spacer region between them, fruiting body formation was significantly delayed and the yield of myxospores was reduced (15). Protein S, therefore, does not seem to be required for differentiation of *M. xanthus* even though it is produced in very large quantities, specifically during development, when the cells are responding to starvation for nutrients. As a result, we have been investigating whether protein S may have a role in the germination of the myxospores, but protein S-deficient spores appeared to be just as viable as wild-type spores under the conditions we used. We also constructed mutants from which the *ops* gene or the spacer region between the two genes has specifically been deleted and found that the *ops* gene mutant showed a delay in fruiting body formation but the spacer region mutant did not. To characterize the product of the *ops* gene, we expressed it in *Escherichia coli* and showed that it can be precipitated with antiserum raised against protein S but that it migrates differently on a sodium dodecyl sulfate (SDS)-polyacrylamide gel. This allowed us to identify it in *M. xanthus* spores. It appears to be synthesized only very late in development after the onset of sporulation.

EXPERIMENTAL PROCEDURES

Bacteria and plasmids. *M. xanthus* FB (DZF1) was used as the wild-type strain. The *tps* deletion mutant (D2) and the *ops-tps* double-deletion mutant (D12) have been described previously (15). The *ops* deletion mutant (D1) and the spacer region deletion mutant (DS) were constructed in a similar manner by using the DNA fragments shown in Fig. 3 and the method of P1 transduction described previously (15, 21). *E. coli* JA221 (*lpp* F' *lacI*q) (18) was used for cloning and expression experiments. *E. coli* C600 (2) was used for phage preparation. Plasmids pREG429 and pBR322::Tn5 were kindly provided by D. Kaiser and D. E. Berg, respectively.

Growth conditions. *M. xanthus* cultures were grown in CYE medium (4), and development

```
         *         *         *         *         *         *         *         *         *         *
GCTGAAGGACCCGTCCCGCAGGACGGAGCTCGGTGCAGCGCTGGAGCCTGGCAGCGAACTGATGCTCCCAGGCCGCCCGCGCGGTGCTGTAGGCGGATTGC        100

         *         *         *         *         *         *         *         *         *         *
TCCTCCCGGAGCGCGGTGCCCAAGCTTCCGGCGGCTTCTGGCAACGAATGGGCCGGGACCGGCTCAAAGGAGGAGAACTGCAATGGCAAACATTACCGTTTT        200
                                                                              MetAlaAsnIleThrValPh

         *         *         *         *         *         *         *         *         *         *
CTACAACGAAGATTTCGGGGGTAAACAGGTCGATCTGAAGCCTGACGAATACAAGCGGGACAAGCTGGAGGCGCTGGGCATCGAGAACAACACCATCAGC        300
eTyrAsnGluAspPheGlyGlyLysGlnValAspLeuLysProAspGluTyrLysArgAspLysLeuGluAlaLeuGlyIleGluAsnAsnThrIleSer

         *         *         *         *         *         *         *         *         *         *
TCGGTGAAGGTGCCGCCTGGCGTGAAGGCTATCCTCTACAAGAACGATGATTTCACCGGCGACCAGATCGAAGTGGTGGCCAATGCCGAGGAGCTGGGCC        400
SerValLysValProProGlyValLysAlaIleLeuTyrLysAsnAspAspPheThrGlyAspGlnIleGluValValAlaAsnAlaGluGluLeuGlyP

         *         *         *         *         *         *         *         *         *         *
CGCTGAACAACAACGTCTCCAGCATCAAGGTCATGTCCGTGCCCGTGCAACCCAGGGCCAGGTTCTTCTACAAAGAGCAGTTCGATGGCAAGGAGGTGGA        500
roLeuAsnAsnAsnValSerSerIleLysValMetSerValProValGlnProArgAlaArgPhePheTyrLysGluGlnPheAspGlyLysGluValAs

         *         *         *         *         *         *         *         *         *         *
CCTGCCTCCTGGTCAGTACACCCAGGCCGAGCTGGAGCGGTACGGCATCGACAACAACACCATCAGCTCGGTGAAGCCGGAGGGCCTGAAGGTCGTCCTA        600
pLeuProProGlyGlnTyrThrGlnAlaGluLeuGluArgTyrGlyIleAspAsnAsnThrIleSerSerValLysProGluGlyLeuLysValValLeu

         *         *         *         *         *         *         *         *         *         *
TTCAAGAACGACAACTTCTCCGCCGGCGACACGCTGTCCGTGACTTCCAACGCCCCCGAGCCTGGGCGCGATGAACAACAACACCTCCAGCATCAGAATCA        700
PheLysAsnAspAsnPheSerAlaGlyAspThrLeuSerValThrSerAsnAlaProSerLeuGlyAlaMetAsnAsnAsnThrSerSerIleArgIleT

         *         *         *         *         *         *         *         *         *         *
CCCCCTGACGCGTGGCGCGCCGGCATTCCAGGTGACGGTCGCTGCCGCGGGGACCGCCACCCTTTCGAGCCCCAGAGACTCTCGAATCACCTGCATCCACG        800
hrPro
         *         *         *         *         *         *         *         *         *         *
ACGGCAGGGGGCGGCCCCACCCCCACTCGGTGACCCCATTCACGCTCACGGCCGAGCTAGTGCCCTGAGCGTCTGCACGCGCTCGTCCCTCCAGTCGAAG        900

         *         *         *         *         *         *         *         *         *         *
ACGGCGAGCGCACGCGCTCGCCTGCGTGAGCGTGTGGAGGCAGGCTCAAGGCCACTTGCCGTACGTGATGGGGCCCCCGGGGTGCGGGCAATTCTGAACACC        1000

         *         *         *         *         *         *         *         *         *         *
TGGCCTTGCCCACGGCATGTGCCGAGGAGACGCGGGCCCGGCAGTAGCCCGCATGACCCGCCCTCAGGCGGATCCGCTCGAGGTCGGCCACCGAACCCCGCA        1100

         *         *         *         *         *         *         *         *         *         *
CCCTGCTGGATCCAGCCCAGGTTGAGCCCGAGGGCCAGGTTCCACGGTCGGTTTCCGGTGACTGGCATCGGCGGCCCTCCTCCCTGGGAGCAACGATGCG        1200

         *         *         *         *         *         *         *         *         *         *
CACGCCGCTCGGACAGGGACAGGCACACGGAGGGGCCCAACTCCGGCTCAGGTGGTCAAGAGTGACGGAGGGCTCTGGCGCGTTGAGAGATGAGCCCCCG        1300

         *         *         *         *         *         *         *         *         *         *
CGCGAAGGCGCGCCGTCGCCCGGCCAGGCTCCCTGTGGCCCCCTGCACTGGGCCAACCAGCTGCTGAAGGACCCGTCCCGCAGGACGGGCTGGTGCAGCG        1400

         *         *         *         *         *         *         *         *         *         *
CTGGAGCCTGGGAGCGAACTGGCGCTCCCAGGCCCGCCCGCGCGGTACTGTGTCACCGTCCGGTGGGCGTCGTGCCGGGAGGCTTGATTGCGGACGAGGA        1500

         *         *         *         *         *         *         *         *         *         *
CCCTGTTCGGCGAGCGCCGCGAAGAGCGAGCTGTTCAGGGACTGGGCTGGAGGACTGTGTCAGGCGGCTCGGCGACGCGTCGACGTTTCCACTCGTGAGG        1600

         *         *         *         *         *         *         *         *         *         *
TAGCAGCTCACCGATGCGTGAGTTGGGATGCGTCTGTACACGCAGCAGCACGTCGGCCATGTATCCTCGGGGTTGACCTTGTTGGCCTCGCAGGTGGCCA        1700

         *         *         *         *         *         *         *         *         *         *
CCAGTGCGTAGAGGCCCGCGAGGTTTTCACCCGCGACCTCGTGGCCGACGAAGAGAAAGTTCGTACGGCCCAGGGCCGCCTTCCGCATCACGCCGGTACA        1800

         *         *         *         *         *         *         *         *         *         *
CCCACGACTGCAACGTGCTCAACCGCACGCCTCGATGCTGGGAAAACTCCTTCTGCGTCAGTCCGCTCGCCTCGAAGGCTTCAGCGACCCGGAACCACTC        1900

         *         *         *         *         *         *         *         *         *         *
CTGCTTCTCCACCGGCTTTGACATTCCCGCCAGGGCCTCGGCGCCCGACTCCTCACTCAACGCCTCTCAGCACGTCTTTGGCCGGACGGATACGGTACTGT        2000

         *         *         *         *         *       * ▼ *         *         *         *         *
AGACGGAᵣTTGCATᵣTTCCGGAGCGCGGTGCCᵣAATGCTᵣTCGGCGGCTTCTGGCAACGAATGGGCCGGGGCCGCTCAACGGAGGAGCACTGCAATGGCAAA        2100
                                                                              MetAlaAs

         *         *         *         *         *         *         *         *         *         *
CATTACCGTTTTCTACAACGAAGACTTCCAGGGTAAACAGGTCGATCTGCCCGCCTGGCAACTACACCCGGGCCCAGTTGGCGGCGCTGGGCATCGAGAAC        2200
nIleThrValPheTyrAsnGluAspPheGlnGlyLysGlnValAspLeuProProGlyAsnTyrThrArgAlaGlnLeuAlaAlaLeuGlyIleGluAsn

         *         *         *         *         *         *         *         *         *         *
AACACCATCAGCTCGGTGAAGGTGCCGCCTGGCGTGAAGGCTATCCTCTACCAGAACGATGGTTTCGCCGGCGACCAGATCGAAGTGGTGGCCAATGCCG        2300
AsnThrIleSerSerValLysValProProGlyValLysAlaIleLeuTyrGlnAsnAspGlyPheAlaGlyAspGlnIleGluValValAlaAsnAlaG

         *         *         *         *         *         *         *         *         *         *
AGGAGTTGGGCCCGCTGAACAACAACGTCTCCAGCATCCGGGTCATATCCGTGCCCGTGCAGCCCAGGGCCAGGTTCTTCTACAAAGAGCAGTTCGATGG        2400
luGluLeuGlyProLeuAsnAsnAsnValSerSerIleArgValIleSerValProValGlnProArgAlaArgPhePheTyrLysGluGlnPheAspGl

         *         *         *         *         *         *         *         *         *         *
CAAGGAGGTGGACCTGCCTCCTGGCCAGTACACCCAGGCCGAGCTGGAGCGGTACGGCATCGACAACAACACCATCAGCTCGGTGAAGCCGCAGGGCCTG        2500
yLysGluValAspLeuProProGlyGlnTyrThrGlnAlaGluLeuGluArgTyrGlyIleAspAsnAsnThrIleSerSerValLysProGlnGlyLeu

         *         *         *         *         *         *         *         *         *         *
GCGGTCGTCCTATTCAAGAACGACAACTTCTCCGGCGACACGCTGCCCGTGAATTCCACGCCCCGACCCTGGGCGCGATGAACAACAACACCTCCACCA        2600
AlaValValLeuPheLysAsnAspAsnPheSerGlyAspThrLeuProValAsnSerAspAlaProThrLeuGlyAlaMetAsnAsnAsnThrSerSerI

         *         *         *         *         *         *         *         *         *         *
TCAGAATCTCCTGACGCGTGGCGCGCCGGCATTCCAGGTGACGGTCCCTGCCGCGGGGACCGCCCCCCTTTCGAGCCCCAGAGACTCTCGAATCACCTCGA        2700
leArgIleSer
```

was induced on CF agar (6) as described previously (7). Germination of spores was carried out on CYE-1.5% agar or A1-0.7% agarose (3).

Spore fractionation. Spores were isolated and purified as described (8). Spore surface proteins were removed by boiling for 5 min in 1% SDS. The proteins inside the spores were released by taking the spores that had been boiled in SDS, washing them twice with distilled water, mixing them with 75- to 150-μm-diameter glass beads (Sigma Chemical Co., St. Louis, Mo.) and just enough water to cover the beads, and sonicating the mixture of spores and beads for 4 min. The beads were then removed by diluting the mixture with 20 volumes of water, vortexing it, allowing the beads to settle, and removing the supernatant. Less than 2% of the spores remained intact after sonication with glass beads, as determined by counting in a Petroff-Hausser chamber. The proteins were precipitated with 10% trichloroacetic acid or antiserum against protein S and were analyzed by SDS-polyacrylamide gel electrophoresis (1). Proteins were visualized by staining with Coomassie blue.

RESULTS AND DISCUSSION

Characterization of protein S. The DNA sequence of the *ops* gene, the *tps* (which codes for protein S), and the 1.4-kilobase spacer region between them is shown in Fig. 1. The DNA sequences of the two genes are 93% homologous, and the homology extends about 100 bases upstream and downstream from the coding regions (10). The amino acid sequences deduced from the DNA sequences of the two genes also show a very high degree of homology (88%). In addition, there is striking internal homology in the protein sequences; they can be divided into four homologous domains, and domains 1 and 3 as well as domains 2 and 4 have particularly extensive homologies (10). This is shown for protein S in Fig. 2. A similar situation exists in calmodulin and the γ-crystallins, two classes of proteins that have been identified only in eukaryotes. In fact, both bovine brain calmodulin and bovine eye lens γ-crystallin II have some homology to protein S. The boxed regions of domains 1 and 3 of protein S (Fig. 2) are very homologous to the proposed Ca^{2+}-binding sequences of bovine brain calmodulin (10); the corresponding regions of domains 2 and 4 have less homology. As predicted by this structure, protein S bound two Ca^{2+} ions per molecule in contrast to the four bound by

calmodulin; the affinities were also lower ($K_d = 1.5 \times 10^{-5}$ and 5×10^{-5}) than those of the calmodulin binding sites (M. Teintze, unpublished data). The homology between γ-crystallin and protein S is more extensive (G. Wistow, L. Summers, and T. Blundell, personal communication), but γ-crystallins are not known to bind Ca^{2+} ions. The functional significance of these similarities is not yet clear, but both the calmodulins and the γ-crystallins are heat-stable proteins like protein S.

Analysis of *ops* and *tps* deletion mutants. In addition to the *tps* deletion and *ops-tps* double-deletion mutants (D2 and D12) described previously (15), we have now constructed mutants in which only the *ops* gene or only the spacer region between the genes has been deleted (D1 and DS; see Fig. 3). When these mutants were induced to develop on CF agar (Fig. 4), it was found that the *ops* deletion mutant (D1) was slightly delayed in fruiting body formation compared with the wild type. The *tps* deletion mutant (D2) and the spacer region mutant (DS; not shown) developed normally, but the double deletion mutant (D12) showed substantially delayed fruiting body formation and a significantly lower yield of myxospores (30% of the number in the other strains). Since all of the open reading frames in the spacer region were deleted in either the DS or D2 mutants which developed normally, the results suggest that it is the *ops* gene which has a role in development and that the very homologous *tps* gene product (protein S) can partially substitute for the *ops* gene product. However, deletion of both genes still allows some cells to complete development, so there is probably some alternative pathway which does not require either of these genes.

As expected, the D1 and DS mutants still synthesized protein S, whereas the D2 and D12 mutants did not (T. Furuichi, unpublished data). Since protein S does not seem to be required for fruiting body formation or sporulation, we investigated whether the protein S-deficient spores of the D2 mutant might be less viable in some way. However, we found no difference in heat and sonication resistance or ability to germinate on CYE agar between D2 and DZF1 spores (T. Furuichi and M. Teintze, unpublished data). Since protein S binds calcium, we also studied germination on minimal plates (A1–0.7% agarose) in the absence and presence of Ca^{2+}. At Ca^{2+} concentrations of 10 μM or less, neither D2 nor DZF1 spores would germinate; as the Ca^{2+} concentration was raised, both types of

FIG. 1. Sequence of the region of the *M. xanthus* chromosome containing the *ops* (upstream) and *tps* (downstream) genes. The boxed sequences show the *tps* gene promoter, and the arrow points to the site of initiation of transcription of the *tps* gene.

Domain 1: Met-Ala-Asn-Ile-Thr-Val-Phe-Tyr-Asn-Glu-Asp-Phe-Gln-Gly-Lys-Gln-Val-Asp-Leu-Pro-Pro-Gly-Asn-Tyr-Thr-Arg-Ala-Gln-Leu-Ala-Ala-Leu-Gly-Ile-$\boxed{\text{Glu-Asn-Asn-Thr-Ile-Ser-Ser-Val-Lys}}$-Val-

Domain 3: Phe-Phe-Tyr-Lys-Glu-Gln-Phe-Asp-Gly-Lys-Glu-Val-Asp-Leu-Pro-Pro-Gly-Gln-Tyr-Thr-Gln-Ala-Glu-Leu-Glu-Arg-Tyr-Gly-Ile-$\boxed{\text{Asp-Asn-Asn-Thr-Ile-Ser-Ser-Val-Lys}}$-

Domain 2: Pro-Pro-Val-Lys-Ala-Tyr-Gln-Asn-Asp-Gly-Phe-Ala-Gly-Asp-Gln-Ile-Glu-Val-Val-Ala-Asn-Ala-Glu-Glu-Gly-Leu-Pro-Leu-Asn-Ala-Ser-Ser-Ile-Arg-Val-Ser-Val-Pro-Val-Gln-Pro-Arg-Ala-Arg-

Domain 4: Pro-Gln-Gly-Leu-Ala-Val-Lys-Val-Leu-Phe-Lys-Asn-Asp-Ala-Phe-Ser-Gly-Ala-Ser-Asp-Ala-Pro-Thr-Leu-Gly-Ala-Met-Asn-Asn-Thr-Ser-Ser-Ile-Arg-Ile-Ser

FIG. 2. Sequence of protein S, showing internal homologies. The boxed sequence represent the most likely Ca^{2+}-binding sites, which are homologous to a proposed Ca^{2+}-binding sequence in bovine brain calmodulin.

spores started to germinate (M. Teintze, unpublished data).

Characterization of the *ops* gene product. Since there was until now no evidence for the presence of the *ops* gene product at any stage of the *M. xanthus* life cycle, we decided to express the gene in *E. coli*. The *Hin*dIII-*Bam*HI fragment containing the *ops* gene (see Fig. 3) was inserted into the inducible expression vector pINIIIA3 (17). The protein produced when the plasmid was induced was immunoprecipitated by antiserum against protein S and migrated with a slightly higher apparent molecular weight than protein S. When spores whose protein S had been removed by treatment with EDTA were incubated with the soluble fraction of *E. coli* containing the *ops* plasmid that had been induced for 1 h, the *ops* gene product was assembled onto the spores (Fig. 5, lane 4). When these spores were then treated with 10 mM EDTA, the *ops* gene product dissociated from the spores (Fig. 5, lane 6). Although the pINIII series vectors have been successfully used to produce very large amounts of many different proteins in our laboratory, the *ops* gene product was produced in only very small quantities. To determine whether the *ops* gene ribosome-binding site did not work well in *E. coli*, the same *Hin*dIII-*Bam*HI fragment was inserted into the vector pINIIIAI (17), which resulted in an *E. coli* ribosome-binding site and initiation codon being 60 bases upstream and in frame with the first codon of the *ops* gene. The result was that the larger read-through product was made in only slightly larger quantities than the product resulting from ribosome binding at the *ops* gene site (M. Teintze, unpublished data). The most likely explanation for the low rate of synthesis, therefore, is the different codon usage (10).

Synthesis and localization of the *ops* gene product in *M. xanthus*. Although the *ops* gene product had not been found on the spore surface (10), the possibility that it might be located inside the spores had not been investigated. Therefore, we looked for the presence of the *ops* gene product on the inside of the spores at various times during development (Fig. 6). Cells were harvested at the indicated times and fractionated into a soluble fraction (supernatant after sonication and centrifugation), a spore surface fraction (supernatant after boiling spores in SDS), and a spore content fraction (after breaking spores open by use of glass beads). The *ops* gene product began to appear on the insides of the spores very late in development (between 41 and 56 h), and the amount increased with time, showing that the spores were still synthesizing protein (Fig. 6). It was not found at any time point in the nonspore or spore surface fractions (not shown). Protein S, on the other hand, was

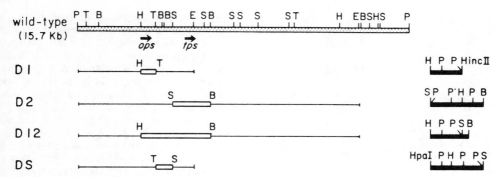

FIG. 3. Construction of deletion mutants. The restriction map of the 15.7-kilobase *Pst*I fragment from the *M. xanthus* chromosome is shown (dotted box). Arrows indicate the positions of the *ops* and *tps* genes. The DNA sequences deleted in the various mutants are shown by empty boxes, and the replacement DNA frgments containing the kanamycin resistance gene from pBR322::Tn5 or pREG429 are shown by the solid boxes to the right. The solid lines show the portions of the *M. xanthus* DNA that were cloned and transduced into the wild-type strain (DZF1) with bacteriophage P1*clr*100CM (20). Homologous recombinations in both arms (double crossover) results in replacement of the appropriate portion of the genome with the kanamycin resistance gene. B, *Bam*HI; E, *Eco*RI; H, *Hin*dIII; P, *Pst*I; S, *Sal*I; T, *Bst*EII.

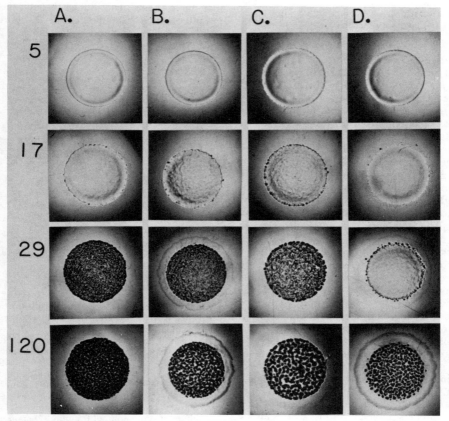

FIG. 4. Development on CF agar plates. Exponentially growing cells were concentrated to about 3,000 Klett units, spotted on CF agar plates, and incubated at 30°C as described previously (7). Numbers at the left indicate the hours elapsed after spotting. Column A, DZF1 (wild type); column B, D1 (*ops* deletion); column C, D2 (*tps* deletion); column D, D12 (*ops-tps* deletion).

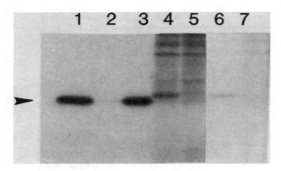

FIG. 5. Assembly of *ops* gene products on spores. Spores were boiled in 1% SDS, and the proteins released were analyzed by SDS-polyacrylamide gel electrophoresis. Treatment before boiling in SDS was as follows. (1) None; (2) spores were extracted with 10 mM EDTA; (3) spores treated as for lane 2 were incubated with protein S and 10 mM Ca^{2+}; (4) spores treated as for lane 2 were incubated in lysate of *E. coli* cells that had been induced to produce the *ops* gene product; (5) spores treated as for lane 4 except without induction of the *ops* gene expression plasmid; (6) supernatant from spores treated as for lane 4 and then extracted with 10 mM EDTA; (7) supernatant from spores treated as for lane 5 and extracted with 10 mM EDTA. Arrow points to protein S.

initially found in the soluble fraction and then appeared on the spore surface, as reported previously (8). However, we also found protein S inside the spores (Fig. 6), where it was still being synthesized late in development along with the *ops* gene product.

A mutant has recently been constructed in which the coding region of the *ops* gene fused to the *tps* gene promoter was inserted into the *M. xanthus* genome. This resulted in the *ops* gene product being synthesized early in development along with protein S and then assembled on the spore surface together with protein S (T. Furuichi, unpublished data). These results suggest that the reason the *ops* gene product is located only inside the spores is that it is normally synthesized only late in development after the cells have completed sporulation. These results are consistent with the expression of β-galactosidase activity very late in development observed in an *ops-lacZ* fusion mutant (5).

Release of protein S during autolysis of developing cells. Since protein S can self-assemble onto myxospores in the presence of Ca^{2+} (8; Fig. 6) and it is known that about 90% of developing cells lyse when the other 10% sporulate (24), we have been investigating whether autolysis is a mechanism by which the protein S that accumulates in the fruiting cells can be released for assembly on the surface of those cells that sporulate. This kind of self-sacrificing social behavior should result in protein S freely

diffusing through the medium, available for assembly on any spore in the vicinity. To test this hypothesis, we investigated whether the spores of the D2 mutant would assemble protein S on their surface during development if they were placed near developing wild-type cells. Spots of either D2 or DZF1 cells that had been developing for 18 h on CF agar plates were overlaid with 0.22-μm-pore membrane filters which had been lying on another CF agar plate with a spot of the other strain developing on top of the filter for the same period of time. This sandwich with the filter separating the two different strains was incubated for another 30 h. The spores were then removed from the tops of the filters, washed, and boiled in 1% SDS. The supernatant was then applied to an SDS-polyacrylamide gel and analyzed for protein S (Fig. 7, lanes 2). An equivalent number of spores of each strain which had been developing on the same plate, without a filter or another strain on top, were treated in the same manner (Fig. 7, lanes 1). The results show that the DZF1 cells produce an excess of protein S over what they assemble on their spores and that this protein S can diffuse through the 0.22-μm filter and be assembled by the D2 spores. *M. xanthus* cells cannot pass through a 0.22-μm pore (16).

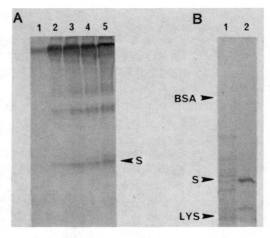

FIG. 6. (A) Synthesis of *ops* gene product. DZF1 cells were spotted on CF agar plates and harvested at: (1) 17 h, (2) 29 h, (3) 41 h, (4) 56 h, (5) 96 h. Spores were isolated, boiled in SDS to remove surface proteins, and broken by using glass beads. The spore content fraction was immunoprecipitated with antiserum against protein S (which also precipitates the *ops* gene product. (B) Content and surface proteins of mature spores. Seven-day-old DZF1 spores were harvested from a CF agar plate. (1) Trichloroacetic acid precipitate of spore content fraction; (2) spore surface fraction. Arrows point to positions of molecular-weight markers. BSA, Bovine serum albumin; S, protein S; LYS, lysozyme.

CONCLUSIONS

The following conclusions can be drawn from the experiments described here. (i) The *ops* gene is expressed in a very late stage of development, and its product accumulates only inside the spore. (ii) The amount of *ops* gene product and the amount of protein S inside the mature spores are of the same order of magnitude. (iii) The amount of protein S on the outer surface of the spores is about an order of magnitude greater than that found inside the spores.

Together with the previously reported observations that 90% of the developing cells lyse (23) and that protein S has no cleavable signal peptide at its amino terminus (10), in contrast to all secreted bacterial proteins that have been studied, these results have led us to propose the following. (i) Neither the *ops* gene product nor protein S is secreted across the cell membrane. (ii) The protein S found assembled on the spore surface is released by the 90% of the cells that lyse during development. This would be consistent with the large quantities of protein S found on the outside of the spore versus those on the inside and the apparent ability of protein S to diffuse freely during sporulation. (iii) The *ops* gene product is produced only in those cells that have been committed to spore formation, a small portion of the total cell population.

Such an "exposure" model would suggest that lysis of a large portion of the developing cells is essential for the sporulation of the others (23). The lysing cells could be providing various important signals or nutrients to the sporeforming cells; this would be an interesting contrast to *Bacillus* sporulation, in which the parent cell nurses the endospore forming inside it. Whether or not this is the case, an interesting question that needs to be addressed is what determines which cells will lyse and which will become spores.

SUMMARY

The *ops* and *tps* genes of *M. xanthus* have about 90% DNA and amino acid sequence homology and are found in the same orientation separated by a spacer region of only 1.4 kilobases. The products of the two genes cross-react immunologically, and both are capable of Ca^{2+}-dependent self-assembly on the surface of myxospores. However, the *ops* and *tps* genes are expressed very differently during the developmental cycle of *M. xanthus*. The *tps* gene is induced early during fruiting body formation on a solid surface, and its product, protein S, is made in large quantities (up to 15% of total protein synthesis). When the cells turn into myxospores, protein S is assembled on the outer surface of the spore. It was also found in much

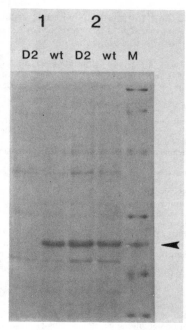

FIG. 7. Transfer of protein S from developing wild-type cells to developing D2 mutant cells. (1) Control spores harvested after 48 h; (2) spores harvested from the tops of filters which had the other strain underneath. D2, *tps* gene deletion mutant; wt, DZF1 wild type; M, molecular-weight markers (phosphorylase b, bovine serum albumin, ovalbumin, carbonic anhydrase, protein S, soy bean trypsin inhibitor, lysozyme). Arrow points to protein S.

smaller quantities inside the spores. The *ops* gene, on the other hand, appears to be induced only very late in development, after the cells have sporulated, since the *ops* gene product was found only inside the spores. Deletion of the *tps* gene had no effect on fruiting body formation or sporulation. Although the mutant spores had no protein S coat, they were still heat and sonication resistant and germinated normally. When the *ops* gene was deleted, fruiting body formation was somewhat retarded compared with the *tps* deletion mutant and the wild type. Deletion of both genes substantially slowed down development and reduced the yield of myxospores. This suggests that the protein S inside the spore may be able to partially fulfill the function of the missing *ops* gene product in the later stages of development. When an *ops* gene under the control of a *tps* gene promoter was inserted into a wild-type strain, the *ops* gene product was synthesized at the same time as protein S and assembled onto the spore surface. Developing *tps* gene deletion mutants assembled protein S onto their spores in the presence of developing wild-type cells, indicating that protein S is released into the medium when sporulation and autolysis occur and is available for assembly on any spore in the vicinity.

This work was supported by Public Health Service grant GM-26843 and National Research Service Award GM-09683 from the National Institute of General Medical Sciences.

LITERATURE CITED

1. **Anderson, C. W., P. R. Baum, and R. F. Gesteland.** 1973. Processing of adenovirus 2-induced proteins. J. Virol. **12**:241–252.
2. **Bachmann, B. J.** 1972. Pedigrees of some mutant strains of *Escherichia coli* K-12. Bacteriol. Rev. **36**:525–557.
3. **Bretscher, A. P., and D. Kaiser.** 1978. Nutrition of *Myxococcus xanthus*, a fruiting myxobacterium. J. Bacteriol. **133**:763–768.
4. **Campos, J. M., J. Geisselsoder, and D. R. Zusman.** 1978. Isolation of bacteriophage MX4, a generalized transducing phage for *Myxococcus xanthus*. J. Mol. Biol. **119**:167–178.
5. **Downard, J. S., D. Kupfer, and D. R. Zusman.** 1984. Gene expression during development of *Myxococcus xanthus*. Analysis of the genes for protein S. J. Mol. Biol. **175**:469–492.
6. **Hagen, D. C., A. P. Bretscher, and D. Kaiser.** 1978. Synergism between morphogenic mutants of *Myxococcus xanthus*. Dev. Biol. **64**:284–296.
7. **Inouye, M., S. Inouye, and D. R. Zusman.** 1979. Gene expression during development of *Myxococcus xanthus*: pattern of protein synthesis. Dev. Biol. **68**:579–591.
8. **Inouye, M., S. Inouye, and D. R. Zusman.** 1979. Biosynthesis and self-assembly of protein S, a development-specific protein of *Myxococcus xanthus*. Proc. Natl. Acad. Sci. U.S.A. **76**:209–213.
9. **Inouye, S.** 1984. Identification of a development-specific promoter of *Myxococcus xanthus*. J. Mol. Biol. **174**:113–120.
10. **Inouye, S., T. Franceschini, and M. Inouye.** 1983. Structural similarities between the development-specific protein S from a Gram-negative bacterium, *Myxococcus xanthus* and calmodulin. Proc. Natl. Acad. Sci. U.S.A. **80**:6829–6833.
11. **Inouye, S., W. Harada, D. Zusman, and M. Inouye.** 1981. Development-specific protein S of *Myxococcus xanthus*: purification and characterization. J. Bacteriol. **148**:678–683.
12. **Inouye, S., Y. Ike, and M. Inouye.** 1983. Tandem repeat of the genes for protein S, a development-specific protein of *Myxococcus xanthus*. J. Biol. Chem. **258**:38–40.
13. **Inouye, S., M. Inouye, B. McKeever, and R. Sarma.** 1980. Preliminary crystallographic data for protein S, a development-specific protein of *Myxococcus xanthus*. J. Biol. Chem. **255**:3713–3714.
14. **Kaiser, D., C. Manoil, and M. Dworkin.** 1979. Myxobacteria: cell interactions, genetic, and development. Annu. Rev. Microbiol. **33**:595–639.
15. **Komano, T., T. Furuichi, M. Teintze, M. Inouye, and S. Inouye.** 1984. Effects of deletion of the gene for the development-specific protein S on differentiation in *Myxococcus xanthus*. J. Bacteriol. **158**:1195–1197.
16. **LaRossa, R., J. Kuner, D. Hagen, C. Manoil, and D. Kaiser.** 1983. Developmental cell interactions of *Myxococcus xanthus*: analysis of mutants. J. Bacteriol. **153**:1394–1404.
17. **Masui, Y., J. Coleman, and M. Inouye.** 1983. Multipurpose expression cloning vehicles in *Escherichia coli*, p. 15–32. *In* M. Inouye (ed.), Experimental manipulation of gene expression. Academic Press, Inc., New York.
18. **Nakamura, K., Y. Masui, and M. Inouye.** 1982. Use of a *lac* promoter-operator fragment as a transcriptional control switch for expression of the constitutive *lpp* gene in *Escherichia coli*. J. Mol. Appl. Genet. **1**:289–299.
19. **Nelson, D. R., and D. R. Zusman.** 1983. Evidence for long-lived mRNA during fruiting body formation in *Myxococcus xanthus*. Proc. Natl. Acad. Sci. U.S.A. **80**:1467–1471.
20. **Rosner, J. L.** 1972. Formation, induction and curing of bacteriophage P1 lysogens. Virology **49**:679–689.
21. **Shimkets, L. J., R. E. Gill, and D. Kaiser.** 1983. Developmental cell interactions in *Myxococcus xanthus* and the *spoC* locus. Proc. Natl. Acad. Sci. U.S.A. **80**:1406–1410.
22. **Takao, T., T. Hitouji, Y. Shimonishi, T. Tanabe, S. Inouye, and M. Inouye.** 1984. Verification of protein sequence by fast atom bombardment mass spectrometry. Amino acid sequence of protein S, a development-specific protein of *Myxococcus xanthus*. J. Biol. Chem. **259**:6105–6109.
23. **Wireman, J. W., and M. Dworkin.** 1975. Morphogenesis and developmental interactions in myxobacteria. Science **189**:516–523.
24. **Wireman, J. W., and M. Dworkin.** 1977. Developmentally induced autolysis during fruiting body formation by *Myxococcus xanthus*. J. Bacteriol. **129**:796–802.

Organization and Transcription of *Anabaena* Genes Regulated During Heterocyst Differentiation

ROBERT HASELKORN,[1] STEPHANIE E. CURTIS,[2] JAMES W. GOLDEN,[1] PETER J. LAMMERS,[1] SANDRA A. NIERZWICKI-BAUER,[1] STEVEN J. ROBINSON,[3] AND NILGUN E. TUMER[1]

Department of Molecular Genetics and Cell Biology, University of Chicago, Chicago, Illinois 60637,[1] Department of Genetics, North Carolina State University, Raleigh, North Carolina 27650,[2] and Department of Botany, University of Massachusetts, Amherst, Massachusetts 01003

Anabaena is one of a number of species of filamentous cyanobacteria that differentiate specialized cells, called heterocysts, at regular intervals along each filament, in response to deprivation of a combined nitrogen source under aerobic conditions. The heterocyst provides an anaerobic local environment for nitrogen fixation, whose ultimate product, glutamine, is transported from the heterocyst to neighboring vegetative cells. The factors that govern the selection of vegetative cells for differentiation, which determines the pattern of heterocyst spacing, remain mostly unknown (6).

The differentiation of a vegetative cell into a heterocyst resembles sporulation in its requirement for the orderly expression of sets of genes. The process differs from sporulation in one significant way: it is irreversible. The heterocyst is a terminal cell that eventually dies, causing the filament to break. Although genetic analysis of heterocyst differentiation is virtually nonexistent, the biochemical changes that characterize the process have been described extensively. Early events, set in motion by nitrogen starvation, include the induction of proteases and the synthesis of new glycolipid and polysaccharide components of the heterocyst envelope. The envelope limits diffusion of gases into the cell, making it possible for hydrogenase and respiratory systems to scavenge oxygen. The proteases degrade a number of vegetative-cell proteins, including the phycobiliproteins which harvest light for the O_2-evolving photosystem II and ribulose biphosphate carboxylase which fixes CO_2. These changes ensure the stability of the oxygen-labile nitrogenase proteins and the flow of reductant exclusively to nitrogenase. Enzymes known to be induced in the heterocyst are nitrogenase, an uptake hydrogenase, the oxidative pentose pathway enzymes, and glutamine synthetase. These changes could be documented because there exist convenient assays for the enzymes mentioned. Global assays based on acrylamide gel electrophoresis of total cell proteins showed at least six classes of proteins distinguished by their time of synthesis in differentiating heterocysts. Other classes were seen only in vegetative cells or in cells growing on ammonia. Finally, a major class of proteins was synthesized in both cell types, but this class represented less than half the total (5).

The big questions about heterocyst differentiation have been posed at the cellular and molecular levels. Concerning the former, a little is known of the rules governing the selection of vegetative cells for differentiation. In an already-differentiated filament growing on N_2, the vegetative cell placed midway between two existing heterocysts is the most likely to differentiate next. That cell is thought to be at the lowest point in a gradient of inhibitor flowing from the heterocysts (17). Glutamine is a good candidate for the inhibitor. In the case of the transition from growth on combined nitrogen to growth on N_2, much less is known about the selection of cells for differentiation. In some species the slightly smaller daughter cell produced by asymmetric division may be selected (16). It is possible that candidate cells have just completed chromosome replication.

At the molecular level, heterocyst differentiation involves the sequential activation and repression of sets of genes. The best models for these processes are found in bacterial sporulation and the lytic development of complex bacteriophages. In both of these examples a major role has been assigned to the transcription apparatus, which appears to be modified sequentially to recognize different promoters at different times. The goal of our work has been to isolate a number of *Anabaena* genes whose expression is regulated up or down in differentiating heterocysts, to determine the structure of the promoters of such genes, and to study their transcription in vitro. Thus far, we have information regarding three genes for nitrogen fixation (*nifH*, *nifD*, *nifK*) encoding the structural components of the nitrogenase complex, which turn on in heterocysts (10, 14); the *rbcL* and *rbcS* genes encoding the large and small subunits of ribulose bisphosphate carboxylase, which turn off in heterocysts (3, 12a); the *psbA* gene family encoding the 32-kilodalton herbicide-binding protein, which continues to be made in heterocysts (3a); and the *glnA* gene encoding glutamine synthetase, whose activity increases severalfold in heterocysts (4, 13, 15). In the course of a broader effort to identify genes expressed differentially in heterocysts, we found that the DNA in the neighborhood of the *nifH*, *nifD*, and *nifK* genes is rearranged during heterocyst differentiation. One of these rearrangements is an excision, analogous to that of bacteriophage lambda escaping from a lysogen, that results in juxtaposition of two *nif* genes that were separated by 11 kilobase pairs (kbp) prior to the rearrangement (J. W. Golden, S. J. Robinson, and R. Haselkorn, manuscript in preparation).

EXPERIMENTAL PROCEDURES

All of the probes and methods used have been fully described in the primary references. The physical map of the region containing the *nifH*, *nifD*, and *nifK* genes in vegetative cell DNA, shown in Fig. 1, is based on Southern hybridization and electron microscope examination of DNA heteroduplexes with cloned *Klebsiella nifHDK* DNA as the reference (10, 14). In addition, the *Anabaena nifH* (12), *nifD* (8), and *nifK* (9) genes have been completely sequenced. The *Anabaena glnA* (15), *rbcL* (3), *rbcS* (12a), and *psbA* (3a) genes were identified and isolated from lambda libraries by use of heterologous probes; these genes have also been sequenced.

Studies of transcription of these genes have been based on RNA prepared from cells grown on ammonia-containing medium or induced for nitrogenase anaerobically, i.e., without heterocyst differentiation. The RNA preparations were characterized by RNA transfer blots, S1 nuclease protection, and primer extension (15). In vitro transcription has utilized RNA polymerase prepared from ammonia-grown cells by published procedures (2). The products of in vitro transcription were characterized either by the size of runoff transcripts or by S1 nuclease protection of the template. The method of breaking heterocysts to provide DNA suitable for Southern blots is to vortex the cells with glass beads, detergent, and phenol-chloroform. This procedure, along with modifications of the lysozyme protocol for isolation of heterocysts (5), will be described in detail elsewhere (Golden et al., in preparation).

RESULTS AND DISCUSSION

We describe here our results on the structure of the promoters of several specific genes regulated during heterocyst differentiation. The *nifH*, *nifD*, and *nifK* genes were chosen as representative genes not expressed in cells growing on ammonia but induced in cells fixing nitrogen. The *rbcL-rbcS* operon is regulated in the opposite sense: on in cells growing on ammonia and off in cells fixing nitrogen. Finally, the *glnA* gene encoding glutamine synthetase is on all the time;

the activity of the enzyme is severalfold higher in heterocysts than in vegetative cells.

In addition to these studies of individual genes whose products are well known, we were interested in broadening the consensus of promoters for genes of each class. To isolate genes in the *nif* class, i.e., genes turned on in heterocysts, we wished to screen a library of recombinant lambda containing *Anabaena* DNA inserts by using total RNA from heterocysts. The procedure for making heterocyst RNA with relatively little physical degradation turned out to be suitable for DNA as well. With high-molecular-weight heterocyst DNA available for the first time, it was possible to determine whether any of the genes whose regulation we had been studying at the transcriptional level were, in addition, reorganized. To date, we have found two rearrangements in the neighborhood of the *nif* genes, but none near *glnA* or the genes coding for photosynthetic functions.

To understand the *nif* gene rearrangement, it is useful to review the physical map of the *nif* genes of *Klebsiella* and of vegetative cells of *Anabaena*, shown in Fig. 1. The upper part of the figure shows the *Klebsiella* genes, organized into eight operons within 20 kbp of DNA (1). This region will suffice to convert *Escherichia coli* to Nif$^+$. The only genes known to be required in addition to these, which are already present in *E. coli*, are *ntrA* and *ntrC*, which code for regulatory factors needed to transcribe the *nif* genes, and an RNA polymerase compatible with these factors (11).

The *nif* genes of *Anabaena* appear to be much more widely dispersed. Within 40 kbp we have been able to locate only four *nif* genes by hybridization with cloned *Klebsiella nif* DNA (14). Most of the neighboring *Anabaena* sequences are not detectably transcribed during nitrogenase induction (14). Even in the regions of homology the maps of *Klebsiella* and *Anabaena* differ. The *nifV/S* gene(s) of *Anabaena* (these genes function in the assembly of nitrogenase) are located on the right of *nifHDK*, whereas they are left of *nifHDK* in *Klebsiella*. Of greater interest is the large intervening sequence between *nifK* and *nifD* in *Anabaena*. These genes code for the two subunits of nitrogenase; they are cotranscribed in *Klebsiella*. Their separation, by non-*nif* DNA that is imperceptibly transcribed, makes no sense.

The *nif* structural genes are not separated in heterocyst DNA. Note in Fig. 1 that the *nifK* gene is located on a 17-kbp *Eco*RI fragment (An207) in vegetative-cell DNA. When heterocyst DNA was cut with *Eco*RI and probed with the *Hind*III fragment 207.8, which is within *nifK*, the only band observed was at 6 kbp. Although other explanations are possible, the

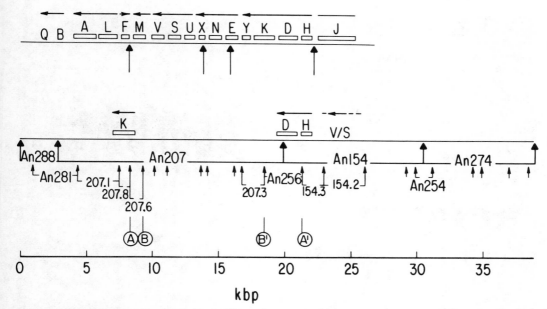

FIG. 1. Comparison of the physical map of the *nif* genes of *Klebsiella* (top) and of *Anabaena* (bottom). Thin arrows indicate *Hind*III sites; thicker arrows, *Eco*RI sites. Boxes correspond to protein sizes. Horizontal arrows indicate the size and orientation of transcription units. For *Klebsiella*, the transcription data are from reference 1. For *Anabaena*, the data are from reference 14 and S. J. RJobinson (unpublished data). The DNA fragment excised during heterocyst differentiation has its left end in segment 207.6 and its right end in segment 256. Since the excised fragment is circular. It contains the new *Hind*III fragment A to A (Golden, unpublished data).

one we tested and confirmed is the excision of 11 kbp between *nifK* and *nifD*. The second rearrangement, for which we show no data, occurs to the right of *nifV/S*. Nothing happens between *nifH* and *nifV/S*. To the right, there is an inversion or deletion that brings a new *Eco*RI site within 6 kbp of the site in *nifD*. No other details of that rearrangement are known.

Returning to the excision between *nifK* and *nifD*, the time course of that process is shown in Fig. 2. For this experiment, samples were taken every 6 h during an aerobic induction of heterocysts, and DNA was prepared from all the cells in each sample. The samples shown on the left of Fig. 2 were cut with *Hind*III and probed with An256, a 2.9 kbp *Hind*III fragment (B' to A' in Fig. 1). The 2.9-kbp band in Fig. 2 is from the 90% of vegetative cells that do not differentiate or rearrange their DNA. The two bands at 1.8 and 2.1 kbp, which replace the 2.9-kbp band quantitatively in heterocysts, correspond to the fragment A to A' (in Fig. 1) formed in the fused chromosome and fragment B to B' formed in the excised DNA. The latter result indicates that the excised DNA is circular.

The same DNA preparations are shown uncut in the right panel of Fig. 2. These gels were probed with 207.3, a *Hind*III fragment within the excised region. Since the DNA was uncut, the appearance of sharp bands in the differentiating

cells also indicates that the excised DNA is circular. The multiple bands are isomeric or polymeric forms of the circle; cut with *Kpn*I, for which there is a single site within the excised region, they all collapse to a single 11-kbp band.

The 2.1- and 1.8-kbp bands detected in *Hind*III digests of heterocyst DNA by An256 were also identified by fragment 207.6 (A to B in Fig. 1), as expected. The 2.1- and 1.8-kbp bands were cloned and sequenced (J. W. Golden, unpublished data). The former comes from the excised circle; the latter, from the fused chromosome. Each junction fragment contains a *nifK* 5' flanking sequence and a *nifD* 3' proximal sequence, separated by an 11-base pair sequence found once in 207.6 and once, directly repeated, in 256. Thus, *nifK* and *nifD* in vegetative-cell DNA are separated by an 11-kbp sequence with an 11-base pair direct repeat at each end. Conservative homologous recombination between the direct repeats leads to excision of the 11-kbp insert, leaving one copy of the repeat in the fused chromosome.

The recombination site in An256 is actually inside the open reading frame corresponding to *nifD* in vegetative-cell DNA (8), which was predicted to code for a polypeptide containing 480 amino acids. After the excision, the new fragment of An207.6 fused onto *nifD* extends the open reading frame to 497 amino acids, of which

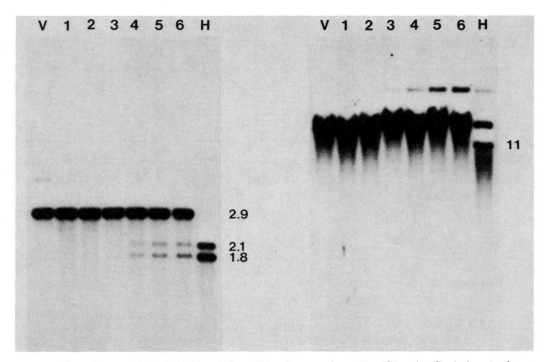

FIG. 2. Time course of the excision of the 11-kbp fragment between *nifK* and *nifD* during *Anabaena* heterocyst differentiation. DNA was prepared from vegetative cells (V), purified heterocysts (H), or whole filaments of a differentiating culture taken at 6-h intervals after transfer to N⁻ medium (lanes 1–6). The set on the left was cut with *Hin*dIII; on the right, the DNA was not cut. The set on the left was probed with nick-translated An256 DNA; the set on the right was probed with 207.3 DNA (see Fig. 1). Sizes are in kbp. The 2.9-kbp band corresponds to An256 itself, in unrearranged chromosomal DNA of vegetative cells. The 2.1- and 1.8-kbp bands correspond to B to B′ and A to A′ in Fig. 1, respectively. On the right side, the heterogeneous band is chromosomal DNA from vegetative cells. The 11-kbp and larger discrete bands are various forms of excised, circular DNA. These convert to a single band at 11 kbp when cut with *Kpn*I, for which there is a single site between *nifK* and *nifD* (14).

the C-terminal 43 residues are new, coded by the *nifK* 5′ flanking sequence. Excision and fusion reduce the intergene region to 199 nucleotides. The sum of the three open reading frames of *nifH*, *nifD*, and *nifK*, together with intergenic regions, a leader of 135 nucleotides, and a 3′ flanking sequence of 100 nucleotides, predicts a transcript of 4.5 kilobases. A transcript of about this size, detected by *nifH*, *nifD*, and *nifK* probes, was found in RNA prepared from pro-heterocysts (Golden, unpublished data). Thus, it appears that one consequence of the excision-fusion is cotranscription of the (new) *nifHDK* operon.

The *nif* genes are regulated by promoter activation. To date, we have determined the sequences of very few promoters, so that the generalizations drawn about transcription control must be regarded as tentative. Nevertheless, among these few promoters the differences in their sequences are striking.

We can compare the promoter structure of three transcripts. The *nifH* transcript, whether fused or not fused, is induced by nitrogen starvation in the absence of O₂; this transcript is not present in cells grown on ammonia (7). The *rbcL* transcript is present in the latter cells but disappears during nitrogen starvation (7). The *glnA* transcript is present in both situations (15). The promoters of these genes, determined by S1 nuclease protection and primer extension with reverse transcriptase (7, 12a, 15), are shown in Fig. 3. Some of these assignments have been confirmed by runoff transcription in vitro (N. E. Tumer, unpublished data).

The *rbcL* promoter resembles the *E. coli* consensus at −10 but is radically different around −35. As expected from this result, the *Anabaena rbcL* gene is not transcribed by *E. coli* RNA polymerase, in vivo or in vitro. However, a preparation of *Anabaena* enzyme from cell grown on ammonia will use the *rbcL* promoter efficiently. Since the same enzyme preparation will also transcribe from an early T4 promoter (about as well as *E. coli* holoenzyme), we suspect that the *Anabaena* enzyme contains several

a

```
     -40              -30              -20              -10         +1
  TTCAAAGAATAACTTATGCCATTTCTTGATATATTGAGAGA
```

b

```
        -40              -30              -20              -10         +1
  CATAACGAACCCATCATGAACACTAATTCTACTGGTTTTTC
        ttgaca(coli)                    tataat(coli)
                ctgg(nif)              ttgca(nif)
```

c

```
          -40              -30              -20              -10         +1
RNA II  CTTTTGTGCAGATGTCGAAAGAAAGGTTAATATTACCTGTA
```

```
          -40              -30              -20              -10         +1
  I  GACTACAAAACTGTCTAATGTTTAGAATCTACGATATTTCA
```

FIG. 3. (a) Nucleotide sequence preceding the start site for transcription of *rbcL-rbcS* mRNA in *Anabaena*. The underlined sequence at −10 corresponds closely to that of a consensus *E. coli* promoter, but the sequence underlined at −35 does not (12a). (b) Nucleotide sequence preceding the start site for transcription of *Anabaena nifHD*. The *E. coli* consensus promoter sequence is shown in lower case letters in the first line below; the *Klebsiella* consensus *nif* promoter sequence is shown in the second line below (l). (c) Top: 5′ flanking sequence of *glnA* gene transcripts observed in vivo in *Anabaena* cells grown on ammonia or in vitro with the use of RNA polymerase from ammonia-grown cells (RNA II) (15; N. E. Tumer, unpublished data). Bottom: 5′ flanking sequence of the major *glnA* gene transcript (RNA I) observed in *Anabaena* induced anaerobically for nitrogenase (15). The underlined sequences for RNA II correspond to a good *E. coli* promoter. The underlined sequence for RNA I correspond to sequences at the same position relative to the *nifD* transcript, above.

species capable of recognizing more than one promoter.

One promoter *not* recognized by the *Anabaena* enzyme in vitro is the *nifH* promoter (7). The 5′ flanking sequence of this transcript is shown in Fig. 3 along with the consensus *nif* promoter sequence of *Klebsiella* (1). Note that *Klebsiella nif* promoters do not have a −35 sequence; instead, they have a −26 sequence. The *Anabaena nif* gene promoter bears some resemblance to the *Klebsiella nif* promoter, but it is equally close to the *rbcL* sequence: four of six matches in the −35 region and four of six matches in the −10 region.

The *glnA* gene transcription is still more complicated. In cells growing on ammonia, this gene is transcribed principally from a promoter 150 base pairs upstream from the gene's coding region. Under nitrogen-fixing conditions, this promoter is shut down and a new one at −90 base pairs is used instead. These promoters are shown as II and I, respectively, in Fig. 3. Promoter II is clearly *E. coli*-like and it is used

by the *E. coli* enzyme in vivo and in vitro, as well as by the *Anabaena* enzyme in vitro. Promoter I has four of six matches in the −35 region with *nifH* and five of six in the −10 region. Consistent with its *nif*-like character, it is not used in vitro by either *E. coli* or *Anabaena* vegetative-cell RNA polymerase.

SUMMARY

The heterocyst is a terminally differentiated cell. In heterocysts, the vegetative-cell functions of photosynthetic oxygen evolution and CO_2 fixation are shut down, to be replaced by cyclic photophosphorylation and oxidative carbon metabolism yielding ATP and reductant for the anaerobic process of nitrogen fixation. Shutting down transcription of one set of genes and turning on others involves modification of the transcriptional machinery, either by replacement of enzyme components needed for promoter recognition or by synthesis of ancillary factors that activate otherwise silent promoters. The *glnA* gene is fitted with multiple promoters to permit its continued expression throughout heterocyst differentiation.

Several physical rearrangements occur in the DNA of the *nif* genes during heterocyst differentiation. One of these transactions is the excision of 11 kbp between *nifK* and *nifD*, resulting in juxtaposition of these two genes and permitting them to be transcribed together.

This research was supported by Public Health Service grant GM 21823 from the National Institute of General Medical Sciences; grant 82-CRCR-1-1056 from the USDA/SEA through the Competitive Research Grants Office; postdoctoral fellowships from the National Institutes of Health to S.E.C., S.A.N.-B., and J.W.G.; postdoctoral fellowships from the American Cancer Society to S.J.R. and N.E.T.; and a postdoctoral traineeship from the National Institutes of Health to P.J.L.

LITERATURE CITED

1. **Beynon, J., M. Cannon, V. Buchanon-Wollaston, and F. Cannon.** 1983. The *nif* promoters of *Klebsiella* have a characteristic primary structure. Cell **34**:665–675.
2. **Burgess, R. R., and J. J. Jendrisak.** 1975. A procedure for the rapid, large-scale purification of *E. coli* DNA-dependent RNA polymerase. Biochemistry **14**:4634–4638.
3. **Curtis, S. E., and R. Haselkorn.** 1983. Isolation and sequence of the gene for the large subunit of RuBP carboxylase from the cyanobacterium *Anabaena* 7120. Proc. Natl. Acad. Sci. U.S.A. **80**:1835–1839.
3a. **Curtis, S. E., and R. Haselkorn.** 1984. Isolation, sequence and expression of two members of the 32 kd thylakoid membrane protein gene family from the cyanobacterium *Anabaena* 7120. Plant Mol. Biol. **3**:249–258.
4. **Fisher, R., R. Tuli, and R. Haselkorn.** 1981. A cloned cyanobacterial gene for glutamine synthetase functions in *E. coli* but the enzyme is not adenylylated. Proc. Natl. Acad. Sci. U.S.A. **78**:3393–3397.
5. **Fleming, H., and R. Haselkorn.** 1974. The program of protein synthesis during heterocyst differentiation in nitrogen-fixing blue-green algae. Cell **3**:159–170.
6. **Haselkorn, R.** 1978. Heterocysts. Annu. Rev. Plant Physiol. **29**:319–344.
7. **Haselkorn, R., D. Rice, S. E. Curtis, and S. J. Robinson.** 1983. Organization and transcription of genes important in *Anabaena* heterocyst differentiation. Ann. Microbiol. (Paris) **134B**:181–193.
8. **Lammers, P. J., and R. Haselkorn.** 1983. Sequence of the *nifD* gene coding for the alpha subunit of dinitrogenase from the cyanobacterium Anabaena. Proc. Natl. Acad. Sci. U.S.A. **80**:4723–4727.
9. **Mazur, B. J., and C.-F. Chui.** 1982. Sequence of the gene coding for the beta subunit of dinitrogenase from *Anabaena*. Proc. Natl. Acad. Sci. U.S.A. **79**:6782–6786.
10. **Mazur, B. J., D. Rice, and R. Haselkorn.** 1980. Identification of blue-green algal nitrogen fixation genes by using heterologous DNA hybridization probes. Proc. Natl. Acad. Sci. U.S.A. **77**:186–190.
11. **Merrick, M. J.** 1982. A new model for nitrogen control. Nature (London) **297**:362–363.
12. **Mevarech, M., D. Rice, and R. Haselkorn.** 1980. Nucleotide sequence of a cyanobacterial *nifH* gene coding for nitrogenase reductase. Proc. Natl. Acad. Sci. U.S.A. **77**:6476–6480.
12a. **Nierzwicki-Bauer, S. A., S. E. Curtis, and R. Haselkorn.** 1984. Cotranscription of genes encoding the small and large subunits of RuP₂ carboxylase in the cyanobacterium *Anabaena* 7120. Proc. Natl. Acad. Sci. U.S.A. **81**:5961–5965.
13. **Orr, J., and R. Haselkorn.** 1982. Regulation of glutamine synthetase activity and synthesis in free-living and symbiotic *Anabaena* spp. J. Bacteriol. **152**:626–635.
14. **Rice, D., B. J. Mazur, and R. Haselkorn.** 1982. Isolation and physical mapping of nitrogen fixation genes from the cyanobacterium *Anabaena* 7120. J. Biol. Chem. **257**:13157–13163.
15. **Tumer, N. E., S. J. Robinson, and R. Haselkorn.** 1983. Different promoters for the *Anabaena* glutamine synthetase gene during growth using molecular or fixed nitrogen. Nature (London) **306**:337–342.
16. **Wilcox, M., G. S. Mitchison, and R. Smith.** 1975. Spatial control of differentiation in the blue-green alga *Anabaena*, p. 453–463. *In* D. Schlessinger (ed.), Microbiology—1975. American Society for Microbiology, Washington, D.C.
17. **Wolk, C. P.** 1967. Physiological basis of the pattern of vegetative growth of a blue-green alga. Proc. Natl. Acad. Sci. U.S.A. **57**:1246–1251.

Approaches to the Study of Cell Differentiation in *Caulobacter crescentus*

AUSTIN NEWTON, NORIKO OHTA, EDWARD HUGUENEL, AND LING-SING CHEN

Departments of Molecular Biology and Biology, Princeton University, Princeton, New Jersey 08544

Caulobacter crescentus is a gram-negative bacterium that was first studied by Roger Stanier and his students at Berkeley in the 1960s (for early review, see 38). It was after this work that the advantages of this organism as a developmental system for investigating both the temporal and spatial regulation of cell differentiation began to be appreciated. The essential features of the life cycle for these studies are the precise timing of developmental events during the cell cycle and the production of two different cell types by asymmetric cell division: the new, motile swarmer cell and the older, nonmotile stalked cell (see Fig. 1A and, for recent reviews, 13, 30).

The dimorphic aspect of the cell cycle is emphasized in Fig. 1A, which shows not only that the two progeny cells are differentiated with respect to cell structure, but also that they are programmed to follow different cell cycles; the stalked cell initiates DNA synthesis immediately after division, whereas the swarmer cell enters a presynthetic gap (G1) required for differentiation into a stalked cell (8). Other studies have shown that the patterns of protein synthesis in the two cells are different as well (1, 6).

For much of the work discussed in this chapter the *Caulobacter* life cycle can be best visualized as occurring in two continuous developmental sequences (Fig. 1B). In one, the stalked cell behaves as a procaryotic version of a stem cell (29) which divides repeatedly to produce new swarmer cells by the assembly of a flagellum, bacteriophage receptor sites, and the pili at one pole of the cell. In the other developmental sequence, the newly divided swarmer cell differentiates into a stalked cell, first by loss of the flagellum and other polar structures, and finally by formation of a stalk; at this point the cell begins chromosome replication and enters the stalked cell division cycle. (Individual cell cycle events are described in 13 and 30.) Although these two developmental sequences can be distinguished at one level because of pleiotropic mutations that independently affect either flagellum assembly or stalk formation (15; J. Sommer and A. Newton, unpublished data), they are not regulated independently of each other; analysis of cell cycle mutants has shown that swarmer cell development, including stalk

formation, requires completion of a late division step(s) in the previous cell cycle (17, 46).

Several features of developmental regulation in *C. crescentus* distinguish it from other systems discussed in this volume. First, a "terminally" differentiated cell type is not formed; the swarmer cell is not some kind of a motile spore, since it is metabolically active and grows in preparation for the initiation of DNA synthesis (45). Second, culture conditions, such as nutrient deprivation, are not required to trigger any of the developmental stages; and third, cell-cell interactions have not been observed to play a role in development. Thus, the cell cycle in the *C. crescentus* cell represents a "stripped down" version of cell differentiation in which the temporal and spatial events are cell autonomous and occur repeatedly as part of a vegetative cell cycle. The rationale for studying developmental regulation in *Caulobacter* is that the mechanisms operative in these cells may provide models for the investigation of similar problems in more complex developmental systems (32). The discussion that follows summarizes some of the approaches that have been used in *C. crescentus* to study the molecular and genetic mechanisms responsible for the temporal and spatial control of development.

EXPERIMENTAL PROCEDURES

Culture conditions and cell synchrony. *C. crescentus* cells were grown in glucose-supplemented minimal salts medium at 30°C as described previously (38). Synchronous swarmer cells were prepared by centrifuging recombinants of strain CB15F on Ludox gradients (17).

S1 nuclease assays. S1 nuclease assays were carried out by the procedure described by Berk and Sharp (4) and elsewhere (L.-S. Chen and A. Newton, manuscript in preparation).

RESULTS AND DISCUSSION

Genetic analysis. In addition to the clearly defined pattern of morphogenesis and the excellent methods of synchronizing swarmer cells for monitoring cell cycle changes (8, 14), a major attraction of *C. crescentus* for developmental work is the increasingly versatile genetics. In the past few years techniques for conjugation

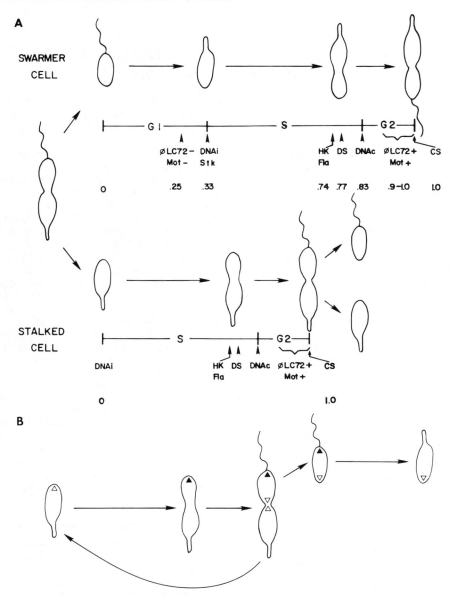

FIG. 1. *C. crescentus* cell cycle. Asymmetric cell division gives the swarmer cell and the stalked cell (A), and repeated division of the stalked cell produces flagellated swarmer cells (B). (A) The cell cycle periods on minimal medium are the presynthetic gap (G1), 60 min; chromosome replication (S), 90 min; and postsynthetic gap (G2), 30 min. Times for events (reviewed in 30) are given as fractions of the cell cycle (+, gain; −, loss); φLC72, adsorption of DNA bacteriophage; Mot, motility; Stk, stalk formation; HK and Fla, initiation of hook protein and flagellin synthesis; DS, division site formation; DNA replication events, see text. (B) The hypothetical organizational centers (open triangles) are laid down late in division and become sites for polar assembly (closed triangles) in the next cell cycle (17; see text).

(10, 31), transduction (12), Tn5 mutagenesis (11), and complementation with the use of stably replicating plasmids (34) have become available. Use of these techniques has allowed Ely and his colleagues to construct a workable genetic linkage map with many nutritional and other markers (2).

Two major areas of genetic and biochemical research interest in *C. crescentus* have been the study of flagellum biosynthesis and control of the cell cycle. Approximately 26 genes required for flagellum formation (*fla* genes) have been identified thus far (20), and 18 of them are clustered in three linkage groups, with the re-

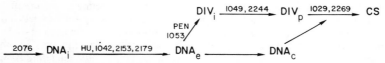

FIG. 2. Organization of gene-mediated steps in the DNA synthetic and cell division pathways of the *C. crescentus* cell cycle. Number designations are for strains with temperature-sensitive mutations in a step required for a DNA synthetic or cell division event (defined in text). Hydroxyurea (HU) reversibly blocks DNA synthesis, and penicillin G (PEN) reversibly blocks initiation of division. (Data from 36.)

mainder scattered on the genetic map (13). We will have more to say about two of these *fla* gene clusters, the flagellin and hook gene clusters, when we discuss the regulation of gene expression.

It is clear from many studies in *C. crescentus* that the timing of developmental events and that of cell cycle events are closely coordinated. To determine whether there is also a functional relationship between the two sets of events, we have isolated temperature-sensitive strains that grow normally at the permissive temperature (30°C) and at the nonpermissive temperature (37°C), but that are blocked at a defined stage of the cell cycle at 37°C (35). Characterization of these mutants has identified genes required for DNA initiation (DNAi), DNA chain elongation (DNAe), initiation of cell division (DIVi), progression of cell division (DIVp), and cell separation (CS). We have now mapped a number of the cell cycle genes on the *C. crescentus* chromosome (A. Newton and L. Kulick, manuscript in preparation) and determined the functional relationship between them in reciprocal shift experiments using the reversible temperature-sensitive mutations in combination with drugs that reversibly block either DNA synthesis or cell division (Fig. 2; 36). Steps mediated by these genes are organized into two dependent pathways, one for DNA synthesis and the other for cell division. As indicated in Fig. 2, the cell division pathway is coupled to the DNA synthetic pathway by the dependence of DIVi and CS on different stages of chromosome replication. Since DNAi is also dependent on completion of the previous round of replication (DNAc), the DNA synthetic pathway is apparently circular (28).

The dependent organization of cell cycle steps accounts for the observed progression of DNA synthetic and cell division events in the *C. crescentus* cell cycle. More interestingly, growth of these mutants at 37°C not only blocks division and eventually leads to the formation of long filamentous cells, but also results in dramatic and characteristic developmental defects (Fig. 3). The three characteristic morphologies of the cell cycle mutants (17, 30) can be described as follows. (i) In DNAe mutants, blockage of DNA synthesis prevents formation of the polar bac-

teriophage φLC72 receptor sites and the assembly of the flagellum; the latter effect first suggested that DNA synthesis is required for flagellar gene expression, a conclusion later confirmed by biochemical experiments (37; see below). (ii) DIVi and DIVp mutants are hydra-like cells whose phenotype is explained by two effects of the cell division block. The mutants do not form new sites for the assembly of surface structures, which are normally laid down at the site of cell division, and the old, polar assembly site is not inactivated by stalk formation (17). Consequently, the cells, which synthesize DNA and express flagellar proteins with normal periodicity at the nonpermissive temperature, assemble successive flagellar filaments at the same cell pole. (iii) CS mutants are highly constricted, filamentous cells which proceed to the last identified stage before cell separation; they also complete the step(s) required to specify the new polar assembly sites for flagellum formation and to initiate stalk formation and swarmer cell maturation (17).

These genetic results clearly implicate steps in the cell division and DNA synthetic pathways in

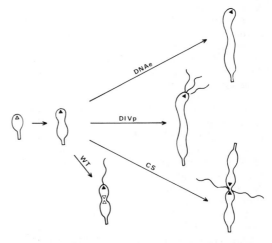

FIG. 3. Effect of blocking cell division on development in *C. crescentus*. The filamentous cells shown are formed after mutants (Fig. 2) are grown for two to three generations at the nonpermissive temperature (see text). Mutants blocked in DIVi and DIVp give similar developmental phenotypes. (Data from 17.)

the regulation of developmental events in *C. crescentus*. They also have suggested approaches to studying the temporal and spatial control of these events, as discussed in the following sections.

Functional differentiation of polar membrane domains. The assembly of the surface structures on the incipient swarmer cell always occurs at the new cell pole, and the mutational analysis outlined above showed that formation of the assembly sites requires completion of a late step(s) in cell division (DIVp). We have proposed that the position of these sites is specified by "organizational centers" that are laid down as part of the division site in the cell cycle before assembly is initiated (17; Fig. 1B). As a consequence of these events, the new cell pole should be stably differentiated from the remainder of the cell envelope. To examine this possibility, we have developed a procedure for the preparation and isolation of membrane vesicles from the flagellated pole of the cell (19).

When *C. crescentus* cells are lysed under the appropriate conditions, membrane vesicles derived from one pole of the cell retain the flagellum, and the filament can be decorated with antiflagellin immunoglobulin. These vesicles can then be separated from the nonflagellated vesicles by binding them to a protein A-Sepharose column and subsequently isolated by elution with glycyl-tyrosine. Two-dimensional gel analysis of the two membrane fractions which had been labeled in vivo with [^{35}S]methionine has revealed a subset of proteins unique to the polar and to the nonpolar vesicles, respectively (see Fig. 3 in reference 20). The association of the flagellins and hook protein exclusively with the polar membrane fraction confirms the efficiency of the separation procedure; other pole-specific proteins might be targeted to this region of the cell envelope in a similar way, or alternatively, they could be laid down at division as part of the organizational center itself. The localized zones of surface array growth observed at the cell division site and at the base of the developing stalk (43) may reflect this same pattern of development. Another example of protein localization within the dividing *C. crescentus* cell is the preferential segregation of the chemotaxis-methylation functions to the motile swarmer cell at division (S. L. Gomez and L. Shapiro, J. Mol. Biol., in press).

An important problem in localized assembly is to identify the stage of translocation at which the subunits of the surface structures are targeted to the polar membrane domain. Pulse-chase experiments have shown that newly synthesized flagellin in *C. crescentus* cells is translocated for assembly in the sequence soluble pool → membrane pool → assembled filament (18). In very short pulse-labeling experiments, followed by radioimmunoassay for labeled flagellin in polar and nonpolar vesicles, almost the entire membrane pool of flagellin was found in the polar vesicles (19). Thus, flagellin is apparently targeted to the cell pole at a relatively early stage of translocation to the membrane.

These results provide direct evidence that the *C. crescentus* cell envelope is both chemically and functionally differentiated into at least two relatively stable domains, here defined by the cell pole and the remainder of the cell envelope. The selective insertion of proteins into different domains of a cell membrane has not been explored extensively, but it is an important element in understanding how positional information is specified. The immunoadsorption procedure described here for the fractionation of antigenically distinct membranes may prove generally useful in addressing this problem.

Organization and expression of flagellar genes on the *C. crescentus* genome. The major flagellar proteins in *C. crescentus* are the 25-kilodalton flagellin A, which is assembled at the distal tip of the filament, the 27.5-kilodalton flagellin B, which forms the proximal portion of the filament (22, 47), and the 70-kilodalton hook protein (41). These proteins are synthesized periodically in the cell cycle at the time of flagellum assembly (23, 37, 42), and experiments with temperature-sensitive cell cycle mutants have shown that blocking DNA synthesis prevents the induction of their synthesis (37, 42). In addition to a possible regulatory role of DNA synthesis, the phenotypes of *fla* mutations reveal a hierarchical control in which expression of the hook protein and flagellin genes is dependent on the expression of other *fla* genes (5, 20, 34, 42). The genetics of what may be a similar regulatory cascade for *fla* gene expression in *Escherichia coli* has been studied in detail (21).

Our studies of *fla* gene expression in *C. crescentus* have focused on the hook gene cluster, which contains the hook protein structural gene (*flaK*). We describe in this section the transcriptional organization of genes in this cluster and experiments designed to examine the level of gene expression by use of S1 nuclease mapping.

Physical mapping of genomic DNA isolated from approximately 50 *fla* mutants induced by Tn5 insertion mutagenesis has localized 19 of the insertions within a 17-kilobase region of the chromosome that contains the hook protein gene (Fig. 4; 34). DNA fragments of various lengths were isolated from wild-type clones covering the hook gene cluster and were subcloned into vector pRK290 (9), which replicates stably in *C. crescentus* cells. The hybrid plasmids were then transferred to the Tn5 insertion mutants and tested for complementation of the nonmotile

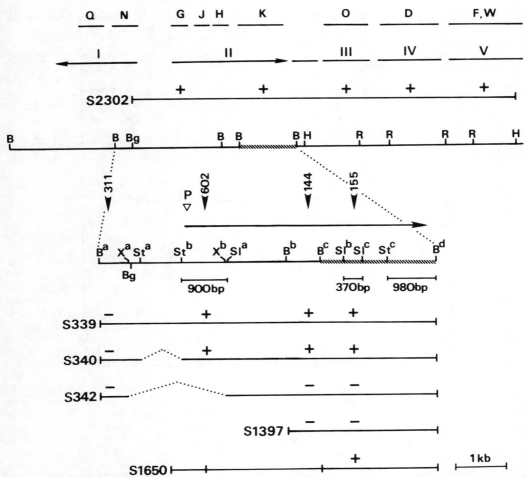

FIG. 4. Physical, genetic, and functional organization of the hook gene cluster in *C. crescentus*. The *fla* or *flb* gene designation and the minimum number of transcriptional units are indicated above the restriction map with an arrowhead showing the direction of transcription (34). The cross-hatched segment of the map is the 2.3-kilobase DNA *Bam*HI fragment containing the *flaK* gene (33), and closed arrows above the map indicate the positions of insertions; P indicates the proposed transcriptional start. DNA fragments subcloned into pRK290 are numbered with an S prefix. S1650 contains the 0.67-kilobase *Bam*HIb-*Bam*HIc fragment plus a tandem duplication of the 2.3-kilobase *Bam*HIc-*Bam*HId fragment. B, *Bam*HI; *Bgl*, *Bgl*II; H, *Hind*III; R, *Eco*RI; X, *Xho*I; St, *Sst*I; Sl, *Sal*I. (Data from 34 and Ohta, Chen, and Newton, unpublished data.)

phenotype. Since Tn5 insertions are polar in *C. crescentus* (34), complementation by this test should occur only when the subcloned fragment carries an intact transcriptional unit, including the promoter and structural genes; a similar approach has been used by Ausubel and co-workers to define transcriptional units in *Rhizobium* (40). The results of these tests indicate that the hook gene cluster contains at least five transcriptional units (I to V; Fig. 4) and that transcription unit II is a polycistronic operon (hook operon) that contains four genes, including *flaK*.

To examine transcription of the hook operon, we have identified two DNA fragments to use in S1 nuclease assays. The direction of transcrip-

tion in the hook operon is known from previous work (33), and the promoter was located within the *Sst*Ib-*Xho*Ib fragment of S339 by making restriction deletions of this subclone and testing them for complementation (Fig. 4). The 370-base pair (bp) *Sal*I-*Sal*I fragment, which is contained entirely within *flaK*, was used to assay temporal regulation of hook protein gene transcription during the cell cycle, and the 900-bp *Sst*I-*Sal*I fragment was used to map the transcriptional start of the operon.

Figure 5A shows the result of using the internal *Sal*I-*Sal*I fragment of the *flaK* gene in the S1 nuclease assay. The fragment was fully protected by an RNA extract from wild-type CB15

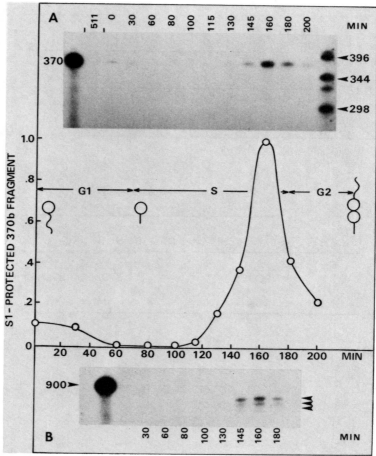

FIG. 5. Transcriptional regulation of the hook operon. (A) Cell cycle regulation of *flaK* gene mRNA was determined by use of the S1 nuclease assay with the 370-nucleotide probe (Fig. 4). (B) The location of the transcriptional start and regulation of the start(s) in the cell cycle were determined by use of the 900-nucleotide probe (Fig. 4); arrows indicate protected DNA fragments (see text). RNA was extracted from synchronous CB15F cells at the times indicated, hybridized to the 5'-end-labeled fragments (24), and then digested with S1 nuclease as described previously (4). Protected DNA fragments were separated on a 5% denaturing polyacrylamide gel. (Data are from Chen and Newton, in preparation.)

cells (Chen and Newton, in preparation), but not by extracts from strain SC511 (see lane 2, Fig. 5A), which is a spontaneous insertion mutant of *flaK* (*flaK155*) that produces no hook protein by radioimmunoassay (34). When RNA samples from a synchronous cell population were assayed, the hook gene transcript was first detected at 130 min in the cell cycle (approximately 0.6 of the swarmer cell cycle) and reached the maximum at 160 min (0.7 of the cell cycle; Fig. 5A). Except for a very low level of the specific RNA present in new swarmer cells, the periodic appearance of the *flaK* gene transcript coincides almost exactly with the pattern of hook protein synthesis determined previously by radioimmunoassay (42). Dot-blot hybridiza-

tion experiments reported by Milhausen and Agabian indicate that the flagellin gene mRNA of *C. crescentus* is also regulated periodically in the cell cycle (26). All of these results are consistent with experiments showing that de novo RNA synthesis is required for the induction of these proteins, and they support the earlier conclusion that periodic expression of the hook protein and flagellin genes is regulated at the transcriptional level (37, 42). Measurements of the rates of specific RNA synthesis are still necessary, however, to eliminate a role of differential messenger stability in regulation of the RNA levels.

S1 nuclease analysis of total in vivo-synthesized RNA was also carried out with the 900-bp

*Sst*I-*Sal*I fragment as the 5'-labeled probe, and at least three protected fragments (800, 770, and 720 nucleotides) were detected (Fig. 5B; Chen and Newton, in preparation). These results are consistent with multiple transcriptional starts, and they also confirm the complementation results which placed the hook operon promoter in the 900-bp fragment (Fig. 4). When the 900-bp probe was used to assay RNA from synchronous cells, a periodic pattern of RNA synthesis was also observed (Fig. 5B). The presence of the three fragments in approximately the same ratio at all stages of expression indicates that, if there are in fact multiple transcriptional starts, they are not used differentially during the cell cycle. Additional S1 mapping and nucleotide sequencing of the hook operon (D. Mullin, L.-S. Chen, and A. Newton, unpublished data) should determine the position and sequence of the promoter for this transcriptional unit.

Complementation analysis also uncovered a previously unidentified promoter at the end of the 2.3-kilobase *Bam*HI-*Bam*HI fragment (crosshatched line in Fig. 4): subclone S1397 does not complement *fla*K mutants, but S1650, which contains a tandem duplication of the 2.3-kilobase *Bam*HI fragment, does (Fig. 4). S1 mapping with the 980-bp *Sst*I-*Bam*HI fragment subsequently confirmed this conclusion by locating a strong promoter approximately 300 bp upstream from the *Bam*HI site (Chen and Newton, in preparation). These results also showed that the promoter is expressed periodically in the cell cycle in late S phase and also in substantial quantities in early G1. Since this promoter is not required for *fla*O expression, an additional, unmapped gene or genes may lie between *fla*K and *fla*O.

Temporal regulation in *C. crescentus* is not exclusively at the level of gene expression, however. Pili are assembled periodically at the flagellated cell pole, but pilin, the 8-kilodalton subunit, is apparently made continuously during most of the stalked cell cycle, and its synthesis is independent of DNA chain elongation (44). Preliminary results also indicate that assembly of pili may require a late cell division step (J. Sommers and A. Newton, unpublished data). Thus, it seems likely that *C. crescentus* cells have evolved a variety of mechanisms to regulate differentiation of their polar structures.

Regulatory interactions of *fla* genes. The requirement of some *fla* gene products for the expression of other *fla* genes has been studied in the *fla*Y-E gene cluster, where mutations in *fla*Y depress expression of the flagellin genes and some of the chemotaxis genes (5), and in the hook gene cluster (34, 42). In the latter case, mutations in *fla*K, the structural gene for the hook protein, or in *fla*J, the gene required for

terminating *fla*K expression at the normal time in the cell cycle (42), reduce flagellin A and flagellin B to undetectable levels (34). The level of a related, but minor 29-kilodalton flagellin (16, 20) is elevated in the *fla*K mutant, however. We conclude from these results that both synthesis of the hook protein and termination of hook assembly are required for activation of the flagellin A and B genes.

The regulatory effect of Tn5 insertions in transcription units III, IV, and V of the hook gene cluster (Fig. 4) is even more striking. These mutations completely abolish expression of the hook operon, as determined by the S1 nuclease assay described in Fig. 5 (Chen and Newton, unpublished data), and of the unlinked flagellin genes, as determined by radioimmunoassay (34). The *fla*O, *flb*D, and *flb*F genes are unusual in this respect because most *fla* mutations in *C. crescentus* only partially depress the levels of flagellin and hook protein gene expression (20). Thus, the 3' transcriptional units of the hook gene cluster could play a central regulatory role in flagellar gene expression. Complementation tests show that these genes act in *trans*, and we propose that their products act positively at the transcriptional level to control expression of the hook operon and the flagellin A and B genes. The *fla*O, *flb*D, or *flb*F products could act directly on the flagellin genes or via the requirement of these products for expression of the hook operon, which is required in turn for expression of the flagellin genes. A speculative model for the interaction between these genes is shown in Fig. 6.

Timing of flagellar gene expression. Experiments in which DNA chain elongation mutants were used have suggested that approximately 0.6 of the chromosome must be replicated before induction of the hook protein and flagellin genes can occur (42). Since this point in the cell cycle corresponds to the time of both flagellar gene expression and replication of the structural genes (T. Lott and A. Newton, manuscript in preparation), the replicating chromosome can be viewed as a "clock" for timing flagellar gene expression and perhaps other developmental events (36). Although this is an attractive working hypothesis, other explanations of the effect of blocking DNA synthesis are also possible, as discussed previously (30).

One way to determine whether replication of a gene is sufficient to induce its expression is to change the time at which the gene is duplicated in the cell cycle. This is now possible by subcloning an intact transcriptional unit containing the gene in pRK290, a vector which replicates throughout the S period in *C. crescentus* (N. Ohta and A. Newton, unpublished data). We describe here the most recent of several such

Hook gene cluster   **Flagellin gene cluster**

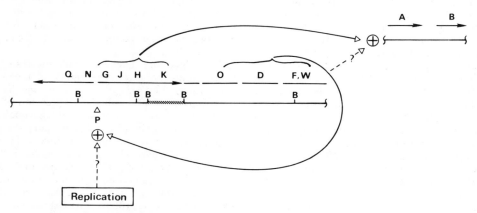

FIG. 6. Working model for regulatory interactions between genes in the hook gene and flagellin gene clusters of *C. crescentus*. These interactions are consistent with the results of complementation tests and assays of flagellin and hook protein gene expression determined in a variety of *fla* mutants (34; see text). The *flaY-E* cluster, which has been cloned independently in the laboratories of Agabian (27) and Shapiro (39), contains genes for flagellin A and the 29-kilodalton flagellin; a second, unlinked cluster apparently contains the flagellin B gene and another copy of the flagellin A gene (P. R. Gill and N. Agabian, Fed. Proc. **42**:1969, 1983).

experiments in which a 15-kilobase genomic fragment containing the hook operon and the three downstream regulatory operons (units III to V) were subcloned. The hybrid plasmid (S2302, Fig. 4) was transferred to the *flaK* mutant SC511, and the rates of flagellin and hook protein synthesis were determined throughout the cell cycle by use of specific radioimmunoassays.

These experiments have shown that the hook protein is overproduced in these strains by a factor of 10-fold, presumably because of the high copy number of genes carried on the plasmid, but that flagellins A and B are synthesized normally (Ohta, Chen, and Newton, in preparation). The expression of *flaK* and that of the flagellin genes are periodic, however, and the times of induction in the cell cycle are almost identical to those observed in wild-type strains by use of radioimmunoassays (42) and S1 nuclease assays (Fig. 5A). Similar experiments with a hybrid plasmid carrying only the hook operon (subclone S339) indicate that the periodicity of transcription is also unaffected by early replication of this segment of the *C. crescentus* chromosome (Ohta et al., in preparation).

These results lead to the interesting conclusion that, although chromosome replication may be necessary for expression of the hook protein gene, replication of the hook operon and adjacent transcriptional units on plasmid pRK290 is not sufficient to initiate early gene expression.

Clearly, an additional component(s) is necessary for the correct timing of this transcriptional event, and in principle any of the genes involved in the hierarchical regulation of the *fla* gene complex could play that role. As suggested previously, the requirement of chromosome replication for gene expression is likely to be superimposed on the hierarchical regulation of the flagellar genes (42). It remains to be proved which steps, or sequence of steps, in this complex network might function as the clock.

A discussion of timing mechanisms in *C. crescentus* should also mention that, depending on the isolation technique and method of analysis, from 6 to 20% of the proteins are synthesized either periodically or in a stage-specific manner during the cell cycle (1, 6, 25). In contrast to the flagellar proteins, most of these proteins are synthesized independently of chromosome replication (M. Clancy and A. Newton, unpublished data). If the synthesis of these proteins is regulated at the transcriptional level, it is interesting to speculate whether changes in the RNA polymerase subunits similar to those described for sporulating *Bacillus subtilis* cells (see articles in this volume) could play a role in altering the transcriptional specificity. This mechanism seems less likely in a system that does not form a stably differentiated cell type; RNA polymerase has been isolated from cells at different stages of the *C. crescentus* cell cycle (3) and from exponential cells in high yield (7) without the observation of multiple species of holoenzyme subunits. Minor RNA polymerase species similar to those described by M. J. Chamberlain (this volume) might have gone undetected, however.

SUMMARY AND CONCLUSIONS

The *C. crescentus* cell cycle is now well established as a model system for studying how developmental programs are translated into temporal and positional events during cell differentiation. Although this problem can be examined in the absence of regulative interactions that are so important in multicellular organisms, development in the "simple" *C. crescentus* cell is still relatively complex, and to be fully understood, it must be analyzed on several levels. The effect of cell cycle mutations has shown that the biosynthesis and correct spatial positioning of surface structures in these cells require completion of specific steps in the DNA synthetic or cell division pathways (17, 36, 42), but to study the mechanisms involved, more information about the synthesis and assembly of the regulated proteins is needed.

The study of the periodic expression of the flagellar genes in the cloned hook operon indicates that these genes are regulated at the transcriptional level (34, Ohta et al., in preparation; Chen and Newton, in preparation). This work also suggests that transcription from the hook operon promoter(s) is under positive regulation by the products of downstream transcriptional units. It remains to be seen whether any of these gene products acts directly with the promoters of the hook or flagellin genes, as hypothesized in Fig. 6. The interactions of genes and gene products in the hook and flagellin gene clusters are probably only part of an extensive hierarchy controlling *fla* gene regulation.

An unanswered question about the regulation of these genes is the identity of the "clock" that times their induction. Although chromosome replication is necessary at a defined stage of the cell cycle for flagellar gene expression (36, 42), replication of the hook operon and three of the associated transcriptional units is apparently not sufficient to determine the time at which these genes are expressed (Ohta et al., in preparation). Additional, as yet unidentified, regulatory interactions must also be involved.

The other half of the developmental puzzle in *C. crescentus* is how the location of the proteins in the cell is specified once they are synthesized. Recent work on membrane lipid synthesis in *C. crescentus* (reviewed in 13) and the coordination of synthesis and assembly of some gene products has suggested that mechanisms controlling when to express gene products could be related to where the products are assembled in the cell

(13). This is an intriguing possibility which should be explored by use of some of the gene probes and membrane isolation techniques now available. As shown by biosynthesis of the pili (44), however, the synthesis and assembly of all structural subunits are not coordinated, and it is possible that the cues for the site of assembly may be specified well before synthesis and the final assembly event (17). Methods for the isolation of flagellated vesicles (19) open one biochemical approach to analyzing those elements of the cell architecture that are important for protein localization.

It seems likely that there is more than a single mechanism in *C. crescentus* for achieving a given type of developmental regulation, as already observed for the biosynthesis of the flagella and the pili. One of the exciting problems in the next few years will be to elucidate and compare the molecular mechanisms responsible for the temporal and spatial control of protein synthesis and localization in these cells.

We thank Evryll Swanson for performing many of the complementation assays and radioimmunoassays discussed.

Work from our laboratory was supported by Public Health Service grants GM22299 and GM25644 from the National Institute of General Medical Sciences.

LITERATURE CITED

1. **Agabian, N., M. Evinger, and G. Parker.** 1979. Generation of asymmetry during development: segregation of type-specific proteins in *Caulobacter*. J. Cell Biol. **81:**123–136.
2. **Barrett, J. T., C. S. Rhodes, D. M. Ferber, B. Jenkins, S. A. Kuhl, and B. Ely.** 1982. Construction of a genetic map for *Caulobacter crescentus*. J. Bacteriol. **149:**889–896.
3. **Bendis, I., and L. Shapiro.** 1973. DNA-dependent RNA polymerase of *Caulobacter crescentus*. J. Bacteriol. **115:**848–857.
4. **Berk, J. A., and P. A. Sharp.** 1977. Sizing and mapping of early adenovirus mRNAs by gel electrophoresis of S1 endonuclease-digested hybrids. Cell **12:**721–732.
5. **Bryan, R. M., M. Purucker, S. L. Gomez, W. Alexander, and L. Shapiro.** 1984. Analysis of the pleiotropic regulation of flagellar and chemotaxis gene expression in *Caulobacter crescentus* using plasmid complementation. Proc. Natl. Acad. Sci. U.S.A. **81:**1341–1345.
6. **Cheung, K. K., and A. Newton.** 1977. Patterns of protein synthesis during development in *Caulobacter crescentus*. Dev. Biol. **56:**417–425.
7. **Cheung, K. K., and A. Newton.** 1978. Polyadenylic acid synthesis activity of purified DNA-dependent RNA polymerase from *Caulobacter*. J. Biol. Chem. **235:**2254–2261.
8. **Degnen, S. T., and A. Newton.** 1972. Chromosome replication during development in *Caulobacter crescentus*. J. Mol. Biol. **64:**676–680.
9. **Ditta, G., S. Stanfield, D. Corbin, and D. R. Helinski.** 1980. Broad host range cloning system for gram-negative bacteria: construction of a gene bank for *Rhizobium meliloti*. Proc. Natl. Acad. Sci. U.S.A. **77:**7347–7351.
10. **Ely, B.** 1979. Transfer of drug resistance factors to the dimorphic bacterium *Caulobacter crescentus*. Genetics **91:**371–380.
11. **Ely, B., and R. H. Croft.** 1982. Transposon mutagenesis in *C. crescentus*. J. Bacteriol. **149:**620–625.
12. **Ely, B., and R. C. Johnson.** 1977. Generalized transduction in *Caulobacter crescentus*. Genetics **87:**391–399.
13. **Ely, B., and L. Shapiro.** 1984. Regulation of cell differen-

tiation in *Caulobacter crescentus*, p. 1–26. *In* R. Losick and L. Shapiro (ed.), Microbial development. Cold Spring Harbor Press, Cold Spring Harbor, N.Y.

14. **Evinger, M., and N. Agabian.** 1977. Envelope-associated nucleoid from *Caulobacter crescentus* stalked and swarmer cells. J. Bacteriol. **132:**294–301.

15. **Fukuda, A., H. Iba, and Y. Okada.** 1977. Stalkless mutants of *Caulobacter crescentus*. J. Bacteriol. **131:**280–287.

16. **Gill, P. R., and N. Agabian.** 1983. The nucleotide sequence of Mr = 28.500 flagellin genes of *Caulobacter crescentus*. J. Biol. Chem. **258:**7395–7401.

17. **Huguenel, E., and A. Newton.** 1982. Localization of surface structures during procaryotic differentiation: role of cell division in *Caulobacter crescentus*. Differentiation **21:**71–78.

18. **Huguenel, E., and A. Newton.** 1984. Evidence that subcellular flagellin pools in *Caulobacter crescentus* are precursors in flagellum assembly. J. Bacteriol. **157:**727–732.

19. **Huguenel, E., and A. Newton.** 1984. Isolation of flagellated membrane vesicles from *Caulobacter crescentus* cells: evidence for functional differentiation of polar membrane domains. Proc. Natl. Acad. Sci. U.S.A. **81:**3409–3413.

20. **Johnson, R. C., D. M. Ferber, and B. Ely.** 1983. Synthesis and assembly of flagellar components by *Caulobacter crescentus* motility mutants. J. Bacteriol. **154:**1137–1144.

21. **Komeda, Y.** 1982. Fusion of flagellar operon genes to lactose genes on a Mu *lac* bacteriophage. J. Bacteriol. **150:**16–26.

22. **Koyasu, S., M. Asada, A. Fukuda, and Y. Okada.** 1981. Sequential polymerization of flagellin A and flagellin B into *Caulobacter* flagella. J. Mol. Biol. **153:**471–475.

23. **Lagenaur, C., and N. Agabian.** 1978. *Caulobacter* flagellar organelle: synthesis, compartmentation, and assembly. J. Bacteriol. **135:**1062–1069.

24. **Maxam, A. M., and W. Gilbert.** 1980. Sequencing end-labeled DNA with base-specific chemical cleavages. Methods Enzymol. **65:**499–560.

25. **Milhausen, M., and N. Agabian.** 1981. Regulation of polypeptide synthesis during *Caulobacter* development: two-dimensional gel analysis. J. Bacteriol. **148:**163–173.

26. **Milhausen, M., and N. Agabian.** 1983. *Caulobacter* flagellin mRNA segregates asymmetrically at cell division. Nature (London) **302:**630–632.

27. **Milhausen, H., P. R. Gill, G. Parker, and N. Agabian.** 1982. Cloning of developmentally regulated flagellin genes from *Caulobacter crescentus* via immunoprecipitation of polyribosomes. Proc. Natl. Acad. Sci. U.S.A. **79:**6847–6851.

28. **Nathan, P., and A. Newton.** 1982. Circular organization of the DNA synthetic pathway in *C. crescentus*. J. Bacteriol. **151:**503–506.

29. **Newton, A.** 1972. Role of transcription in the temporal control of development in *Caulobacter crescentus*. Proc. Natl. Acad. Sci. U.S.A. **69:**447–451.

30. **Newton, A.** 1984. Temporal and spatial control of *Caulobacter* cell cycle, p. 51–75. *In* P. Nurse and E. Streiblora (ed.), The microbial cell cycle. CRC Press, Boca Raton, Fla.

31. **Newton, A., and E. Allebach.** 1975. Gene transfer in *Caulobacter crescentus*: polarized inheritance of genetic markers. Genetics **80:**1–11.

32. **Newton, A., M. A. Osley, and B. Terrana.** 1975. *Caulobacter crescentus*: a model for the temporal and spatial control of development, p. 442–452. *In* D. Schlessinger (ed.), Microbiology—1975. American Society for Microbiology, Washington, D.C.

33. **Ohta, N., L.-S. Chen, and A. Newton.** 1982. Isolation and expression of cloned hook protein gene from *Caulobacter crescentus*. Proc. Natl. Acad. Sci. U.S.A. **79:**4863–4867.

34. **Ohta, N., E. Swanson, B. Ely, and A. Newton.** 1984. Physical mapping and complementation analysis of transposon Tn5 mutations in *Caulobacter crescentus*: organization of transcriptional units in the hook gene cluster. J. Bacteriol. **158:**897–904.

35. **Osley, M. A., and A. Newton.** 1977. Mutational analysis of development control in *Caulobacter crescentus*. Proc. Natl. Acad. Sci. U.S.A. **74:**124–128.

36. **Osley, M. A., and A. Newton.** 1980. Temporal control of the cell cycle in *Caulobacter crescentus*: roles of DNA chain elongation and completion. J. Mol. Biol. **138:**109–128.

37. **Osley, M. A., M. Sheffrey, and A. Newton.** 1977. Regulation of flagellin synthesis in the cell cycle of *Caulobacter*: dependence on DNA replication. Cell **12:**393–400.

38. **Poindexter, J. S.** 1964. Biological properties and classification of the *Caulobacter* group. Bacteriol. Rev. **28:**231–295.

39. **Purucker, M., R. Bryan, K. Amemiya, B. Ely, and L. Shapiro.** 1982. Isolation of a *Caulobacter* gene cluster specifying flagellum production by using non-motile Tn5 insertion mutants. Proc. Natl. Acad. Sci. U.S.A. **79:**6797–6801.

40. **Ruvkun, G. B., V. Sundaresan, and F. M. Ausubel.** 1982. Direct transposon Tn5 mutagenesis and complementation analysis of *Rhizobium meliloti* symbiotic nitrogen fixation genes. Cell **29:**551–559.

41. **Sheffrey, M., and A. Newton.** 1979. Purification and characterization of a polyhook protein from *Caulobacter crescentus*. J. Bacteriol. **138:**575–583.

42. **Sheffrey, M., and A. Newton.** 1981. Regulation of periodic protein synthesis in the cell cycle: control of initiation and termination of flagellar gene expression. Cell **24:**49–57.

43. **Smit, J., and N. Agabian.** 1982. Cell surface patterning and morphogenesis: biogenesis of a periodic surface array during *Caulobacter* development. J. Cell Biol. **95:**41–49.

44. **Smit, J., and N. Agabian.** 1982. *Caulobacter crescentus* pili: analysis of production during development. Dev. Biol. **89:**237–247.

45. **Terrana, B., and A. Newton.** 1975. Pattern of unequal cell division and development in *Caulobacter crescentus*. Dev. Biol. **44:**380–385.

46. **Terrana, B., and A. Newton.** 1976. Requirement of a cell division step for stalk formation in *Caulobacter crescentus*. J. Bacteriol. **128:**456–462.

47. **Weissborn, A., H. M. Steinman, and L. Shapiro.** 1982. Characterization of the proteins of the *Caulobacter crescentus* flagellar filament: peptide analysis and filament organization. J. Biol. Chem. **257:**2066–2074.

AUTHOR INDEX

SUBJECT INDEX

278